# Studies on Large Branchiopod Biology and Aquaculture

# Developments in Hydrobiology 64

*Series editor*

H. J. Dumont

# Studies on Large Branchiopod Biology and Aquaculture

*Edited by*

D. Belk, H.J. Dumont and N. Munuswamy

*Reprinted from Hydrobiologia, vol. 212 (1991)*

Kluwer Academic Publishers
Dordrecht / Boston / London

**Library of Congress Cataloging-in-Publication Data**

Studies on large branchiopod biology and aquaculture / edited by D.
Belk, H.J. Dumont, N. Munuswamy.
p. cm. -- (Developments in hydrobiology ; v. 64)
Most of the papers were presented at the Euphyllopod Symposium
held at the State University of Ghent, Ghent, Belgium, August 9-12,
1989.
Includes index.
ISBN 0-7923-1169-8 (acid-free paper)
1. Branchiopoda--Congresses. 2. Aquaculture--Congresses.
I. Belk, D. (Denton) II. Dumont, H. J. (Henry J.) III. Munuswamy,
N. IV. Euphyllopod Symposium (1989 : State University of Ghent)
V. Series.
QL444.B8S78 1991
595.3'2--dc20 91-11197

ISBN 0-7923-1169-8

Published by Kluwer Academic Publishers,
P.O. Box 17, 3300 AA Dordrecht, The Netherlands.

Kluwer Academic Publishers incorporates
the publishing programmes of
D. Reidel, Martinus Nijhoff, Dr W. Junk and MTP Press.

Sold and distributed in the U.S.A. and Canada
by Kluwer Academic Publishers,
101 Philip Drive, Norwell, MA 02061, U.S.A.

In all other countries, sold and distributed
by Kluwer Academic Publishers Group,
P.O. Box 322, 3300 AH Dordrecht, The Netherlands.

*Printed on acid-free paper*

# Contents

*Hydrobiologia* **212**: ix, 1991.
*D. Belk, H. J. Dumont & N. Munuswamy (eds), Studies of Large Branchiopod Biology and Aquaculture.*


# Preface

This book contains a collection of papers dealing with various aspects of the biology and aquaculture of the 'true phyllopods' as envisioned by Sars (1890). Today these members of the class Branchiopoda are grouped into four orders by Fryer (1987). They are the Anostraca, the Notostraca, and the two conchostracan orders Spinicaudata and Laevicaudata. The majority of the included papers were presented at the Euphyllopod Symposium convened at the State University of Ghent in Ghent, Belgium 9–12 August 1989. In addition, other relevant manuscripts that came to the attention of the editors during the peer review and manuscript improvement process are included.

The 1989 Euphyllopod Symposium in Ghent was the first international meeting ever devoted exclusively to the large branchiopod crustaceans. It was conceived and made a reality by the hard work of Prof. Dr Henri Dumont. Researchers with an interest in phyllopods from Australia, Belgium, Brazil, France, India, Ireland, Italy, Mexico, South Africa, Spain, United States of America, and Yugoslavia enjoyed a highly productive, exciting, and stimulating four days. Since there were no concurrent sessions, all attendees had an opportunity to hear each paper and join in the discussions. It was the delegate's unanimous hope that other international euphyllopod symposia be held in the future.

The Euphyllopd Symposium participants extend their special thanks to Henri Dumont for bringing us together. We also recognize the importance of Christine De Walsche's work as Symposium Secretary and offer her our sincere appreciation. We thank the State University of Ghent for use of university facilities and especially for the very economical living accommodation provided to visiting delegates. We thank the Institute of Ecology and Zoogeography for hosting the symposium. We are very appreciative of the financial support provided by the Belgian National Science Fund and by the Ministry of the Flemish community, Department of Higher Education and Research. This support was vital. We thank Kluwer Academic Publishers for agreeing to publish the proceedings of our meeting in the Developments in Hydrobiology series.

DENTON BELK
*San Antonio, Texas, USA*
*15 December 1990*

*Hydrobiologia* **212**: x–xii, 1991.
*D. Belk, H. J. Dumont & N. Munuswamy (eds), Studies of Large Branchiopod Biology and Aquaculture.*

# List of participants

ABREU-GROBOIS, F. A.
I.C.M.L.-U.N.A.M.
Apartado postal 811
Mazatlan Sinaola 82000
Mexico

AHL, Jonna
St.-Ambrose University
Department of Biology
Davenport, IO 52803–2898
USA
*Present address:*
University of Hartford
College of Basic Studies
West Hartford, CT 06117
USA

ALONSO, Miguel
University of Barcelona
Department of Ecology
Barcelona 08028
Spain

APPLETON, C. C.
University of Natal
Department of Zoology & Entomology
Pietermaritzburg 3201
Natal
South Africa

BARELLI, Maria Gaetana
University la Sapienza
B.A.U Department
Viale dell Universita 32
Rome 00185
Italy

BELK, Denton
Our Lady of the Lake University of San Antonio
411 S.W. 24th Street
San Antonio, TX 78285
USA

BRENDONCK, Luc
R.U.G.
Department of Agriculture
9000 Ghent
Belgium

COHEN, Rosa Graciela*
University of Buenos Aires,
Facultad de Ciencias exactas y Naturales
Buenos Aires 1428
Argentina

COUTTEAU, Peter
Artemia Reference Center
State University of Ghent
Rozier 44, 9000 Ghent
Belgium

CRIEL, Godelieve
R.U.G.
Department of Anatomy
K.L. Ledeganckstraat 35
9000 Ghent
Belgium

DE MATTHAEIS, E.
University of Rome 'La Sapienza'
Department of Animal and Human Biology
Viale dell Universita 32
Rome 00185
Italy

DE WALSCHE, Christine
R.U.G.
Department of Ecology & Zoogeography
K.L. Ledeganckstraat 35
9000 Ghent
Belgium

DUMONT, Henri
R.U.G.
Department of Ecology & Zoogeography
K.L. Ledeganckstraat 35
9000 Ghent
Belgium

FANFANI, Alberto
University 'La Sapienza'
Department of Animal and Human Biology
Viale dell Universita 32
Rome 00185
Italy

FUGATE, Michael
University of California
Department of Biology
Riverside, CA 92521
USA

GRAINGER, J. N. R.
University of Dublin
Trinity College
Department of Zoology
Dublin 2
Ireland

GUCLIEMINOTTI, Giorgio
University 'La Sapienza'
B.A.U. Department
Viale dell Universita 32
Rome 00185
Italy

HONTORIA, F.
Instituto de Acuicultura de torre de la sal
Ribera de Cabanes
12595 Torre de la Sal
Castellon
Spain

JAUME, D.
University of Barcelona
Facul. Biologia
Department of Ecology
Barcelona 08028
Spain

LAVENS, Patrick
Artemia Reference Center
Rozier 44
B-9000 Ghent
Belgium

MAEDA-MARTINEZ, Alejandro
Universidad Juarez del Estado De Durango
Department of Investigacion
Escuela superior de Biologia
Apartado postal 329-B
Gomez Palacio
Durango
Mexico

* Did not attend in person.

MARINCEK, Magdalena
University of Belgrade
Faculty of Biology
Department of Zoology
Belgrade 11 000
Yugoslavia

MERTENS, Johan
R.U.G.
Department of Ecology & Zoogeography
K.L. Ledeganckstraat 35
9000 Ghent
Belgium

MITCHELL, S. A.
University of the Orange Free State
Department of Zoology
P.O. Box 339
9300 Bloemfontein
South Africa
*Present address:* Water Research Commission
P.O.B. 824
Pretoria 0001
South Africa

MUNUSWAMY, N.
University of Madras
Department of Zoology
Life Science Building
Quindy Campus
Madras 600 025
India

MURA, Graziella
University 'La Sapienza'
Department of Animal and Human Biology
Viade dell Universita 32
00185 Rome
Italy

NAVARRO, J. C.
Instituto de Acuicultura de torre de la sal
Ribera de Cabanes
12595 Torre de la Sal
Castellon
Spain

PERINI, Virgilio
University 'La Sapienza'
B.A.U. Department
Viale dell Universita 32
00185 Rome
Italy

PETROV, Brigita
University of Belgrade
Faculty of Biology
Department of Zoology
Belgrade 11 000
Yugoslavia

PILLA, E. J.
*Present address:*
Univ. of Swansea
Dept. of Genetics
Singleton Park
Swansea
UK

PRASAD, Durga
Nagarjuna University
Department of Zoology
Nagarjuna Nager GUNTUR 522 510
Andhra Pradesh
India

SASSAMAN, Clay
University of California
Department of Biology
Riverside, CA 92521
USA

SEAMAN, M. T.
University of the OFS
Department of Zoology-Entomology
Bloemfontein 9300
South Africa

THIÉRY, Alain
Université D'Avignon
Faculté des Sciences
Departement de Biologie Animale
Hydrobiologie
Avignon 84 000
France

WILLIAMS, W. D.
University of Adelaide
Department of Zoology
GPO Box 498
Adelaide
S.A. 5001
Australia

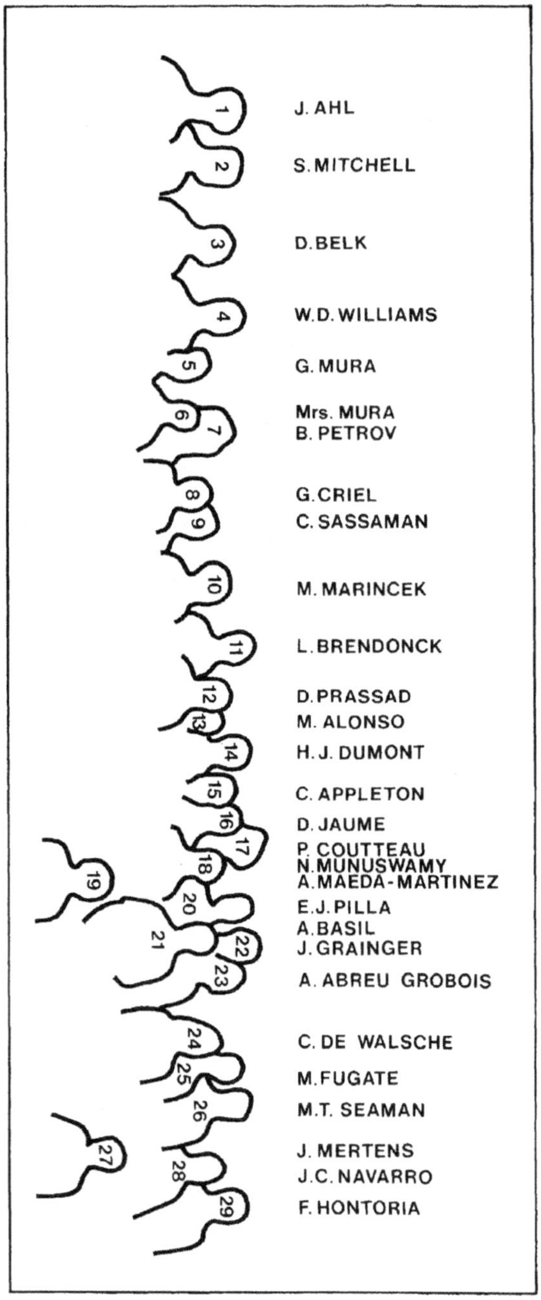
J. AHL
S. MITCHELL
D. BELK
W.D. WILLIAMS
G. MURA
Mrs. MURA
B. PETROV
G. CRIEL
C. SASSAMAN
M. MARINCEK
L. BRENDONCK
D. PRASSAD
M. ALONSO
H. J. DUMONT
C. APPLETON
D. JAUME
P. COUTTEAU
N. MUNUSWAMY
A. MAEDA-MARTINEZ
E.J. PILLA
A. BASIL
J. GRAINGER
A. ABREU GROBOIS
C. DE WALSCHE
M. FUGATE
M.T. SEAMAN
J. MERTENS
J.C. NAVARRO
F. HONTORIA

*Hydrobiologia* **212**: 1–10, 1991.
*D. Belk, H. J. Dumont & N. Munuswamy (eds), Studies on Large Branchiopod Biology and Aquaculture.*



# The growth rate and growth efficiency of *Streptocephalus macrourus* (Crustacea, Anostraca) cultured on microalgae

S.A. Mitchell*
*Department of Zoology and Entomology, University of the Orange Free State, PO Box 339, Bloemfontein 9300, South Africa; *present address: Water Research Commission, PO Box 824, Pretoria 0001, South Africa*

*Key words:* harvesting, production *Streptocephalus macrourus*, Anostraca, livestock waste, microalgae, biomass

## Abstract

Affordable biological technology for the reclamation of wastes and water of the waste streams from intensive livestock units is important in a country short of water. This study tested the concept of reclamation of waste by *Streptocephalus macrourus* (Crustacea: Anostraca) from the effluent of a high rate algal pond processing livestock wastes. *S. macrourus* showed a growth efficiency of 39% to 74% when fed optimal rations and cultured at densities between 10 and 400 $l^{-1}$. The maximum daily growth rates (0.15–0.21) approximate the growth rates of cladoceran or rotifer cultures managed for maximal biomass production. *S. macrourus*' ability to withstand crowding enabled the production from *S. macrourus* cultures (up to 91.8 mg dry mass $l^{-1}$ $d^{-1}$, or 1241 mg wet mass $l^{-1}$ $d^{-1}$) to exceed production recorded from cladoceran or rotifer cultures. Temperature influenced growth rate, with the highest growth rate occurring at 24 °C. The dilution rate of continuously fed cultures influenced growth rate, with the optimum dilution rate tested being 10 ml organism$^{-1}$ $d^{-1}$. Mass mortality occurred when organisms were held at a density of 4000 $l^{-1}$. *S. macrourus* is able to convert algae grown on livestock waste efficiently into anostracan biomass, and is able to give a very high daily production.

## Introduction

The currently recommended practice for the treatment of wastes from intensive livestock units in South Africa is to lagoon the effluent and allow it to evaporate (Funke *et al.*, 1984). A motivating factor for the revision of this policy in a semi-arid country such as South Africa is that water is in short supply and so wherever possible it should be upgraded for further use. Another reason for reviewing this policy, and one rated by Carpenter (1978) as a world-wide research research priority, is the recovery of nutrients from waste. Lincoln *et al.* (1986) estimated that up to 66.7% of the nitrogen in the feed fed to an intensive pig unit was lost in the waste stream. In their system 50% of the waste nitrogen was reclaimed.

One problem with the treatment of waste by algal mass culture is the removal of the suspended solids (either algal of bacterial) from the algal pond effluent. Various authors (Milligan *et al.*, 1980; Schluter & Groeneweg, 1981; Dinges, 1982; Proulx & de la Noue, 1985; a.o.) have proposed the use of filter feeding invertebrates as a second stage in the system which, under correct management, can filter the solids from the waste stream and can be harvested more easily than the algae. These invertebrates may then be used either as a protein supplement in the livestock unit, as a live food in aquaculture or for some other such application. An advantage of the use of such organisms as a protein supplement is the fact that the protein quality is comparable with other animal proteins rather than with plant proteins (e.g. Gallagher & Brown, 1975). The incorporation of these organisms grown on recycled nutrients into animal feeds which require a protein of animal origin to give the requisite amino acid spectrum will reduce the requirement for the input of protein from a 'new' source.

An important consideration when harnessing a natural food chain for the recovery of nutrients from waste waters is the efficiency with which the nutrients contained in the waste are converted into recoverable biomass. The loss of energy at each step in the food chain makes it desirable to keep the food chain short and efficient.

This study investigated the efficiency with which the anostracan *Streptocephalus macrourus* was able to convert algal biomass cultured in waste water into anostracan biomass under various management regimes.

## Methods and materials

### *Algal culture*

Algae for use in the study were cultured in outdoor ponds with a capacity of 500 l as described by Walmesley & Shillinglaw (1984), and agitated by airlift pumps as described by Mitchell (1986). However, during this study the ponds were neither covered not heated. One quarter of the culture volume was replaced daily with new medium consisting of cattle waste leachate (1 g dry mass cattle waste $l^{-1}$) and urea (50 mg $l^{-1}$) to provide nitrogen, as only solid cattle waste was used and this was deficient in nitrogen.

### *Estimation of algal dry mass*

The dry biomass of algae in the cultures was estimated twice weekly by filtering (0.45 $\mu$m pore size) and weighing the ovendried (24 h at 105 °C) algae (three replicates). A range of dilutions was prepared from the same algal suspension, and the optical density of each dilution was measured. The relationship between the dry mass of algae and the optical density of algae in suspension could then be calculated for the range of dilutions prepared. The optical density of the algal culture was measured daily, and the biomass estimated from the predetermined relationship between dry mass and optical density. From these figures the daily ration for each treatment was calculated.

### *Estimation of dry mass of S. macrourus*

Twice weekly the dry mass of the organisms was estimated from the total length (mm) by the formula

$$Y = e^{-2.6412 + 0.2201*X}$$
$$r = 0.9695$$

where $X$ is the length (mm), $Y$ is the dry mass (mg).

This relationship is valid for organisms between 2 and 18 mm long. The dry mass of organisms 19 mm or longer is underestimated.

The relationship between wet mass and dry mass of *S. macrourus* was as follows:

$$\text{Wet mass} * 0.0739 = \text{dry mass (SD 0.006)}.$$

### *Experimental design*

Two different methods of culture management were used. In the first, organisms were fed the required ration of algae once daily on a semi-continuous basis. In the second, medium containing the appropriate algal biomass was pumped into the culture continuously at the appropriate rate by a peristaltic pump, and the culture effluent was drained continuously to maintain the desired volume.

*A. Semi-continuously fed cultures*

A matrix (in triplicate) of treatments covering both a range of densities of organisms (10, 50, 100, 200, 300 and 400 organisms $l^{-1}$) and a range of feeding rations (daily rations of algae as estimated dry biomass at 0.11, 0.21, 0.32, 0.47 and 0.70 of estimated dry biomass of *S. macrourus*) were conducted in 200 ml of medium in 250 ml erlenmeyer flasks. The medium in each flask was aerated through a pasteur pipette. The appropriate volume of medium was replaced by fresh medium containing the daily ration for the treatment. As the organisms in the higher density and higher ration treatments approached sexual maturity it was necessary to concentrate the medium as the algal culture effluent contained insufficient biomass to provide the required ration.

The mortality in each replicate was recorded daily and the ration adjusted to the number (and so total biomass) of organisms in the replicate concerned. The ration for each replicate of each treatment was recalculated twice weekly after the estimated of the biomass in the particular flask. This experiment was carried out at room temperature of 20–22 °C.

A second experiment was conducted at a range of constant temperatures to ascertain the effect of temperature on the growth rate. In these, organisms were grown in medium kept in constant temperature chambers at 12 °C, 16 °C, 20 °C, 24 °C and 29 °C and exposed to a diel cycle of 12 h light. These organisms were fed a ration of 0.5 of the body mass daily.

*B. Continuously fed cultures*

Two experiments were conducted using continuously fed cultures. In the first, organisms were held in 200 ml of medium at a density of 1000 $l^{-1}$ and fed a daily ration of 0.217 of the *S. macrourus* dry biomass in each treatment. The food for the treatments was suspended in 0.5 l, 1 l and 2 l daily (corresponding to 2.5 ml, 5 ml and 10 ml per organism respectively), and pumped through the respective treatments at this rate.

In the second experiment, the *S. macrourus* were also held in 200 ml of medium, but at densities of 1000, 2000 and 4000 organisms $l^{-1}$. In this experiment, the medium was passed through the cultures at a rate of 10 ml daily per organism (corresponding to the highest of the three rates used in the first fed experiment). The organisms in this experiment were fed a ration of 0.656 of their estimated body mass daily.

Where necessary in the following calculations, the number of surviving organisms in the treatment under consideration was taken into account.

*Calculation of growth rate*

The daily growth rate was calculated (using the estimate for dry mass) by the following formula

$$\ln(X_t/X_0) = mt \text{ (Pirt, 1975)}$$

where $X_t$ is the dry mass at time $t$, $X_0$ the dry mass at the beginning of the period, $t$ the time in days and $m$ the specific daily growth rate.

*Calculation of growth efficiency*

The growth efficiency of the organisms was calculated from the formula

$$\frac{\text{mass of food consumed} * 100}{\text{mass increase of organisms}}$$

and was expressed as a percentage.

*Calculation of daily production*

The production of the organisms was derived from the mass increase during the period concerned multiplied by the number of organisms in the replicate concerned.

## Results

*The effect of rations on survival of S. macrourus to sexual maturity*

In the semi-continuous cultures survival was low in the treatments receiving a ration of 0.11 of their body mass daily, particularly in the treatments containing organisms at 50 $l^{-1}$ or less (survival

<5%). The survival at higher densities varied from 16% to 42%. Survival was higher in treatments receiving a daily ration of 0.21 of their body mass, with mean survival to sexual maturity varying between 34% and 75% per treatment. Mean survival of organisms fed at 0.32 varied between 60% and 100% with individual duplicates of each treatment showing survival in excess of 80%. Most treatments which received rations of 0.47 or 0.70 showed survivals in excess of 80%.

Survival of organisms in continuously fed cultures showed a different pattern. There was >95% survival in each treatment of the experiment fed at 0.217 of their body mass daily. This was substantially higher than the survival recorded for organisms fed semi-continuously at this level. Survival amongst organisms fed at 0.656 of their body mass daily varied between treatments. Survival in the treatment with 1000 organisms $l^{-1}$ was 75.4%. Survival in the treatment containing organisms at 2000 $l^{-1}$ was lower at 59.7% and survival in the treatment containing organisms at 4000 $l^{-1}$ was 4.5%. Most of the mortality in this latter treatment occurred after day 19. Maximum survival of organisms was obtained at daily rations between 0.32 and 0.7 of body mass for the semi-continuously fed treatments and at 0.217 of body mass for those in the continuously fed treatments. Densities up to 400 $l^{-1}$ had no apparent influence on survival, but densities of 2000 and 4000 $l^{-1}$ had an appreciably negative influence on survival.

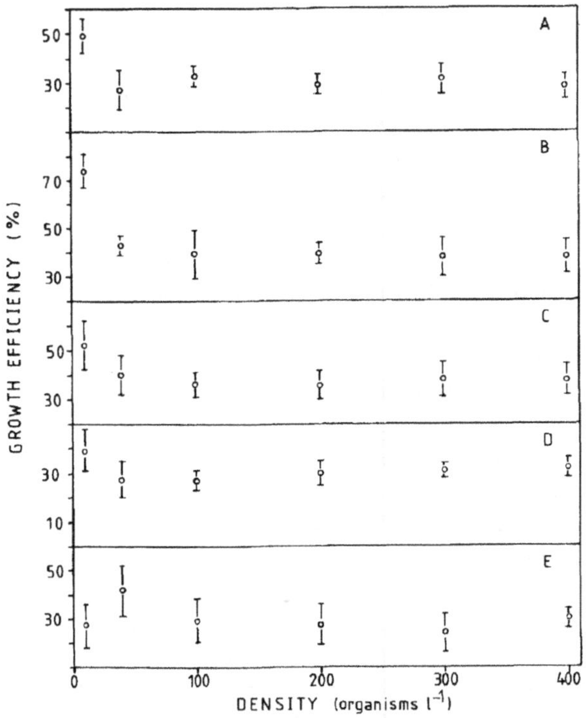


*Fig. 1.* The mean growth efficiency (±95% confidence limits) of *Streptocephalus macrourus* when held at different densities and fed various rations expressed as a proportion of their body mass.
A – 0.7
B – 0.47
C – 0.32
D – 0.21
E – 0.11

*The effect of rations on growth efficiency*

The growth efficiency (GE) of *S. macrourus* varied with management and was significantly higher at all densities (2 way ANOVA and Tukey's Test) in the semi-continuously fed experiment when organisms were fed 0.47 of their body mass daily (Fig. 1B). The GE was generally higher when organisms were not crowded, the exception being amongst organisms facing incipient starvation (ration of 0.11 of body mass daily). The maximum GE (74%) was obtained when organisms were fed a ration of 0.47 body mass daily and held at a density of 10 organisms $l^{-1}$ (Fig. 1B). The GE obtained from organisms fed this ration decreased to 39% at 100 organisms $l^{-1}$ and the mean GE for organisms held at densities between 100 and 400 $l^{-1}$ was 39.4%. The GE for organisms in treatments fed at 0.32 of body mass daily was significantly lower than that of organism fed at 0.47, but significantly higher than that of organisms fed 0.7; 0.21 or 0.11 of their body mass daily (Fig. 1C). The mean GE of organisms fed a ration of 0.7 body mass daily was not significantly different from that of organisms fed at 0.21 or 0.11 of their body mass daily (2 way ANOVA and Tukey's Test, Figs. 1 and 2).

The GE of organisms in continuously fed treatments fed at 0.217 was strongly influenced by

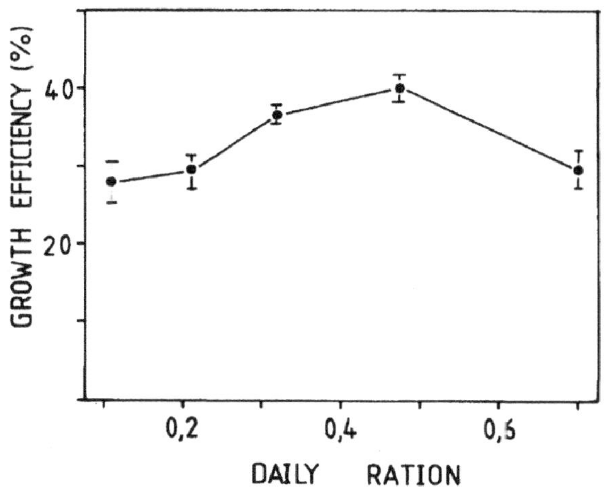


*Fig. 2.* The mean growth efficiency (±95% confidence limits) of *Streptocephalus macrourus* cultured at densities between 50 and 400 organisms $l^{-1}$ when fed various rations.

dilution rate. Organisms in the culture in which the daily ration was carried in 2.5 ml of medium per organism showed a GE of 16.6%. This increased to 31.5% when daily ration was carried in 5 ml per organism and to 34.8% when daily ration was carried in 10 ml per organism. The GE of the organisms held at 1000 $l^{-1}$ (27.7%), 2000 $l^{-1}$ (33.8%) and 4000 $l^{-1}$ (32.0%) showed little variation.

The optimum daily ration tested was 0.47 of body mass. Organisms fed a daily ration of 0.32 of body mass gave a similar GE to those fed at 0.47, but rations of higher or lower proportions of body mass gave lower GEs than those obtained from organisms fed rations between 0.32 and 0.47 of body mass daily. In continuously fed experiments, a higher dilution rate increased GE, but crowding did not effect GE of the surviving organisms even when mortality was substantial.

### *The effect of rations on growth rate*

The main difference between growth efficiency and growth rate was that although organisms fed a ration of 0.70 of their body mass daily showed a reduced growth efficiency, the growth rate was slightly higher at this ration that at the other rations tested, although it was only significantly higher than that of organisms fed at 0.21 and 0.11 of their body mass daily (2 way ANOVA and Tukey's Test; Fig. 3 & 4). The mean growth rate of organisms fed a ration of 0.7 body mass daily and held at densities between 50 and 400 $l^{-1}$ was 0.158 which is similar to the mean growth rate of organisms fed at 0.47 body mass daily (0.144) for the same range of densities (Fig. 4). Organisms fed at rations of 0.11 and 0.21 of body mass daily not only showed substantially lower growth rates (0.033) for organisms fed at 0.11 body mass daily and 0.052 for organisms fed at 0.21 body mass daily than organisms fed at higher levels, but they also took longer to reach sexual maturity and did so at a smaller size (Table 1). The time taken by organisms to reach sexual maturity at higher

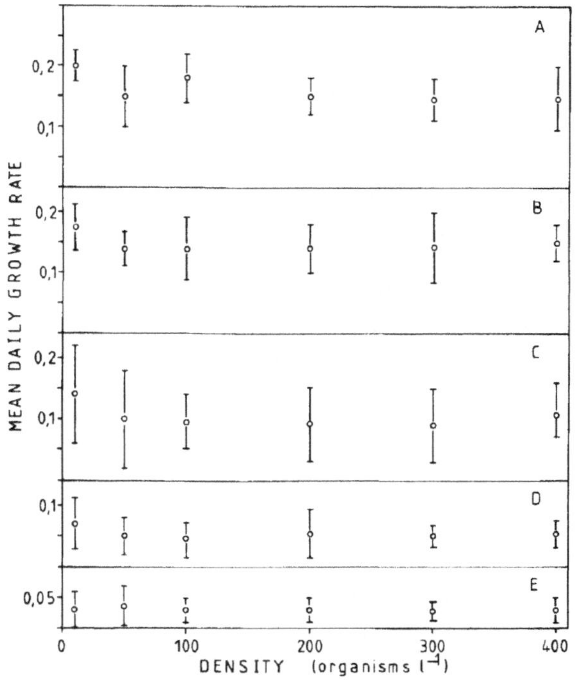


*Fig. 3.* The men daily growth rates (±95% confidence limits of *Streptocephalus macrourus* when held at different densities and fed various rations expressed as a proportion of their body mass.
A – 0.7
B – 0.47
C – 0.32
D – 0.21
E – 0.11

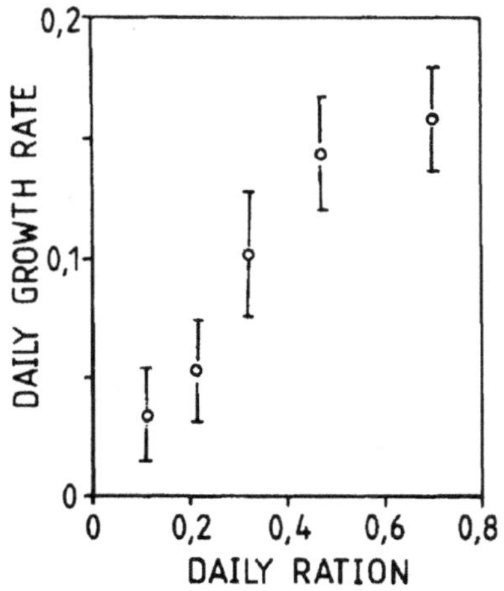


*Fig. 4.* The mean growth rate (±95% confidence limits) of *Streptocephalus macrourus* cultured at densities between 50 and 400 organisms $l^{-1}$ when fed various rations.

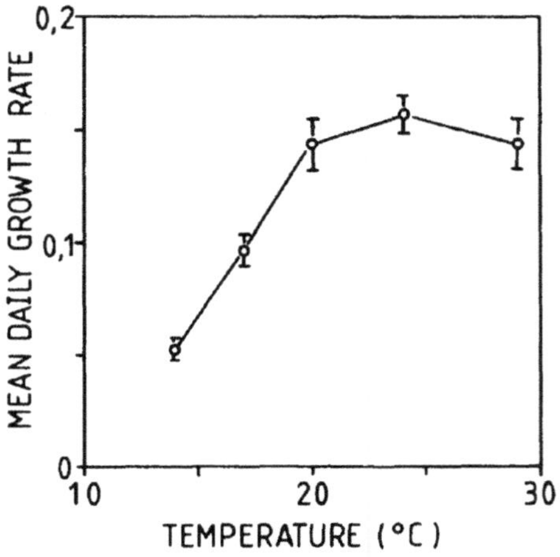


*Fig. 5.* The mean daily growth rates (±95% confidence limits) of *Streptocephalus macrourus* when grown in constant temperatures ranging from 12 °C to 29 °C, and fed a ration of 0.5 of the biomass in the culture daily.

rations did not vary between treatments, although organisms held at higher densities grew slightly more slowly than those held at lower densities.

The growth rates of organisms grown at a constant temperature of 12 °C was 0.052 $d^{-1}$ (Fig. 5). The daily growth rates increased linearly to 0.143 at 20 °C and reached a maximum of 0.155 at 24 °C. At 29 °C the daily growth rate (0.141) was similar to that at 20 °C (Fig. 5).

The growth rates of organisms which were fed continuously at varying dilution rates varied from 0.03 in the treatment diluted at 2.5 ml per organism to 0.06 in the treatment diluted at 10 ml per organism. This range of growth rates is within the range shown by the organisms in the semi-continuously fed cultures at a similar ration, but growth rate at the highest dilution rate was double that at the lowest. The mean daily growth rates of the organisms in the treatments fed continuously but held at different densities varied from 0.11 to 0.14, which is lower than was shown by organisms fed similar rations on a semi-continuous basis.

*Table 1.* The time taken for *Streptocephalus macrourus* to reach sexual maturity when fed at various rations, expressed as a proportion of their body mass daily.

| Ration | Time (days) to sexual maturity | Length (mm) at sexual maturity |
|---|---|---|
| 0.11 | 50 | 10.5–11.5 |
| 0.21 | 50 | 14.5–15.5 |
| 0.32 | 25–30 | 17.0–18.0 |
| 0.47 | 25–30 | 19.0–22.0 |
| 0.70 | 25–30 | 19.0–22.0 |

The highest growth rates were obtained at the highest rations in the semi-continuously fed experiment. In the continuously fed experiments organisms fed at a ration of 0.217 body mass daily showed growth rates similar to the semi-continuously fed experiment. However, organisms fed at 0.656 of their body mass daily showed lower growth rates than did organisms which were fed a similar ration semi-continuously.

### *The effect of rations on production*

The mean daily production of semi-continuously fed organisms fed any particular ration increased linearly with increasing density with the correlation coefficient $r^2$ varying from 0.886 for organisms fed a ration of 0.11 to 0.98 for organisms fed a ration of 0.47. However, mean daily production of organisms fed at rations of 0.11 and 0.21

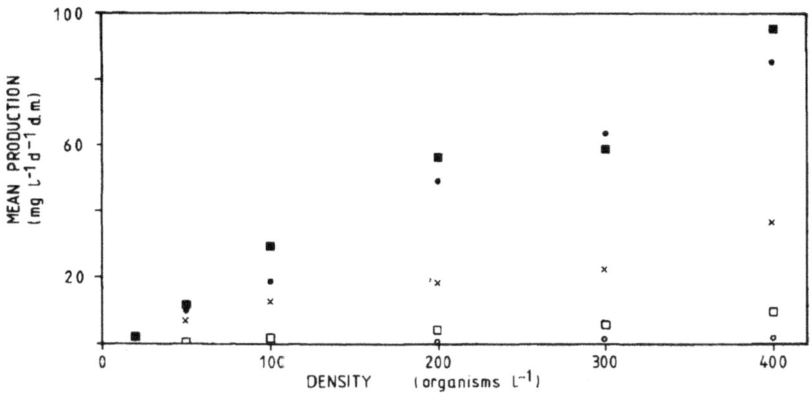


*Fig. 6.* The mean daily production of *Streptocephalus macrourus* when held at different densities and fed various rations expressed as a proportion of their body mass.

| | Ration | Correlation ($r^2$) |
|---|---|---|
| ■ | – 0.7 | 0.96 |
| ● | – 0.47 | 0.98 |
| × | – 0.32 | 0.92 |
| □ | – 0.21 | 0.96 |
| ○ | – 0.11 | 0.89 |

of body mass daily were low, being 1.5 mg dry mass $l^{-1} d^{-1}$ and 9.7 mg dry mass $l^{-1} d^{-1}$ respectively at a density of 400 organisms $l^{-1}$ after mortality had been taken into account (Fig. 6). Mean daily production was higher amongst organisms fed a ration of 0.32 of body mass daily (36.7 mg dry mass $l^{-1} d^{-1}$ at 400 organisms $l^{-1}$), but this was substantially lower than obtained from organisms fed a ration of 0.47 (86.3 mg dry mass $l^{-1} d^{-1}$ at 400 organisms $l^{-1}$) and 0.70 (91.8 mg dry mass $l^{-1} d^{-1}$ at 400 $l^{-1}$) (Fig. 6).

The production of organisms fed continuously but at various dilution rates was strongly influenced by the dilution of the culture medium. The culture diluted at 2.5 ml per organism $d^{-1}$ showed a mean daily production of 12.7 mg dry mass $l^{-1} d^{-1}$, and this increased to 32.3 mg dry mass $l^{-1} d^{-1}$ for organisms in the culture diluted at 5 ml per organism $d^{-1}$ and 45.5 mg dry mass $l^{-1} d^{-1}$. The organisms fed continuously at a ration of 0.656 body mass daily gave a higher production than those fed at a ration of 0.217 body mass daily in spite of suffering heavier mortality. A mean daily production of 105.8 mg dry mass $l^{-1}$ was obtained from the treatment containing 1000 organisms $l^{-1}$, and the treatment containing 2000 organisms $l^{-1}$ gave a mean daily production of 128.3 mg dry mass $l^{-1}$. The treatment containing organisms at 4000 $l^{-1}$ gave a mean daily production of 168.4 mg dry mass $l^{-1} d^{-1}$ before the heavy mortality which began three days before the survivors laid their first eggs.

## Discussion

Once the basic metabolic requirements of *S. macrourus* have been met, the growth efficiency remains near maximal over a range of rations before decreasing again. The growth efficiency of the rotifer *Brachionus calyciflorus* showed a similar response to varying rations (Mitchell, 1986). At rations which did not fully meet the basic metabolic requirements of the *S. macrourus*, both growth rate and growth efficiency were low with an increased mortality. At rations higher than required for optimal growth efficiency, a substantial decline in growth efficiency is associated with a relatively small increase in growth rate.

The response of *S. macrourus* to variations in rations is comparable to that of fish rather than to that of other filter feeding invertebrates with shorter life cycles such as rotifers or cladocerans. The latter organisms live in food limited equilibrium with the environment, and growth efficiency remains near optimal over a relatively wide range of food concentrations. The production from cultures of rotifers or cladocerans varies with the ration. However, organisms which have longer life cycles, such as *S. macrourus*, are unable to respond to short-term changes in ration by changes in reproductive rate. The response is manifested by changes in the growth rate and growth efficiency instead. Thus, it is important that the daily ration fed to this category of organisms approximates that which will give the best return. This was the problem faced by Milligan *et al.* (1980) when using *Artemia* (Anostraca) to clear algae and bacteria from a hypersaline waste stream. The solution in this study was found with a polyculture of *Artemia* and the salt water rotifer (*Brachionus plicatilis*), where *Artemia* numbers were kept more or less constant, and short term variations in biomass of bacteria and algae in the waste stream were compensated for by variations in the numbers of *B. plicatilis* in polyculture.

Incipient starvation affected organisms in three ways. The first effect was a reduction in growth rate by an amount proportional to the reduction of ration below the optimum. This response was apparent even in organisms fed a ration sufficient to give a growth efficiency approaching the maximum obtained. As the ration was decreased further, the first trend continued, but the organisms took longer to reach sexual maturity. The third response to reduced rations was mortality. The percentage mortality not only increased with decreasing rations, but also with decreasing density of organism at any ration. This may indicate that there is a threshold concentration of algal biomass below which, in terms of energy, feeding becomes uneconomic for the organism. This aspect of *S. macrourus* feeding needs further investigation.

If organisms are not fed correctly, growth rate and growth efficiency may be suboptimal. The level of ration which gave the best combination of survival, growth efficiency and growth rate was 0.47 of the organisms' body mass daily. While growth efficiency was highest at low densities of organisms, production was low. Growth efficiency and mean daily growth rate varied little between densities of 50 to 400 organisms $l^{-1}$ in semi-continuously fed cultures, and production increased linearly with numbers of organisms $l^{-1}$ in cultures managed in this manner. Reeve (1963) obtained peak growth efficiencies from *Artemia* cultures fed at low levels (his units of measurement were not comparable to those used in this study), with growth efficiencies of up to 79% being recorded between hatching and sexual maturity. This is similar to the highest growth efficiency (74%) recorded from *S. macrourus* for the same period of growth.

The growth rates of the organisms held at $1000\,l^{-1}$ and fed continuously at 0.217 of their body mass daily were comparable to those obtained in semi-continuously fed cultures at lower densities. However, growth rates increased with increasing dilution of ration to the point that the growth rate obtained at the highest dilution was double that obtained at the lowest dilution. This, taken together with a similar increase in the growth efficiencies over the same dilutions, indicates that conditions were better for the organisms in treatments diluted at higher rates. However, organisms fed continuously at 0.656 of their estimated body mass daily used the ration less efficiently than those fed semi-continuously on a similar ration.

The highest growth efficiency achieved by *S. macrourus* when held at densities between $100\,l^{-1}$ and $400\,l^{-1}$ (ca. 40%) is similar to the highest shown by various species of cladocerans under optimal conditions in both wild and cultured populations growing on similar diets (39.3% Lei & Armitage, 1980; 40%, Tarifeno-Silva *et al.*, 1982; 40%, Proulx & de la Noue, 1985). The figure of 39–40% is higher than the maximum obtained for either *Brachionus plicatilis* (24.7%, Hirata, 1979; 25%, James *et al.*, 1983) or

*B. calyciflorus* (25%, Boraas, 1983; 24.5%, Mitchell, 1986) when grown on similar diets.

The growth rates shown by *S. macrourus* were lower than the highest recorded for rotifers. Maximum production was obtained from a culture of the rotifer *Brachionus calyciflorus* at a population growth rate of 0.21 $d^{-1}$ (Mitchell, 1986) and 'near maximum' numbers of *B. rubens* were obtained at a population growth rate of 0.25 $d^{-1}$ (Groeneweg & Schluter, 1981). Both of these studies were conducted in the presence of excess food, and both found that the organisms were unable to maintain a growth rate of 0.5 $d^{-1}$. Boraas (1983), on the other hand, worked with *B. calyciflorus* in a food limited system and obtained maximum production at a population growth rate of 1.296 $d^{-1}$. Mitchell & Richmond (1987) recorded a maximum growth rate of 1.26 $d^{-1}$ for *B. plicatilis* under conditions of low density and excess food. By comparison, the growth of cichlid fish is lower (*Tilapia rendalli*; 0.036 $d^{-1}$ at 28 °C (Caulton, 1982) and *Oreochromis mossambicus*; 0.046 $d^{-1}$ at 24 °C (Gaigher & Krause, 1983).

While the potential growth rate of *S. macrourus* is lower than that of rotifers or cladocerans under certain conditions, the potential production is higher because the organism can be held at higher densities than rotifers or cladocerans. Maximum figures obtained for the output of *B. calyciflorus* from mass culture were 25 mg dry mass $l^{-1}$ $d^{-1}$ (Boraas, 1983) and 19.2 mg dry mass $l^{-1}$ $d^{-1}$ (Mitchell, 1986). Ivleva (1973) quoted yields of between 30 and 100 mg wet mass (ca. 3–10 mg dry mass) $l^{-1}$ $d^{-1}$ for *Daphnia magna* under mass culture conditions, and Dinges (1982) recorded a yield of 11.2 mg wet mass (ca. 1.1 mg dry mass) $l^{-1}$ $d^{-1}$ for *Daphnia* harvested from waste water stabilisation ponds. The highest yield reported by Ivleva (1973) is approximately 50% of the rotifer yields obtained by Boraas (1983) or Mitchell (1986). Proulx & de la Noue (1985) reared *D. magna* in a flow through system, and realised a mean yield of 432 mg wet mass (ca. 43.2 mg dry mass) $l^{-1}$ $d^{-1}$. This is substantially higher than the yields of rotifers referred to above, and the authors feel that this harvest was below optimal. They suggest that a harvest rate of 35% every 3 days (11.7% $d^{-1}$) would give optimal results. If the cultures were harvested daily as done by Mitchell (1986) and not every third day, the harvest volume could possibly be increased still further before a decrease in yield occurred. These authors reported an increase in juvenile mortality as a result of crowding at the higher densities.

The yields discussed above were considerably lower than that obtained from *S. macrourus*. In the semi-continuously fed experiment at a density of 400 organisms $l^{-1}$, mean productions of 86.3 mg dry mass $l^{-1}$ $d^{-1}$ (ration – 0.47) and 91.8 mg dry mass $l^{-1}$ $d^{-1}$ (ration – 0.7) (1168 and 1242 mg wet mass $l^{-1}$ $d^{-1}$ respectively) were recorded over the growth period. The mean daily production realised from the continuously fed cultures was higher, although the mean production per individual organism was lower at the greater densities. The severe mortality which occurred just before the onset of sexual maturity at a density of 4000 $l^{-1}$ indicated that mortality is apparently density dependant, in contrast to the observation made by Williams (1985) or temporary pool fauna. However, the biomass at which mortality occurred was much higher than would normally be found in the field, and was also much higher than the biomass at which juvenile mortality was recorded amongst *Daphnia magna* by Proulx & de la Noue (1985).

The results of this investigation indicate that *S. macrourus* is able to efficiently convert the algal biomass in the effluent from a high rate algal pond into anostracan biomass. *S. macrourus* is tolerant to crowding, and so is also able to yield substantially higher productions than rotifers or cladocerans under mass culture conditions. The optimal survival and production will be obtained from a *S. macrourus* culture containing organisms at a density of between 400 and 1000 $l^{-1}$ and fed at a ration of approximately 0.47 of the organisms' estimated dry mass daily.

**Acknowledgements**

I thank Mrs. E. Phillips for competent technical assistance and Prof. P.H. Hewitt for allowing me to use the constant temperature chambers. This work was funded by the Water Research Commission, Pretoria.

## References

Boraas, M. E., 1983. Population dynamics of food limited rotifers in two stage chemostat culture. Limnol. Oceanogr. 27: 546–563.

Carpenter, R. L., 1978. Aquaculture as an alternative waste water treatment. In Pastakia C. M. R. (ed.), Fish farming and wastes. Report of the proceedings of an international conference held at the University College of London, 4–5 January 1978. London: 110–115.

Caulton, M. S., 1982. Feeding, metabolism and growth of tilapias: some quantitative considerations. In Pullin R. S. V. & R. H. Lowe-McConnell (eds), The biology and culture of tilapias. ICLARM. Manila, Phillipines: 157–183.

Dinges, R., 1982. Natural systems for water pollution control. van Nostrand Reinhold Environmental Engineering Series. New York. 252 pp.

Funke, J. W., J. G. Knoesen & J. C. Venter, 1984. Code of practice for the handling of manure from intensive animal feeding units. Water Research Commission, Pretoria. 42 pp.

Gaigher, I. G. & J. B. Krause, 1983. Growth rates of mosambique tilapia (*Oreochromis mossambicus*) and silver carp (*Hypophthalmichthys molitrix*) without artificial feeding in floating cages in plankton-rich waste water. Aquaculture 31: 361–367.

Gallagher, M. & W. D. Brown, 1975. Composition of San Francisco Bay brine shrimp (*Artemia salina*). J. agric. Food. Chem. 23: 630–632.

Groeneweg, J. & M. Schluter, 1981. Mass production of freshwater rotifers on liquid wastes. II. Mass production of *Brachionus rubens* Ehrenberg in 1838 in the effluent of high rate algal ponds used for the treatment of piggery waste. Aquaculture 25: 25–33.

Hirata, H., 1979. Rotifer culture in Japan. In Styczynska-Jurewicz E., T. Backiel, E. Jaspers & G. Persoone (eds), Cultivation of fish fry and its love food. European Mariculture Society special publication No. 4. Bredene, Belgium: 361–376.

Ivleva, I. V., 1973. Mass cultivation of invertebrates, biology and methods. Israel program for scientific translations. Jerusalem. 148 pp.

James, C. M., M. Bou-Abbas, A. M. Al-Khars, S. Al-Hinty & A. E. Salman, 1983. Production of the rotifer *Brachionus plicatilis* for aquaculture in Kuwait. Hydrobiologia 104: 77–84.

Lei, C-H. & K. B. Armitage, 1980. Ecological energetics of a *Daphnia ambigua* population. Hydrobiologia 70: 133–143.

Lincoln, E. P., B. Koopman, L. O. Bagnall & R. A. Nordstedt, 1986. Aquatic system for food and feed production from livestock wastes. J. Agric. Engng. Res. 33: 159–169.

Milligan, D. J., J. A. Quick, S. E. Hill, J. A. Morris & R. J. Hover, 1980. Sequential use of bacteria, algae and brine shrimp to treat industrial waste water at pilot plant scale. In Persoone G., P. Sorgeloos, O. Roels & E. Jaspers (eds), The Brine Shrimp Artemia Vol. 3. Universa Press, Wetteren, Belgium: 193–206.

Mitchell, S. A., 1986. Experiences with outdoor semi-continuous mass culture of *Brachionus calyciflorus* Pallas (Rotifera). Aquaculture 51: 289–297.

Mitchell, S. A. & A. Richmond, 1987. The use of rotifers for the maintenance of monoalgal mass cultures of *Spirulina*. Biotechn. Bioeng. 30: 164–168.

Pirt, J. S., 1975. Principles of cell and microbe cultivation. Blackwell Scientific Publications. Oxford.

Proulx, D. & J. de la Noue, 1985. Harvesting *Daphnia magnum* grown on urban tertiarily-treated effluents. Wat. Res. 19: 1319–1324.

Reeve, M. R., 1963. Growth efficiency in *Artemia* under laboratory conditions. Biol. Bull. 125: 133–145.

Schluter, M. & J. Groeneweg, 1981. Mass production of freshwater rotifers on liquid wastes. I. The influence of some environmental factors on population growth of *Brachionus rubens* Ehrenberg 1838. Aquaculture 25: 17–24.

Tarifeno-Silva, E., L. Y. Kawasaki, D. P. Yu, M. S. Gordon & D. J. Chapman, 1982. Aquacultural approaches to recycling of dissolved nutrients in secondarily treated domestic wastewaters. II. Biological productivity of artificial foodchains. Wat. Res. 16: 51–57.

Walmsley, R. D. & S. N. Shillinglaw, 1984. Mass algal culture in outdoor plastic-covered minipond systems. Ann. appl. Biol. 104: 185–197.

Williams, W. D., 1985. Biotic adaptations in temporary lentic waters, with special reference to those in arid and semi-arid regions. Hydrobiologia 125: 85–110.

Hydrobiologia 212: 11–17, 1991.
D. Belk, H. J. Dumont & N. Munuswamy (eds), Studies on Large Branchiopod Biology and Aquaculture.




# Culturing Artemia (Tuticorin strain) in organic and agricultural wastes at different salinities

J. Antony Basil[1] & G.T. Pandian[2]
[1]*Department of Biochemistry, School of Biological Sciences, Madurai Kamaraj University, Madurai 625 021, India*; [2]*Department of Zoology, Aditanar College of Arts and Science, Tiruchendur, Virapandianpatnam p.o., S. India*

*Key words:* Artemia (Tuticorin strain), organic waste, agricultural waste, salinity, growth, fecundity

## Abstract

Artemia sp (Tuticorin strain) was cultured at a density of 250 individuals $l^{-1}$ at 35‰, 45‰, 60‰, 75‰ salinity using five combinations of groundnut oil cake, decayed cabbage leaves, single superphosphate and Baker's yeast as feed. Effects on survival, growth, and fecundity were noted.

## Introduction

In India, Artemia occurs (Fig. 1) naturally in the salt ponds of Maharastra (Kulkarni, 1953), Sambhar lake of Rajasthan (Baid, 1958), Tamil Nadu (Royan *et al.*, 1970; Basil *et al.*, 1985) and along Sourastra and Kutch coasts of Gujarat (The Gulf of kutch) (Royan, 1979). The annual cyst production from these natural areas is not sufficient to meet India's demand. Recycling agricultural and organic waste for mass production of Artemia is therefore an obvious priority.

*Fig. 1.* Salt producing areas along the coast line of India and locations of Artemia occurrence.

## Material and methods

*Artemia salina* cysts were collected from salt pans of Tuticorin, South India. They were hatched under laboratory conditions and reared to adults.

In our experiments, highly saline water from Tuticorin was added to natural sea water, raising salinity to 45‰; 60‰ and 75‰. Rearing experiments (in duplicate) were conducted in a 12 litre rectangular plastic trough, in open air, and in the shadow. To this, we added 10 days treated biogas slurry (cow dung), 20 days treated poultry litter and agricultural waste products like groundnut oil cake and 10 days treated decayed cabbage leaves. Samples containing biogas slurry, poultry litter, cabbage leaves, and combinations were analysed for bacteria, following the method of Lakshmanan *et al.* (1971).

In addition, single super phosphate and Baker's

*Table 1.* Feed Combination of organic and agricultural waste products in culturing Artemia ($g.l^{-1}$).

| Combination | Poultry litter | Biogas slurry | Groundnut oil cake | Decayed cabbage leaves | Single super phosphate | Baker's Yeast |
|---|---|---|---|---|---|---|
| A | 1.5 | 1.5 | – | 0.25 | – | – |
| B | 1.5 | 1.5 | – | 0.25 | 0.05 | – |
| C | 1.5 | 1.5 | 0.15 | 0.25 | – | 0.025 |
| D | 1.5 | 1.5 | 0.15 | 0.25 | 0.05 | – |
| E | 1.5 | 1.5 | 0.15 | 0.25 | 0.05 | 0.025 |

yeast were used. The combinations (A–E) used are given in Table 1. The culture containers were enriched with A–E 3 days before the inoculation of *Artemia* nauplii. The length of the nauplii was measured before inoculation (initial length). Growth and survival were recorded every third day by taking 5 animals (preserved in 10% formaldehyde) from each container for measuring. To study survival three samples were collected and counted every third day. For fecundity, the number of eggs was counted by dissecting the brood pouch under a binocular microscope.

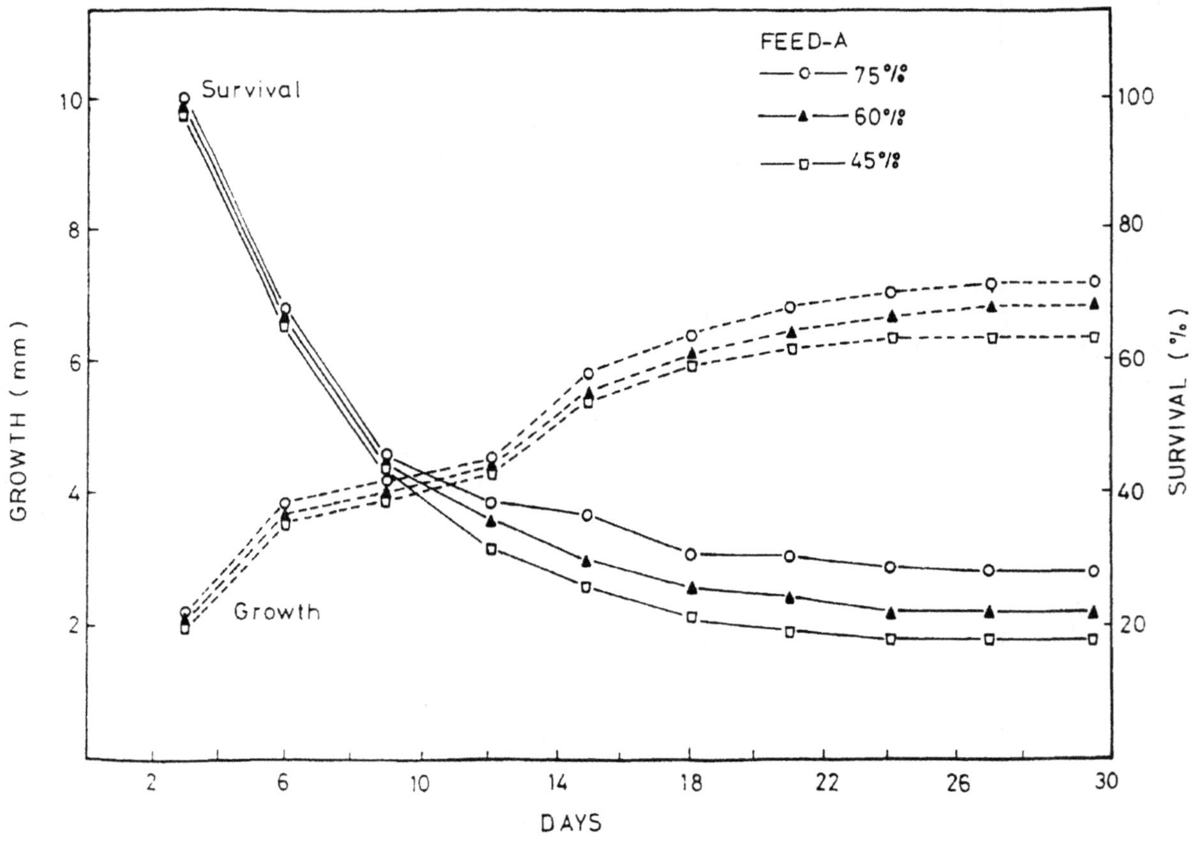


*Fig. 2A*

*Table 2.* Occurrence of bacteria in the organic waste materials (Cowdung, Poultry manure, Cabbage and Combination).

| Bacteria | Organic waste materials | | | |
|---|---|---|---|---|
| | Cowdung | Poultry Manure | Cabbage | Combination |
| *Escherichia coli* | + | – | + | – |
| *Streptococcus bous* | + | – | + | – |
| *Clostridium Spp.* | + | – | – | + |
| *Lactobacillus Spp.* | – | + | – | + |

## Results

Bacteria: Screening revealed four types of bacteria in the substances. Poultry manure contained one type (*Lactobacillus sp.*), Cowdung and cabbage contained three types. The mixture of the three organic wastes contained all four (Table 2).

## Survival

Figure 2 (A–E) illustrates Artemia survival under three salinities and five feed combinations.

Combination 'A' was poorest. Maximum survival was only 28%. Combination 'E' gave maximum survival (92%). Besides food combinations,

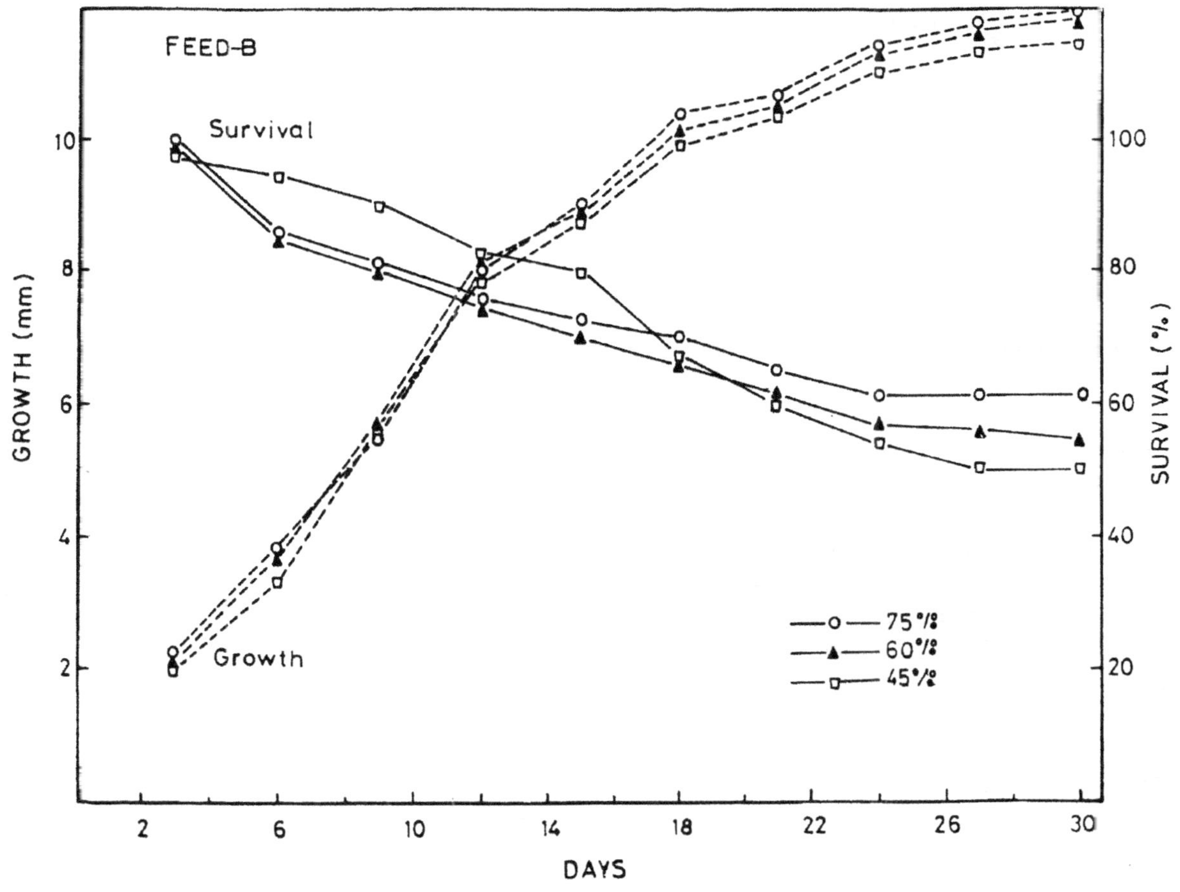


*Fig. 2B*

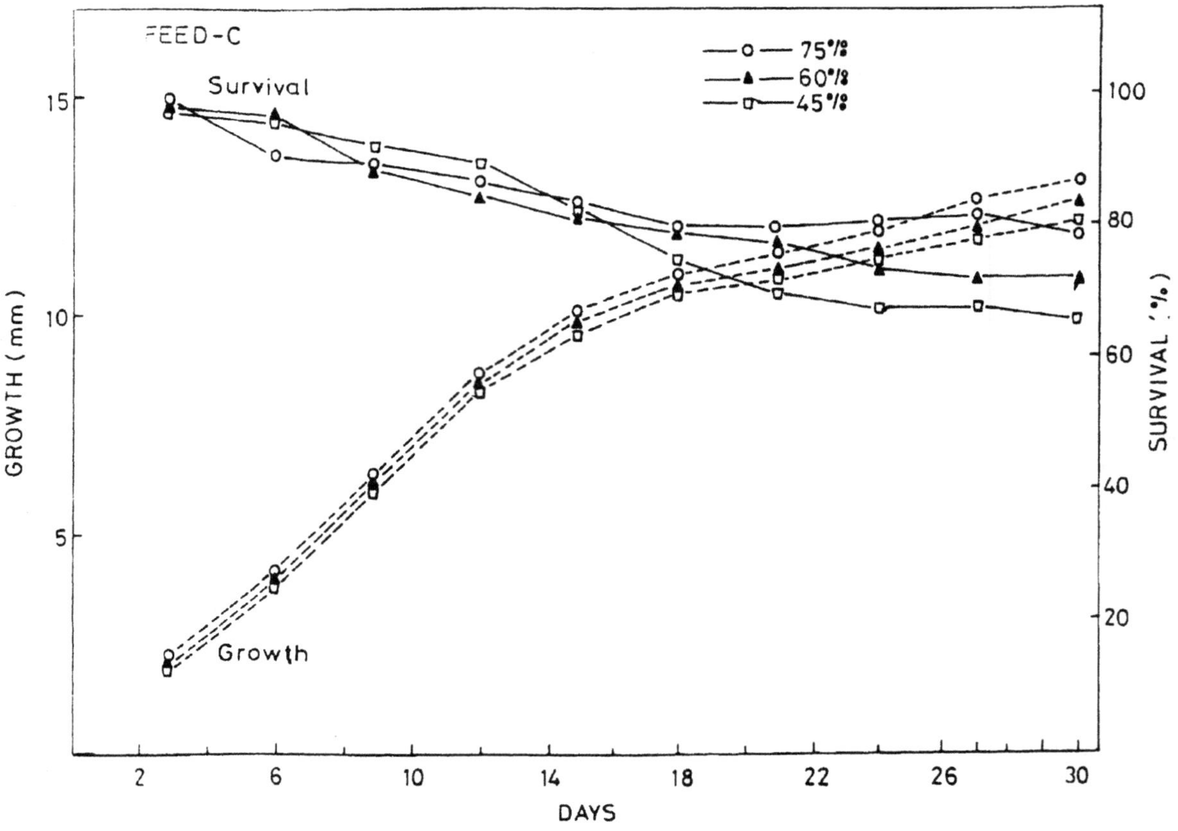


*Fig. 2C*

highest salinity gave maximum survival, but no statistical significance could be ascertained.

## Growth

In all cases (Fig. 2), growth was sigmoidal. Combination 'A' supported the lowest growth; (maximum length 7.1 mm after 30 days). All other food combinations supported better growth, but differences between B–E could not be statistically ascertained.

## Maturity (Table 3)

Under combination 'A' no individual reached maturity, even after 30 days. Maturation time was 15 to 18 days in the other combinations.

*Table 3.* Maturity and fecundity

| Salinity | Feed combination | | | | |
|---|---|---|---|---|---|
| | A | B | C | D | E |
| 45‰ | | | | | |
| Maturation (days) | Nil | 18 | 18 | 18 | 15 |
| Fecundity (Nos/brood) | Nil | 62 | 80 | 86 | 101 |
| 60‰ | | | | | |
| Maturation (days) | Nil | 18 | 15 | 15 | 15 |
| Fecundity (Nos/brood) | Nil | 73 | 95 | 97 | 120 |
| 75‰ | | | | | |
| Maturation (days) | Nil | 18 | 15 | 15 | 15 |
| Fecundity (Nos/brood) | Nil | 81 | 103 | 107 | 130 |

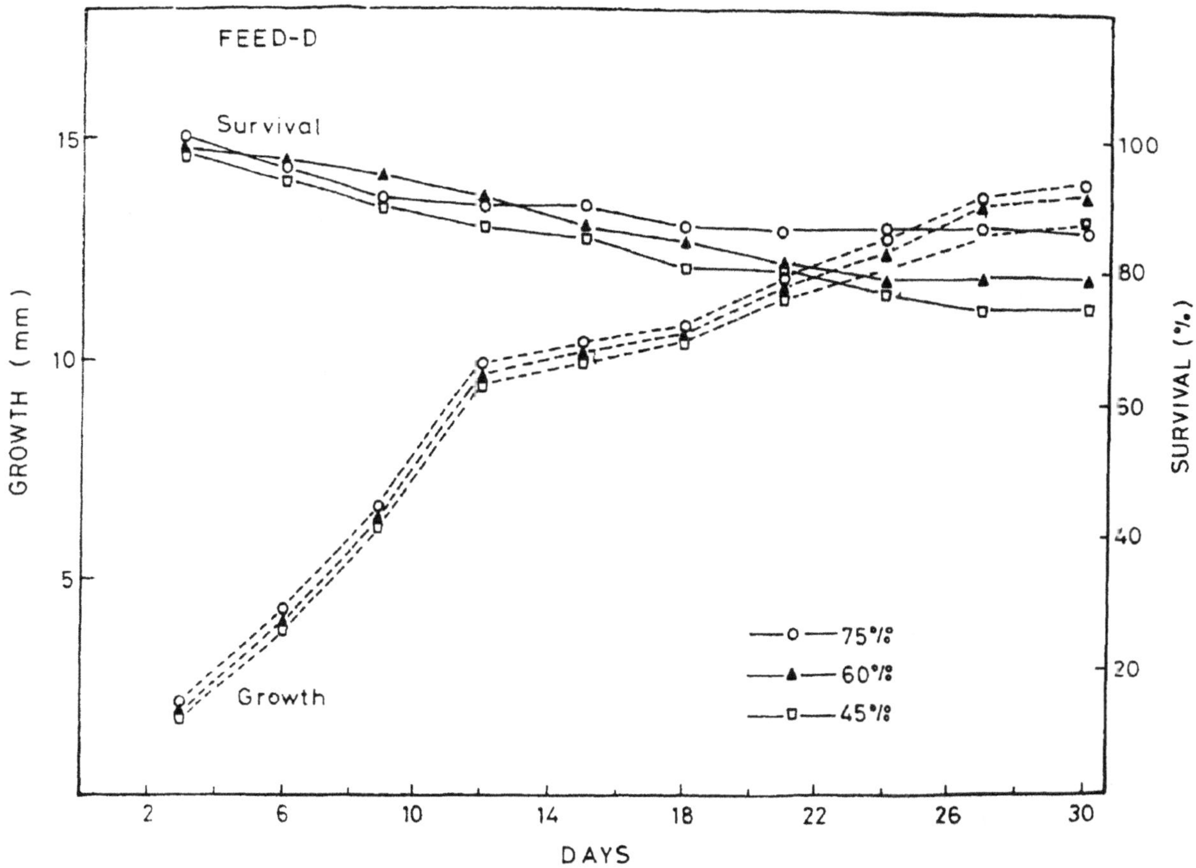


*Fig. 2D*

**Fecundity (Table 3)**

Under combination 'A' fecundity was nil in all salinities. Fecundity was satisfactory in all other combinations and is suggestive of an increase B → < E, with the higher fecundity at the higher salinity. These differences were not statistically testable, however.

**Discussion**

Utilization of organic wastes for the culture of *Artemia sp.* (Bombay strain) has been initiated by Dwivedi *et al.* (1980) by feeding them with pig dung supplemented with yeast, single superphosphate, and groundnut oil cake. Basil *et al.* (1985) cultured Artemia (Tuticorin strain) by using cowdung, poultry manure, and cabbage leaves. Basil *et al.* (1989) used cowdung (biogas slurry) as a nutrient.

The purpose of using these materials is to reduce culture cost. In Dwivedi's system, survival was about 60% when pig dung with yeast was used at 45‰ salinity. Survival increased at 75‰ salinity. Basil *et al.* (1989) recorded 90% survival at 60‰ salinity. The survival of Artemia therefore clearly has a salinity optimum, although this was not shown in a statistically significant way in the present contribution.

The nature of the organic waste materials is even more important. When organic and the agricultural waste materials alone are used, survival is less than 30–50% (Dwivedi, 1980; Basil, 1989).

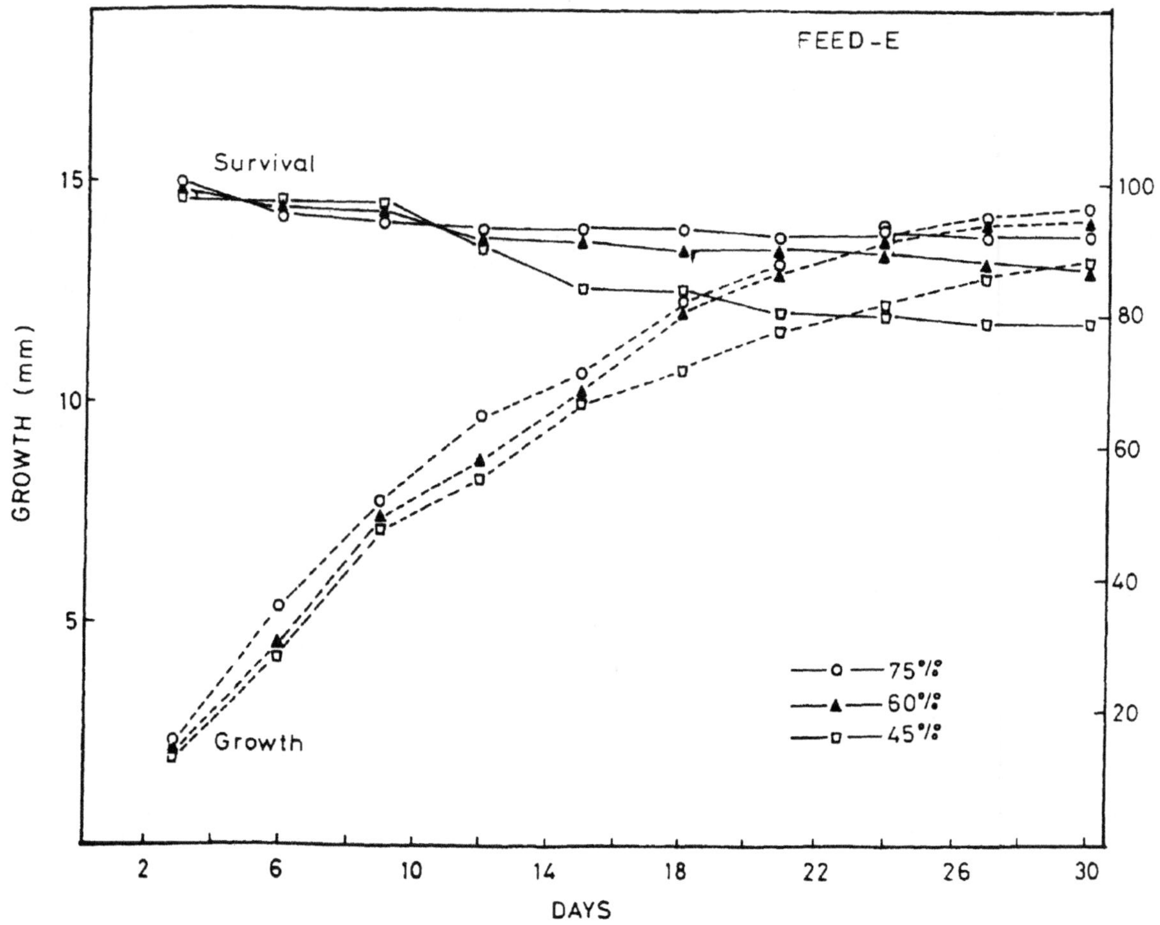


*Fig. 2E*

Survival increases when organic waste materials are supplemented with single superphosphate (Table 1). The growth, maturity, and fecundity respond in the same way as survival. This indicates that, besides salinity, a balanced diet such as B, C, D and E (preferably E) is a prerequisite for successful mass production of *Artemia*.

## Acknowledgements

A.B. thanks Prof. J. Jayaraman, Head, Dept. of Biochemistry, Madurai Kamaraj University, for encouragement and facilities for research, and Prof. M. Lakshmanan, Vice chancellor of Madurai Kamaraj University for encouragement and help. G.T.P. thanks the Principal and Management of Aditanar College of Arts and Science, for encouragements.

## References

Baid, I. C., 1958. Occurrence of Artemia salina in Sambar lake, Rajasthan. Curr. Sci. 27: 58–59.

Balasundaram, C. & A. K. Kumaraguru, 1987. Laboratory studies on the growth and reproduction of Artemia (Tuticorin strain), 331–338. Artemia Research and its Applications. Vol 3. Ecology, Culturing, Use in Aquaculture (P. Sorgeloos, D. A. Bengtson, W. Decleir & E. Jaspers (eds). Universa Press, Wetteren, Belgium. 556 pp.

Basil, J. A., D. Raja David Premkumar, A. P. Lipton & M. P. marian, 1987. Preliminary studies on the culture of

Artemia using renewable organic wastes, 275–278. Artemia Research and its Applications. Vol 3. Ecology, Culturing, Use in Aquaculture (P. Sorgeloos, D. A. Bengtson, W. Decleir & E. Jaspers (eds). Universa Press, Wetteren,, Belgium. 556 pp.

Basil, J. A., D. Raja David Premkumar, A. P. Lipton & M. P. Marian, 1987. Artemia in the salt pans of Vedaranyam, Southern India. Artemia Research and its Applications. Vol 3. Ecology, Culturing, Use in Aquaculture (P. Sorgeloos, D. A. Bengtson, W. Decleir & E. Jaspers (eds). Universa Press, Wetteren, Belgium. 556 pp.

Basil, J. A., Kannan A. Sathasiva, P. Israel, G. Mathuram & D. Selvarani. Culture of Artemia using biogas slurry (cowdung), agricultural wastes and synthetic artemia meal – Book of Abstracts. p. 59 in Aquaculture '89 conference, Los Angeles, Ca, Feb. 12–16, 1989.

Dwivedi, S. N., S. K. R. Ansari & M. Q. Ahmed, 1980. Mass culture of Brine Shrimp under controlled conditions in cement pools at Bombay, India. The brine shrimp Artemia. Vol. 3.: 175–183. Ecology, culturing, use in Aquaculture. G. Persoone, P. Sorgeloos, O. Roels & E. Jaspers (eds). Universa Press, Wetteren, Belgium, 456 p.

Lakshmanan, M., M. Kunthala Jayaraman, J. Jayaraman & A. Gnanam, 1971. Laboratory experiments in Microbiology and Molecular Biology. Higginbothams Ltd, Madras.

Kulkarni, C. V., 1953. Occurrence of brine shrimp Artemia salina in Bombay. J. Bombay Nat. Hist. Soc. 51: 951–952.

Royan, J. P., P. Navaneethakrishnan & A. M. Selvaraj, 1970. Occurrence of Artemia salina in Southern India. Curr. Sci. 39: 14.

Royan, J. P., 1979. Occurrence of Artemia species in the Gulf of Kutch, Mahasagar. Bull. natl. Inst. Oceanog. 12: 271–272.

*Hydrobiologia* **212**: 19–20, 1991.
*D. Belk, H. J. Dumont & N. Munuswamy (eds), Studies on Large Branchiopod Biology and Aquaculture.*



# A rapid, non-stressing technique for measuring growth in Anostraca and other delicate, medium-sized aquatic invertebrates

Christine De Walsche & Henri J. Dumont
*Institute of Animal Ecology, State University of Ghent, K.L. Ledeganckstraat 35, 9000 Ghent, Belgium*

*Key words:* video technique, Anostraca, method

## Abstract

We describe a rapid and elegant method for measuring aquatic animals in the millimeter to centimeter range, using video recording.

## Introduction

Repeated measurements of individual animals, such as needed in studies on somatic growth, are stressful, and injuries inflicted during the treatment may seriously affect the results. This is particularly true of agile swimmers, with a delicate body structure, like the Anostraca.

The method described here is aimed at reducing the stress experienced by the test animals to a minimum, while retaining the advantages of accuracy, repeatability, and speed.

## Description of the method (Fig. 1)

The recording device needed consists of a standard stereoscopic microscope (with a trinocular head) (1), provided with a C-mount, onto which a video camera (2) is mounted. The video camera is coupled, either directly to a video-printer (Sony video graphic printer UP 811) (4), or indirectly, via a video-recorder (3) and a TV-monitor (5).

Animals to be measured are transferred to petri dishes, placed on a transparant calibrated grid (in the case of our studies with various Anostraca, we used standard transparant millimeter paper), and illuminated from above (fiber light source) as well as from below. The animals are filmed for a few seconds, either while freely swimming about in 3–4 mm of water (larvae and small animals), or in a minimum of water (adults and animals in the centimeter range), and subsequently returned to their culture vessel.

The video tape can afterwards be reviewed by searching for the best images in terms of sharpness and contrast. These images are then video-printed (Sony thermographic paper UPP 110 S) and measured with a curvimeter. The background grid provides a standard for conversion of lengths measured to real lengths, such that recordings can be done using a zoom-objective.

## Discussion

We found this method to give quite satisfactory results, even though the present resolution of the video-printing procedure is still rather poor (Fig. 2). The latter disadvantage is, however, compensated by the gain in time (thus a less stressing treatment for the animals), the possibility to store images indefinitely, and the drastic reduction in

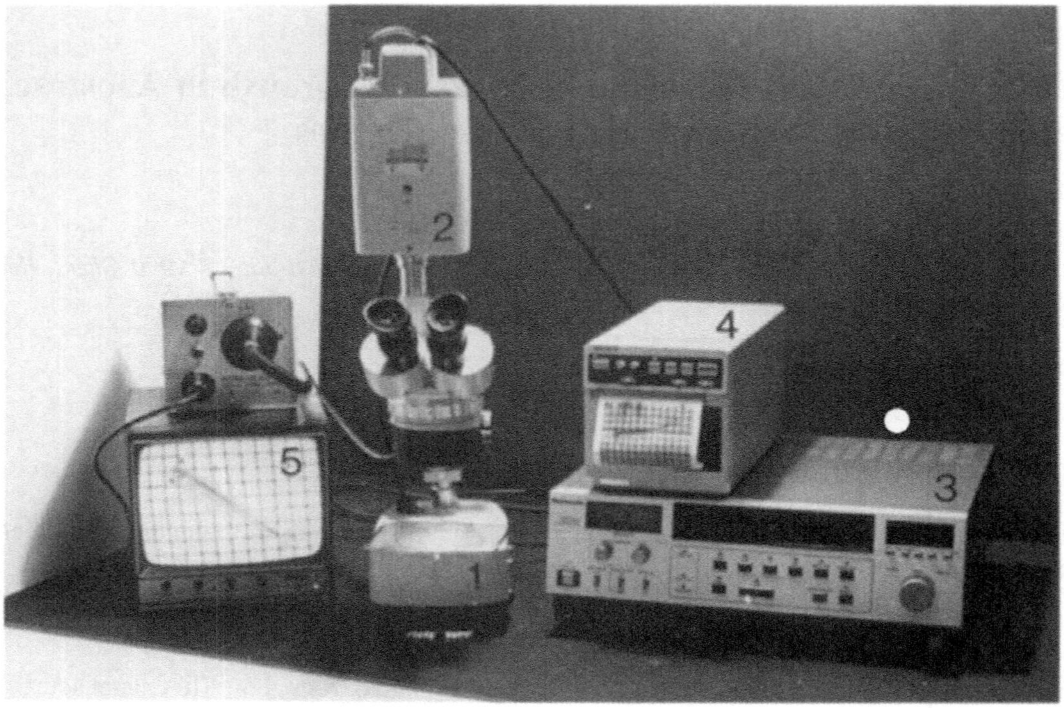


*Fig. 1.* An overview of the recording unit with (1) stereoscopic microscope, (2) video camera, (3) video-recorder, (4) video-printer, and (5) TV-monitor.

mortality among test animals when compared to other methods, such as the direct measurement of animals on wet filter paper.

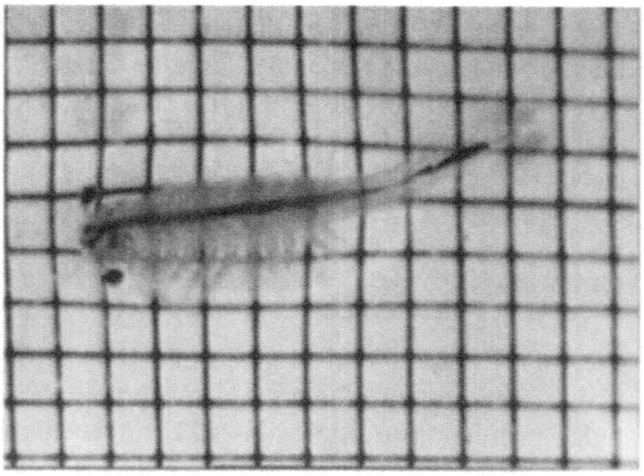

*Fig. 2.* An example of a video print of a specimen of *Thamnocephalus platyurus*, against a calibrated grid (spacings 1 mm). Length of the animal: 9.02 ± 0.10 mm ($m \pm s$; $n = 3$).

*Hydrobiologia* **212**: 21–26, 1991.
*D. Belk, H. J. Dumont & N. Munuswamy (eds), Studies on Large Branchiopod Biology and Aquaculture.*



# Observations on temperature optimum, cyst production, and survival of *Streptocephalus proboscideus* (Frauenfeld, 1873) (Crustacea: Anostraca), fed different diets

Christine De Walsche, Johan Mertens & Henri J. Dumont
*Institute of Animal Ecology, State University of Ghent, K.L. Ledeganckstraat, 35, 9000 Ghent, Belgium*

*Key words:* Anostraca, *Streptocephalus*, aquaculture, diets, bacteria, blue-green bacteria


## Abstract

Somatic growth in *S. proboscideus*, fed *Chlorella vulgaris* increased with temperature reaching a plateau after about 8 to 11 days at between 26 and 31 °C. Survival was best below 29 °C. Fertility (the number of cysts produced per female) and survival, tested at 27 °C, demonstrated some variability as a function of the composition of four different diets. However, large variances and few replicates make exact evaluation impossible.

It was found that *S. proboscideus* can successfully be cultured, using a variety of food sources. Only pure *Spirulina platensis* gave unsatisfactory results in terms of low cyst production and high mortality. Filamentous blue-green algae, because of their possible toxicity, should preferably be excluded from formulated diets.


## Introduction

Successful laboratory culture of fairy shrimps requires control over factors such as the chemical composition of the culture medium, size and shape of the culture vessels, temperature, light regime, and food.

If in addition, one wishes to explore the possibility of long-term, large-sized cultures, with an aim at aquacultural applications, economic factors come into play as well. Among other things, a low-cost, easily accessible food source should be available. In contrast with *Artemia* culturing (Dobbeleir *et al.*, 1980; Johnson, 1980; Lavens *et al.*, 1987; Sick, 1976), where much research has been devoted to the development of cheap food sources, live as well as inert, little work has been done on the approximately 250 other species of Anostraca. *Streptocephalus proboscideus*, a non-selective filter feeder, which naturally occurs in ephemeral rainpools in North-East Africa, is dependent on a wide array of suspended food particles, such as bacterial clumps and algae. Live animal prey is also consumed, in which case size and shape of a food item are less important than its escape capabilities (swimming mode and speed) (Mertens *et al.*, 1990).

The aim of the present note is twofold: to define a temperature 'comfort zone' for the culturing of *S. proboscideus*, and to initiate testing the quality of a series of low-cost inert diets on survival and fertility of the species.

Moore (1957) noted that *Streptocephalus seali*, fed yeast, or a combination of yeast and *Chlamydomonas intermedia*, grew and matured much more rapidly than when fed algae or *Paramecium multimicronucleatum* only. Bearing in mind the wide feeding spectrum mentioned above, we chose to use mixed diets as well, with bacteria, cyanobacteria, and yeast as components. We tested four formulated diets which had a commercially available cyanobacterial fraction in common, in addition to a pure diet of the cyanobacterium.

## Material and methods

Specimens used in our experiments were obtained from continuous laboratory bulk cultures. Cysts for the *S. proboscideus* cultures were isolated from dry mud, collected in rainpools along the Blue Nile at the village of Al Gedid ('the new one'), c. 90 km south of Khartoum, Sudan, in December 1985. This is the classical habitat studied by Rzoska (1961).

### *Temperature optimum*

The choice of the temperature used in the dietary experiments (rounded to 27 °C; actually 26.9 ± 0.3 °C) was made after exploring the performance of animals (in terms of growth, maturation time, and survival) across the interval of 20–31 °C which seemed *a priori* reasonable in view of the tropical origin of the species. Within this interval, we investigated five temperatures: 20.9 ± 0.8 °C ($m \pm s$; $n = 15$); 23.5 ± 0.2 °C; 26.4 ± 0.2 °C; 29.3 ± 0.3 °C; 31.3 ± 1.2 °C. Temperatures were thermostatically controlled to the nearest degree centigrade; inevitable small fluctuations occurred, however, and therefore daily measurements to the nearest 0.1 °C were made, and averaged across the entire period of experimentation. Per temperature, 20 Instar I larvae were used. This moderately low number was intended to exclude a possible influence of crowding on individual performance. These larvae were reared in 2.5 l aquaria containing a constant volume (100 ml) of Cairns medium (Cairns, 1969) *per animal*. The aquaria were subjected to a 14 : 10 h (l : d) light regime, continuously aerated by means of a compressor, and connected to a Watson-Marlow peristaltic pump (202 U/1) which provided a constant supply of *Chlorella vulgaris*. It has long been known (e.g. Ryther, 1954) that growth, reproduction, and filtering rate in filter feeders are inhibited by substances released by old algal cultures. To minimize such influences, we only used algae from rapidly growing cultures, less than 5 days old. Algal cells were concentrated by centrifuging, and washed in distilled water before being pumped into the shrimp cultures. Their concentration in the cultures was kept constant to c. 100.000 cells ml$^{-1}$ (controlled with a Bürker counter). This was adequate for survival, growth, and maturation of the shrimps (their intestine was always well filled), but not for a high cyst production. Cyst production was, however, not a target variable in this part of the experiment.

Growth and survival were followed on a daily basis during 16 and 22 days respectively. A non-stressing method was used: video images of the animals were recorded, printed, and length measurements from brain to tip of cercopodes were taken from these prints using a curvimeter (De Walsche & Dumont, 1991). Shrimps were videotaped over transparent calibrated 1 mm grid in order to allow measurements to be converted to real size.

### *Dietary experiments*

After hatching in bulk cultures, a maturation period of 14 d was allowed before shrimps were used in experiments. Each experiment was performed in triplicate (each data point in the figures represents the mean of three trials), and involved five males and five females, transferred to 2.5 litre aquaria containing 1.5 l of dechlorinated tap water (conductivity 700 ± 1 μS), which was periodically changed (10% every two days).

The total duration of an experiment was 30 days, of which the first six days were omitted from subsequent analysis, and considered as an acclimation phase of the animals to their new environment. Continuous aeration, and thus suspension of the food particles, which were added twice a day, was provided by a compressor. All aquaria were subjected to a 14 : 10 h (l : d) light regime. To facilitate comparison with Fig. 1, only the results up to day 22 are shown on Fig. 5.

Surviving animals were counted and cysts harvested and counted at 2 day intervals. At each occasion the number of harvested cysts was divided by twice the number of surviving females, to obtain an estimate of cysts produced per female per day.

Cyanobacteria, and combinations of bacteria, cyanobacteria, and yeast were used as food sources. The fact that the intestine of the shrimps was coloured according to their respective diet, provided evidence that the food particles offered were ingested. Further, all shrimps in a control experiment, to which no food was added, starved to death within seven days, confirming that the shrimps were able to digest the food administered. As sources of bacterial matter we used: 1. Horse-manure extract: horse dung was incubated in tap water for one day in a container equipped with a 120 $\mu$m Trislot sieve (Trislot Systems, Waregem, Belgium); the extract was pumped out through the sieve, and again filtered over a 80 $\mu$m sieve. 2. A Spray-dried cell cream of technical grade *Corynebacterium lilum*, used in glutamic acid fermentation by James Allan & Sons, Stockton, California. 3. Active silt: a cloudy suspension of bacteria grown on the effluent of a dairy factory, made available by the Faculty of Agronomy, State University of Ghent. The flocculent nature of the material is a result of the mixing of several bacterial species. The taxonomic composition of the bacteria in this active silt is an industrial secret. 4. *Spirulina platensis*: spray-dried, powdered cyanobacterial colonies produced by Ein Yahav Nature Products, Negev desert. The size of the powdered particles (measured to the nearest 1 $\mu$m) ranged from 10 to 130 $\mu$m. The active silt and the dung extract contained bacteria with cell sizes ranging from 0.2 to 5 $\mu$m. The raw *Corynebacterium* cell-cream contained particles in the following size range (provided by the producer): 9% larger than 105 $\mu$m; 7.5% between 75 and 105 $\mu$m; 51% between 75 and 45 $\mu$m, and 40% smaller than 45 $\mu$m. All food mixtures were sieved over an 80 $\mu$m sieve after homogenising with an Ultra-Turrax T25 homogeniser, and made up to formulated diets as follows (all amounts given in dry weight):

Diet 1: A mixture of horse manure extract (0.033 g $l^{-1}$), *S. platensis* (0.1 g $l^{-1}$), and *S. cerevisiae* (baker's yeast) (0.013 g $l^{-1}$).

Diet 2: A mixture of *C. lilum* (0.07 g $l^{-1}$), *S. platensis* (0.1 g $l^{-1}$), and *S. cerevisiae* (0.013 g $l^{-1}$).

Diet 3: diet 2 with in addition a suspension of clay particles. The clay and *C. lilum* particles were mixed in a ball mill.

Diet 4: active silt (0.033 g $l^{-1}$), *S. platensis* (0.1 g $l^{-1}$), *S. cerevisiae* (0.013 g $l^{-1}$).

In addition, a tripled concentration (0.3 g $l^{-1}$) (to compensate the lack of yeast and bacteria) of *S. platensis* was tried out as a monodiet. Experimental conditions were equal to those described for the other diets.

## Results

*Temperature optimum (Figs. 1 & 2)*

Survival became *quasi* constant after c. 4 days (Fig. 1) and somatic growth after c. 8 days (Fig. 2). Therefore, growth experiments were arrested and interpreted after 16 d of observation.

Survival was good throughout the temperature

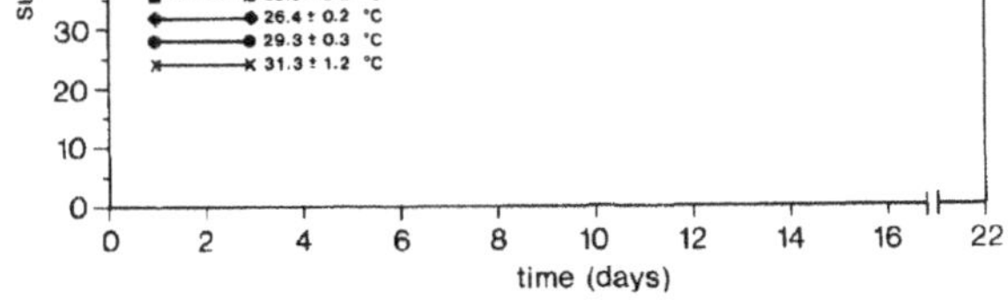


*Fig. 1.* Survival of *S. proboscideus* as a function of time, kept at different temperatures, fed *C. vulgaris*. (1) = animals lost by manipulative error. Overlapping points are shown as unfilled symbols.

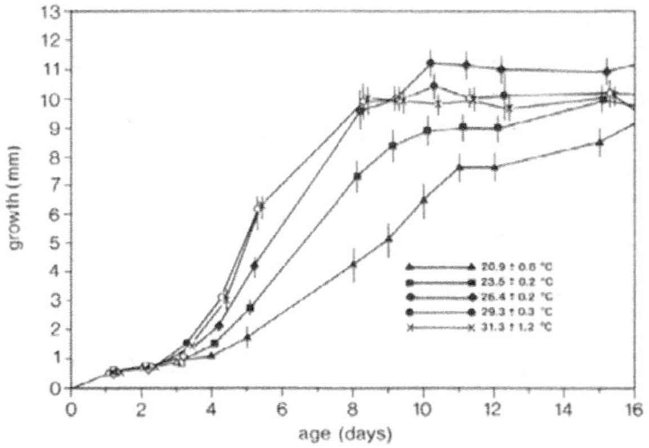


*Fig. 2.* Growth (± S.E.) of *S. proboscideus* kept at different temperatures, fed *C. vulgaris*. Overlapping points are shown as unfilled symbols.

range tested, and varied from 45 to 80% after 22 d. For technical reasons, the experiment was not replicated. However, the positioning of the curves in Fig. 1 strongly suggests that the effects of temperature on survival were minimal under the conditions of the experiment.

Maturation time, here defined as the time necessary for a minimum of 50% of the females to produce a first clutch of cysts, was also calculated. Females kept at 31.2 ± 1.2 °C, and those kept at 29.3 ± 0.3 °C reached maturity after 8 days, while in females reared at 26.4 ± 0.2 °C this took 9 days. Female shrimps, kept at 23.5 ± 0.2 °C needed 15 days to produce their first clutch of cysts; none of the females reared at 20.9 ± 0.8 °C produced a clutch even after the 16th day of observation.

Somatic growth (Fig. 2) followed a sigmoid pattern, with the growth curves converging almost simultaneously to a plateau at a length of 10–11 mm for animals reared at 31.2 ± 1.2 °C, 29.3 ± 0.3 °C, and 26.4 ± 0.2 °C. Animals kept at 23.5 ± 0.2 °C and 20.9 ± 0.8 °C grew distinctly more slowly.

We conclude that the temperature interval 27–31 °C is a 'comfort zone' for rearing *S. proboscideus*. For economic reasons (heating costs), we chose to conduct all subsequent experiments near the lower boundary of this zone, i.e. 27 °C.

*Cyst production as a function of diet (Figs. 3 & 4; Table 1)*

Changes in cyst production for five feeding programs across four weeks (30 d) of observation are shown in Fig. 3 (curves smoothed by using 3 day moving averages). The figures in Table 1 represent the mean production of cysts per day per female (row 1) maximum average clutch size per female

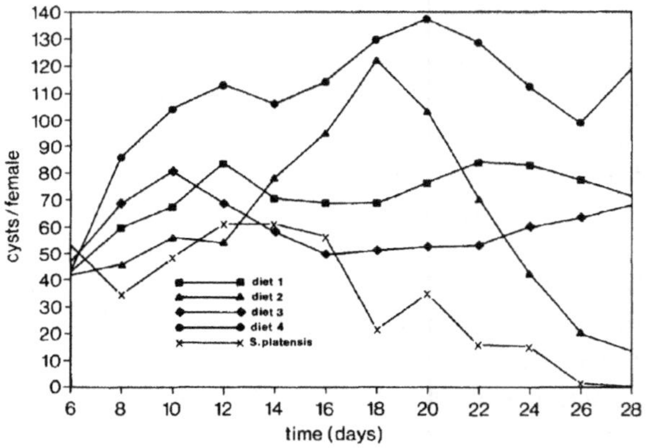


*Fig. 3.* Cyst production by *S. proboscideus* as a function of time and diet (smoothed by using moving averages). Each point represents the average value of three replicas.

*Table 1.* Row 1: Average number of cysts per female per day (± S.D.) between day 6 and 30. Row 2: Maximum average clutch size per female per diet during the 30 d of observation. Row 3: Mean total number cysts per female, produced whithin a period of 30 d.

| | diet 1 | diet 2 | diet 3 | diet 4 | Spir. |
|---|---|---|---|---|---|
| cyst.♀$^{-1}$.d$^{-1}$ | 71.2 ± 17.7 | 57.2 ± 42.6 | 61.3 ± 13.8 | 110.0 ± 35.3 | 30.4 ± 37 |
| max*.♀$^{-1}$.d$^{-1}$ | 91 | 125 | 83 | 151 | 130 |
| total fertility | 1947 | 1576 | 1634 | 2903 | 1158 |

max* = maximum average clutch size

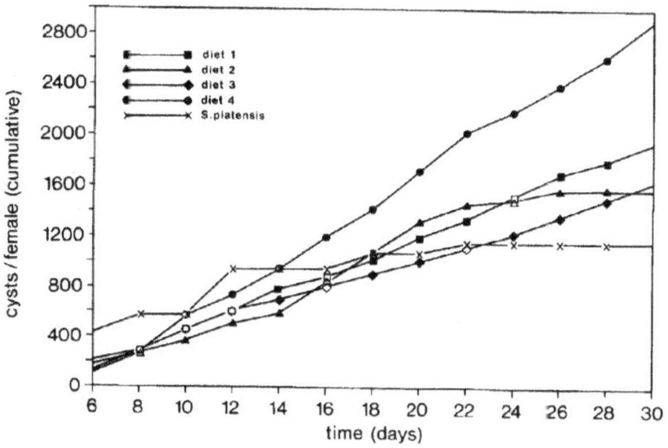


*Fig. 4.* Cumulative cyst production curve of *S. proboscideus*, as a function of diet. Each point represents the average value of three replicas. Overlapping points are shown as unfilled symbols.

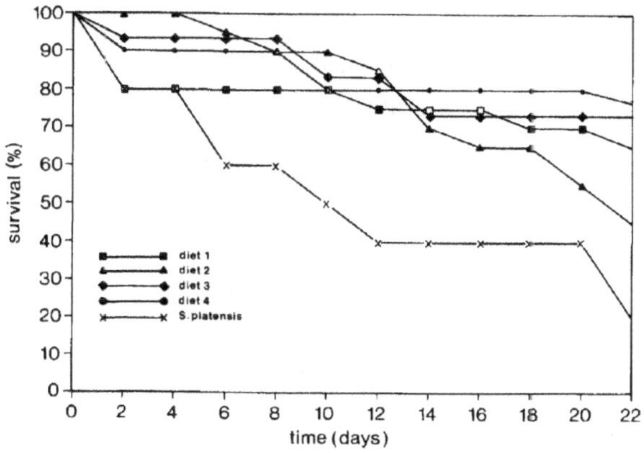


*Fig. 5.* Survival of *S. proboscideus* as a function of time and diet. Each point represents the average value of three replicas. Overlapping points are shown as unfilled symbols.

*Table 2.* Average percent survival ($\pm$ S.D.) (sexes pooled) after 10, 22, and 30 days.

| | diet 1 | diet 2 | diet 3 | diet 4 | Spir. |
|---|---|---|---|---|---|
| 10 d. | 80.0 ± 11.5 | 90.0 ± 5.8 | 83.3 ± 6.6 | 80.0 ± 20.0 | 50.0 |
| 22 d. | 66.7 ± 20.8 | 46.7 ± 25.0 | 73.3 ± 28.9 | 76.7 ± 32.1 | 20.0 |
| 30 d. | 60.0 ± 5.8 | 16.7 ± 5.8 | 73.3 ± 16.7 | 66.7 ± 18.6 | 0.0 |

per day (row 2), and mean total number of cysts produced per female (row 3), for each diet. Figure 4 shows the cumulative curves of cyst production.

### *Survival as a function of diet (Fig. 5; Table 2)*

Best survival at 22 days (average 76.7%) was obtained with an active silt diet, followed by a *C. lilum* diet to which clay was added (average 73.3%), and by the horse manure diet (average 66.7%). Only an average of 46.7% of the shrimps survived the 22 d of feeding on *C. lilum* without clay. Survival of shrimps fed on monodiet of *S. platensis* was the worst: only 20% survived 22 d of observation, and none were alive after 28 days.

## Discussion and conclusion

Our results are encouraging as far as the possibility of long-term culturing of *S. proboscideus* is concerned. Mortalities in the cultures, if substantial, are acceptable, and only slight temperature effects, in the case of *Chlorella*, and dietary effects, in the case of formulated diets (compared to *Chlorella*) (Figs. 1, 5), could be detected. Only the *Spirulina* monodiet was totally unsatisfactory, leading to early, progressive mortalities, that wiped out the entire culture before the end of the experiment. As far as cyst production goes, the *Spirulina* monodiet was also, consistently, the poorer one (Table 1, Fig. 3). Addition of various non-cyanobacteria and/or clay usually improved production (Fig. 5), but the number of replica's (3) was too low to allow any discrimination between diets (all pairwise t-tests not significant at $p < 0.05$). This does not mean that differences between diets did not exist, but only that large variances and few replicates do not permit discrimination on this point.

The fact that pure *Spirulina* gave the poorest results may be due to the toxic nature of this and other cyanobacteria (Gorham, 1964; Gentille, 1971). In this context, it is noteworthy to attract attention to the fact that Anostraca are rarely, if ever, found in association with blooms of filamentous cyanobacteria, which should, therefore, preferably be excluded from formulated diets.

## References

Cairns, J., 1969. Fish bioassays – Reproducibility and rating. Rev. Biol. 7: 7–12.

De Walsche, C. & H. J. Dumont, 1991. A rapid non-stressing technique for measuring growth in Anostraca and other delicate, medium-sized aquatic invertebrates. In Belk, D., H. J. Dumont & N. Munuswamy (eds), Studies on Large Branchiopod Biology and Aquaculture. Developments in Hydrobiology 64. Kluwer Academic Publishers, Dordrecht: 19–20. Reprinted from Hydrobiologia 212.

Dobbeleir, J., N. Adam, E. Bossuyt, E. Bruggeman & P. Sorgeloos, 1980. New aspects of the use of inert diets for high density culturing of brine shrimp. In Persoone, G., P. Sorgeloos, O. A. Roels & E. Jaspers (eds), The brine shrimp Artemia. Vol. 3. Ecology, Culturing, Use in Aquaculture. Universa Press, Wetteren Belgium: 165–174.

Gentille, J. H., 1971. Blue-green and green algae toxins. In S. Kadis (ed.), Microbial toxins V. 7. Academic: 27–65.

Gorham, P., 1964. Toxic algae. In D. F. Jackson (ed.), Algae and Man. Plenum Press, pp. 307–336.

Johnson, P. T., 1980. Evaluation of various diets for optimal growth and survival of selected life stages of Artemia. In Persoone, G., P. Sorgeloos, O. A. Roels & E. Jaspers (eds), The brine shrimp Artemia. Vol. 3. Ecology, Culturing, Use in Aquaculture. Universa Press, Wetteren Belgium: 185–192.

Lavens, P., A. De Meulemeester & P. Sorgeloos, 1987. Evaluation of mono- and mixed diets as food for intensive Artemia culture. In Sorgeloos, P., D. A. Bengtson, W. Decleir & E. Jaspers (eds), Artemia research and its application. Vol. 3. Universa Press, Wetteren, Belgium.

Mertens, J., N. Munuswamy, C. De Walsche & H. J. Dumont, 1990. On predatory tendences in the feeding ecology of the fairy shrimp *Streptocephalus proboscideus* (Frauenfeld, 1873) (Crustacea: Anostraca). Hydrobiologia 198: 119–123.

Moore, G. W., 1957. Studies on the laboratory culture of Anostraca. Trans. am. microsc. Soc. 76: 159–173.

Ryther, J. H., 1954. Inhibitory effects of phytoplankton upon the feeding of *Daphnia magna* with reference to growth, reproduction and survival. Ecology 35: 522–533.

Rzoska, J., 1961. Observations on Tropical Rainpools and General Remarks on Temporary Waters. Hydrobiologia 17: 265–286.

Sick, L. V., 1976. Nutritional effect of five species of marine algae on the growth, development, and survival of the brine shrimp Artemia salina. Mar. Biol. 35: 69–78.

*Hydrobiologia* **212**: 27–37, 1991.
D. Belk, H. J. Dumont & N. Munuswamy (eds), *Studies on Large Branchiopod Biology and Aquaculture.*



# A model for growth of *Artemia franciscana* cultures based on food ration-dependent gross growth efficiencies

F.A. Abreu-Grobois, R. Briseño-Dueñas, M.A. Herrera & M.L. Malagón
*Estación 'Mazatlán', Instituto de Ciencias del Mar y Limnología, Universidad Nacional Autónoma de México, Apartado Postal 811, Mazatlán, Sinaloa 82000, Mexico*

*Key words:* *Artemia*, culture strategies, yields, feeding, growth efficiency, sequential harvesting

## Abstract

Laboratory cultures of *Artemia franciscana* grown under batch regimes at constant temperatures (28 °C) and salinity (35 g $l^{-1}$), three initial food concentrations (0.1, 0.4 and 1 M cells $ml^{-1}$), various daily food rations (0.1–9 M *Dunaliella tertiolecta* cells $Artemia^{-1}$), and different population densities (1–16 ind $ml^{-1}$) were used to develop a model of population growth. Growth rates and gross growth efficiencies ($K_1$) were largely independent of population densities and initial food concentrations but determined by age and daily amount of food ingested. While maximum growth rates were found with the highest rations, $K_{1\ max}$ peaked at rations of 0.5 million cells $d^{-1}$ and decreased at feeding levels above this. A contour plot showing the trend relating $K_1$ to *Artemia* size and ingestion rate in combination and was used to model growth in analogous controlled feeding conditions. Computer simulations using this model paralleled published results of final 15-day average individual sizes of *Artemia*. Optimal results for near constant food utilization are predicted for high initial population densities (100 *Artemia* nauplii $ml^{-1}$) and daily culls of enough animals to equilibrate food demand with food availability. This strategy could permit a range of *Artemia* sizes harvested, maximize final individual sizes and retain high total yields (> 1.2 kg dry wt $l^{-1}$). Effects of different culture strategies are discussed.

## Introduction

Interest in the mass rearing of *Artemia* has developed as it may be cultured at high densities capable of producing yields of up to 20 kg wet weight $m^{-3}$ in 14 days of culture (Sorgeloos, 1985). Applications of these are manifold: the cultures can be performed under controlled conditions, year-round, irrespective of climate (Lavens *et al.*, 1985); *Artemia* can be manipulated in size and nutritional quality to be utilized as carrier for various nutritional, prophylactic and therapeutic components to cultured predators (Léger *et al.*, 1987); cultures can be employed to harvest, and transform to useful biomass, marine microalgae grown in sewage (McShan *et al.*, 1974), artificial upwelling (Tobias *et al.*, 1979), fish/shrimp pond effluents (Hughes-Games, 1971) or industrial treated wastewaters (Milligan *et al.*, 1980); and, the culture systems can be vertically integrated to supply fresh live food of various sizes to hatchery, nursery, or maturation departments of fish or shrimp farm complexes (Lavens *et al.*, 1985).

Adequate dimensioning of intensive culture systems according to application needs requires accurate information on the effects of rearing conditions on the growth dynamics of the organisms.

For *Artemia*, data are available on how growth and survival are affected by strain and species differences (Vanhaecke & Sorgeloos, 1980), environmental variables (Baid, 1963; Reeve, 1963a; Vanhaecke *et al.*, 1984; Von Hentig, 1971; Wear & Haslett, 1986), and food quality (Sick, 1976), quantity (Nimura, 1980; Mason, 1964) and concentration (Brune & Anderson, 1984; Reeve, 1963a, b).

In applications where live algae are used to rear *Artemia*, food production constitutes a major cost. Although this outlay can be offset by employing algae as part of nutrient recycling systems, conversion efficiency is an important consideration both for economic reasons as well as to be able to predict biomass yields.

In spite of the amount of work on *Artemia*, published results on the effects of feeding levels on its growth are contradictory. One study, on *Artemia* fed with *Chlamydomonas* sp. (Nimura, 1980), demonstrated negative effects of overfeeding on both growth and conversion efficiencies with no food concentration effects. Working with *Phaeodactylum tricornutum*, Reeve (1964a) found low growth efficiencies at both low and high food concentrations (concomitant with low and high ingestion levels, respectively) and peaks at intermediate values. Contrary to this, Brune & Anderson (1984), also using *P. tricornutum*, found only increasing conversion efficiencies with increased food concentrations although variations were observed during the growth of the shrimps.

The present work studied the effects of controlled feeding of *D. tertiolecta* on the growth of *Artemia franciscana*, and on its gross food conversion efficiency, $K_1$ (Conover, 1964), during development from metanauplius to adult over a two week period. These data were employed to model population growth dynamics under analogous mass culture conditions. *A. franciscana* was chosen because most quantitative research and development of aquacultural systems has been done with this species, and partly as it has been used as a standard for comparison of growth performances of other strains and species of the genus (Vanhaecke & Sorgeloos, 1980). As its test food, the marine phytoflagellate *Dunaliella tertiolecta* was used since as it has been used widely as food for *Artemia* and, besides being plastic to chemical and physical conditions and easy to cultivate to extremely high densities (7–8 g dry wt $l^{-1}$) (Spectorova *et al.*, 1982), this alga can be grown to high protein cell contents under high nutrient régimes (Fábregas *et al.*, 1986), a quality correlated with highest rates of growth in *Artemia* (Sick, 1976).

## Materials and methods

*Artemia franciscana* cysts were provided by the Artemia Reference Center, State University of Ghent, Ghent, Belgium. They were harvested in June, 1985 from a population recently innoculated in Kenya with San Francisco Bay strain cysts (P. Vanhaecke, pers. comm.). Cysts were decapsulated and hatched using standard methods (Sorgeloos *et al.*, 1986). Nauplii were transferred, after rinsing, to experimental setups which contained 0.45 $\mu$m filtered natural sea water (35–37 g $l^{-1}$) at 28 ± 1 °C. *Artemia* individual dry weights were calculated from individual

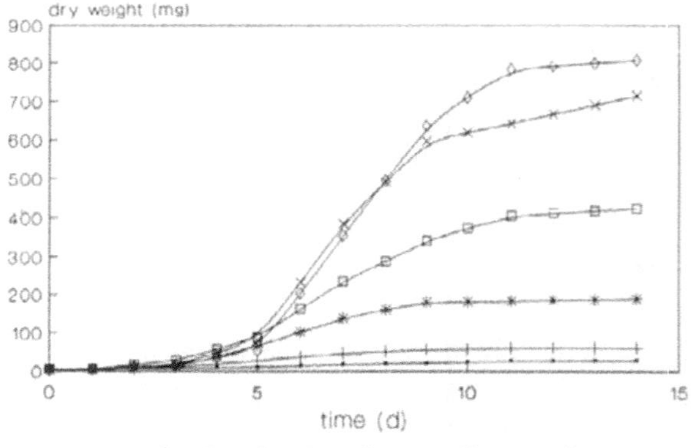


*Fig. 1.* Growth curves of *A. franciscana* under controlled feeding. Treatments 1–6 received a constant ration of food (*D. tertiolecta*): aprox. 10, 25, 50, 100, 300, and 900 $\mu$g d. wt. of algae cells ind$^{-1}$ d$^{-1}$, respectively, at an initial concentration of 0.4 M cells ml$^{-1}$. Treatments 7 and 8 (not shown) but with growth curves identical to t5 received respectively the same quantity of food as t5 but at initial concentration of 0.1 and 1 M cells ml$^{-1}$. Duncan's m.r.t. on final sizes: t6 = t5 > t4 > t2 = t1.

lengths using a second-degree polynomial regression of Reeve's log-transformed data (1964a: 137, Fig. 1) ($F = 1630$; $P < 0.001$):

individual dry weight ($\mu$g) =

$$10^{(-2.53 - 1.63*\log(len) + 0.81*(\log(len))^2)} \times 10^3 \qquad (1)$$

where *len* = length of the individual (mm) measured from the anterior margin of the head to the anus. Unless otherwise specified, all dry weight measures are expressed in $\mu$g. Dry weight-length correlations of our own material, using the same method as for algae (see below), confirmed the fit of this equation. Measurements of the length of *Artemia* individuals were performed with animals narcotized with minimal exposure to chloroform-saturated sea water under a microscope containing a calibrated ocular graticule. Previous experiments demonstrated that handling had no significant effects on growth and survival.

*Dunaliella tertiolecta* cultures were maintained under the same temperature and salinities, with constant dim lighting provided by 30 W 'Daylight' fluorescent lamps. For feeding experiments, algae cultures were harvested serially after not more than 5 days of growth, counted with a haemocytometer and diluted to the desired concentrations with fresh filtered sea water. Dry weights of the algae were obtained after filtering, in triplicate, measured volumes of haemocytometer-counted cultures, onto pre-weighed 0.45 $\mu$m fiberglass membrane filters, rinsing with isotonic ammonium formate, and drying at 65 °C for 48 h. One million cells of *D. tertiolecta* were calculated to weigh 102 ($\pm$9.2) $\mu$g dry wt.

For the controlled feeding experiment (experiment 1), 8 treatments (t1–t8) were established with at least 50 organisms each and at least one replicate each. Treatments 1–6, respectively received 0.1 (10.2 $\mu$g dry weight), 0.25 (25.5 $\mu$g), 0.5 (50 $\mu$g), 1.0 (102 $\mu$g), 3 (306 $\mu$g), and 9 (918 $\mu$g) million (M) algae cells individual$^{-1}$ day$^{-1}$ at an initial ($t = 0$) concentration of 0.4 M cells ml$^{-1}$. Treatments 7 and 8 were also fed 3 M cells d$^{-1}$, as in t5, but at initial concentrations of 0.1 and 1 M cells ml$^{-1}$, respectively. Actual feeding levels deviated by no more than 10% from schedule and these rations were taken into account for computations. Feeding régimes in each treatment were sustained throughout the experiment by preserving the ratio of number of animals to volume of medium and conserving the (initial) algae concentration. Total volumes were modified only to adjust for mortality. Controls were kept with algae and no *Artemia* to correct for possible algae growth or precipitation. Daily grazing rates were calculated using standard equations (Omori & Ikeda, 1984) from algae cell concentrations at the start and at the end of each feeding period.

For experiment 2, inverted plastic bottles of 75 ml working capacity with 120 $\mu$m plankton net windows were placed on a floating raft on a 100 l *D. tertiolecta* culture subjected to constant light and an air-driven laminar flow. The design permitted the free passage of actively growing algae cells from the surrounding medium through the treatment bottles. Algae concentrations were adjusted when necessary to 0.4 M cells ml$^{-1}$ by volume replacements with fresh nutrient medium, Because of the large volume of algae culture, plant growth always exceeded or was equal to the feeding requirements of the *Artemia*. Each bottle, in replicate, was inoculated with 1, 2, 4, 8, or 16 *A. franciscana* nauplii per ml$^{-1}$. Mortality was adjusted by raising the bottle on the raft to reduce the working capacity and maintain the population density.

## Results

### *Controlled feeding (Experiment 1)*

#### *Growth with different rations*

Growth with rations ranging from 0.1 (t1) to 9.0 (t6) M cells d$^{-1}$ *Artemia*$^{-1}$ produced significantly different sizes ($F = 136.3$, $P < 0.001$) after 14 days of culture (Fig. 1). Maximum sizes were observed in t6 *Artemia* with mean length 10.03 mm ($\pm$0.72), equivalent to 807 $\mu$g dry weight (d.w.). Final sizes and growth curves in treatments 5,7 and 8 (t7–8 not shown in the Fig.)

were not significantly different (Duncan's multiple range test) from each other or from treatment 6. Their mean final sizes were 9.62 ($\pm$0.85, equivalent to 714 $\mu$g), 9.68 ($\pm$0.71, 727 $\mu$g), and 9.30 ($\pm$0.95, 647 $\mu$g) mm, respectively. Differentiation in size between treatments became evident on day 5 when fastest growth rates were observed, particularly for treatments 3–8. At this point the *Artemia* attained sizes of about 3.5 mm (50 $\mu$g d.w.). After day 7, treatments 5–8 differentiated in size from the rest.

### *Maximum ingestion rates*

An approximation for the maximum daily ingestion rate (i.r. $_{max}$) was derived (Fig. 2) using data for *Artemia* at different developmental stages raised in treatment 6. These organisms had, at least, 0.1 M cells ml$^{-1}$ remaining at the end of each daily experimental period. This concentration has been reported to correspond to the threshold concentration of *D. tertiolecta* above with the i.r. of *Artemia* remains constant and maximal (Reeve, 1964a). The regression equation ($r^2$ = 0.76) obtained was:

$$\text{i.r.}_{max} = 609.5 \times (1 - e^{(-0.075 \times \text{d.w.})})\ \mu\text{g d}^{-1} \tag{2}$$

where d.w. is the individual dry weight, in $\mu$g, of *Artemia*.

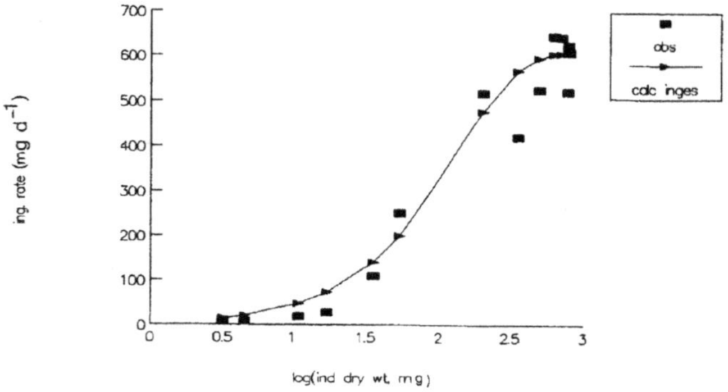


*Fig. 2.* Regression curve and observed maximum daily ingestion rates (i.r. $_{max}$) derived from t6 for *A. franciscana* of different sizes feeding on *D. tertiolecta* cells.

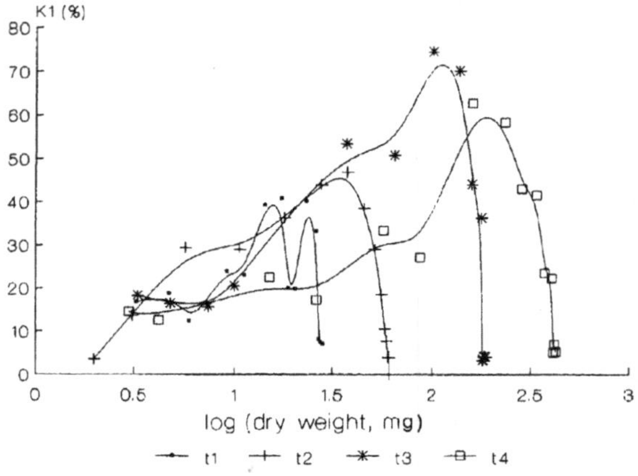


*Fig. 3.* $K_1$ values for *A. franciscana* feeding on *D. tertiolecta* in treatments 1–4 and their trend curves.

### *Gross food conversion coefficients ($K_1$)*

Because *Artemia* in the different treatments grew at different rates, for meaningful analysis of $K_1$ values it was necessary to compare the results of the different treatments on the basis of individual sizes (expressed in $\mu$g d.wt.) rather than on a time scale. There was a general trend for low $K_1$ values at the beginning of growth (from the second day after hatching when $K_1$ determinations were begun). These values of $K_1$ ranged from 3.5% (t2) to 20.8% (t7) on day 2 and rose, with the development of the animals (Figs. 3 and 4), to one or more maxima. $K_1$ values dropped towards low

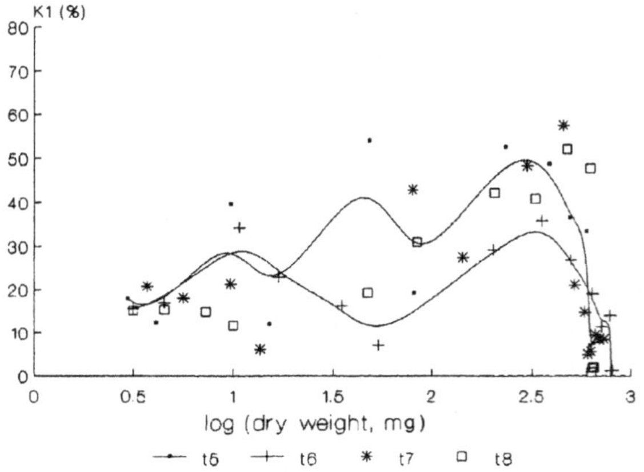


*Fig. 4.* $K_1$ values for *A. franciscana* feeding on *D. tertiolecta* in treatments 5–8 and their trend curves. Trends for t7 and t8 not shown for clarity.

values as growth slowed down at sizes corresponding to limits of the different feeding levels. With the lowest rations (t1 and 2), maximum sizes obtained (3.06 and 4.09 mm, or 28.3 and 61.1 $\mu$g, respectively) were, presumably, those at which the energy obtained from grazing was only sufficient for the maintenance of these particular biomasses. Up to treatments with rations of about 0.5 million cells or 53 $\mu$g d.w. $ind^{-1}\ d^{-1}$ (t3), $K_1$ maxima ($K_{1max}$) were observed to increase with increasing rations. Beyond the ration for t3, with the highest $K_{1max}$ value (74.5%), $K_{1max}$ results for the remaining treatments diminished with increasing rations. $K_{1max}$ for treatments 5, 7 and 8, which were offered identical rations but at different concentrations all resulted in similar values and not significantly different from those of t6 (Fig. 4). Although the $K_{1max}$ values appeared to fall with the increasing rations in t4–t8, maximal (final) sizes obtained increased with greater food availability. Maximal sizes, in log($\mu$g), are observable in the figures coinciding with a tendency for the $K_1$ curves to fall towards 0, i.e. a trend toward zero growth with positive ingestion values.

*Ration-related survival rates*

Daily survival among the eight treatments averaged ~90%. Final survival appeared to be related to ration levels, with maximal survivals obtained for intermediate feeding levels (t4 and t5) and being reduced both at low and at high rations (Fig. 5; 85 ± 5%, 86 ± 7%, 66 ± 10%, 58 ± 13%, 72 ± 8%, respectively for t4–t8). Lower survivals for treatments 1–3 were significantly different from t4–t6 (Duncan's multiple range test, $P < 0.05$). The regression equation obtained for the relationship was:

$$\text{survival} = e^{(-0.21 + 0.36*\log(\text{rat}) - 0.12*\text{rat})} \quad (3)$$

where *rat* = daily ration, in millions of *D. tertiolecta* cells $ind^{-1}\ d^{-1}$ ($r^2 = 0.913$).

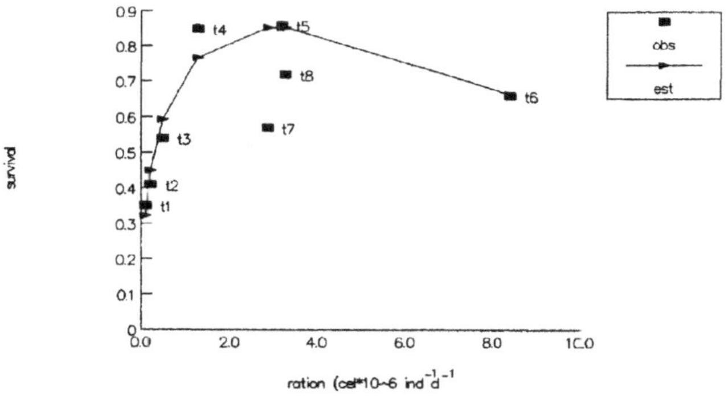


*Fig. 5.* Final survival and daily ration for *A. franciscana* under controlled feeding conditions (see text for regression curve).

*Combined effect of ingestion rates and size of animals of $K_1$*

The previous results suggested a simultaneous relationship between *Artemia* dry weight, their ingestion rates and the associated $K_1$ values. It is important to consider that it is not the ration (amount of food offered) but the quantity which is actually ingested daily which has bearance on the results. In this experiment the available rations simply placed different upper ceilings to the amount of algae available in the different treatments.

A contour plot of the combined effects of individual dry weights (i.d.w.) and ingestion rates (i.r., both in logarithms) or $K_1$ values is presented in Fig. 6. From this representation it is clear how optimal gross food conversion efficiencies are associated with particular combinations of i.d.w. and i.r.. Not all combinations are, of course, possible as upper limits to the ingestion rate capacity exist for each *Artemia* weight class as defined by equation (2). A ridge of high $K_1$'s extends diagonally across the contour plot describing, for each weight class the ingestion rates associated with highest $K_1$. Highest overall $K_1$ values are associated with a combination of i.d.w. of about 100 $\mu$g (2 in log scale) and an i.r. of about 50 $\mu$g d.w. (1.7 in logs). Higher order polynomial regressions were employed to evaluate the interactive effects of i.r. and i.d.w. on determining $K_1$. The results (Table 1) demonstrate significant qua-

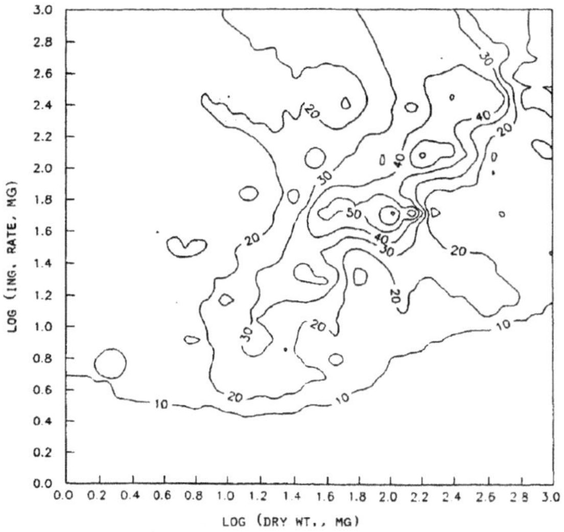


*Fig. 6.* Contour plot in lines of equal $K_1$ values for combinations of dry weight (in logs of $\mu$g, abscissa) and daily ingestion rate (in $\mu$g dry weight $d^{-1}$) for *A. franciscana* feeding controlled rations of *D. tertiolecta*.

*Table 1.* Results of the polynomial regression and its associated ANOVA for the effects of ingestion rates and individual dry weights on $K_1$.

| Variable | Estimated coefficient | s.e. | $t$ |
|---|---|---|---|
| log(i.d.w.) | – 5.4 | 36.7 | – $0.56^{ns}$ |
| log(i.r.) | – 19.9 | 35.8 | – $0.15^{ns}$ |
| log(i.d.w.)* log(i.r.) | – 298.0 | 79.6 | – 3.75*** |
| $(\log(i.d.w.))^2$ | 170.9 | 42.7 | 4.01*** |
| $(\log(i.r.))^2$ | 180.6 | 57.9 | 3.12** |
| $(\log(i.d.w.))^2$* log(i.r.) | 49.6 | 14.7 | 3.37** |
| $(\log(i.r.))^2$* log(i.d.w.) | 50.1 | 12.2 | 4.09*** |
| $(\log(i.d.w.))^3$ | – 58.1 | 16.6 | – 3.51** |
| $(\log(i.d.w.))^3$ | – 56.5 | 13.3 | – 4.25*** |
| $R^2 = 0.859$ | | | |

Probability levels for 2-tailed tests:
$^{ns}$ not significant at 0.05
* $P < 0.01$; ** $P < 0.001$; *** $P < 0.01$

ANOVA

| Source | SS | DF | MS | F |
|---|---|---|---|---|
| regression | 70450.667 | 9 | 7827.853 | 35.721*** |
| residual | 24981.524 | 114 | 219.136 | |

dratic and cubic interactions between the variables. However, although both 2nd and 3rd degree equations were obtained with multiple correlation $r^2$ values $> 0.7$ and statistically significant coefficients, the visual fit of the contour plot was considered adequate for our purposes and the modelling of growth in *Artemia* can be derived manually from the contour lines.

*Density effects on growth and yields (Experiment 2)*

The design of experiment 2 permitted the evaluation of the effects of density *per se* on the growth dynamics of *Artemia*, independent of any side-effects, such as exposure to increasing concentrations of excretory products. Food levels maintained in the culture tank, were considered adequate and permitting optimal access to maximal high rations.

Growth was maximal and not significantly different (Duncan's multiple range test, $P > 0.05$) between treatments with densities of more than 2 *Artemia* $ml^{-1}$ (Fig. 7, reaching sizes (9.4–9.80 mm, 657–754 $\mu$g i.d.w.) comparable to those in the controlled feed experiment at the highest food levels. Observed growth, however, was significantly reduced in treatments with the lowest density.

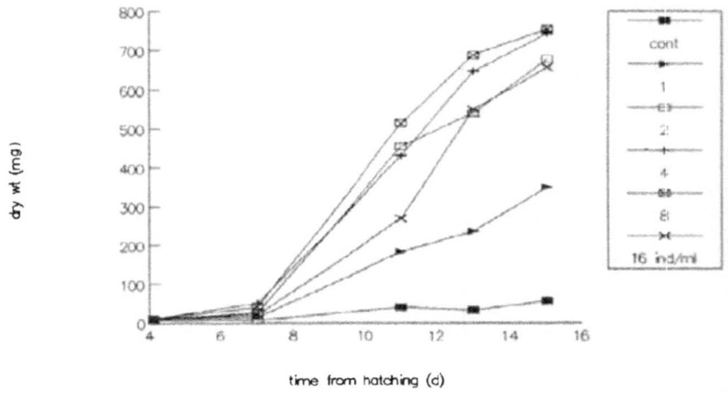


*Fig. 7.* Increases in weight for *A. franciscana* reared at various densities (1–16 ind $ml^{-1}$) in cages within a *D. tertiolecta* culture held at near constant concentrations of Ca 0.4 M cells $ml^{-1}$. Cont = control, *Artemia* held in same containers but with no access to flood. (weight values converted from length measurements by regression curve, see text).

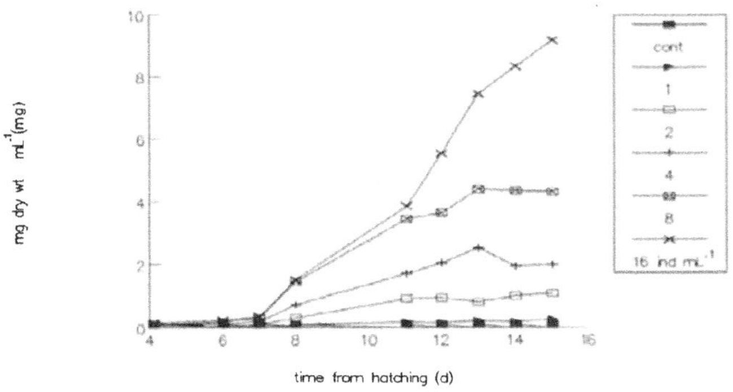


*Fig. 8.* Estimated biomass (survival x mean dry weight $ml^{-1}$) for *A. franciscana* data shown n Fig. 7.

Survival among treatments varied and appeared to be inversely correlated with density. Maximal final survival (87% ± 2%) was obtained for 16 ind $ml^{-1}$, with intermediate densities giving lower survivals (72% ± 5% and 74% ± 8%, resp.), but higher than those for 1–2 ind $ml^{-1}$ (60% ± 9% and 48% ± 7%, resp.). The lower growth and survival in the least dense cultures might be explained by the observation that the shrimp in these cultures were unable to keep the plankton net clean by their feeding actions, probably leading to reduced food passage and inefficient flushing of the culture containers.

The observed yields in this experiment (Fig. 8), calculated by multiplying the density of the final cultures by the mean dry weights (derived from regression equation 2), is suggestive of the potential for systems where *Artemia* can be grown at maximal rates (with an optimal access to food) and at densities greater than 8 *Artemia* $ml^{-1}$. Highest yields were found at the highest density, where survival was highest and individual sizes were among the highest. Thus, an estimated yield of 9.2 mg d.w. $ml^{-1}$ was obtained at the end of 15 days of culture.

## Discussion

The different growth rates recorded in the controlled feeding experiment, corresponded closely to previous studies under similar laboratory conditions (Nimura, 1980; Vanhaecke & Sorgeloos, 1980), particularly with Mason's (1963) who used almost identical feeding rations and the same species of algae as food.

Values for $K_1$ not only are comparable to those found for this organism (Gibor, 1957; Nimura, 1980; Reeve, 1963a; Sushchenya, 1970) but also are determined by an interaction of size of the organism (in terms of i.d.w.) and its i.r.. Consideration of these two parameters simultaneously for *Artemia* was first reported by Brune & Anderson (1984) although they used a different species of phytoplankton, flow-through culture systems and reported only increasing $K_1$ values with increased food ingestions with no apparent fall either side of an overall $K_{1max}$ such as we found. Negative effects on $K_1$ by high levels of food ingestion has been found in this organism (Reeve, 1963a) and another crustacean, *Daphnia* pulex (Richman, 1958). Reeve (*loc. cit.*) explained his observations by noting that at the lower ingestion rates food passage is slower and digestion more efficient. Sushchenya (1970) assumed this phenomenon to be universal and applied Paloheimo & Dickie's (1965, 1966) equations derived from growth and feeding in fish to fit growth curves from Richman's *D. pulex* data.

Changes in the growth efficiency values of *Artemia* associated with development have already been reported although the magnitudes of the $K_1$ values differ markedly between some of the studies (cf. Mason (1964) vs. Brune & Anderson (1984)). Although the different results could be caused by differences in measuring techniques, nutritional value of different algae species or culture systems, particular stages in the development of *Artemia* are associated with critical inflection points in the reported $K_1$-growth curves. One such critical point corresponds to individuals of ca 100 $\mu$g i.d.w. (Figs. 3, 4), equivalent to ca 5 mm lengths, where dips in the values were found regardless of the ration. This critical point is also visible in Brune & Anderson (1984: 112, Fig. 3) and could be associated with some delicate stage in the development of the organism. Regions of $K_{1max}$ for treatments with the highest growth rates (t4–t8) occurred in organisms of ca 180–344 $\mu$g i.d.w. (6–7.5 mm lengths) which correspond to stages where the *Artemia* is fully mature (Schrehardt, 1987) and the specific filtration rate tends towards its lowest values (Brune & Anderson, 1984).

The above elements allow us now to develop a model for population growth, based on the associations between $K_1$ and i.r.-i.d.w. during growth, but also accomodating the observation that as the individual weights approach maximal values, $K_1$ tends towards zero. The desirability of this feature has been discussed by Silvert & Pauly (1987).

A mass balance equation can be established for the growth rate of an animal:

$$G = I - T \qquad (4)$$

where $G$ is the energy of growth, $I$ is the energy of the food intake, and $T$ is the energy of metabolic expenditure which includes respiration, excretion, molting and, in this form of the equation, the portion of non-assimilated food.

Increments in weight of individuals over any short period of time, $dt$, can be expressed as:

$$dw/dt = i - t \qquad (5)$$

where the same parameters are considered in units $d^{-1}$ and $i$ is equivalent to i.r.

As the gross growth coefficient, $K_1$, is defined as:

$$K_1 = dw/dt \times 1/\text{i.r.} \qquad (6),$$

data of i.r. and $K_1$ for any particular *Artemia* size class would allow us to predict growth rate by the reciprocal of this equation. The effect of metabolic expenditures are not considered here except as gross contributions through the $K_1$ values, for it can be shown that the expression for $K_1$ is related to total metabolic expenditure $t$ in equation (5):

$$K_1 = 1 - \text{t/i.r.} \qquad (7)$$

In our case, we can calculate $K_1$ values for particular combinations of i.d.w. and i.r. using a method analogous to Brune & Anderson (1984) from Fig. 6 for discrete time intervals, such as single days.

When considering population growth, mortality must also be taken into account. Although the results of both experiments showed mortality rates $>20\%$ over 14 d culture periods in the worst cases, these probably resulted from the frequent handling involved and extremes in conditions. Typical algae-fed mass rearings of *Artemia*, which we are attempting to model (e.g. Tobias *et al.*, 1980), 2-weekly survival rates $>90\%$ are common. For this reason, and considering the high survival or our treatments in near-optimal conditions, a daily survival rate of 0.991 was considered appropriate, yielding a final survival at day 14 of 0.88. Density-dependent effects were considered negligible in the light of our density experiments up to 16 ind $ml^{-1}$.

Thus, the growth in day $i$ of an individual *Artemia* under culture can be approximated as:

$$\Delta w_i = K_1(\text{i.r.}_{\cdot i}, x_i) * \text{i.r.}_{\cdot i} + x_i\ (\Delta t) \qquad (8)$$

where, $\Delta w_i$ = average weight gain ($\mu$g) for $i$th day.
i.r.$_{\cdot i}$ ($\mu$g) = i.r.$_{\cdot i\max}$ if the food ration available is $\geq$ i.r.$_{\cdot i\max}$, or
= ration available otherwise. The ration available is calculated by dividing the amount of food offered by the number of *Artemia* in the culture.
$x_i$ ($\mu$g) = mean i.d.w.
$K_1$ = 1% gross growth efficiency obtained from the contour plot for the particular $x_i$ and the estimated i.r.$_{\cdot i}$.

The overall population growth for the $i$th day will thus be:

$$\Delta W_i = \Delta w_i + N_i\ (\Delta t) \qquad (9)$$

where, $N_i$ (the total number of *Artemia* in the culture) = $N_{i-1} * 0.991$.

Beginning with an *Artemia* hatching weight of 1.8 $\mu$g, increments of average individual and population growth can be modelled following day increments and selecting the $K_1$ values corresponding to the projected $x_i$ and i.r. values, the latter of which will depend on the food availabilities under consideration.

The result of one such simulation appears in Fig. 9, where two theoretical regimes were modelled. Culture I growth curve is derived from feeding levels according to demand calculated from the i.r.$_{max}$ equation. Culture II has a ceiling limit of 163 $\mu$g *Artemia*$^{-1}$ d$^{-1}$ (approx. 1.6 M cells of *D. tertiolecta* d$^{-1}$) to parallel Mason's data (1963: 144, Fig. 3) and our own results in experiment 1. The shape and growth rates are similar to the observed results.

While the fit to observed animal growth demonstrates the soundness of our model, its greatest value lies in providing a capacity to evaluate alternative rearing strategies depending on requirements and resources and to dimension for adequate food production. For example, fastest growth rates and hence overall production rates in a given time period can be expected from demand feeding, *albeit* at lower efficiency levels. On the other hand, highest efficiency, under systems analogous to ours, would be predicted when rationing feed according to levels corresponding to ridges of high $K_1$ shown in Fig. 6.

The food requirements of the population will always rise with the growth of the individuals,

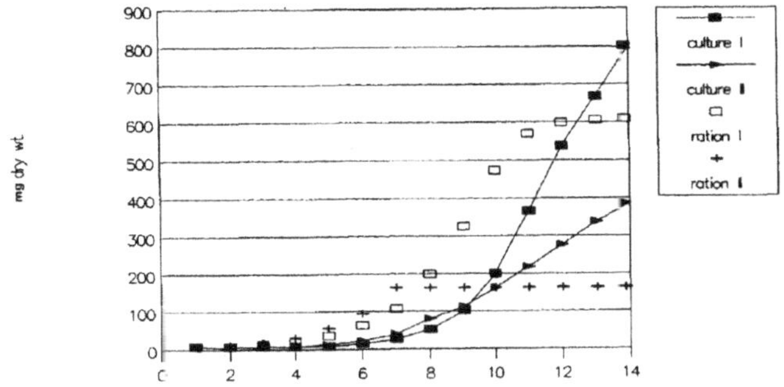


*Fig. 9.* Simulated growth curves for two culture conditions. Culture I is projected as receiving food according to demand (levels indicated as ration I), while culture II is modelled as receiving amounts limited by a ceiling of 163 $\mu$g d$^{-1}$ ind$^{-1}$ (ration II).

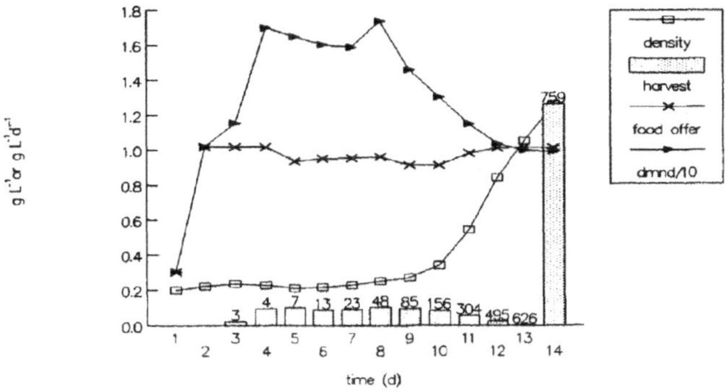


*Fig. 10.* Simulation of a proposed strategy for *Artemia* culture with sequential partial culls (see text). Food requirements (dmnd in figure) were calculated from an estimation of the global i.r. $_{max}$ values if no harvesting was performed. Demand data are 1/10 the calculated values in the graph only after day 2. Mass values are g d.w. Figures above bars represent estimated mean i.d.w. of the culled *Artemia*.

such that at the final stages for *Artemia*, the food demand which must be met becomes more than 10 times that found at the metanaupliar stage. Considering that algae production systems work at their most efficient when steady and continuous harvesting is maintained, the irregular demand of food by a single *Artemia* culture at best would hinder the optimization of the plant cultures. In Fig. 10 a projected run of an alternative strategy for mass culture of *Artemia* which we consider could aleviate this problem is provided. Considering a given upper daily limit to the amount of food available of 1 g d.w. per liter of *Artemia* culture (taking into account the results of Spectorova *et al.*, 1982 less than 1 l of algae culture per day would be necessary to obtain this amount of food) the problem is to maximize both total yields and final average i.d.w. after 14 d of batch culture.

The suggested scheme would involve starting cultures with 100,000 nauplii l$^{-1}$, corresponding to a biomass density of <0.4 g d.w. l$^{-1}$ not high enough to affect the quality of the medium. Food would then be offered on demand (filled triangles in Fig. 10) until the food requirements of the growing *Artemia* would reach the pre-established ceiling depending on the growth rate according to the feeding regime. At this point daily culls would begin, removing a number of *Artemia* equivalent (in terms of the sum of their individual i.r.$_{max}$

values) to the projected daily food 'deficits' (bars in Fig. 10, in g $l^{-1}$). As this procedure would be continued, a series of harvests would be obtainable, each containing a graded selection of individual *Artemia* sizes (figures above bars, in $\mu$g d.w.). Not only could the projected final mean i.d.w. of the last harvest be expected to be among the highest attainable in the best algae-fed conditions after 14 d culture (Ca 750 $\mu$g d.w.) and the final harvest yield greater than 1 g $l^{-1}$ (assuming 0.991 daily survival) but, additionally, an Ca 1 g $l^{-1}$ could be available in the form of over 90000 metanauplii and juveniles ranging in graded individual sizes from 3 to 600 $\mu$g d.w.

Although this mode of cultivating *Artemia* has not been reported and thus we cannot compare our projected results with real data, it is an extrapolation of conditions not unlike those found under our experiments. In particular, the projected standing biomass under culture, Ca 0.2 g $l^{-1}$ d.w. or 2 kg $m^{-3}$ in live weight, is considered normal for intensive operations. The last harvest projected yield, equivalent to 13 kg $m^{-3}$, is, if anything, conservative if one considers that 20 kg $m^{-3}$ have already been achieved (Sorgeloos, 1985).

While the data discussed here may only reflect results of cultures using the same species of phytoplankton under the same culture conditions and methodology, the approach is applicable to any system. Culture systems utilizing flow-through conditions are much more efficient than batch-cultures for achieving high yields (Lavens *et al.*, 1985) yet, although control of feeding is performed in terms of input and output algae concentrations, the individual ingestion rates and hence the growth efficiencies can still be predicted and manipulated by methods such as that described here. This Brune & Anderson (1984) have already demonstrated with methods analogous to ours. However, they did not find negative effects on conversion efficiencies by the highest ingestion rates. Whether their observations are characteristic to flow-through conditions, and reduction of $K_{1max}$ with maximal feeding is associated to batch cultures remains to be studied. However, although Mitchell (1991), using the freshwater anostracan *Streptocephalus macrourus*, found increasing $K_1$ values with increasing ration levels in a flow-through system, he also observed a reduction at the highest rations.


## Acknowledgements

This work was financed by grant PCECBNA-030216 from the Mexican National Council for Science and Technology (CONACYT) to one of the authors (FAAG).


## References


Baid, I. C., 1963. The effects of salinity on growth and form of *Artemia salina* (L.). J. exp. Zool. 153: 279–283.

Brune, D. E. & T. H. Anderson, 1984. The application of process kinetics for predicting optimum performance of continuous brine shrimp culture. J. World Maricul. Soc. 15: 108–119.

Conover, R. J., 1968. Zooplankton- life in an nutritionally dilute environment. Am. Zool. 8: 107–118.

Conover, R. J., 1964. Food relations and nutrition of zooplankton. Proc. Symp. exp. mar. Ecol., occ. Pub. no. 2, Graduate School of Oceanography, Univ. Rhode Island, USA.: 81–91.

Fábregas, J., C. Herrero, J. Abalde, R. Llao & B. Cabezas, 1986. Biomass production and biochemical variability of the marine microalga *Dunaliella tertiolecta* (Butcher) with high nutrient concentrations. Aquaculture 53: 187–199.

Gibor, A., 1957. Conversion of phytoplankton to zooplankton. Nature 179: 1304.

Hughes-Games, W. L., 1971. Growing the Japanese oyster (*Crassostrea* gigas) in subtropical seawater fish ponds. I. Growth rate, survival and quality index. Aquaculture 11: 217–229.

Lavens, P., P. Baert, A. De Meulemeester, E. van Ballaer & P. Sorgeloos, 1985. New developments in the high density flow-through culturing of brine shrimp *Artemia*. J. World Maricul. Soc. 16: 498–508.

Léger, P., D. A. Bengtson, P. Sorgeloos, K. L. Simpson & A. D. Beck, 1987. The nutritional value of *Artemia*: a review. In P. Sorgeloos, D. A. Bengtson, W. Decleir & E. Jaspers (eds), *Artemia* Research and its Applications, 3. Universa Press, Wetteren, Belgium: 356–372.

Mason, D. T., 1963. The growth response of *Artemia* salina (L) to various feeding regimes. Crustaceana, 5: 138–150.

McShan, M., N. M. Trieff & D. Grajcer, 1974. Biological treatment of wastewater using algae and *Artemia*. J. Wat. Pollut. Cont. Fed. 46: 1742–1750.

Milligan, D. J., J. A. Quick, S. E. Hill, J. A. Morris & R. J.

Hover, 1980. Sequential use of bacteria, algae and brine shrimp to treat industrial wastewater at pilot plant scale. In G. Persoone, P. Sorgeloos, O. A. Roels & E. Jaspers (eds), The Brine Shrimp *Artemia*, 3. Universa Press, Wetteren, Belgium: 193–206.

Mitchell, S. A., 1991. The growth rate and growth efficiency of *Streptocephalus macrourus* (Crustacea, Anostraca) cultured on microalgae. In Belk, D., H. J. Dumont & N. Munuswany (eds), Studies on Large Branchiopod Biology and Aquaculture. Development in Hydrobiology 64. Kluwer Academic Publishers, Dordrecht: 1–10. Reprinted from Hydrobiologia 212.

Nimura, Y., 1980. Retarded growth of *Artemia* salina by overfeeding. Bull. Jap. Soc. sci. Fish. 46: 681–687.

Omori, M. & T. Ikeda, 1984. Methods in Marine Zooplankton Ecology. John Wiley & Sons, N.Y. 332 pp.

Paloheimo, J. E. & L. M. Dickie, 1966. Food and growth of fishes. III. Relations among food, body size and growth efficiency. J. Fish. Res. Bd Can. 23: 1209–1248.

Paloheimo, J. E. & L. M. Dickie, 1965. Food and growth of fishes. I. A growth curve derived from experimental data. J. Fish. Res. Bd Can. 22: 521–542.

Reeve, M. R., 1963a. Growth efficiency in *Artemia* under laboratory conditions. Biol. Bull. 125: 133–145.

Reeve, M. R., 1963b. The filter-feeding of *Artemia*. I. In pure cultures of plant cells. J. exp. Biol. 40: 195–205.

Richman, S., 1958. The transformation of energy by *Daphnia pulex*. Ecol. Monographs. 28: 273–291.

Schrehardt, A., 1987. A scanning electron-microscope study of the post-embryonic development of *Artemia*. In P. Sorgeloos, D. A. Bengtson, W. Decleir & E. Jaspers (eds), *Artemia* Research and its Applications, 1. Universa Press, Wetteren, Belgium: 5–32.

Sick, L. V., 1976. Nutritional effect of five species of marine algae on the growth, development, and survival of the brine shrimp *Artemia salina*. Mar. Biol. 35: 69–78.

Silvert, W. & D. Pauly, 1987. On the compatibility of a new expression for gross conversion efficiency with the von Bertalanffy growth equation. Fish. Bull. 86: 139–140.

Sorgeloos, P., P. Lavens, P. Léger, W. Tackaert & D. Versichele, 1986. Manual for the culture and use of brine shrimp *Artemia* in aquaculture. State University of Ghent, Ghent, Belgium, 319 pp.

Sorgeloos, P., 1985. Potentials of converting microalgae into brine shrimp *Artemia*. Arch. Hydrobiol. Beih. Ergebn. Limnol. 20: 147–149.

Spectorova, L. V., O. I. Goronkova, L. P. Nosova & O. N. Albitskaya, 1982 (for 1981). High-density culture of marine microalgae- promising items for mariculture. I. Mineral feeding regime and installations for culturing *Dunaliella tertiolecta* Butch. Aquaculture 26: 289–302.

Sushchenya, L. M., 1970. Food rations, metabolism and growth of crustaceans. In J. H. Steele (ed) Marine Food Chains, University of California Press, Berkeley and Los Angeles, CA, USA.: 127–141.

Tobias, W. J., P. Sorgeloos, E. Bossuyt & E. Roels, 1979. The technical feasibility of mass-culturing *Artemia salina* in the St. Croix 'Artificial Upwelling' Mariculture System. In J. W. Avault (ed.) Proc. 10th Ann. Meeting WMS. Louisiana State University, Baton Rouge, LA, USA: 203–214.

Tobias, W. J., P. Sorgeloos, O. A. Roels & B. A. Sharfstein, 1980. International Study on *Artemia*. XIII. A comparison of production data of 17 geographical strains of *Artemia* in the St. Croix Artificial Upwelling-Mariculture System. In G. Persoone, P. Sorgeloos, O. A. Roels & E. Jaspers (eds), The Brine Shrimp *Artemia*, 3. Universa Press, Wetteren, Belgium: 383–392.

Vanhaecke, P., S. E. Sidall & P. Sorgeloos, 1984. International study on *Artemia*. XXXII. Combined effects of temperature and salinity on the survival of *Artemia* of various geographical origins. J. exp. mar. Biol. Ecol. 80: 259–275.

Vanhaecke, P. & P. Sorgeloos, 1980. International study on *Artemia*. XIV. Growth and survival of *Artemia* larvae of different geographical origin in a standard culture test. Mar. Ecol. Prog. Ser. 3: 303–307.

Von Hentig, R., 1971. Einfluss von Salzgehalt und Temperatur auf Entwicklung, Wachstum, Fortpflanzung und Energiebalanz von *Artemia salina*. Mar. Biol. 9: 145–182.

Wear, R. G. & S. J. Haslett, 1986. Effects of temperature and salinity on the biology of *Artemia franciscana* Kellogg from Lake Grassmere, New Zealand. 1. Growth and mortality. J. exp. mar. Biol. Ecol. 98: 153–166.

*Hydrobiologia* **212**: 39–44, 1991.
*D. Belk, H. J. Dumont & N. Munuswamy (eds), Studies on Large Branchiopod Biology and Aquaculture.*



# Some aspects of *Artemia* biology affected by cestode parasitism

F. Amat, A. Gozalbo, J.C. Navarro, F. Hontoria & I. Varó
*Instituto de Acuicultura de Torre de la Sal, 12595 Ribera de Cabanes, Castellón, Spain*

*Key words:* *Artemia*, parasitism, survival, castration

## Abstract

*Artemia* individuals from a wild population, bearing cestode cysticercoids (Hymenolepididae) were kept in the laboratory during four months to observe influences of the parasite on the host biology. No differences were found between parasitized and unparasitized shrimps regarding final adult size. The higher survival of parasitized animals suggests that parasitism is not pathogenic. The most dramatic effect was host castration. Parasitized individuals showed higher total lipid levels, probably linked to carotenoid pigments conveying a red color to parasitized shrimps.

## Introduction

During a research program on the ecology of *Artemia* populations along the Spanish Mediterranean coast (Amat *et al.*, 1988), Cestode cysticercoids (Hymenolepididae) were observed parasitizing the hemocoele of brine shrimps.

Many aquatic and terrestrial crustaceans play an important role as intermediate hosts for parasites of fishes, birds and mammals (Moore, 1984; Fischer, 1968). Brine and fairy shrimps are among these crustaceans (Maksimova, 1973, 1974, 1976; Jarecka, 1984; Codreanu & Codreanu-Balcescu, 1978; Gabrion *et al.*, 1982). However, there is little information about the effects of parasites on the biology of Anostraca.

In a separate paper (Amat *et al.*, in preparation), a first account about parasitism in *Artemia* along the Spanish Mediterranean coast was given, with special attention to its prevalence and the diversity of anatomical regions parasitized by cysticercoids. This report outlines several effects of parasitism on *Artemia* biology, and shows how these ultimately favor the parasite (Courtney & Christensen, 1987).

## Materials and methods

A wild population of parthenogenetic *Artemia* (Bonmati saltern, Santa Pola, Spain) was sampled on October 27, 1988 (S = 130‰, T = 23 °C). After sample processing, as described elsewhere (Amat *et al.*, in preparation), we found 18.10% of the females to be parasitized with 1.06 cysticercoids/female.

Parasitized (200) and unparasitized (200) females were randomly selected and placed in 4 glass carboys (100 each) filled with 2 l of brine (S = 90‰) and fresh *Dunaliella salina* culture, provided with gentle air bubbling, and kept at room temperature (20–22 °C) for a normal daylight period. Brine and algae were renewed twice a week.

Experimental populations were observed once every 10–12 days. Surviving brine shrimp were removed, lightly narcotized, counted, and measured under a dissecting microscope. Special attention was paid to the reproductive state attained by unparasitized females and each was classified as: Immature (quiet ovaries), ovulating (oocites moving along the ovaries or filling the oviducts), or ovated (embryos, naupliae, or cysts filling the ovisac). Parasitized females attaining mature size were also examined for condition of the reproductive system. Females were returned to the flasks after recovering from the anaesthetic.

Eight observations were performed. The first one evaluated the starting population.

Some samples were kept frozen until analysis for biochemical composition (Gozalbo & Amat, 1988). Results were statistically treated by means

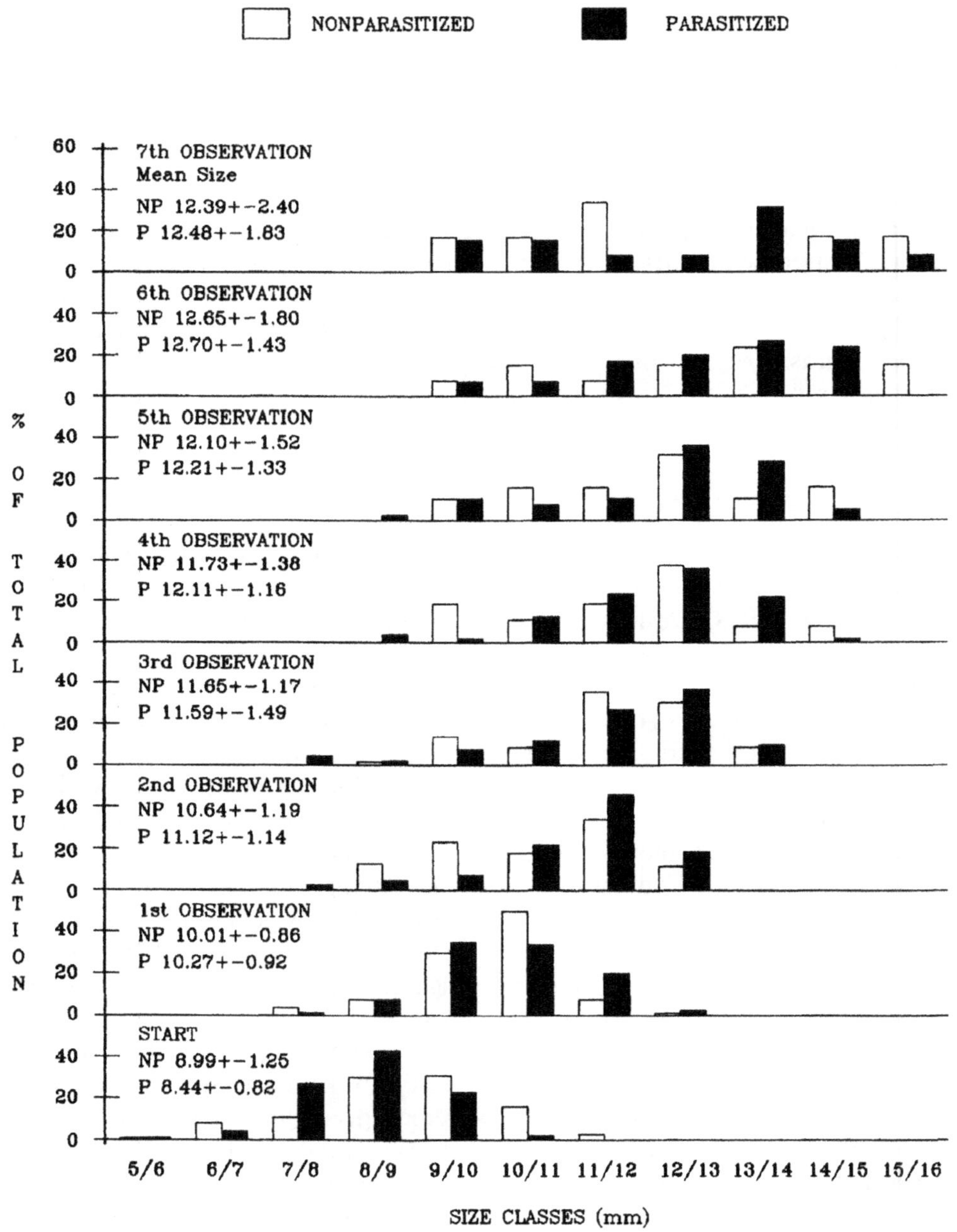


*Fig. 1.* Frequency (%) distribution of female sizes in both experimental groups (parasitized and nonparasitized) from starting population (ST) up to the final (7th) observation, with specification of mean population size (mm).

of a two tailed t-test (Sokal & Rohlf, 1981) in order to detect significant differences ($p \leq 0.05$) between parasitized and unparasitized females.

## Results

### *Brine shrimp size*

There are well documented cases of remarkable size increases in parasitized individuals compared to unparasitized ones in the same population (Mueller, 1963; Stark, 1965; Pearre, 1976). A difference in shape displayed by parasitized *Artemia*, reddish color, and larger size was reported by Amat *et al.* (in preparation) in wild populations, where healthy and mature females attained 7–11 mm, and parasitized 9–14 mm. This was not the case among our experimental populations. Figure 1 relates to size changes in both groups. Despite the increase of mean size by about 4 mm, there was no difference in size between parasitized and nonparasitized individuals.

### *Survival*

Pathogenicity is an important factor in the compatibility of host and parasite (Calentine, 1967). Figure 2 suggests that this kind of parasitism is not pathogenic, because of the higher survival displayed by parasitized shrimps after the second observation. Survival was twofold after the fourth and three times greater from then to the end of the experiment (approximate mean age of surviving females 110 days). It is of interest to note that until the 6th observation the biggest parasitized female (14.4 mm) bore 11 cysticercoids, a value not surpassed in the available literature.

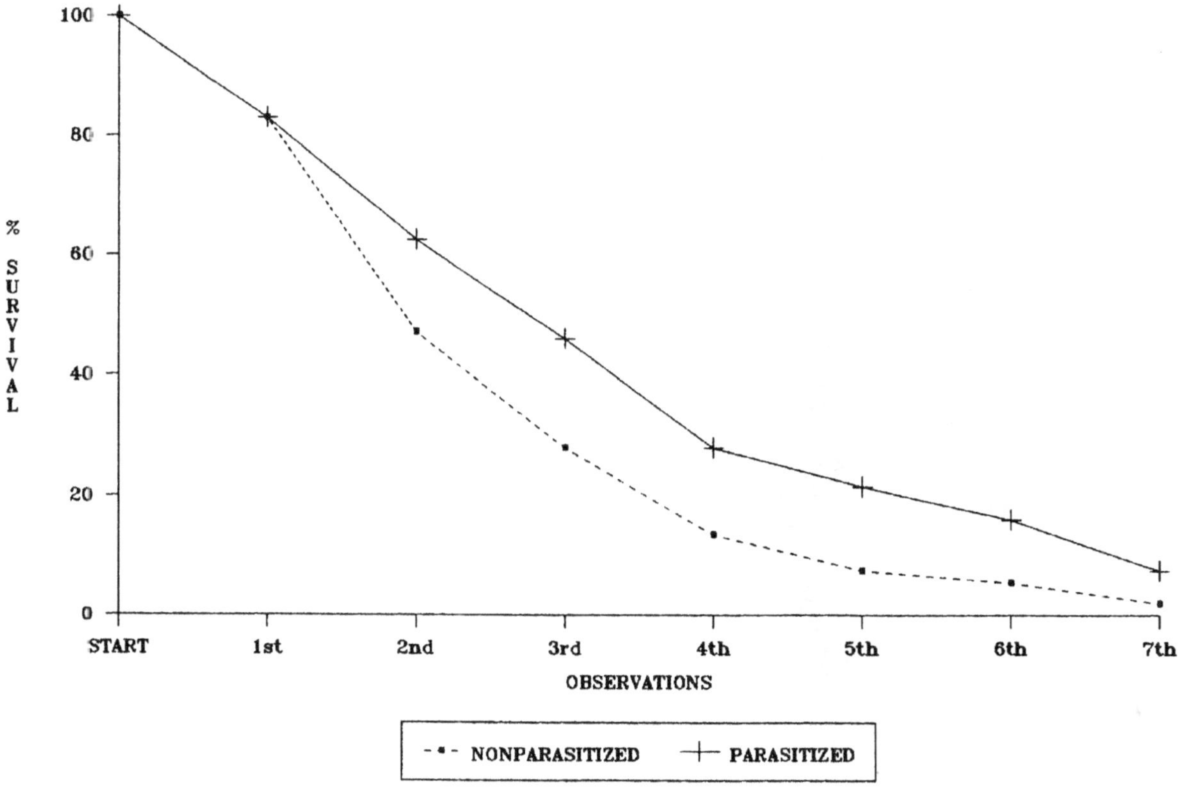


*Fig. 2.* Mean percent survival in parasitized and nonparasitized female experimental groups from the starting population (ST) up to the final (7th) observation.

*Fecundity*

The presence of parasites can produce severe changes in the host's reproductive capabilities, including total inhibition, through hormonal and mechanical interference (Munro, 1953; Hynes, 1955; Stark, 1965; Muzzall & Rabalais, 1975a; Keymer, 1980; Moore, 1983; Maema, 1986; Hurd & Arma, 1987; Hurd *et al.*, 1987). Figure 3 shows the gradual decrease of immature females

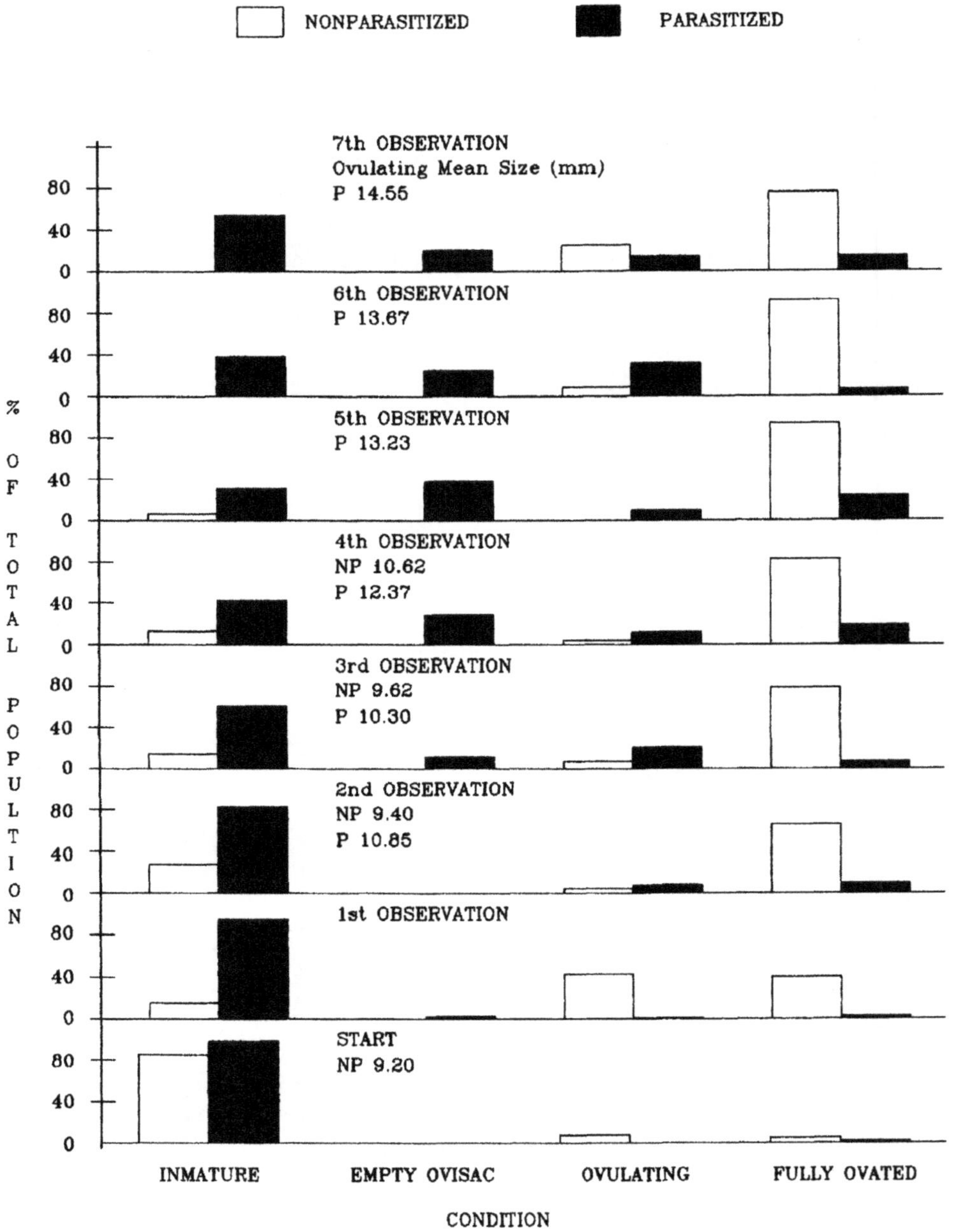


*Fig. 3.* Frequency (%) distribution of different reproductive conditions in both parasitized and nonparasitized female experimental groups from the starting population (ST) up to the final (7th) observation, with specification of ovulating female mean size (mm).

*Table 1.* Values obtained for the main biochemicals analised (as % dry weight) in parasitized and unparasitized brine shrimp *Artemia*.

| | Nonparasitized | | Parasitized | t-value | t-prob |
|---|---|---|---|---|---|
| | Immature | Mature | | | |
| Lipids | 9.49 ± 1.09 | 9.42 ± 0.59 | 11.13 ± 0.76 | − 2.33 | 0.04 |
| Phospholipids | 6.69 ± 0.05 | 6.68 ± 0.96 | 5.12 ± 0.71 | 1.63 | 0.13 |
| Cholesterol | 1.40 ± 0.19 | 0.98 ± 0.03 | 1.33 ± 0.09 | − 0.77 | 0.46 |
| Glycogen | 16.69 ± 1.00 | 16.84 ± 0.98 | 21.63 ± 2.62 | − 4.10 | 0.002 |

in the unparasitized population. They were actively ovulating at 9.20–9.60 mm mean size and ovated at 10–11 mm, displaying a mean reproductive output of about 80 embryos or cysts per brood.

The population of parasitized females always displayed high levels of immature specimens. They usually reached mature size without ovulating. When they did ovulate, their scarce broods produced only 10–12 embryos or cysts. Most of them slightly expanded the ovisac but there was no oocyte immigration. Parasitized females ovulated weakly while increasing in size from 10.85 to 14.55 mm.

### *Biochemical composition*

The main biochemicals analysed and the values obtained are detailed in the Table 1. A t-test indicated that there were no significant differences ($p \leq 0.05$) between the results obtained for unparasitized immature and mature females, while data from the unparasitized animals, when compared with those from parasitized individuals gave significant differences pointing to higher lipid and glycogen levels in parasitized females.

## Discussion

Brown & Thompson (1986) and Moore (1983) stated that parasites can provoke severe changes in shape, size, and behavior of intermediate hosts, aiming to make them more conspicuous to predators. Bright red color displayed by parasitized *Artemia* is an example of this mechanism. However, we did not find differences in size between healthy and parasitized specimens in our experimental populations. In opposition to what usually happens in wild populations where, probably because of the higher longevity of parasitized brine shrimps, several older and bigger individuals coexist with unparasitized individuals that are smaller. It may be that in these parasitized individuals higher total lipid levels, assuming that red pigments like canthaxantin or echinenone add to total lipids, provide buoyancy and protection against solar radiations. (Margalef, pers. com.).

According to Courtnay & Christensen (1978), the life cycle of an intermediate host will not be seriously hampered if there exists a good compatibility with its parasite. In our study, there was little pathology associated with the relationship. In fact, host longevity increased. This strategy usually allows the host to live long enough for the parasite to reach the infective stage and be transferred to the final host. In this case, there was no evidence for transfer to a final host because the parasite appeared always as a cysticercoid, seldom with active strobilation.

Finally, in agreement with Munro (1953), Hynes (1955), Schmidt (1964), Kennedy (1972), Baudoin (1975), and Moore (1983) we consider it possible that benefits accrued to a parasite castrating or otherwise reducing host reproduction may include, besides increased host viability, an increase in energy available because of reduced host reproductive expenditure. As an example of this, we observed significantly higher glycogen levels in parasitized females, although it is not possible to state if this glycogen comes from pathophysiological effects induced by the parasite or if it is stored in the cysticercoids prior to its investment in the final parasitic form following metamorphic development.

## Acknowledgements

The authors are grateful to Mrs. Ma. Nieves Sanz who kindly helped in samples processing. This work was supported by a grant CAICYT 83/002. J.C. Navarro was supported by the 'Generalitat Valenciana' and F. Hontoria by the 'Excelentisima Diputación de Castellón'.

## References


Amat, F., F. Hontoria, J. C. Navarro & A. Gozalbo, 1988. Bioecologia de *Artemia* de la laguna de La Mata (Torrevieja, Alicante). Actas Coloq. Luso-Espanhol. Ecol. Bacias Hidrog. Rec. Zool.: 17–24.

Amat, F., M. P. Illescas & J. Fernandez. Brine shrimp *Artemia* from Spanish Mediterranean salterns parasitized by *Flamingolepis liguloides* (Cestoda, Hymenolepididae) cysticercoids. (in preparation).

Baudoin, M., 1975. Host castration as a parasitic strategy. Evolution 29; 335–352.

Brown, A. F. & D. B. A. Thompson, 1986. Parasite manipulation of host behaviour: acanthocephalans and shrimps in the laboratory. J. Biol. Educ. 20: 121–127.

Calentine, R. I., 1967. Larval development of four caryophyllaeid cestodes. J. Iowa Acad. Sci. 72: 418–424.

Codreanu, R. & D. Codreanu-Balcescu, 1978. The occurrence in *Artemia salina* L. (Crustacea Anostraca) from Rumania of a peculiar cysticercoid larva belonging to avian hymenolepididae (Cestoda). Proc. IVth. Int. Cong. Parasit. Ser. B: 39 pp.

Courtney, C. C. & B. M. Christensen, 1987. Host-parasite relationships of caryophyllaeid cestodes and aquatic oligochaetes: I. Host longevity and parasite intensity. J. Parasitol. 73: 1124–1132.

Fischer, H., 1968. The life cycle of *Proteocephalus fluviatilis* Bangham (Cestoda) from smallmouth bass, *Micropterus dolomieui* Lacépède. Can. J. Zool. 46: 569–580.

Gabrion, C., G. MacDonald & V. Boy, 1982. Dynamique des populations larvaires du cestode *Flamingolepis liguloides* dans une population d'*Artemia* en Camargue. Acta Oecolog. 3: 273–293.

Gozalbo, A. & F. Amat, 1988. Composición bioquímica de biomasas silvestres de *Artemia*. Investigación Pesquera 52: 375–385.

Hynes, H. B. N., 1955. The reproductive cycle of some British freshwater Gammaridae. J. anim. Ecol. 24: 352–387.

Hurd, H. & C. Arma, 1987. *Hymenolepis diminuta*: Effect of infection upon the patency of the follicular epithelium in the intermediate host, *Tenebrio molitor*. J. Int. Pathol. 49: 227–234.

Hurd, H., J. C. Mercer & A. E. Munn, 1987. The effect of *Hymenolepis diminuta* upon ecdysteroid activity in the haemolymph of the intermediate host, *Tenebrio molitor*. Parasit. Res. 74: 198–199.

Jarecka, L., 1984. Development of *Hymenolepis arctowskii* Jarecka et Ostas, 1984 (Cestoda, Hymenolepididae) in the intermediate host *Branchinecta gaini* Daday (Branchiopoda) of the Antarctic. Acta Parasit. Pol. 29: 337–342.

Kennedy, C. R., 1972. The effect of the cestode *Caryophyllaeus laticeps* upon production and respiration of its intermediate host. Parasitology 59: 783–794.

Keymer, A. E., 1980. The influence of *Hymenolepis diminuta* on the survival and fecundity of the intermediate host, *Tribolium confusum*. Parasitology 81: 405–421.

Maema, M., 1986. Experimental infection of *Tribolium confusum* (Coleoptera) by *Hymenolepis diminuta* (Cestoda): host fecundity during infection. Parasitology 92: 405–412.

Maksimova, A. P., 1973. Branchiopoda as the intermediate hosts of Hymenolepididae. Mat. Int. Conf. Hymenolep. W. 14–16: 82–85.

Maksimova, A. P., 1974. Branchiopods (Branchiopoda, Anostraca), intermediate hosts of cestodes Fam. Hymenolepididae. Parazitologiya 7: 349–352.

Maksimova, A. P., 1976. A new cestode, *Fimbriarioides tadornae* sp. N. from *Tadorna tadorna* and its development in the intermediate host. Parazitologiya 10: 16–24.

Moore, J., 1983. Parásitos que cambian el comportamiento de su patrón. Investigación y Ciencia 94: 58–65.

Moore, J., 1984. Altered behavioral responses in intermediate hosts. An acanthocephalan parasite strategy. Am. Nat. 123: 572–577.

Mueller, J. F., 1963. Parasite induced weight gain in mice. In T. C. Chang (Ed.) Some biochemical and immunological aspects of host-parasite relationships. Ann. N. Y. Acad. Sci. 113: 217–233.

Munro, W. R., 1953. Intersexuality in *Asellus aquaticus* L. parasitized by a larval acanthocephalan. Nature 172: 313.

Muzzall, P. M. & F. C. Rabalais, 1975a. Studies on *Acanthocephalus jacksoni* Bullock. 1962. II An analysis of the host-parasite relationship of larval *Acanthocephalus jacksoni* in *Lirceus lineatus* (Say). Proc. Helm. Soc. Wash. 42: 35–38.

Pearre, S., 1976. Gigantism and partial parasitic castration of chaetognata infected larval trematodes. J. mar. biol. Ass. U.K. 56: 503–513.

Schmidt, G. D., 1964. Life cycle and development of *Prosthorhynchus formosus* (Van Cleave, 1918) Travassos, 1926. an acanthocephalan parasite of birds. Dissertation. Colorado St. Univ. Disserts. Fort Collins. Col. USA.

Sokal, R. R. & F. J. Rohlf, 1981. Biometry. W. H. Freeman & Co. New York, 859 pp.

Stark, G. T. C., 1965. *Diplocotyle* (Eucestoda), a parasite of *Gammarus zaddachi* in the estuary of the Yorkshire Esk. Britain. Parasitology 55: 415–420.

*Hydrobiologia* **212**: 45–59, 1991.
D. Belk, H. J. Dumont & N. Munuswamy (eds), *Studies on Large Branchiopod Biology and Aquaculture.*



# Life history and interspecies relationships of *Chirocephalus diaphanus* Prévost and *Tanymastix stagnalis* (L.), (Crustacea, Anostraca) inhabiting a group of mountain ponds in Latium, Italy

Graziella Mura
*Department of Animal and Human Biology, University La Sapienza, Rome, Italy*

*Key words:* Anostraca, coexistence, life history

## Abstract

Life histories of two coexisting Anostracan species, inhabiting a group of mountain ponds in Latium (Pantani di Forca Canapine) are described and discussed, with particular regard to hatching requirements, growth rate, fecundity and life span.

Unlike previously assumed (Mura, 1985), the cysts of the two species *Tanymastix stagnalis* and *Chirocephalus diaphanus* seem to respond to the same hatching stimuli, since their nauplii coexist and can be separated by morphological differences.

Significant differences recorded as to size, time in attaining sexual maturity, fecundity and life span, are in favour of a niche separation by size and by time. In particular, size differences are supposed to reflect significant differences also at filtering apparatus level, thus allowing prey selection by size.

## Introduction

Available literature on the subject (Broch, 1969, 1988; Brown & Carpelan, 1971; Donald, 1983; Geddes, 1973; Horne, 1967; Huggins, 1976; Lynch, 1964; Prophet, 1959, 1963; Wiman, 1978), records the coexistence of two or more Anostracan species, either sharing a pond at different times of the year depending on climatic events, or (in most cases) attaining sexual maturity at different moments and co-occurring in the adult stage for a certain period according to their specific tolerance ranges for environmental factors. Attention focuses mainly on the relative abundance of each species in relation to different environmental conditions characterizing the study years. No information is provided on the early life of the species.

A detailed description and discussion was given by Bernice (1972), Moore (1959, 1963) and Nourisson (1961), though concerning different situations. The former, discussing interspecies relationships between co-occurring *Branchinella kugenumaensis* and *Streptocephalus dichotomus*, records the two species as temporally separated. The nauplii of the first species appear earlier after the rainfall than *S. dichotomus* larvae, collected 3–7 days later.

A difference in hatching requirements was assumed to explain this phenomenon, even though it is unclear how the nauplii of the two species can be told apart, since no comments or illustrations are provided in the text concerning morphological features.

As to the case presented by Moore (1959), individuals of *Eubranchipus moorei* (misidentified as *E. holmani*) and of *S. seali* co-occurring in mixed populations were approximately at the same stage of development. He inferred therefore that the species had coexisted since the initial filling of the ponds. In this instance difference in hatching time was probably not the factor separating the two populations. A differential early growth rate was suggested to play the major role. Here too, however, no comments on early larval morphology of the involved species are provided, and the author did not identify larvae smaller than 5–6 mm.

Nourisson (1961), discussed a case of sympatry of *Chirocephalus diaphanus*, *Branchipus schaefferi* and *Tanymastix lacunae* in ponds of the Camargue (South France). He assumed a difference in hatching requirements for the three species to explain the presence of mixed populations of different sizes, thus inferring they were of different ages. No distinction was made at the larval level and the involved species were identified after reaching adult size in an aquarium.

In the above literature, two factors important in coexistence are identified: timing in attaining sexual maturity and an association of species which usually differ in environmental requirements (temperature, ionic composition of the medium, salinity etc.), one being more favoured by local conditions than the other. One point remains to be confirmed, however: is the observed succession in the presence of the adults due to a different hatching time depending on specific requirements, or is simultaneous hatching followed by a different early growth rate? An answer to this question could probably be provided by a morphological analysis of the early larval stages collected soon after filling of the pond.

In this paper the results of field studies dealing with *Chirocephalus diaphanus* Prévost and *Tanymastix stagnalis* (L.) in pasture pools of the Central Appennine in Latium, at 1600 m above sea level, are discussed in an attempt to better understand Anostracan species associations.

The coexistence of these two species has been

*Fig. 1.* The study site, pond P7, Pantani di Forca Canapine.

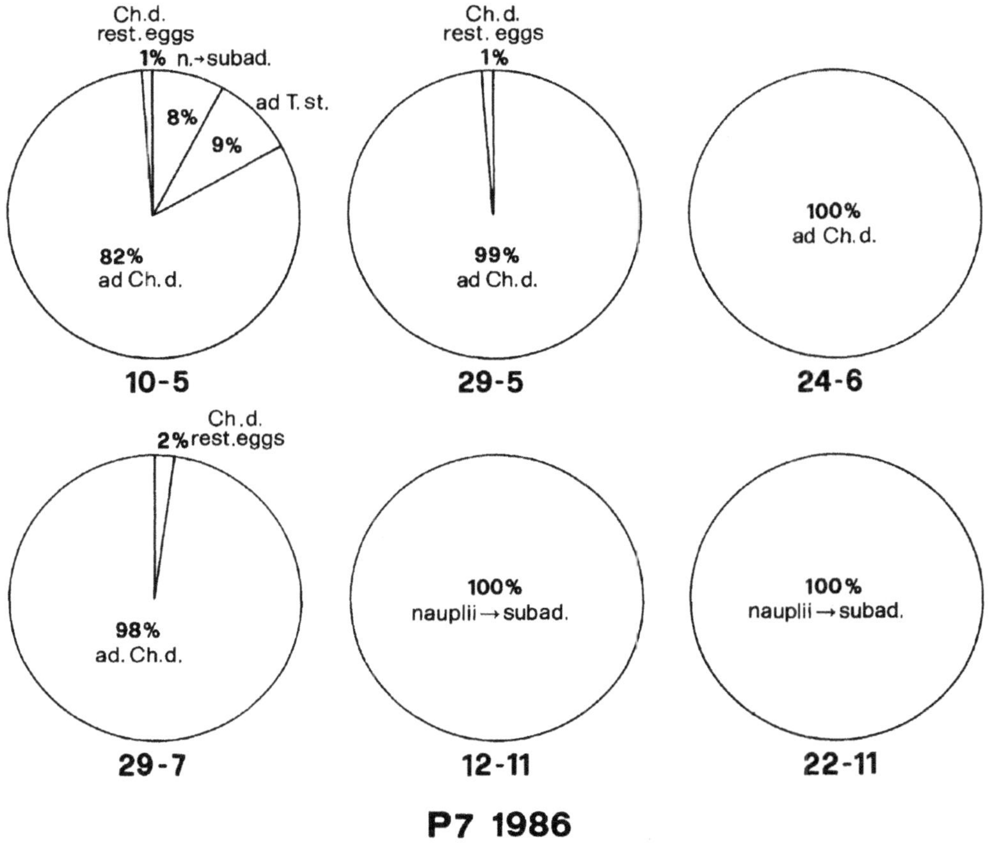


*Fig. 2.* Frequency of sampling during 1986.

recorded previously (Mura, 1985). The ecological conditions and the hydrological cycle of the ponds were described by Mura, 1987.

## Materials and methods

One of the previously described ponds (Pantani di Forca Canapine) (Mura, 1987), identified as P7 (Fig. 1), was studied during three years (1986–1988). It was periodically sampled, from the beginning of snowmelt, in late March or early April, until summer drying, and again, from filling by rainfall in autumn until the onset of winter, when severe climatic conditions prevented me from reaching the study area.

In 1986 the frequency of sampling was fortnightly (Fig. 2). It was weekly in 1987, in order to obtain more detailed information (Fig. 3). A further study was performed in 1988, again at 15 day intervals (Fig. 4). Collections of fairy shrimps were made with a plankton net of 150 $\mu$m mesh size.

Anostracan larvae were measured from head to anus with an eyepiece micrometer in a Zeiss microscope and classified for developmental stages according to Nourisson (1964). Total length of adults was determined under a dissecting microscope, from the tip of the head to the tip of the cercopods, excluding the terminal setae (Moore, 1955). At each sampling the number of males and females was counted. To assess clutch size, individuals with full ovisacs were isolated and allowed to drop their eggs. Care was taken to avoid females with only partly full egg sacs, since these could have shed part of their eggs (Hildrew, 1985).

The diameter and total number of eggs per

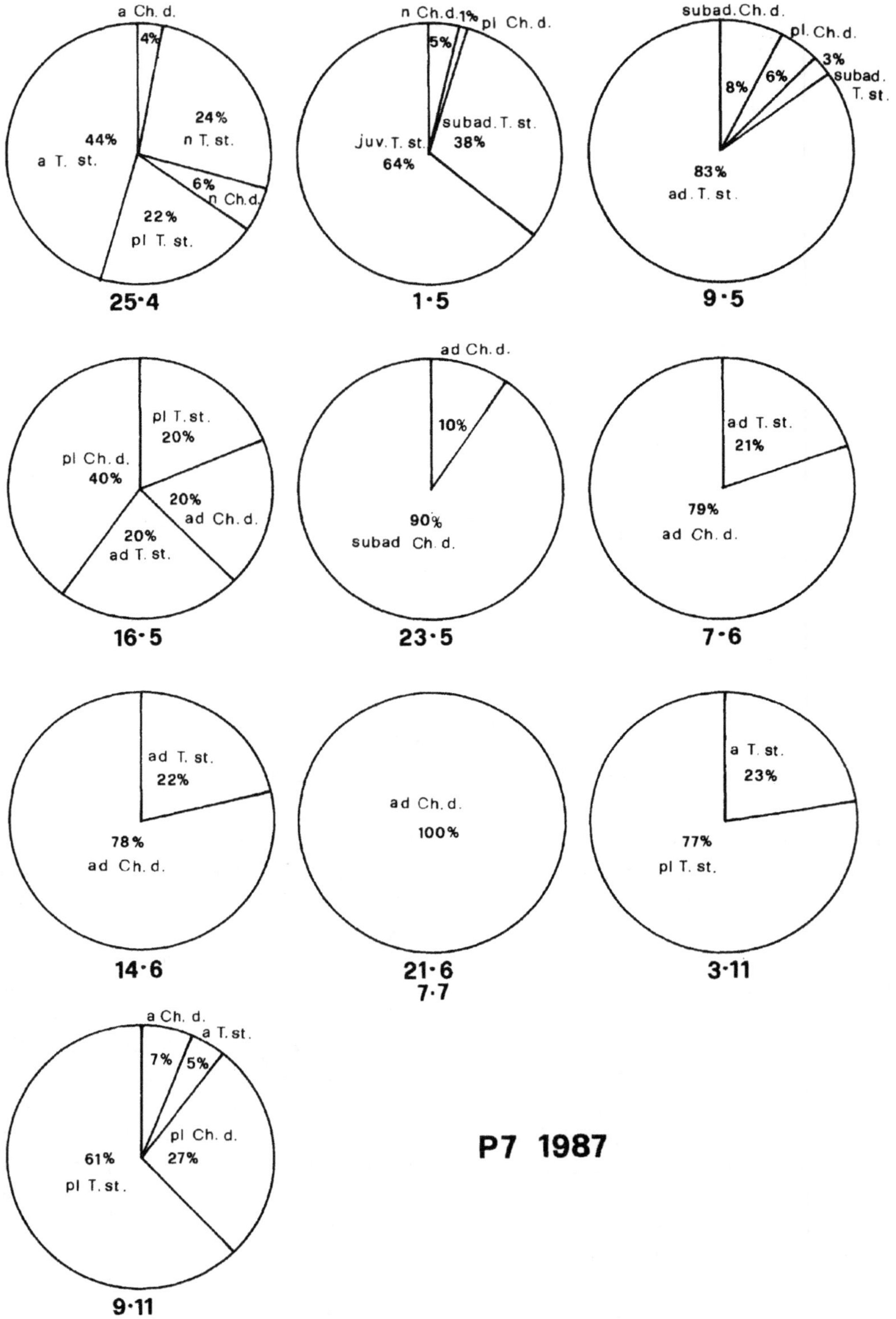


*Fig. 3.* Frequency of sampling during 1987. *n* = nauplii, *a* = metanauplii, juv. = juvenes, subad. = just differentiated males and females, ad. = sexually mature individuals.

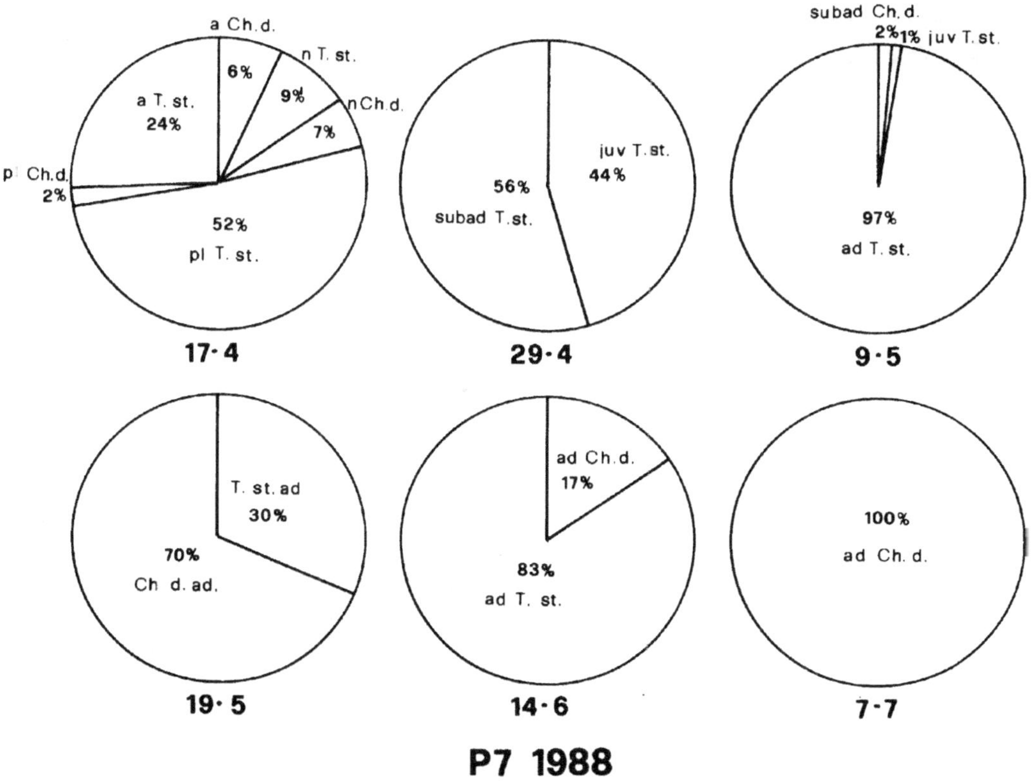


*Fig. 4.* Frequency of sampling during 1988. *n* = nauplii, *a* = metanauplii, (rudiments of thoracic segments evident), pl = advanced metanauplii (rudiments of thoracic appendages evident).

clutch was determined for each sampled female. Sex of post-larval forms was determined from the shape of genitalia and second antennae.

Water and air temperatures were recorded with a standard laboratory thermometer; pH measurements were performed with a field pH meter. Meteorological information was provided by the Istituto Centrale di Statistica. (ISTAT).

In order to obtain information on the egg-laying habits of the two species, soil samples were taken at intervals along the main diameter of the pond. After sieving, cysts were counted and their percentages in each sample determined (Fig. 5).

## Results

Seasonal fluctuations in meteorological conditions, pH and temperature during the study years are illustrated in Figs. 6–9.

### *Hatching*

As cited elsewhere (Mura, 1987), the ponds usually flood twice a year: in early spring, by snowmelt, and in autumn, by rainfall. The Anostracan populations thus exhibit bivoltine cycles depending on precipitation. Marked differences in incubation periods of the cysts were observed correspondingly in the spring cycle and in the autumn cycle of the ponds.

During the years of observations, in fact, I recorded hatching a few days (2–3) after autumnal flooding, and after about 20 days from the beginning of snowmelting, in early spring. Winter inspections of the ponds were almost impossible for several reasons (distance from Rome, severe meteorological and walking conditions at that altitude, etc.).

In spring, hatching began towards the second half of April, depending on water temperature.

In 1986 my first visit was unfortunately too late

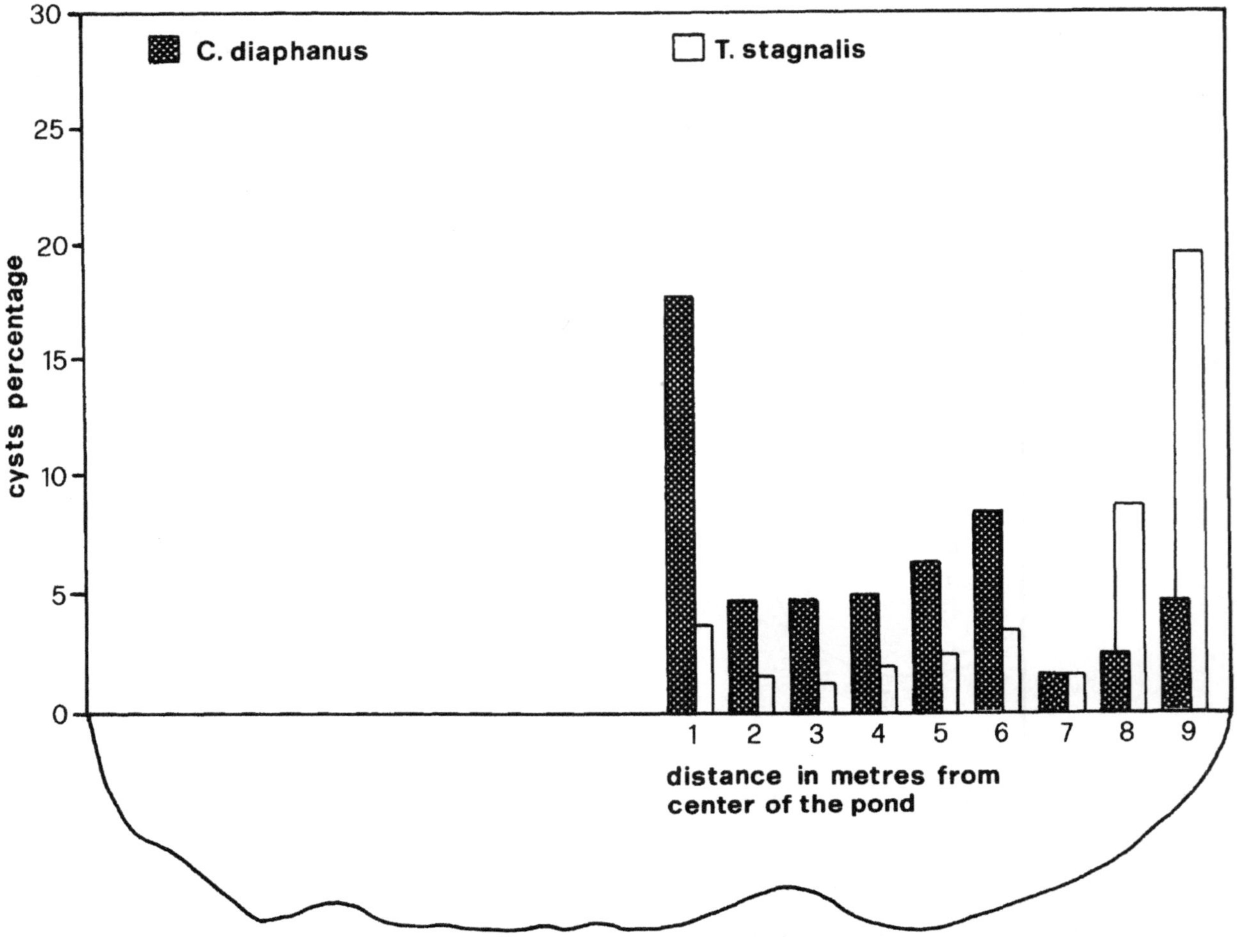


*Fig. 5.* Pattern of oviposition of *T. stagnalis* and *Ch. diaphanus* in the pond studied.

to detect early larval forms. In 1987 and 1988 I could assess that both species hatch as a true nauplius. Nauplii of *C. diaphanus* and of *T. stagnalis* were collected on April 25th in 1987, and on April 17th in 1988, together with older larvae (Figs. 3–4).

From the sampling of 1987, it appeared that the hatching period seems more restricted for *T. stagnalis* than for *C. diaphanus*, since nauplii of the latter were still collected 7 days after the nauplii of the former had disappeared, on the 1st of May.

In 1988, no nauplii of either species were collected after the first sampling.

During autumn 1987, the first larvae were collected on 3 November (Fig. 3) and all were identified as *T. stagnalis*; no nauplii were present, all were older larvae (mean size = 0.8 mm). On the subsequent sampling, however, both species occurred (Fig. 3), though no nauplii were collected (mean size of *T. stagnalis* = 1.3 mm, mean size of *C. diaphanus* = 1.1 mm). After this date snow and ice invaded the ponds. The pattern was more or less the same in 1986 (Fig. 2). In 1988, no consistent flooding took place (Fig. 9).

Separation of the larvae by developmental stage (Nourisson, 1964; Di Luzio, 1977) and analysis of their morphological differences, enabled me to identify the larvae of the two species. As indicated in Figs. 3–4, *C. diaphanus* and *T. stagnalis* larvae were present in each sample, in different percentages. Mean sizes of

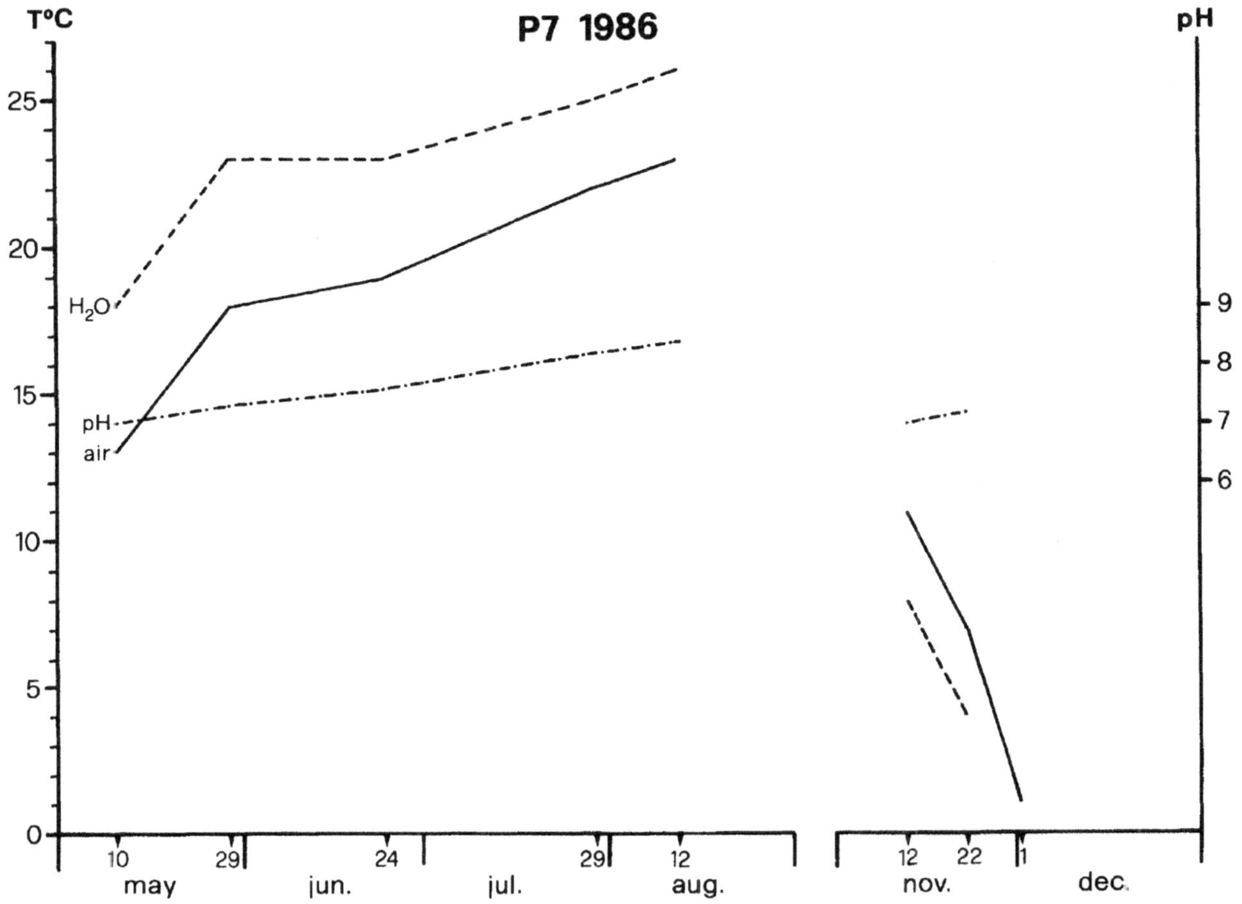


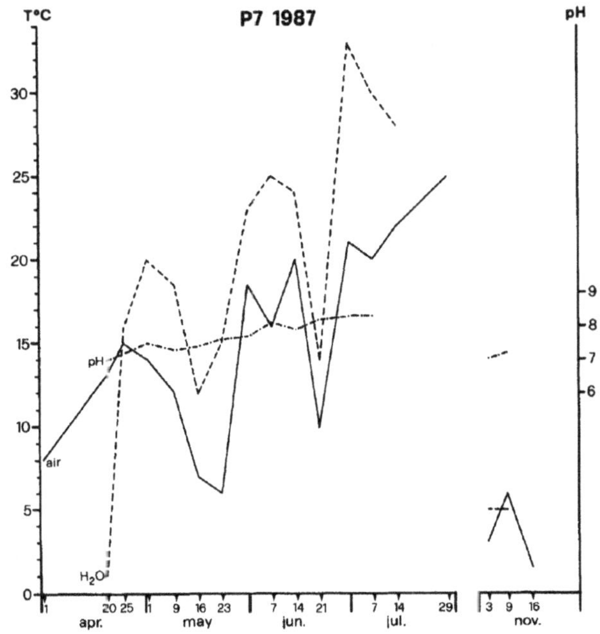


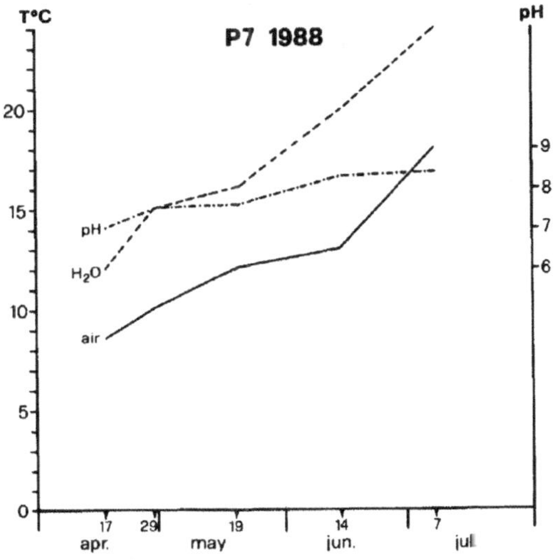


*Figs. 6, 7, 8.* Annual fluctuations of pH and temperature in 1986, 1987 and 1988.

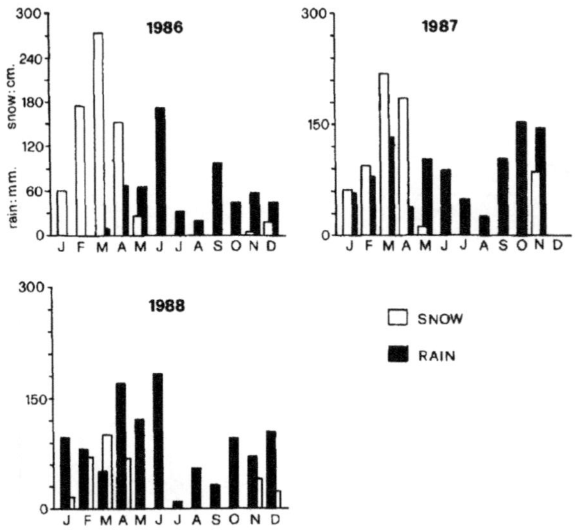


Fig. 9

the nauplii (0.458 mm for *C. diaphanus* and 0.420 for *T. stagnalis*) showed significant differences, but striking differences were detectable on the first antennae and the furca. *Tanymastix* larvae have a short AI and their furca appears late during development of furca (Figs. 10–11). Student's T test for body size and AI length of the nauplii are presented in Table 1.

*Table 1.* Statistical evaluation of biometric data on body size and antennule length for nauplii of *T. stagnalis* and *Ch. diaphanus*.

| | | | |
|---|---|---|---|
| A – Body size ($n$ = 41) | | | |
| | *Ch. diaphanus* | | *T. stagnalis* |
| range | 0.400–0.544 | | 0.366–0.624 |
| mean | 0.458 | | 0.421 |
| *D* | 0.0072 | | 0.1169 |
| *Df* | | 80 | |
| *V* | | 0.0024 | |
| *c* | | 0.0486 | |
| *t* | | 3.494 | |
| $p < 0.002$ | | | |
| B – Length of first antennae ($n$ = 41) | | | |
| | *Ch. diaphanus* | | *T. stagnalis* |
| range | 0.176–0.272 | | 0.016–0.144 |
| mean | 0.219 | | 0.096 |
| *D* | 0.0269 | | 0.0207 |
| *Df* | | 80 | |
| *V* | | 0.0595 | |
| *c* | | 0.0244 | |
| *t* | | 22.899 | |
| $p < 0.002$ | | | |

### *Development*

Simple means, ranges, and growth curves based on collections made in 1987 and 1988 are illustrated in Figs. 12–13.

The basic pattern of growth is similar, and both species show a decrease in growth rate after they attain sexual maturity (eggs in the oviducts, full seminal vesicles). Nonetheless, differences in growth rates and maturation times between the species exist. In *T. stagnalis* in 1987, sexual differentiation began on 1 May and was completed on 9 May, 14 days after hatching. The first sexually mature individuals were collected on 16 May (21 days after birth). Until this date, increase in average length was rapid (0.314 mm $d^{-1}$), but then a decrease was observed (0.196 mm $d^{-1}$) until the end of life. In 1988 the pattern was similar. Sexual maturity was attained in 22 days, on 9 May. Growth rate correspondingly was 0.490 mm $d^{-1}$ and decreased after this date to 0.130 mm $d^{-1}$.

In 1987, *C. diaphanus* began to differentiate sexually on 9 May and completed this process 21 days after hatching, on 16 May. First sexual mature individuals were collected 7 days later, on 23 May (28 days after hatching). The growth rate corresponded to 0.467 mm $d^{-1}$ initially and decreased to 0.282 mm $d^{-1}$ after sexual maturation. In 1988, sexually mature individuals were recorded 32 days after hatching, on 19 May. Growth rate was 0.503 mm $d^{-1}$ until this date, and then decreased to 0.255 mm $d^{-1}$ until the end of the life cycle.

### *Reproduction and life span*

I also noticed differences in the sizes at which species differentiate and attain sexual maturity.

In 1987, minimum size at sexual differentiation was 4.8 mm for *T. stagnalis* and 6.8 mm for

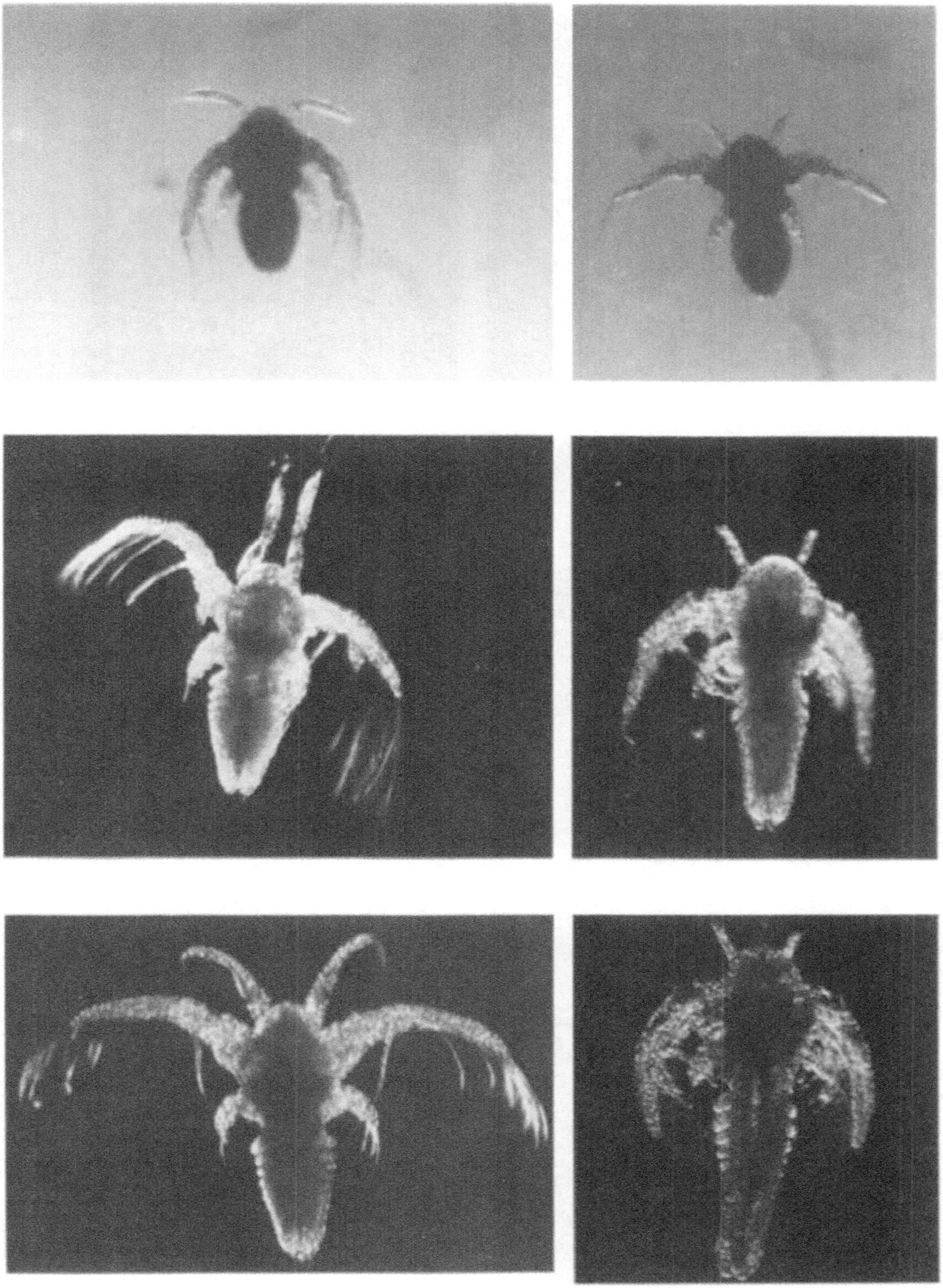

*Fig. 10.* Morphological differences in early larval stages of *T. stagnalis* (a–c) and *Ch. diaphanus* (d–f).

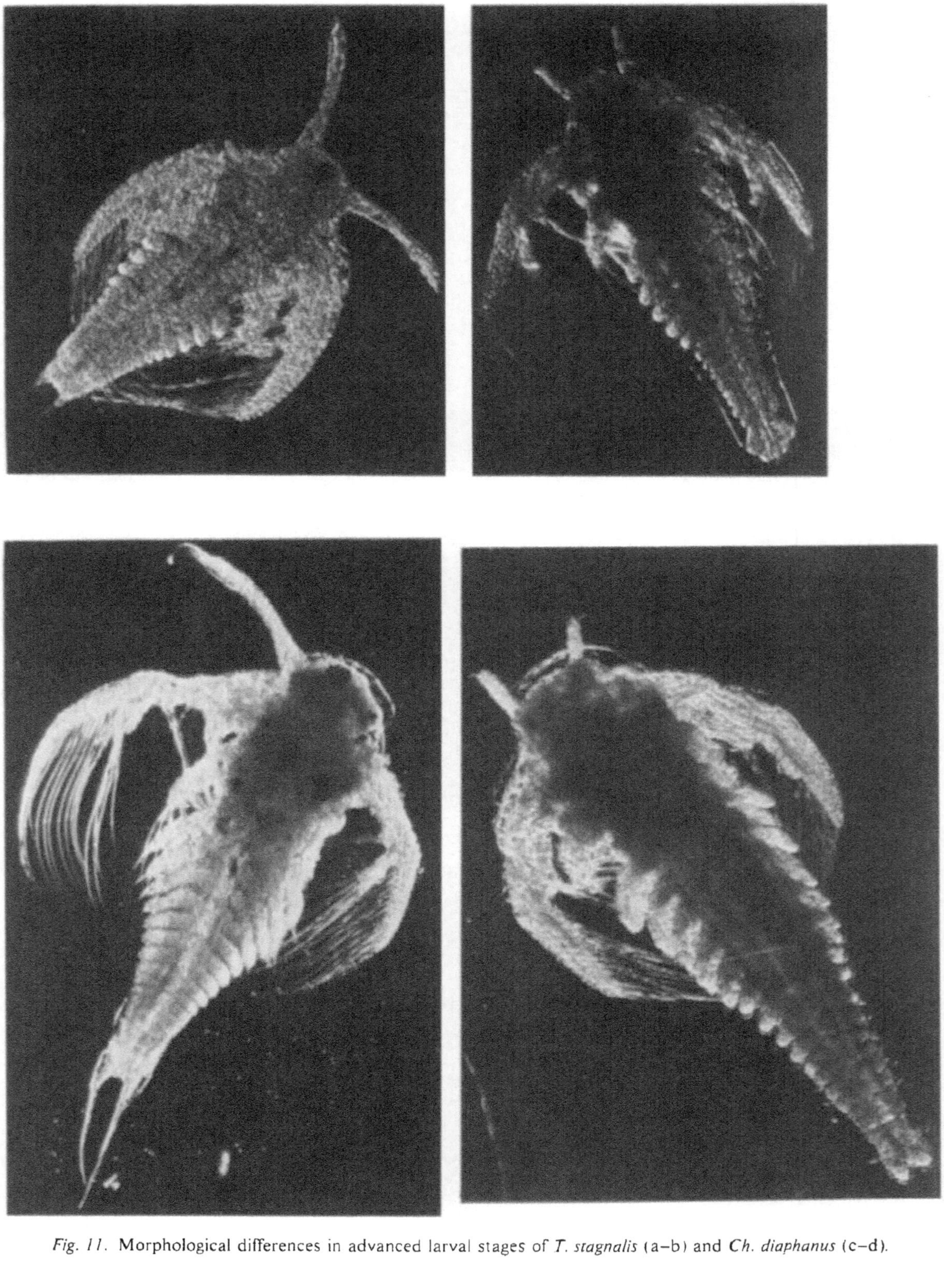

*Fig. 11.* Morphological differences in advanced larval stages of *T. stagnalis* (a–b) and *Ch. diaphanus* (c–d).

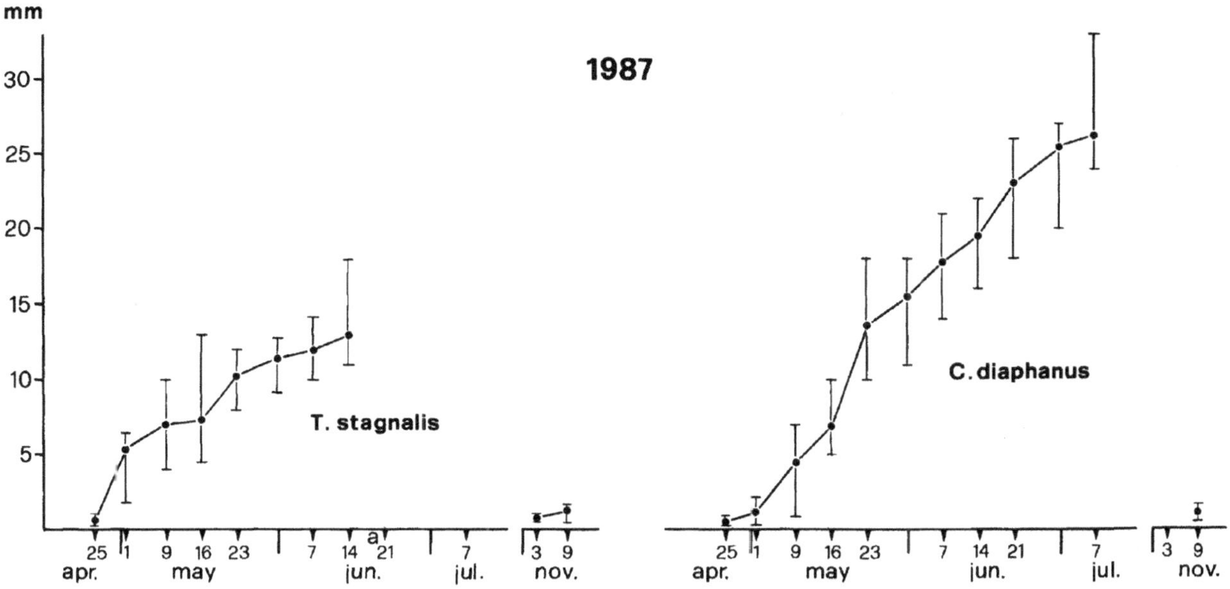


*Fig. 12.* Growth rates of the two studied species in 1987.

*C. diaphanus*. In 1988, the former differentiated at a minimum size of 6 mm, and the latter at 11.4 mm.

Egg production began in females at lengths of 8 mm for *T. stagnalis* and of 12 mm for *C. diaphanus* in 1987. In 1988, comparable data were 9.9 mm and 13 mm, respectively. The reproductive period lasted about 4 weeks in *T. stagnalis* and 6 weeks in *C. diaphanus* in 1987, and 5 weeks and 7 weeks in 1988.

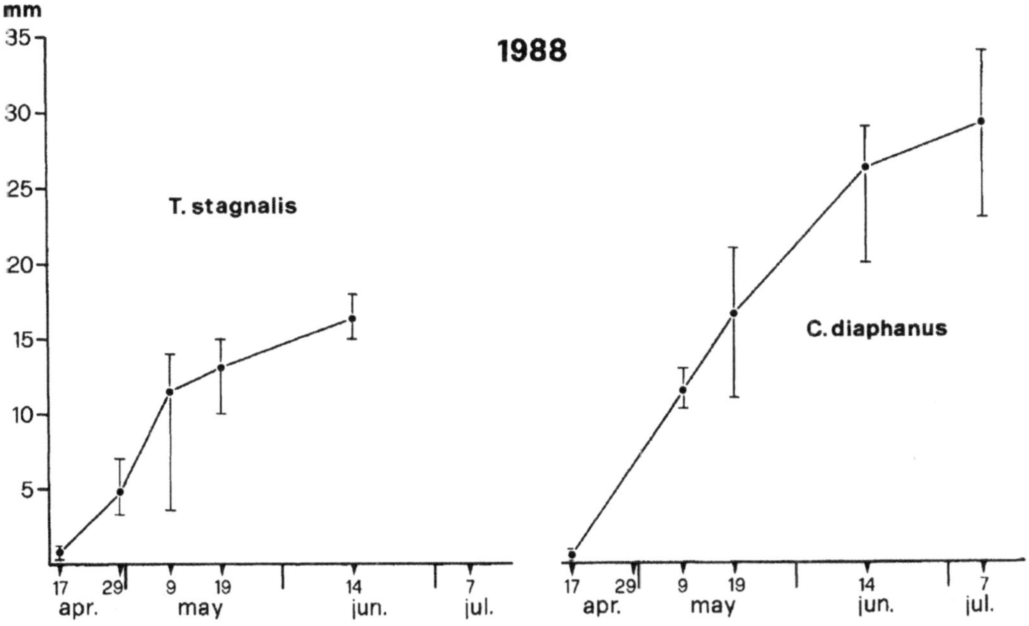


*Fig. 13.* Growth rates of the two studied species in 1988.

In Fig. 14, egg counts of 60 females of *T. stagnalis* and of 122 females of *C. diaphanus*, all having a full complement of eggs, are plotted against length. A positive relationship exists, in both cases, between clutch size and body length

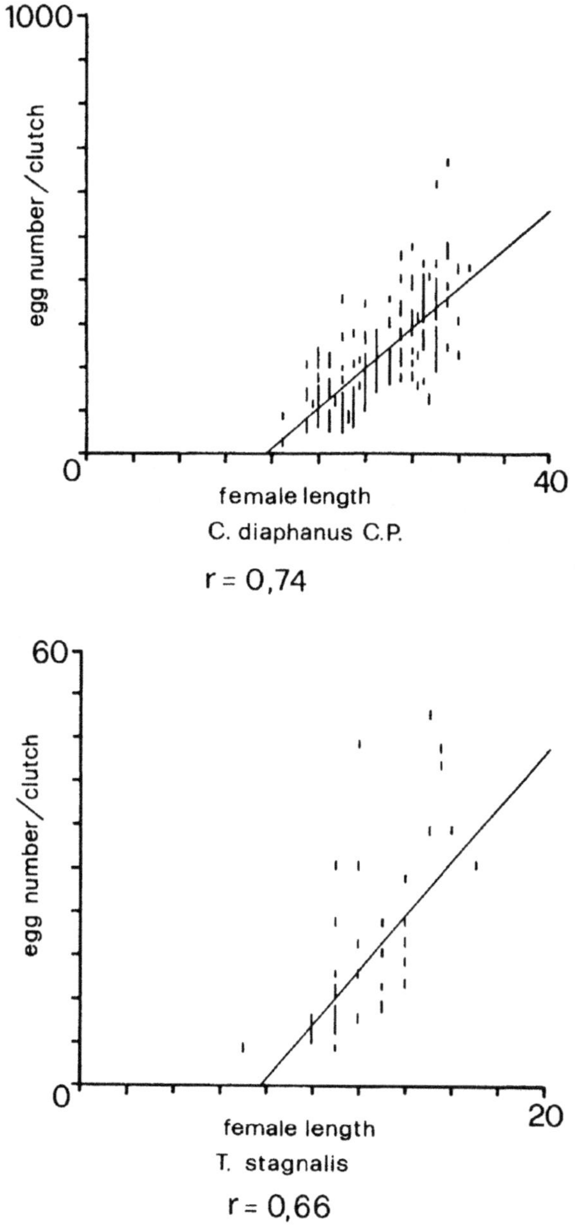


*Fig. 14.* Relationship between clutch size and total length in 60 females of *T. stagnalis* and 122 females of *Ch. diaphanus*.

($r = 0.66$ for *T. stagnalis* and $r = 0.75$ for *C. diaphanus*).

Average maximum clutch size was 44 for *T. stagnalis*, corresponding to a size of 15 mm, and 671 for *C. diaphanus*, corresponding to a size of 32 mm.

Life span is shorter in *T. stagnalis* (50 days in 1987 and 58 days in 1988) than in *C. diaphanus* (73 days in 1987 and 81 days in 1988). Marked differences exist also in the maximum size reached by the two species (18 mm in 1987 and in 1988 for *T. stagnalis*; 33 mm in 1987 and 34 mm in 1988 for *C. diaphanus*).

## Discussion and conclusions

The above results are similar to the case discussed by Moore (1959).

Hatching of the co-occurring species seems to occur at the same time or within a narrow time interval (hours). The nauplii of both species (Fig. 3) were occurring since the beginning of the samplings, mixed with older larvae. Both populations consisted of individuals hatched within a period of a few days. Hatching in *C. diaphanus*, lasting 6–7 days, seemed to last longer than in *T. stagnalis*, unless the first hatching of the latter began slightly earlier than my first sampling.

The early coexistence of nauplii of *T. stagnalis* and *C. diaphanus* suggests that the cysts of the two species respond to the same hatching stimuli, even though larvae and adults show different tolerances to environmental factors. These different responses could be the reason for the difference in relative abundance of the two species.

The hatching temperature seems to be 3–7° in both spring and autumn, the pattern in the years of study was similar, even though differences in thermal fluctuations were noticed between years.

Hatching normally occurs after a rise of temperature, combined with the thawing of the bottom mud (in spring), or after a drop combined with extensive rains (in autumn).

Whilst the spring cycle was nearly regular during the study, the autumn cycle was irregular, depending on climatic events. Often rain was not sufficient to fill the ponds; if hatching took place in such conditions, the larvae were killed. This was the case in 1986 and in 1987, while in 1988 I could not even find larvae.

Occasional samplings in earlier years gave different results; in 1980 (November 25th) I found adults of *T. stagnalis* together with subadults of – presumably – *C. diaphanus*. In 1981 (October 15th), both species were found. *T. stagnalis* was abundant, fully mature with egg-bearing females, but of *C. diaphanus* only few young adult, non-ovigerous females were mixed with many sub-adults. About one month later (on November 23rd), under 20 cm of ice, I found fully mature individuals of both species, but their relative abundance was reversed. In 1985 (November 13rd and December 4th), I found again larvae of an undetermined species (at the time I had not yet decided to examine larval morphology), but on the following visit all the ponds were fully invaded by ice.

These field observations show that thermal conditions affect the two species differently, and to a different extent. *T. stagnalis* is favoured by low temperatures and shows more rapid maturation, reaching sexual maturity at a smaller size. By contrast, *C. diaphanus* is more tolerant of higher temperatures, lives longer and reached larger sizes. As a consequence, total fecundity was different, being dependent not only on the number of eggs per clutch, but also on the total number of clutches during a female's life.

Nonetheless, differences were observed in number of eggs per clutch, and in clutch size between species.

My observations confirm that *T. stagnalis* is characterized by a lower fertility (Freiner & Grüttner, 1984; Grainger, 1981) than *C. diaphanus* (Lake, 1969).

Both species exhibit a positive relationship between body length and clutch size, as recorded for *Branchinecta gigas* (Daborn, 1973, 1975), if we consider those size classes which correspond to maximum fecundity (15 mm for *T. stagnalis*, 32 mm for *C. diaphanus*). From these sizes on, there is some indication that a decrease in egg production occurs, but this has to be confirmed by further observations. The pattern of egg production in *T. stagnalis* and *C. diaphanus* could perhaps be due to female senescence (Daborn, 1976, 1977; Linder, 1960).

As to the tolerance limits for temperature, and in spite of the low optimum documented in the literature (Freiner & Grüttner, 1984; Garreau de Loubresse, 1965; Grainger, 1981; Maier & Tessenow, 1983), I noticed that *T. stagnalis* can survive temperatures as high as 20–22 °C, providing the rise is not sudden. This was confirmed in 2 years of study.

During 1987, May was characterized by a mean temperature of 4–5 °C; in June we assisted to a rise of 12 degrees in 14 days, followed by a sudden drop from 20 °C to 7 °C in two days. In 1988, May had a mean temperature of 8.5 °C; at the beginning of June, in 3 days, a drop of 13 degrees occurred (from 16 °C to 3 °C) followed by a new rise from 3 °C to 13 °C in only one day. After these two sudden temperature shifts, *Tanymastix* had disappeared. Data from previous field studies on this species (Freiner & Grüttner, 1983) are in agreement with these observations.

Conversely, due to its broad tolerance range (Hall, 1961; Lake, 1969; Nourisson, 1961; Nourisson & Aguesse, 1961), all these fluctuations were well survived by *C. diaphanus*, which lived on until the end of July, when temperature reached values as high as 30 °C or more, in each year.

These differences in temperature tolerance could be indirectly confirmed by the distribution pattern of the eggs of the two species on the bottom of the ponds (Fig. 5).

Starting sexual maturation earlier, *T. stagnalis* begins to lay eggs at lower temperatures, before the contraction of the ponds, and decreases its production as evaporation progresses with increasing temperature. This probably explains the higher percentage of cysts in the peripheral areas. To the contrary, the major concentration of cysts of *C. diaphanus* occurs in the central area, which is the last to dry up. During the study, the ratio between the two species fluctuated following more of less the same pattern (Figs. 3 and 4), *T. stagnalis* being more abundant at the beginning of the pond's cycle while *C. diaphanus* prevailed towards the end.

Considering that the two species have always been seen together since the discovery of the Latium ponds in 1976, it seems unreasonable to interpret their co-occurrence as a process of colonization and extinction followed by recolonization (Donald, 1982). Rather, it seems that the two species colonized these basins from nearby areas, thanks to their easy dispersability and viability of their cysts. It is worthwhile to note that within a radius of about 50 km from the study site, several mountain ponds are inhabited by the one or by the other. All of these ponds are usually visited by migrating cattle (sheep and horses).

As far as I can infer from my results, hatching conditions for the species pair are closely related, if not identical, unlike what I previously suspected (Mura, 1985). Nonetheless, they seem to avoid competition. The differences recorded as to size, time of sexual maturation, and life span (mostly due to different responses to environmental conditions) suggest that a niche separation exists. Perhaps, body size differences reflect a difference in filtering apparatus, and hence a difference in prey size.

## Acknowledgements

I thank Dr Giuseppe Bleiner for his assistance in computer graphics, Mr Fulvio Ferrara for helping in photographical documentation, and Dr D. Belk for help in improving the manuscript. This research was performed with financial support of M.P.I. 60%.

## References


Bernice, R., 1972. Ecological studies on *Streptocephalus dichotomus* Baird. Hydrobiologia 39: 217–240.

Broch, E. S., 1965. Mechanism of adaptation of the fairy shrimp *Chirocephalopsis bundyi* Forbes to the temporary pond. Mem. Cornell Univ. Exp. Sta. 392: 3–48.

Broch, E. S., 1988. Osmoregulatory patterns of adaptation to inland astatic waters by 2 species of fairy shrimps *Branchinecta gigas* Lynch and *Branchinecta mackini* Dexter. J. Crust. Biol. 8: 383–391.

Brown, L. & L. H. Carpelan, 1971. Egg hatching and life history of a fairy shrimp *Branchinecta mackini* Dexter (Crustacea Anostraca) in a Mohave desert playa (Rabbit dry lake). Ecology 52: 41–54.

Daborn, G. R., 1973. Community structure and energetics in an argillo trophic lake with special reference to the giant fairy shrimp *Branchinecta gigas* Lynch. Ph. D. Thesis Univ. Alberta Edmonton, 168 pp.

Daborn, G. R., 1975. Life history and energy relations of the giant fairy shrimp *Branchinecta gigas* Lynch 1937 (Crustacea Anostraca). Ecology 56: 1025–1039.

Daborn, G. R., 1976. The life cycle of *Eubranchipus bundyi* (Forbes) (Crustacea: Anostraca) in a temporary vernal pond of Alberta. Can. J. Zool. 54: 193–201.

Daborn, G. R., 1977. The life history of *Branchinecta mackini* DExter (Crustacea: Anostraca) in an argillotrophic lake of Alberta. Can. J. Zool. 55: 161–168.

Dexter, R. W., 1967. Annual changes in populations of Anostraca Crustacea. Proc. Symp. on Curstacea. Part II: 568–576.

Di Luzio, R. M., 1977. Osservazioni morfologiche e biologiche su una popolazione d'alta quota di *Tanymastix stagnalis* (Phyllopoda, Anostraca) del Piano di Raccollo, Campo Imperatore, L'Aquila. Tesi Sperimentale, Fac. Scienze, L'Aquila. 142 pp.

Donald, D. B., 1983. Erratic occurrence of Anostracans in a temporary pond: colonization and extinction or adaptation to variations in annual weather? Can. J. Zool. 61: 1492–1498.

Freiner, D. & O. Grüttner, 1984. Der Eichener Kiemenfusskrebs. Natur Mus. 114: 274–286.

Garreau de Loubresse, N., 1965. Caractéres les plus apparents de la reproduction de *Tanymastix lacunae* G. en élevage. C. R. Acad. Sc. Paris 260: 1751–1754.

Geddes, M. C., 1973. Salinity tolerance and osmotic and ionic regulation in *Branchinella australiensis* and *B. compacta* (Crustacea Anostraca). Comp. Biochem. Physiol. 45: 559–569.

Geddes, M. C., 1981. The brine shrimp *Artemia* and *Parartemia* comparative physiology and distribution in Australia. Hydrobiologia 81: 169–179.

Grainger, J. N. R., 1981. The effect of constant and changing temperatures on the period from egg hatching to egg laying in *Tanymastix stagnalis*. J. Therm. Biol. 6: 353–356.

Hall, R. E., 1961. On some aspects of the natural occurrence of *Chirocephalus diaphanus* Prévost. Hydrobiologia 17: 205–217.

Hildrew, A. G., 1985. A quantitative study of the life history of a fairy shrimp (Branchiopoda: Anostraca) in relation to the temporary nature of its habitat, a Kenian rainpool. J. Anim. Ecol. 54: 99–110.

Horne, F., 1967. Effects of physical-chemical factors on the distribution and occurrence of some Southeastern Wyoming Phyllopods. Ecology 48: 472–477.

Huggins, D. G., 1976. The sympatric occurrence of three species of Eubranchiopoda in Douglas County. Kansas Southw. Nat. 20: 557–578.

Lake, P. S., 1969. The effect of temperature on growth, longevity and egg production in *Chirocephalus diaphanus* Prévost. (Crustacea: Anostraca). Hydrobiologia 33: 342–351.

Linder, H. J., 1959. Studies on the freshwater fairy shrimp *Chirocephalopsis bundyi* (Forbes). 1 Structure and histochemistry of the egg shell formation. J. Morphol. 107: 259–282.

Lynch, J. E., 1964. Packard's and Pearse's species of *Branchinecta*: analysis of a nomenclatural involvement. Am. Midl. Nat. 71: 466–488.

Maier, G. & U. Tessenow, 1983. *Tanymastix stagnalis*: Vorkommen im Hannoverschen Wendland und Befunde zur Larvalentwicklung (Crustacea: Anostraca). Abh. Naturwiss. Ver. Hamburg 25: 351–355.

Moore, W. G., 1955. The life history of the spiny-tailed fairy shrimp in Louisiana. Ecology 36: 176–184.

Moore, W. G., 1959. Observations on the biology of the fairy shrimp *Eubranchipus holmani*. Ecology 40: 399–403.

Moore, W. G., 1963. Some interspecies relationships in Anostraca populations of certain Louisiana ponds. Ecology 44: 131–139.

Mura, G., 1985. Preliminary report on the sympatric occurrence of two species of fairy shrimps (Crustacea: Anostraca) in some temporary ponds from Italy. Riv. Idrobiol. 24: 73–82.

Mura, G., 1987. Osservazioni faunistico-ecologiche su di un gruppo di pozze di alpeggio dei Monti Reatini. Riv. Idrobiol. 26: 123–130.

Nourisson, M., 1959. Evolution d'une mare temporarire de haute altitude à *Chirocephalus stagnalis* Shaw (Crustacé Phyllopode) C. r. Acad. Sci. Paris 248: 3052–3054.

Nourisson, M., 1961. Existence d'une intervalle de température favorable au développement des œufs asséchés de *Chirocephalus stagnalis* Shaw (Crustacé Phyllopode). C. R. Acad. Sci. Paris 253: 1994–1996.

Nourisson, M., 1964. Recherches écologiques et biologiques sur le Crustacé Branchipode *Chirocephalus diaphanus* Prévost. Etude expérimentale du déterminisme de l'œuf. Thèse. Univ. Nancy: 154 pp.

Nourisson, M. & P. Aguesse, 1961. Cycle annuel des Phyllopodes d'une mare temporaire de Camargue. Bull. Soc. Zool. France 86: 754–762.

Prophet, C., 1959. A winter population of *Streptocephalus seali* Ryder inhabiting a roadside ditch in Lyon County, Kansas. Trans. Kans. Acad. Sci. 62: 153–161.

Prophet, C., 1963. Physical-chemical characteristics of habitats and seasonal occurrence of some Anostraca in Oklahoma and Kansas. Ecology 44: 798–801.

Wiman, F. H., 1978. Speciation in the fairy shrimp genus *Streptocephalus*. Ph. D. Thesis Univ. Wisconsin, Madison. 130 pp.

Hydrobiologia 212: 61–66, 1991.
D. Belk, H. J. Dumont & N. Munuswamy (eds), Studies on Large Branchiopod Biology and Aquaculture.



# Influence of temperature on hatching of eggs of *Lepidurus couesii* (Crustacea, Notostraca)

B.J. Hann & Blair Lonsberry
*Department of Zoology, University of Manitoba, Winnipeg, Manitoba, R3T 2N2, Canada*

*Key words:* Notostraca, tadpole shrimps, temporary ponds, Crustacea, adaptation, egg hatch

## Abstract

Tadpole shrimps (Notostraca) occur sporadically in temporary ponds and their survival there depends largely upon the drought-resistant eggs they produce. Environmental conditions conducive to hatching of eggs of *Lepidurus couesii* were investigated in the laboratory. Almost all eggs hatched best at 20 °C, whether desiccated for a short or long period, with a prior freezing shock or without such a shock, with intact shells. Eggs that endured a longer period of desiccation and eggs with abraded shells displayed a more equivocal response to different temperatures for hatching. Long-term hatchability of eggs was demonstrated. Time required for successful hatching was shortest at 20 °C, with no discernible difference between 17 °C and 25 °C. Both short- and long-term survival of populations of the species appears to be fostered by the adaptive response to temperature shown by the eggs.

## Introduction

Of the many microcrustaceans able to colonize temporary ponds (Cladocera, Copepoda, Ostracoda, Conchostraca, Anostraca, Notostraca) in the north temperature zone, the tadpole shrimps (Notostraca) occur most sporadically. This belies the antiquity of the group, living fossils known since the Carboniferous (Fryer, 1988), their worldwide distribution, and their considerable number and biomass in ponds when they are found. All of the crustacean groups inhabiting temporary water bodies have evolved means to survive the periods of desiccation and freezing experienced between flooding of the habitat. The production of resistant or diapause eggs that can withstand these harsh conditions has enabled successful invasion and colonization of these habitats, characterized by extremely variable and unpredictable environmental conditions (Longhurst, 1955).

*Lepidurus couesii* Packard, one of four species of Notostraca that occur in Canada, is broadly but sporadically distributed throughout the Holarctic (Hartland-Rowe, 1965). Like other microcrustaceans, including notostracans, *L. couesii* produces resistant eggs. Little is known regarding the hatching process or about the environmental conditions present in temporary ponds that may stimulate hatching. From studies on a number of notostracan species, desiccation of the eggs for varying periods of time has been found to promote subsequent hatching (Carlisle, 1968; Hempel-Zawitkowska & Klekowski, 1968; Takahashi, 1977a, b). Temperature and salinity of the water appear to be important in achieving optimal egg hatch (Horne, 1967, 1971; Scott & Grigarick, 1979), as is photoperiod (Takahashi, 1975). Freezing has also been reported to induce egg hatch (Braswell, 1967), and osmotic conditions may be relevant (Hempel-Zawitkowska, 1969).

Long-term persistence of a population of notostracans in a temporary pond depends strongly on two factors. Firstly, hatching of eggs should occur in response to appropriate stimuli which should signal, with reasonable predictability, that adequate conditions will prevail in the habitat to permit individuals to reach maturity and reproduce successfully. Secondly, an egg bank should accumulate in the sediments of the pond to provide an auxiliary and supplemental source of eggs in the event of catastrophe to the current population (such that no new infusion of eggs results in a given year), or prolonged periods of unfavourable conditions (e.g. drought).

This study examines the conditions conducive to hatching of eggs of *L. couesii* and assesses their role in the success and persistence of notostracans in temporary ponds from both short-term and evolutionary perspectives.

## Materials and methods

Eggs used in the hatching experiments were obtained from two populations of *L. couesii*, one inhabiting an old borrow pit pond (50° 10′ N, 98° 6′ W) near the University of Manitoba field station (Delta Marsh), SE of Lake Manitoba, and the second found in a roadside pond in SW Winnipeg, MB (49° 85′ N, 97° 20′ W). Eggs (designated old eggs) were obtained from the population in the borrow pit pond during laboratory experiments involving reproductive adults in June 1986 (Straszynski, 1986), and were stored dry in vials, at room temperature, and in darkness. Eggs (designated new eggs) were collected from a roadside pond which contained a viable population in spring 1988. The eggs were collected in September 1988 by removing the dry vegetation (*Sphagnum* sp.) and topmost centimeter of soil from the deepest location in the pond. The vegetation and soil were flooded in small amounts, the eggs were carefully removed and immediately redried so flooding of the eggs during experimental treatment would occur synchronously.

Eggs were observed daily for 12 days using a dissecting microscope to determine time of first rupture of the egg shell (with the embryo retained within the hatching membrane), and time of emergence of the embryo from the hatching membrane as a first instar larva.

To assess the influence of temperature on hatching success of eggs, experiments were carried out using groups of eggs treated as follows: (1) new eggs (dry, room temperature storage), (2) new eggs (dry, room temperature storage, 0 degrees for one week prior to experiment), (3) old eggs (dry, room temperature storage, intact shells), and (4) old eggs (dry, room temperature storage, abraded shells).

Experiments were performed on all four treatment groups of eggs in three controlled environment chambers at three temperatures: 17 degrees C, 20 degrees C, and 25 degrees C, all with photoperiod 14L : 10D. In all experiments, eggs were flooded with synthetic pond water (Hebert & Crease, 1980), and kept submersed during the course of the experiment by placing them under small plastic petri dishes immersed within a large petri dish of synthetic pond water. Ten eggs were enclosed under each small petri dish to permit ready observation of the eggs and to monitor progress of individual eggs during the experiment. At least 30 (and up to 60) eggs were used in each treatment in each experiment.

Eggs that did not hatch in the trials at 17 °C, 20 °C and 25 °C were dried, rewetted, and observed for supplemental hatching at 20 °C.

Temperature effects on hatching success were examined by comparison of cumulative percentage hatch of eggs obtained for each of the four egg treatment groups at each of three temperatures using a two-sample Kolmogorov-Smirnov test (Sokal & Rohlf, 1981). The influence of temperature on time to hatch for each egg treatment group was determined using ANOVA (Sokal & Rohlf, 1981).

The first experiment was carried out in November 1988 (with additional observations on frozen eggs in February 1989). The second experiment to examine the impact of increased duration of desiccation on hatch success at different temperatures took place in June 1989. Results of the November,

February and June trials were examined for significant differences in cumulative percentage hatch using the Kolmogorov-Smirnov test, and data were pooled as no significant differences were detected. Thus, experiment 1 consisted of pooled results from the November and February trials, and experiment 2 of results of the June trials.

## Results

### *Experiment 1*

New eggs (unfrozen) had a significantly ($p < 0.05$) higher cumulative hatch at 20 °C compared to eggs hatched at 17 and 25 °C (Fig. 1). No significant differences were obtained between eggs hatched at 17 and 25 °C. The highest total cumulative hatch occurred at 20 °C (78.6%), followed by 17 °C (44.4%) and 25 °C (26.7%). Hatch initiation began on the third day after flooding for both 20 °C and 25 °C, and on the fourth day at 17 °C.

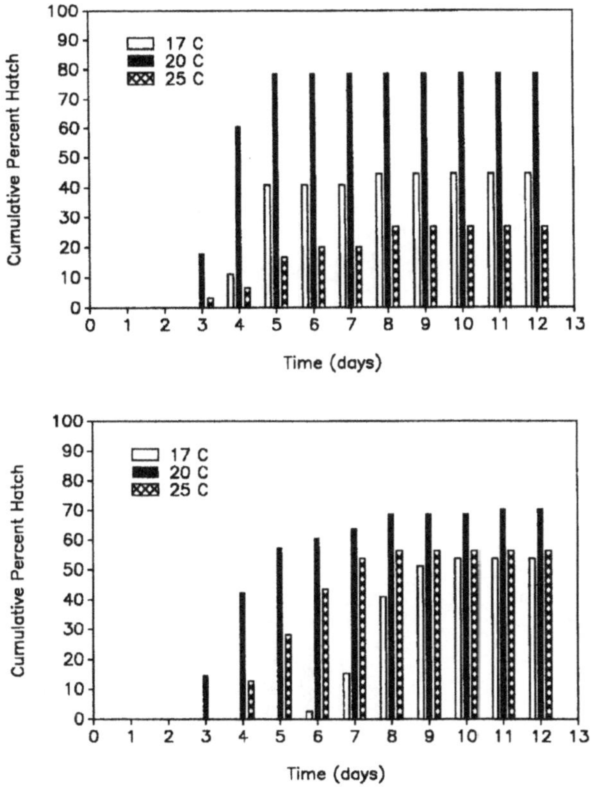


*Fig. 1.* Cumulative percentage hatch of new eggs (experiment 1) over time at three temperatures for unfrozen (top) and frozen (bottom) eggs.

Although the highest total cumulative hatch of new eggs (frozen) was observed at 20 °C (70.5%) as for the unfrozen new eggs, there were significant ($p < 0.05$) differences in cumulative hatch among all three temperatures for this egg treatment group (Fig. 1). Hatch initiation began on the third day at 20 °C, on the fourth day at 25 °C, but not until the sixth day at 17 °C.

Comparison of frozen with unfrozen new eggs at each temperature revealed no significant differences ($p > 0.05$) in cumulative hatch of eggs at 20 °C, but discernible differences were observed at 17 °C and 25 °C. At 17 °C and 25 °C, higher total cumulative hatch was achieved with frozen eggs compared with unfrozen eggs.

Old eggs (with intact shells, unfrozen) hatched most successfully at 20 °C, with cumulative percentage hatch significantly higher than at 17 °C and 25 °C (Fig. 2). Initiation of hatch began on the third day at 20 °C and 25 °C, but not until day 5 at 17 °C.

There were no significant differences in total percentage hatch at 20 °C among new eggs (unfrozen), new eggs (frozen), and old eggs (intact shells, unfrozen). Considerably lower hatch success at 20 °C was obtained with the old, abraded eggs (Fig. 2). There were no significant differences in cumulative percentage hatch of old (abraded) eggs among temperatures. Unhatched eggs from the three temperature treatments showed slight response to rewetting at 20 °C with 19% hatching successfully.

Eggs of *L. couesii* began to hatch as early as three days after flooding, with earliest hatching occurring at the warmer temperatures (except with the old, abraded eggs). Significant differences in number of days to hatch were detected among temperature treatments ($p < 0.001$); both frozen and unfrozen eggs held at 20 °C hatched earlier than at either 17 °C or 25 °C.

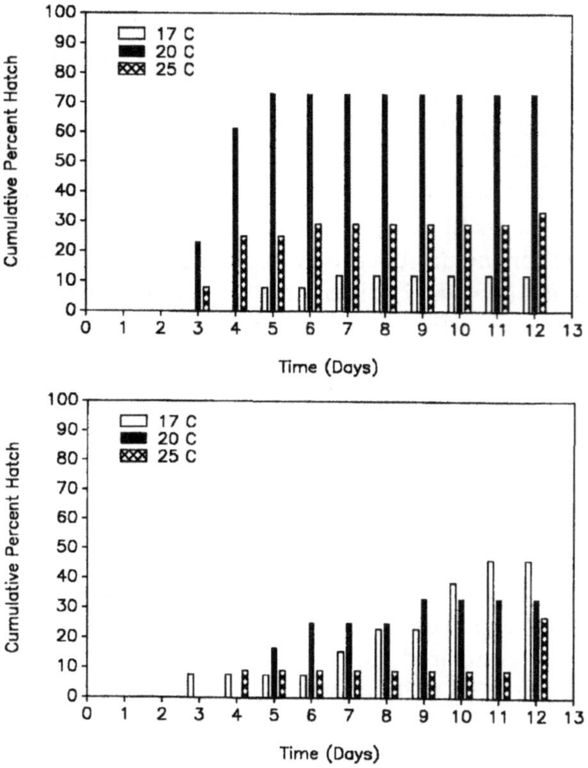


*Fig. 2.* Cumulative percentage hatch over time at three temperatures of old eggs with intact shells (top) and abraded shells (bottom).

*Experiment 2*

No significant differences were found in cumulative percentage hatch among temperatures for either the frozen or unfrozen eggs (Fig. 3). Unfrozen eggs had the higher total percentage hatch at all temperatures. As in experiment 1, the maximum total percentage hatch of frozen eggs occurred at 20 °C, whereas for the unfrozen eggs in experiment 2, maximum hatch occurred at 25 °C. Hatch initiation commenced on day 3 for both groups of eggs at all three temperatures.

Unfrozen eggs hatched significantly sooner at 20 °C ($p < 0.001$), whereas there was no difference in hatching time detected among temperature treatments for frozen eggs.

## Discussion

The optimal temperature for hatching of eggs of *L. couesii* with intact shells, and short- and long-term desiccation, was 20 °C. Scott & Grigarick (1979) found that 22 °C produced maximum hatch percentage for *Triops longicaudatus* (Leconte), a species which is primarily southern in distribution, widespread in western USA and Mexico (Hartland-Rowe, 1965). The distribution of *L. couesii* is more northern, found in Alberta, Saskatchewan, Manitoba, northwestern USA, and USSR (Edmondson, 1959, Hartland-Rowe, 1965). The close correspondence between the temperature of maximum egg hatch and environmental temperature typically experienced by eggs of *T. longicaudatus* (Scott & Grigarick, 1979) contrasts sharply with that of *L. couesii* with a maximal egg hatch success rate at 20 °C, but for

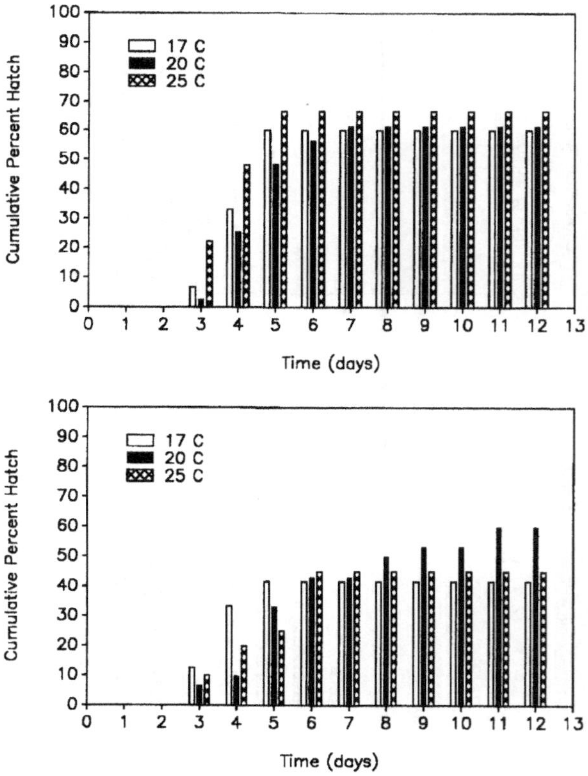


*Fig. 3.* Cumulative percentage hatch of new eggs (experiment 2) over time at three temperatures for unfrozen (top) and frozen (bottom) eggs.

which egg hatch must begin as soon as ice melts in the northern ponds.

Although 20 °C produced maximum hatch of eggs for new eggs, the marked difference among temperatures was observed only for unfrozen eggs. Those eggs that had been frozen were induced to hatch with considerable success (> 50%) at all 3 temperatures. This suggests that the threshold or window for induction of hatching may be less restrictive following a freezing shock, akin to that experienced during over-wintering in a frozen pond. The influence of duration of freezing on hatch success is unknown.

The longer period of desiccation of eggs prior to hatch induction in experiment 2 (9 months compared with 3 months) appeared to moderate the influence of temperature on hatch success. The similar response rate at all three temperatures for both frozen and unfrozen eggs suggests that the eggs may respond to the rewetting stimulus at a broad range of temperatures given a prior prolonged desiccation period. However, this result differs noticeably from that obtained for the old eggs from the Delta ponds where there was a marked temperature effect even after 2 years of desiccation. The extent of inter-populational variation in response to temperature and duration of drying remain to be examined.

The marginal increment in hatch success following repeated drying and reflooding of eggs at 20 °C confirms the pattern observed for *T. cancriformis* (Fox, 1949; Hempel-Zawitkowska, 1967) and *T. longicaudatus* (Takahashi, 1977). As in the previous studies, even with repeated rewetting, 100% hatch was not achieved. The viability of the unhatched eggs was not determined. However, the incomplete hatch of egg batches even from individual females under all conditions could be interpreted as an adaptation to the highly variable environment in which successive opportunities to hatch, in conjunction with partial hatch induction, may be the optimal evolutionary strategy.

The old eggs that had been dried for 2 years hatched at 20 °C as successfully as those new eggs dried for several weeks or months. If the old eggs had been integrated into the 'egg bank' in the sediments of the pond, they would have been subjected to additional freezing treatment over the 2 years. Nonetheless, the continued high hatch success of old eggs suggests that they could remain viable and capable of hatching for longer than one season, thereby contributing to the long-term persistence of the population.

The 10–50% hatch success of the abraded, old eggs with damaged shells, though lower than for all other groups of eggs, is remarkable in that such low quality eggs were capable of hatching at all after 2 years of desiccation. Thus, the long-term probability of successful hatching of eggs in the egg bank, even those subject to the abrasive and adverse storage conditions in the sediments, is expected to be reasonably good. However, further experiments with eggs stored *in situ* in the pond for varying periods of time are needed to confirm this prediction.

## Acknowledgements

This study was funded by Natural Sciences and Engineering Research Council of Canada operating grants to BJH.

## References

Braswell, A. L., 1967. Preincubation treatment required for hatching of *Lepidurus apus*. Utah Acad. Proc. 44: 344–352.

Carlisle, D. B., 1968. *Triops* (Entomostraca) eggs killed only by boiling. Science 161: 279–280.

Edmondson, W. T., 1959. Fresh-water biology, 2nd edition. John Wiley & Sons, N.Y., 1248 pp.

Fox, H. M., 1949. On *Apus*: its rediscovery in Britain, nomenclature and habits. Proc. Zool. Soc. Lond. 119: 693–702.

Fryer, G., 1988. Studies on the functional morphology and biology of the Notostraca (Crustacea: Branchipoda). Phil. Trans. r. Soc., Lond. 321: 27–124.

Hartland-Rowe, R., 1965. The Anostraca and Notostraca of Canada with some new distribution records. Can. Field Nat. 79: 185–189.

Hebert, P. D. N. & T. J. Crease, 1980. Clonal coexistence in *Daphnia pulex* (Leydig): another planktonic paradox. Science 207: 1363–1365.

Hempel-Zawitkowska, J., 1967. Natural history of *Triops cancriformis* (Bosc.). Zool. pol. 17: 173–239.

Hempel-Zawitkowska, J., 1969. Hatchability of eggs of *Triops cancriformis* (Bosc.) in solution of chlorides and sulfates. Pol. Arch. Hydrobiol. 16: 105–114.

Hempel-Zawitkowska, J. & R. Z. Klekowski, 1968. The influence of desiccation at different air humidities on hatchability of *Triops cancriformis* (Bosc.) eggs. Pol. Arch. Hydrobiol. 15: 183–189.

Horne, F. R., 1967. Effects of physical-chemical factors on the distribution and occurrence of some Southeastern Wyoming phyllopods. Ecology 48: 472–477.

Horne, F. R., 1971. Some effects of temperature and oxygen concentration on phyllopod ecology. Ecology 52: 343–347.

Longhurst, A. R., 1955. Evolution in the Notostraca. Evolution 9: 84–86.

Scott, S. R. & A. A. Grigarick, 1979. Laboratory studies of factors affecting egg hatch of *Triops longicaudatus* (Leconte) (Notostraca: Triopsidae). Hydrobiologia 63: 145–152.

Sokal, R. R. & F. J. Rohlf, 1981. Biometry, 2nd edition. W. H. Freeman & Co., N.Y., 859 pp.

Straszynski, E., 1986. *Lepidurus couesii* Packard (Crustacea, Notostraca) in temporary ponds in southern Manitoba: sexual dimorphism and behavioural interactions. Delta Marsh Field Station Annual Report 21: 98–106.

Takahashi, F., 1975. Effect of light on the hatching of eggs in *Triops granarius* (Notostraca: Triopsidae). Envir. Control in Biol. 13: 29–33 (in Japanese with English summary).

Takahashi, F., 1977a. Pioneer Life of the Tadpole Shrimps. *Triops* spp. (Notostraca: Triopsidae). Appl. Ent. Zool. 12: 104–117.

Takahashi, F., 1977b. High water temperature prevents egg hatching in *Triops granarius* (Notostraca: Triopsidae). Appl. Ent. Zool. 12: 196–197.

*Hydrobiologia* **212**: 67–76, 1991.
*D. Belk, H. J. Dumont & N. Munuswamy (eds), Studies on Large Branchiopod Biology and Aquaculture.*



# Indirect effects of the fairy shrimp, *Branchipus schaefferi* and two ostracod species on *Bacillus thuringiensis* var *Israelensis*-induced mortality in mosquito larvae

Leon Blaustein* & Joel Margalit
*Laboratory for Biological Control, Department of Biology, Ben Gurion University, P.O. Box 653, Beer Sheva 84105, Israel;*present address and offprint requests to: Mitrani Center for Ecology, Blaustein Institute for Desert Research, Ben Gurion University, Sede Boker 84990, Israel*

*Key words: Aedes aegypti*, Anostraca, *Bacillus thuringiensis* var *israelensis*, biological control, *Branchipus schaefferi*, indirect effects, mosquitoes, Ostracoda

## Abstract

*Bacillus thuringiensis* var *israelensis* (BTI), when ingested by mosquito larvae, is highly toxic to them. Many other aquatic invertebrates feed on bacteria but, in general, BTI is not toxic to them. We tested in the laboratory the hypothesis that certain crustaceans indirectly benefit mosquito larvae by reducing mortality caused by BTI. We presumed the mechanism to be ingestion of the bacteria by the crustaceans resulting in a lower concentration available to the mosquito larvae.

Mortality of *Aedes aegypti* larvae exposed to BTI was reduced in the presence of the fairy shrimp, *Branchipus schaefferi* and the ostracod, *Cypridopsis vidua* (only during summer trials and not autumn trials for the latter species) but was not reduced in the presence of the ostracod, *Heterocypris incongruens*. By contrast, *H. incongruens* preyed upon infected, though still-moving, larvae. Feeding on the bacteria by the crustaceans may not be an important mechanism; our data indicates that the crustaceans did not reduce BTI in the water. Moreover, *B. schaefferi*, introduced into water and then removed prior to the introduction of BTI and mosquitoes, also reduced mosquito mortality.

The mechanism for the protective effect of *B. schaefferi* and *C. vidua* is unknown. However, these results suggest that the abundance of certain organisms co-occurring with mosquito larvae may partially explain why the effective concentration of BTI varies among habitats.

## Introduction

The indirect effects of one species on the interaction between two other species in aquatic systems has received increased attention in recent years (see papers in Kerfoot & Sih, 1987). Many types of indirect effects have been defined (reviewed by Miller & Kerfoot, 1987). Some of the more commonly studied indirect effects include keystone predation (predation on a competitive dominant indirectly benefits the inferior competitor; *sensu* Paine, 1980) and apparent competition (when the presence of an alternative prey causes an increase in a predator which eventually causes an increase in predation on a second prey species; *sensu* Holt, 1977).

Indirect effects have also been considered in management of mosquito populations. The introduction of larvivorous fishes into rice fields for mosquito control may actually result in an increase in mosquitoes if the loss of natural control by invertebrate predators which are eaten by the fishes outweigh the direct consumption by the fishes on the mosquitoes (Hoy *et al.*, 1972; Bence, 1988). The presence of alternative prey has also been shown to reduce mosquito mortality by general predators (Bence, 1988; Chesson, 1989; Blaustein, 1990). Blaustein & Karban (1990) showed that the presence of mosquitofish, *Gambusia affinis* indirectly caused an increase in the development rate of mosquito larvae. Presumably, mosquitofish, by reducing the numbers of other organisms sharing the same trophic levels as mosquitoes, reduced competition.

In this paper, we assess another potential indirect effect on mosquitoes – whether the presence of various crustaceans indirectly affects mortality of mosquito larvae caused by the bacterium, *Bacillus thuringiensis* var *israelensis* (BTI). Several bacteria – BTI (Goldberg & Margalit, 1977) and *Bacillus sphaericus* (Singer, 1973; Ramoska *et al.*, 1977) – have high mosquito larvicidal activities and are now used worldwide for mosquito control. The mosquito larvae ingest these bacteria which are toxic to them. A favorable feature is that both BTI (reviewed in Margalit *et al.*, 1985) and *B. sphaericus* (e.g., Mulligan *et al.*, 1977) are not toxic to most nontarget organisms. Many crustaceans co-occur with mosquito larvae and are suspected to have a high overlap in diet. Some mosquito larvae feed exclusively by filter feeding and ingest phytoplankton, bacterioplankton and microscopic planktonic animals (e.g., Hinman, 1930; Walker *et al.*, 1988). Many cladocerans (e.g., Gophen *et al.*, 1974; Lampert, 1974) and fairy shrimp (e.g., Dimentman, 1976) also filter feed on these food items and thus should have a high overlap with filter-feeding mosquito larvae. Other mosquito larvae browse on substrates in addition to filter feeding. Both some ostracods (Barnes, 1974) and some fairy shrimp (Modlin, 1982) utilize both feeding modes and hence should have a high overlap in diet with these mosquitoes. Thus, these animals and mosquitoes may compete for food.

Our *a priori* hypothesis was that these animals, by ingesting BTI, should decrease the concentration of this bacteria in the water and thus reduce mortality in mosquito larvae. Rettich (1986) tested this hypothesis in the laboratory with *Daphnia curvirostris* and found that high densities of this cladoceran reduced mortality by BTI in *Aedes cantans* larvae. This has not been tested for other organisms which might share the same trophic levels as mosquitoes. Here, we report on experiments assessing how BTI-induced mortality in the mosquito, *Aedes aegypti* are affected by two ostracod species and one fairy shrimp species in the laboratory.

## The crustaceans

One ostracod, *Cypridopsis vidua* (O.F. Muller, 1776) (form: parva), was collected from a permanent pool at Ein Akev in the Negev Desert of Israel. This species has been described largely as a detritivore (Cannon, 1925). *C. vidua* was shown to alter the algal species composition and to eliminate rotifers from laboratory mesocosms (Diner *et al.*, 1986) but its impact on bacteria has not been assessed. We found it associated with various mosquito species along the edges of the pool. Only adults were used.

We collected the second ostracod, *Heterocypris incongruens* (Ramdohr, 1808), from temporary waters of Ein Avdat in the Negev Desert and from an abandoned swimming pool in Beer Sheva. Its feeding on bacteria has not been assessed. Only adults were used.

The fairy shrimp tested here, *Branchipus schaefferi*, is a species found in Europe, Asia and Africa. In Israel, it occurs in rain pools within the Southern Coastal Plain and the Judean and Negev Desert (Dimentman, 1976). Based on the intersetular distance for all age classes, particles of the size of BTI (2–3 microns) and BTI spores (1–1.5 microns) should be well- within its food size range (Dimentman, 1976). It co-occurs with mosquitoes in the Negev Desert (Dimentman & Margalit, 1981). We collected specimens in March, 1988 from a temporary pool near Lahav, Israel for the first two experiments. For the remaining experiments, specimens were obtained by flooding (with distilled water) soil collected during October from this same pool and providing yeast as food. Both immatures and adults were assessed.

## Methods

### *BTI-Induced mortality*

Field-collected crustaceans were acclimated to laboratory conditions for at least 24 hours. The basic experimental design was as follows. Plastic cups containing 100 ml water served as the test arenas. Aged tap water (tap water which sat in the laboratory for 24 hours) was used for the ostracods. A mixture of half field water combined with half aged tap water was used for the fairy shrimp experiments. *Aedes aegypti* larvae from a colony were used in all of the experiments. The feeding behavior of this mosquito in general and on BTI in particular, has been well-documented (e.g., Khawaled *et al.*, 1988). Twenty *A. aegypti* third instar larvae were added to each cup. Preliminary experiments showed that the crustaceans did not affect mortality of healthy mosquito larvae in the absence of BTI.

In each experiment, a treatment containing mosquito larvae but no crustaceans, served as a control. We decided *a priori* to discard the results of any experiments in which greater than 5% mortality occurred in the control cups. Mortality in the control cups was never greater than two of sixty mosquito larvae (3.3%) and generally was 0% in the results presented. Hence, mortality in mosquitoes in the absence of BTI was essentially zero and is not reported in the figures and tables. Other cups contained BTI plus various densities of one crustacean species beginning with zero. The BTI powder, Duphar-Biochem's unformulated product, was placed into a homogenizer with water to break up aggregates. Serial dilutions were then made to arrive at the desired concentrations. The concentrations of BTI used in the various experiments are presented with the results.

For *C. vidua*, six experiments were conducted. For most of the experiments, ostracod densities of 0, 50, 100 and 200 per 100 ml were tested and the ostracods, mosquitoes and BTI were added simultaneously. In an additional treatment, the mosquitoes were added two hours after the introduction of 100 *C. vidua* and the bacteria.

Two experiments were conducted with *H. incongruens*. Densities of 0, 20, 40 and 80 per 100 ml were added four hours prior to the introduction of the mosquito larvae and bacteria. Another treatment consisted of a simultaneous introduction of the bacteria, mosquito larvae and 80 *H. incongruens*.

Six experiments were conducted with *B. schaefferi*. The first two experiments compared BTI-induced mortality in the presence and absence of five adult fairy shrimp (length ca 14 mm) introduced four hours prior to the introduction of the mosquitoes and BTI. Subsequent experiments examined the mosquito-bacteria interaction using fairy shrimp of various sizes (length: 7 to 14 mm) and densities ranging from zero to as many as five. Again, the fairy shrimp were added four hours early. We also included another treatment of simultaneous introductions of three fairy shrimp with the mosquito and bacteria in one experiment.

We conducted three experiments to determine if fairy shrimp, introduced prior to the introduction of BTI and mosquitoes but removed before adding the BTI and mosquitoes, had an effect. Ten fairy shrimp were added to cups containing 100 ml of aged tap water. Other cups contained only the tap water. After six hours, the fairy shrimp were removed by pouring the water through a strainer. The BTI and mosquitoes were then added.

Three replicates for each level of each treatment were used for most experiments. The response measured was mortality after 24 hours under conditions of 25 °C and a light : dark cycle of 14 : 10 unless otherwise stated.

### *Crustacean feeding on BTI*

Identical cups and water used in the above experiments were also used for preliminary experiments to indirectly assess crustacean feeding on BTI by measuring the BTI concentrations in the water.

The presence or absence of BTI was crossed with the presence or absence of the crustacean (100 individuals for *C. vidua*, 80 for *H. incongruens* and 3 for *B. schaefferi*). BTI concentrations of $10^{-6}$ mg powder per ml were used for the ostracod species and $2 \times 10^{-6}$ mg per ml for the fairy shrimp. Two replicate cups were used for the two treatments without BTI while three were used for the two treatments with BTI. At the initiation of the experiment, two replicate samples of 1 ml each were taken from each cup at mid depth after vigorous shaking. The samples were then heat-shocked for 10 minutes at 70 °C to kill non-spore forming bacteria. After vigorous shaking, 0.1 ml from each sample were plated onto a culture dish containing LB medium (Miller, 1972). After four hours (and also 24 hours with *B. schaefferi*), the procedure was repeated. The petri dishes were incubated at 25° and the bacterial colonies were counted after 48 hours. Identification of BTI colonies was based solely on visual inspection of the colony and microscopic inspection of bacilli from individual colonies but not on toxicity. Thus, we refer to our counts as '*B.t. israelensis*-like' colonies. However, the very low numbers of *B.t. israelensis*-like colonies resulting from the water-only treatment and the water-plus-fairy shrimp treatment ensured that colonies of *Bacillus* species other than BTI incorporated into our counts were negligible.

## Results

Table 1 summarizes the comparison between mosquito mortality by BTI in the absence and presence of 100 *C. vidua* which were introduced simultaneously with the bacteria and two hours before the mosquito larvae. A strong treatment × time interaction occurred ($p < 0.001$, Table 1b). The presence of this ostracod species significantly reduced the larvicidal activity of BTI during the initial experiments (mid-to-late summer). However, later during the autumn, *C. vidua* had no effect. Ostracods, added simultaneously with both the mosquitoes and bacteria had a similar, though not as strong, effect (Fig. 1).

We did not find evidence that this ostracod fed on BTI. Among the BTI treatments, the number of *B.t. israelensis*-like colonies per culture plate was not significantly different in the presence and in the absence of the ostracods which were collected during late October (student's $t = 0.50$, $p = 0.67$; Fig. 2a). We tried to collect *C. vidua* the following summer and autumn for additional experiments but it had strangely disappeared from the habitat in which it was previously so abundant.

*H. incongruens*, introduced simultaneously with the bacteria and four hours prior to the introduction of the mosquitoes, did not reduce mortality in mosquito larvae. In fact, mortality tended to be greater in the presence of ostracods though not significantly (ANOVA: $F_{5,17} = 2.91$, $p = 0.060$; Fig. 3). A second experiment indicated that *H. incongruens* may have had a negative effect on mosquito larvae. Assessment of treatment differences was not possible after 24 hours in this

*Table 1a.* Summary of *B.t.i.*-Induced mortality of *Aedes aegypti* larvae after 24 hours (mean and standard error) in the absence and presence of 100 *Cypridopsis vidua* with a two hour delay in adding *Aedes*. P-values are from Student's t-test.

| Date | Mean no. larvae killed (S.E.) | | *P*-value |
|---|---|---|---|
| | Ostr. absent | Ostr. present | |
| 17 Aug. 1988 | 4.67 (0.33) | 0.00 | 0.006 |
| 21 Aug. 1988 | 4.00 (0.58) | 0.00 | 0.002 |
| 8 Sep. 1988 | 2.67 (0.88) | 0.00 | 0.039 |
| 10 Oct. 1988 | 1.33 (0.88) | 0.00 | 0.205 |
| 14 Oct. 1988 | 3.00 (1.16) | 3.33 (1.86) | 0.886 |
| 21 Oct. 1988 | 18.00 (0.58) | 17.67 (1.20) | 0.815 |

*Table 1b.* Analysis of variance from data above (first transformed to natural logs).

| Source of variation | d.f. | M.S. | F | *P*-value |
|---|---|---|---|---|
| Ostracod | 1 | 6.88 | 49.62 | <0.001 |
| Date | 5 | 5.21 | 112.01 | <0.001 |
| Interaction | 5 | 0.86 | 6.20 | <0.001 |
| Error | 24 | 2.31 | | |

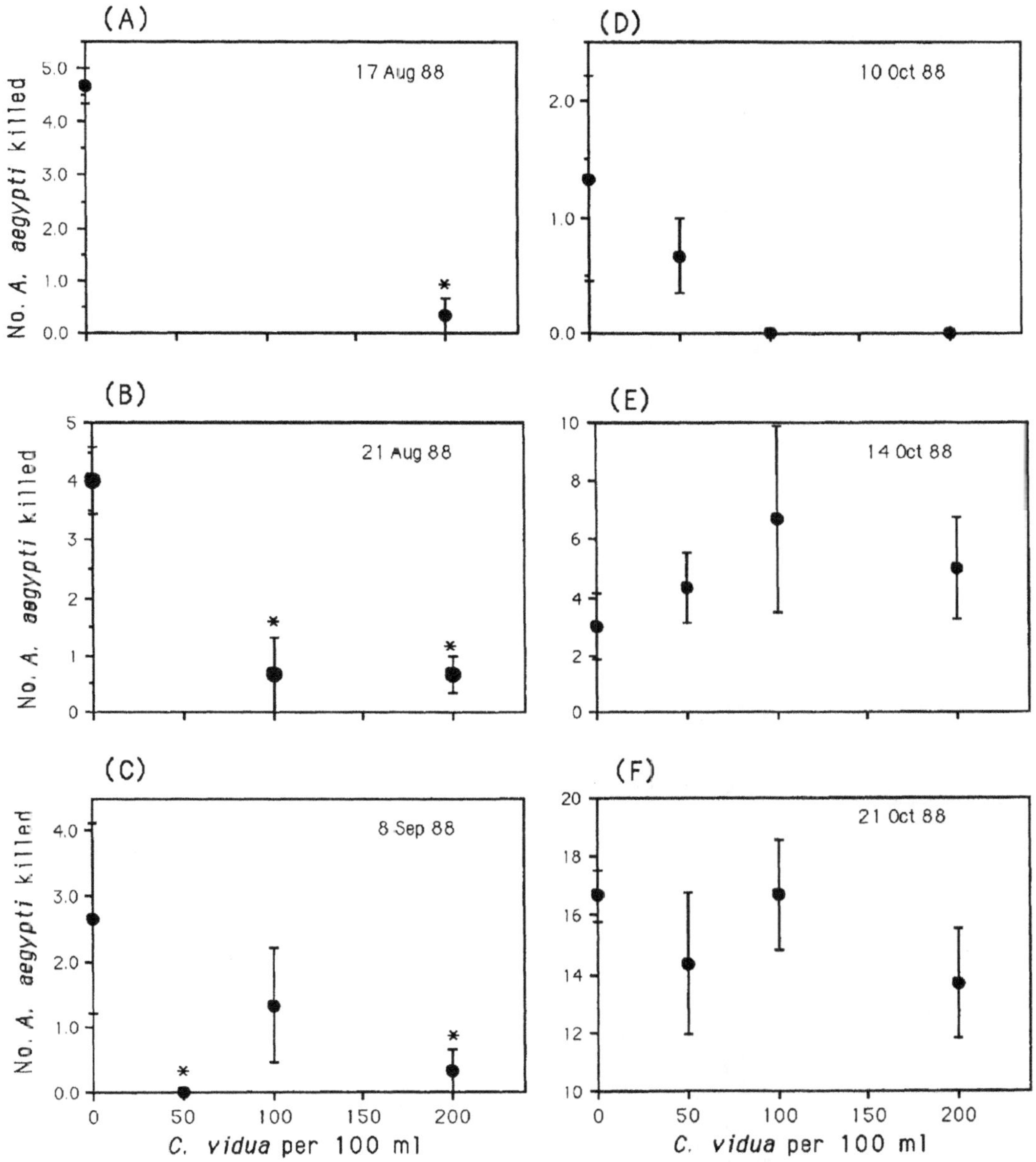


*Fig. 1.* Mortality of *Aedes aegypti* larvae as a function of *Cypridopsis* density at BTI concentrations of $5 \times 10^{-7}$ mg per ml for the first five dates and $10^{-6}$ mg per ml for the last date. Ostracods, mosquitoes and BTI were added simultaneously. Error bars are $\pm 1$ standard error. Asterisks on figure indicate which densities were significantly different (Scheffe's F-test) from treatment containing no ostracods. Mortality in controls (absence of BTI) was negligible and is not depicted.

experiment because 100% mortality occurred. However, we also made observations during the first three hours and observed ostracods preying on mosquito larvae which were exposed to the BTI but were still mobile (Fig. 4).

The number of *B.t. israelensis*-like colonies per culture plate was not significantly different in the presence and in the absence of 80 *H. incongruens* (Student's $t = 2.095$, $p = 0.104$; Fig. 2b). In fact, numbers tended to be greater in the presence of the ostracods.

Fairy shrimp reduced mortality by BTI to

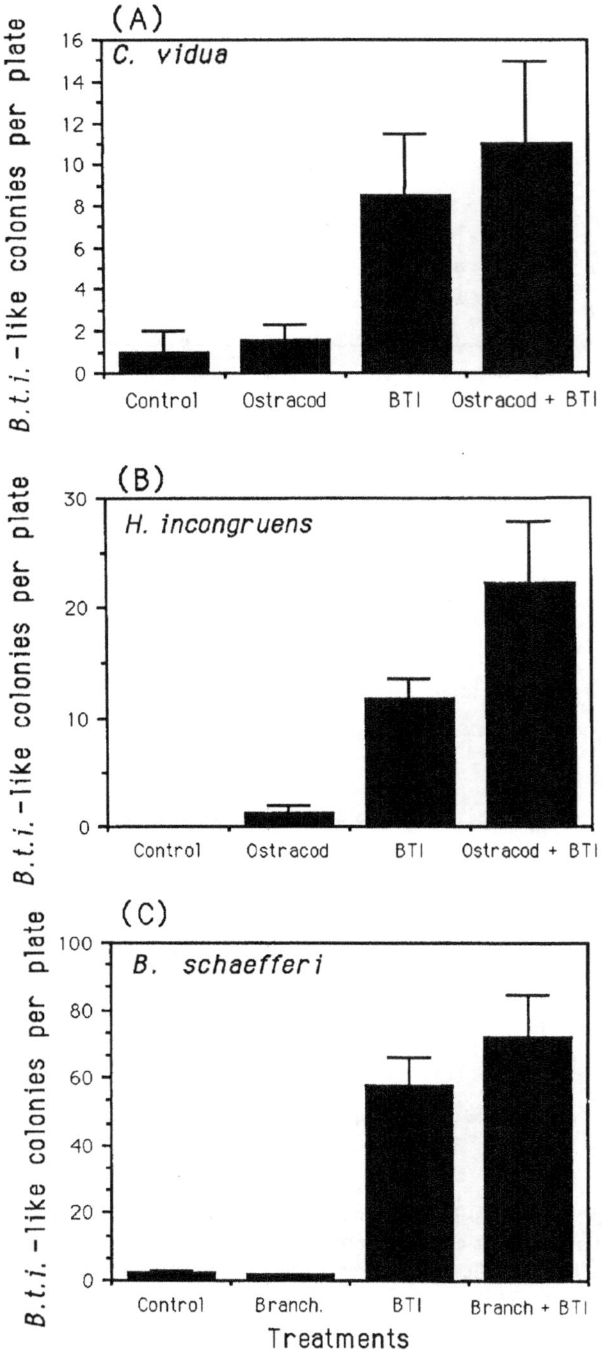


*Fig. 2. Bacillus thuringiensis israelensis*-like colonies per culture plate from 0.1 ml of water after four hours in the presence and absence of the crustaceans: a) *Cypridopsis vidua* (BTI concentration = $10^{-6}$ mg per ml), b) *Heterocypris incongruens* (BTI concentration = $10^{-6}$ mg per ml), and c) *Branchipus schaefferi* (BTI concentration = $2 \times 10^{-6}$ mg per ml). Error bars are ±1 S.E.

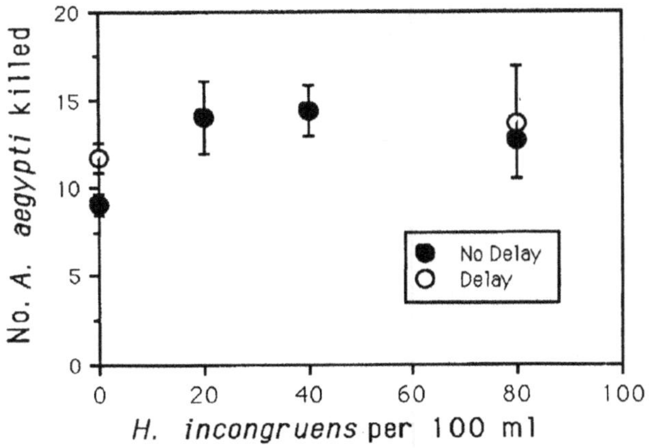


*Fig. 3.* Mortality of *Aedes aegypti* larvae as a function of *Heterocypris incongruens* density in the presence of BTI (concentration = $10^{-6}$ mg per ml). Dark symbols indicate that ostracods were added four hours prior to adding mosquito larvae. Open symbol indicates simultaneous introduction. Error bars are ±1 S.E. Mortality in controls (absence of BTI) was negligible and is not depicted.

*A. aegypti* larvae in all of the experiments (Table 2, Fig. 5). In the first two experiments with *B. schaefferi*, 5 fully-matured fairy shrimp, introduced four hours prior to introducing the mosquito larvae, sharply reduced BTI-induced mortality in the larvae (Table 2). Subsequent experiments demonstrated that less than five individu-

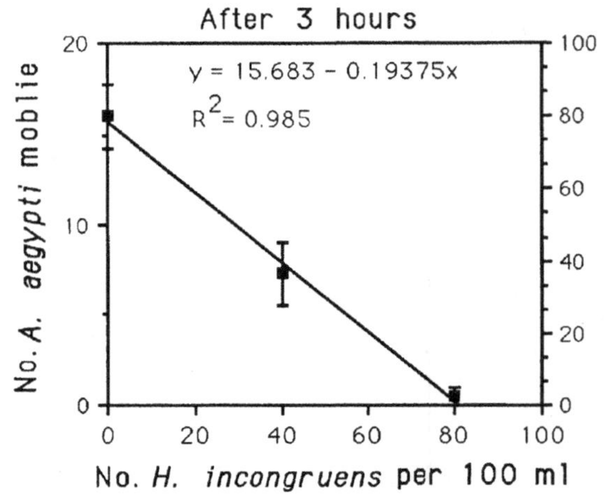


*Fig. 4.* Number of *Aedes aegypti* larvae still alive and mobile as a function of *Heterocypris incongruens* density four hours after introducing the mosquito larvae. The BTI concentration used was $2 \times 10^{-6}$ mg per ml. Mortality in controls (absence of BTI) was negligible and is not depicted.

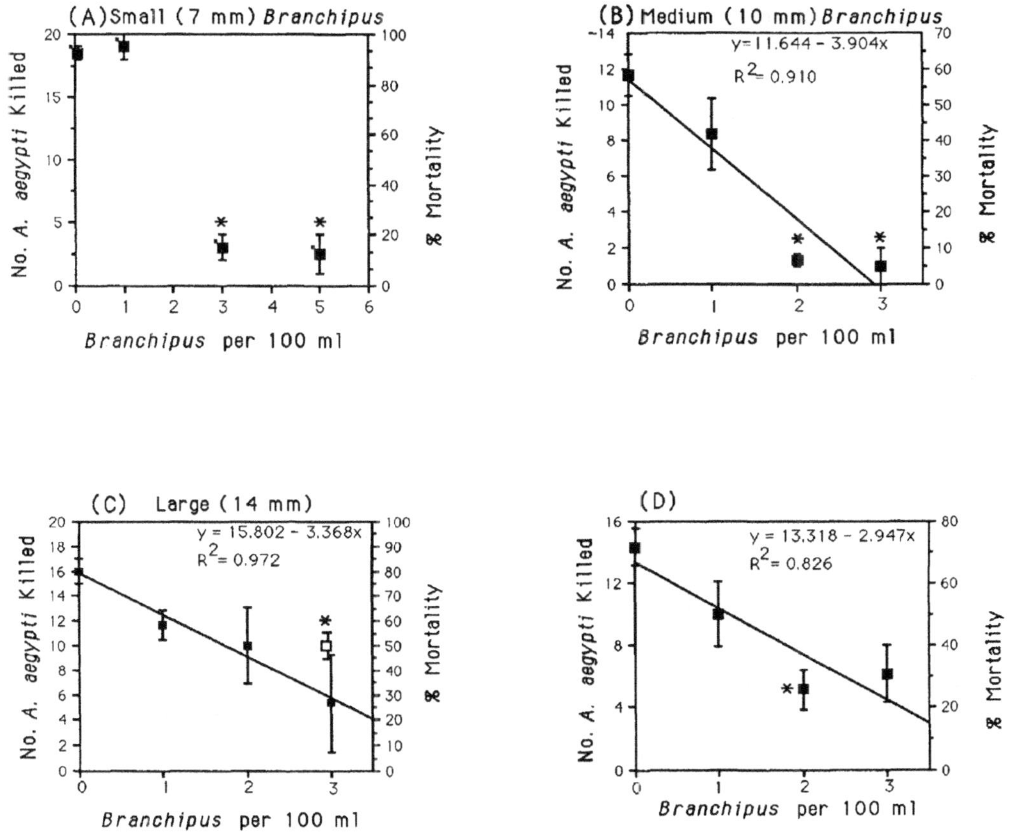


*Fig. 5.* Mortality of *Aedes aegypti* larvae as a function of *Branchipus schaefferi* density in the presence of BTI (concentration of $2 \times 10^{-6}$ mg per ml). Dark symbols are for organisms added with a four hour delay. Open symbol is with simultaneous introduction. Data were subjected to both regression and ANOVA. Asterisks on figure indicate which densities were significantly different (Scheffe's F-test) from treatment containing no ostracods. Error bars are ±1 standard error. Mortality in controls (absence of BTI) was negligible and is not depicted.

*Table 2.* *B.t.i.*-induced mortality of *Aedes aegypti* after 24 hours in the absence and presence of five *Branchipus schaefferi*. *B.t.i.* and *Branchipus* were introduced four hours before *Aedes*. *P*-values are from student's t-tests.

| Treatment | *A. aegypti* larvae killed | | | | | |
|---|---|---|---|---|---|---|
| | Rep 1 | Rep 2 | Rep 3 | Mean (S.E.) | Percent | *P*-value |
| Experiment 1 | | | | | | |
| *B.t.i.* | 20 | 20 | 20 | 20.0 | 100.0% | <0.001 |
| *B.t.i.* + *Branch.* | 5 | 3 | 2 | 3.3 (0.88) | 16.7% | |
| Experiment 2 | | | | | | |
| *B.t.i.* | 20 | 20 | 20 | 20.0 | 100.0% | <0.01 |
| *B.t.i.* + *Branch.* | 12 | 7 | 13 | 10.7 (1.86) | 53.3% | |

*Table 3a.* *B.t.i.*-induced mortality of *Aedes aegypti* larvae after 24 hours in water preconditioned and not preconditioned by exposure to *Branchipus schaefferi.* Ten *Branchipus* were introduced for six hours and then removed upon introduction of *B.t.i.* and *Aedes.*

| Treatment | *A. aegypti* larvae killed | | | | |
|---|---|---|---|---|---|
| | Rep 1 | Rep 2 | Rep 3 | Mean (S.E.) | Percent |
| Experiment 1 | | | | | |
| *B.t.i.* | 7 | 10 | – | 8.5 (1.47) | 42.5% |
| *B.t.i.* + *Branch.* | 2 | 3 | – | 2.5 (0.49) | 12.5 % |
| Experiment 2 | | | | | |
| *B.t.i.* | 11 | 9 | 10 | 10.0 (0.58) | 50.0% |
| *B.t.i.* + *Branch.* | 8 | 6 | 6 | 6.7 (0.67) | 33.4 % |
| Experiment 3 | | | | | |
| *B.t.i.* | 6 | 5 | 5 | 5.3 (0.40) | 26.7% |
| *B.t.i.* + *Branch.* | 2 | 3 | – | 2.5 (0.49) | 12.5 % |

*Table 3b.* Analysis of variance on above data.

| Source of variation | d.f. | M.S. | F-ratio | *P*-value |
|---|---|---|---|---|
| Treatment | 1 | 59.21 | 49.19 | <0.001 |
| Experiment | 2 | 27.12 | 22.53 | <0.001 |
| Interaction | 2 | 3.13 | 2.60 | 0.129 |
| Error | 9 | 1.20 | | |

als also reduced mortality (Fig. 5). Both small (length = 7 mm), medium (length = 10 mm) and large (length = 14 mm) fairy shrimp, tested separately, sharply reduced BTI-induced mortality (Fig. 5a, b, c). Three fairy shrimp added simultaneously with the mosquito larvae and BTI also caused a reduction in mosquito mortality ($t = 4.715, p = 0.009$; Fig. 5c). An assessment of the effects of sex ($F = 1.449, p = 0.274$) and size (medium vs large fairy shrimp) ($F = 0.776, p = 0.421$) showed no significant differences. These different sizes and sexes were then combined to give six replicates per density and the effect of *B. schaefferi* density was consistent with previous tests (Fig. 5d).

We did not demonstrate that the fairy shrimp reduced the concentration of BTI in the water after four hours in two trials and after 24 hours in the second trial (Student's t-tests, $p >> 0.05$ for all tests; first trial results shown in Fig. 2c).

In all three experiments, mortality was consistently lower in cups which previously contained fairy shrimp but removed prior to the addition of BTI and mosquitoes than in cups which never contained fairy shrimp (Table 3). Overall, mortality was significantly reduced after 24 hours from 39.4% without fairy shrimp to 21.4% with prior conditioning of the medium (Table 3b). We also noted mortality after 72 hours for the last two experiments and the differences were considerably greater; mortality for experiments 2 and 3 respectively averaged 75% and 68.4% without fairy shrimp and 40% and 37.5% with prior introduction of fairy shrimp.

## Discussion

The results clearly and consistently demonstrated that *B. schaefferi*, in some as yet unknown way, positively affected the survival of mosquito larvae exposed to BTI. The variability in the magnitude of the differences among experiments may have been due to various factors such as the amount of alternative food in the water and slight differences in the age of the mosquitoes. This reduction in BTI-induced mortality is consistent with the findings of Rettich (1986) who demonstrated the same type of indirect positive effect of *Daphnia curvirostris* on mosquito larvae. He did not assess whether changes in the BTI concentration of the water occurred. This proposed mechanism (reduction of BTI by the crustaceans) to explain our results is not supported by our preliminary feeding experiments. This is rather surprising in light of the fact that *B. schaefferi* is capable of feeding on particles of this size based on its intersetular distance (Dimentman, 1976).

Because the number of BTI-like colony forming units tended to be higher (though not significantly) rather than lower in the presence of all three crustaceans, these results suggest that perhaps the crustaceans excrete something which causes the BTI to replicate. This is not the likely explanation to our results as the samples were plated after four hours of exposure to the crustaceans and it takes approximately six hours for the spores to replicate once introduced into the water (e.g., Khawaled *et al.*, 1988). It is possible that the BTI is ingested but the spores are excreted back into the water unharmed. It is also possible that the BTI introduced into the water are in aggregates which are in some way broken up by the crustaceans. Since an aggregate of many BTI spores and a single isolated spore will both produce a single colony forming unit in a culture dish, breaking up of aggregates by crustaceans could mask a reduction, via feeding, of the BTI.

We demonstrated that only prior conditioning of the medium by fairy shrimp is needed to reduce larval mortality. Hence, another mechanism in addition to, or instead of, feeding appears to be involved Metabolites, excreted by the fairy shrimp, may in some way inactivate the endotoxin. It is possible that the fairy shrimp excrete substances that are not ingested by the mosquito larvae but that the bacteria adhere to these substances making the BTI unavailable to the mosquito larvae. Further experiments are needed to determine the mechanism. Lastly, the excreted materials by the fairy shrimp may serve as an alternative food for the mosquito larvae. Availability of alternative foods of mosquito larvae have been shown to reduce the efficacy of larvicidal bacteria (Ramoska & Pacey, 1979; Aly, 1983).

*C. vidua* also reduced BTI-induced mortality, at least during the summer. As with the fairy shrimp, feeding upon the bacteria does not appear to explain the result. We are unable to determine why the ostracods had no effect later in the season. Further experiments are needed with this organism to determine if there is a seasonal effect and if, like the fairy shrimp, its mere presence and then removal prior to introducing BTI and mosquitoes will also have an effect. All results indicate that *H. incongruens* has no positive effect on *A. aegypti* survival and, in fact, may increase mortality in those larvae ingesting a sublethal dose. This finding is similar to that of Weiser (1976) who found *H. incongruens* to attack and feed on 'still-moving' *Chironomus plumosus* larvae which were infected with a fungus, *Coelomomyces chironomi*.

The design of our experiment precluded other potential indirect effects that crustaceans may have on BTI-induced mortality in mosquito larvae. Our laboratory experiments did not take into account the amount of suspended organic matter. The presence of organic particles in the water reduces the efficacy of BTI (reviewed in Margalit & Dean, 1985). Some crustaceans may actually cause the reverse effect seen in the current paper by feeding on the organic matter in the water. Alternatively, the crustaceans, by feeding on alternative foods of the mosquito larvae, may result in greater consumption of BTI by the mosquitoes which would enhance the larvicidal efficacy of BTI. The sum of all of these potential indirect effects may need to be taken into account to determine the effect of these crustaceans on the efficacy of BTI.

Although the mechanisms involved remain to be determined, our results show that the presence of certain crustaceans may indirectly benefit mosquito larvae in the presence of a pathogenic bacteria. This adds to a growing body of literature demonstrating the many types of indirect effects and their importance in community dynamics. It is also of applied interest; when applying bacteria such as BTI for mosquito control, the effective dose may be a function of the abundance of certain crustaceans and other animals.

## Acknowledgements

We are grateful to Kamal Khawaled, Arieh Zaritsky, Zeev Barak and Chanan Dimentman for discussions and to K. Khawaled, A. Zaritsky and Delton Belk for comments on the manuscript. We are also grateful to Koen Martens for identifications of the ostracod species and to Chanan Dimentman for the fairy shrimp species.

## References

Aly, C., 1983. Influence of feeding behavior of *Aedes aegypti* on the effectiveness of *Bacillus thuringiensis* var *israelensis*. Bull. Soc. Vector Ecol. 8: 94–100.

Barnes, H. G., 1974. Invertebrate Zoology. Saunders, Philadelphia, 970 pp.

Bence, J. R., 1988. Indirect effects and biological control of mosquitoes by mosquitofish. J. Appl. Ecol. 25: 505–521.

Blaustein, L., 1990. Evidence for predatory flatworms as organizers of zooplankton and mosquito community structure in rice fields. Hydrobiologia, 199: 179–191.

Blaustein, L. & R. Karban, 1990. Indirect effects of the mosquitofish, *Gambusia affinis*, on the mosquito, *Culex tarsalis*. Limnol. Oceanogr., 35: 767–771.

Cannon, H. G., 1925. On the feeding mechanism of a freshwater ostracod, *Pioncypris vidua (O.F. Muller)*. J. linn. Soc. Lond., Zool. 36: 325–335.

Chesson, J., 1989. The effect of alternative prey on the functional response of *Notonecta hoffmani*. Ecology 70: 1227–1235.

Dimentman, Ch., 1976. Distribution and biology of Anostraca (Crustacea) in temporary pools in Israel and factors affecting it. PhD dissertation, Hebrew University, 140 pp.

Dimentman, Ch., 1981. The rain pool ecosystems of Israel: Geographical distribution of freshwater Anostraca (Crustacea). Israel J. Zool. 90: 1–15.

Dimentman, Ch. & J. Margalit, 1981. Rainpools as breeding and dispersal sites of mosquitoes and other aquatic insects in the Central Negev Desert. J. Arid Envir. 4: 123–129.

Diner, M. P., E. P. Odum & P. F. Hendrix, 1986. Comparisons of the roles of ostracods and cladocerans in regulating community structure and metabolism in freshwater microcosms. Hydrobiologia 133: 59–63.

Goldberg, L. J. & J. Margalit, 1976. Bacterial spore demonstrating rapid larvicidal activity against *Anopheles* sergentii, *Uranotaenia unguiculata*, *Culex univittatus*, *Aedes aegypti* and *Culex pipiens*. Mosq. News 37: 355–358.

Gophen, M. & W. Geller, 1974. Filter mesh size and food particle uptake by Daphnia. Oecologia 64: 408–412.

Hinman, E. H., 1930. A study of the food of mosquito larvae (Culicidae). Am. J. Hyg. 12: 238–270.

Holt, R. D., 1977. Predation, apparent competition and the structure of prey communities. Theor. Pop. Biol. 12: 197–229.

Hoy, J. B., E. E. Kauffman & A. G. O'Berg, 1972. A large scale field test of *Gambusia affinis* and Chlopyriphos for mosquito control. Mosq. News 32: 162–171.

Kerfoot, W. C. & A. Sih, 1987. Predation: direct and indirect effects on aquatic communities. University Press of New England, 386 pp.

Khawaled, K., Z. Barak & A. Zaritsky, 1988. Feeding behavior of *Aedes aegypti* larvae and toxicity of dispersed and of naturally encapsulated *Bacillus thuringiensis* var *israelensis*. J. Invertebr. Zool. 52: 419–426.

Lampert, W., 1974. A method for determining food selection by zooplankton. Limnol. Oceanogr. 19: 995–998.

Margalit, J. & A. Dean, 1985. The story of *Bacillus thuringiensis* var *israelensis* (*B.t.i.*). J. Am. Mosq. Control Ass. 1: 1–7.

Margalit, J., C. Pascar-Gluzman, H. Bobroglio, A. Barak, & L. Lahkim-Tsror, 1985. Biocontrol of mosquitoes in Israel. In: Laird, M. & J. W. Miles (eds), Integrated mosquito control strategies, Vol. 2, pp. 361–364. Academic Press.

Miller, J. H., 1972. Formulas and recipes. In: Miller, J. H. (ed.), 'Experiments in molecular genetics' Cold Spring Harbor, New York, p. 431.

Miller, T. E. & W. C. Kerfoot, 1987. Redefining indirect effects. In: Kerfoot, W. C. & A. Sih (ed.), Predation: direct and indirect effects on aquatic communities. University Press of New England, pp. 33–37.

Modlin, R. F., 1982. A comparison of two *Eubranchipus* species (Crustacea, Anostraca). Am. Midl. Nat. 107: 107–113.

Mulligan, F. S., C. H. Schaefer & T. Miura, 1978. Laboratory and field evaluation of *Bacillus sphaericus* as a mosquito control agent. J. Econ. Ent. 71: 774–778.

Paine, R. T., 1980. Food webs: linkage, interaction strength and community infrastructure. J. anim. Ecol. 49: 667–685.

Pennak, R. W., 1978. Freshwater invertebrates of the United States. 2nd edition. John Wiley & Sons, New York, 803 pp.

Ramoska, W. A. & C. Pacey, 1979. Food availability and period of exposure as factors of *Bacillus sphaericus* efficacy on mosquito larvae. J. Econ. Ent. 27: 523–525.

Ramoska, W. A., S. Singer & R. Levy, 1977. Bioassay of three strains of *Bacillus sphaericus* on field collected mosquito larvae. J. Invert. Pathol. 30: 151–154.

Rettich, F., 1986. Our experience with the use of *Bacillus thuringiensis* H-14 in mosquito control in Czechoslovakia. Dipterologica bohemoslovaca IV: 129–133.

Singer, S., 1973. Insecticidal activity of recent bacterial isolates and their toxins against mosquito larvae. Nature 244: 110–111.

Walker, E. D., E. J. Olds & R. W. Merritt, 1988. Gut content analysis of mosquito larvae (Diptera: Culicidae) using DAPI stain and epifluorescence microscopy. J. Med. Ent. 25: 551–554.

Weiser, J., 1976. The intermediary host for the fungus *Coelomomyces chironomi*. J. Invertebr. Pathol. 28: 273–274.

*Hydrobiologia* **212**: 77–82, 1991.
*D. Belk, H. J. Dumont & N. Munuswamy (eds), Studies on Large Branchiopod Biology and Aquaculture.*



# The biology of *Tanymastix stagnalis* (L.) and its survival in large and small temporary water bodies in Ireland

J. N. R. Grainger
*Department of Zoology, Trinity College, Dublin 2, Ireland*

*Key words: Tanymastix*, Anostraca, biology, survival strategy, climate, temporary waters

## Abstract

*Tanymastix stagnalis* is known from seven locations in Ireland. These range from small to large temporary water bodies. It was earlier shown that the best hatching and survival of eggs and young stages occurred in the presence of soil and at temperatures between about 10° and 15°. Some eggs hatched after one drying period, others after up to at least 4 drying and wetting periods. It was found from field observations that the habitat appeared after a minimum of 200 mm of rain in a 2-month period. This occurred on average more than once per year, mostly between June and February. A computer model was constructed using field data, and meteorological data (rainfall and temperature) covering 100 years, to decide how secure the organism was in two contrasting habitats. This showed that in both habitats there was little likelihood of extinction from metorological causes. There was, however, the possibility of destruction of the smaller habitats either through drainage or conversion to permanent ponds through compaction of the soil by heavy farm animals. The distribution and the possibility of harvesting the organisms is briefly discussed.

## Introduction

The Anostracan *Tanymastix stagnalis* (L.) was first discovered in Ireland in 1974 by Young (1975, 1976) and a further six stations were located by Grainger (1976). It is the only Anostracan known to inhabit Ireland. The biology of *Tanymastix* and the factors controlling the hatching of the resting eggs have been described by Loubresse (1965a, b), Al-Tikrity (1979) and Al-Tikrity & Grainger (1990). The European distribution of the species has been well documented by Freiner & Gruttner (1984) and there have been further records by Mura (1985).

The species is found in large and small temporary waterbodies in West Ireland. Its general biology will not be recounted here but it is important to mention the following points:

1. The species is essentially a cool water species which does not flourish in temperatures greater than 20°.
2. The resting eggs can survive a number of immersions in water and then successfully hatch. Al-Tikrity & Grainger (1990) found that out of 100 eggs some could (in the presence of soil) hatch following 4 periods of wetting and drying.

The aim of the present paper is to examine the likelihood of survival of the organism in two contrasting locations in Ireland in which it has been found. All locations in which the species has been found dry out completely during a dry period.

## Materials and methods

The two sites chosen are near Kilcolgan, Co. Galway, which is about 15 km S of Galway city. One, (Site A), is located about 3 km E of Kilcolgan and consisted of a series of pools in a large undulating field. Some of these join together after heavy rain. This is located on the Irish National Grid at Grid reference M 472170. The smaller pools were about 5 to 10 m in diameter and normally flooded to a depth of no more than about 3 m.

The second location (Site B) is about 6 km from Site A and is about 2 km S of Kilcolgan. It is a long elongated depression which when full was at least 0.5 km long, and up to 60 m wide. The depth varied considerably but could reach 8 m after heavy rain. Its location on the National Grid is M 431143.

In the present field study rainfall rates were determined locally at 24 hr intervals using a portable raingauge. Changes in water height were measured using bamboo sticks pushed into the water of the temporary habitats and the changes in water level were measured by using a rule and making pencil marks on the sticks after 24-hour periods.

Rainfall measurements are available from 1944 to 1978 from Kilcolgan.

Accurate daily air temperature data are available for University College, Galway from 1966 to the present.

*Table 1.* Mean evaporation from Class A pans in Galway from 1969 to 1978.

| Month | Evaporation ($cm \cdot month^{-1}$) |
|---|---|
| JAN | 0.67 |
| FEB | 1.25 |
| MAR | 3.24 |
| APR | 5.47 |
| MAY | 7.76 |
| JUNE | 9.57 |
| JULY | 9.02 |
| AUG | 7.23 |
| SEPT | 4.49 |
| OCT | 2.48 |
| NOV | 0.91 |
| DEC | 0.70 |

Direct measurements of evaporation using a Class A pan are available from University College, Galway from 1969 to the present. These were averaged (to 1978) as monthly means and these are shown in Table 1. The present analysis finished with the 1978 data in all cases.

In order to get a broader picture over a longer period of time the rainfall and temperature data from Birr (about 80 km away) was used. This is continuous from 1873 but the present analysis finished in 1978 so the relevant data is up to that year.

The Kilcolgan monthly rainfall data were correlated with the Birr data from 1944 to 1978. The correlations were positive and very good ($r = 0.8502$; $n = 420$; $P < 0.05$). The method used to determine the relation between them was that described by Pearson (1901). This gave:

$$Y = A + B \cdot X$$

where $Y$ = Kilcolgan monthly rainfall; $X$ = Birr monthly rainfall;
$A = 24.46357$;

$B = 0.81457$.

Similarly the Galway mean monthly air temperatures were correlated with Birr air temperatures from 1966 to 1978. The correlation coefficient was also very good ($r = 0.9963$; $n = 156$; $P < 0.05$). The equation relating the two was:

$$Y = A + B \cdot X$$

where $Y$ = Galway monthly mean air temperature;

$X$ = Birr mean monthly air temperature;

$A = 0.90466$;

$B = 0.96629$.

Calculations and modelling were done on an Apple II + microcomputer and the programs used are all available on application to the author.

**Results and discussion**

1976 and 1977 were very dry summers in Ireland. It was observed in 1976 that following 210 mm of rain over a two-month period in the autumn, immediately following the dry spell, water appeared in all the habitats where *Tanymastix* is found including Sites A and B. In 1977, also following a dry summer, it was observed that 200 mm of rain over a two-month period caused the habitat to appear in all sites (including A and B). This was therefore taken as the minimum necessary for the habitat to appear. How long the habitat remained dependent upon physical factors which could be measured or estimated.

Table 2 shows the occurrence of appearance of the habitat for the Kilcolgan area over the period 1873 to 1978. Over these 105 years there were at least 187 floodings. This amounts on average to 1 to 2 per year and these are mostly distributed between June and January.

The next question concerns the maintenance of the habitat once it appears. This is determined by rainfall, drainage and evaporation.

The effects of rainfall on the level of water in stations A and B were estimated in the early months of 1978 using the method described. Once the habitat appeared the surrounding land is largely saturated and so the rainfall has an exaggerated effect on the level of the water. Each 1 mm of rain falling during a 24 hour period resulted in the level of the water in site A rising by an average of 29.8 mm. (3 measurements). This I have called the rainfall multiplier.

In Site B the equivalent result was 49.3 (3 measurements).

Drain rate was measured at each site over 24 hour periods during winter when there was no rain and this gave the following results:

Site A 7.6 cm $d^{-1}$
Site B 26.9 cm $d^{-1}$

These results are the means of 3 measurements. There seemed to be no relation between drain rate and water level.

The calculation of the change in water level once the habitat appears is shown in Table 3. $T$ is the actual second month (of a two-month period) during which > 200 mm rain is estimated to have fallen.

The problem now arises of whether development of the eggs which hatch can be completed before the habitat disappears. The relation between duration to egg-laying and various constant temperatures has been described in Grainger (1981). It was shown that while development under changing temperature conditions does have a significant effect on the duration of development the effect is a rather small one. This means that developmental factions can be used in roughly estimating when total development is completed when temperatures are changing.

Using the relation mentioned above, the Birr mean monthly temperatures have been used to get an estimate of the Galway (and hence Kilcolgan)

*Table 2.* Occurrence of two-month periods at Kilcolgan in which rainfall was > 200 mm.
(adjusted Birr data 1873–1978 inclusive)

| | |
|---|---|
| JAN | 12 |
| FEB | 2 |
| MAR | 6 |
| APR | 4 |
| MAY | 7 |
| JUNE | 13 |
| JULY | 32 |
| AUG | 18 |
| SEPT | 26 |
| OCT | 20 |
| NOV | 28 |
| DEC | 19 |
| TOTAL = 187 | |

The first month is the one which is counted.

*Table 3.* Change in water level.

Water level at start of month $T + 1$ = L cm
Water level at start of $T + 2 = L + (R \cdot X) - E - (D \cdot M)$

$R$ = monthly rainfall (cm)
$X$ = rainfall multiplier
$E$ = mean monthly evaporation (cm)
$D$ = drain rate ($cm \cdot day^{-1}$)
$M$ = number of days in month
$T$ = second month in actual two-month period

mean air temperatures. Pybus & Pybus (1980) have shown that there is a good relation been air and shallow lake water temperatures in the Galway area. This is not at all surprising since most days in W Ireland have a considerable period of cloud cover. The average cloud cover is about 71% of the sky. There is also a relatively small number of annual hours of bright sunshine (Keane, 1986). In the present study area it is about 1300 hours. Thus it is possible to get an approximate estimate of water temperature of relatively shallow water bodies such as are considered here from air temperatures. Table 4 shows a comparison of Birr data adjusted for Galway for two months January and June. Very low and very high temperatures are unusual. These conditions would seem to be ideally suited for *Tanymastix*.

The main computer analysis was now carried out according to the scheme in Table 5. The Birr monthly rain and mean air temperature meteorological data (adjusted for Galway and Kilcolgan) were used together with the measured rainfall multipliers and drain rates for Sites A and B. Since the initial water level when the habitat appears (at the beginning of month T + 1) is not known the program must be run for various initial water levels (L). The results are given in Tables 6 and 7. It was thought best to use six successive failures to lay eggs as constituting an extinction since in experiments using 100 eggs some still hatched following four wettings and dryings. Using a very large number of eggs would, most likely, have produced some eggs which would survive 6 immersions.

It would appear that in Site A *Tanymastix* should be relatively secure but that survival in Site B would seem to be slightly more precarious and would require normal flooding to depths of about 400 cm. This in fact usually appears to be the case.

Factors other than climate, however, must be taken into account particularly in smaller ponds.

*Table 4.* Mean monthly air temperature frequency for Galway.
(adjusted Birr data 1873 to 1978 inclusive)

| °C | Jan | June |
|---|---|---|
| −2– 0 | 1 | 0 |
| 0– 2 | 2 | 0 |
| 2– 4 | 10 | 0 |
| 4– 6 | 51 | 0 |
| 6– 8 | 37 | 0 |
| 8–10 | 5 | 0 |
| 10–12 | 0 | 2 |
| 12–14 | 0 | 49 |
| 14–16 | 0 | 53 |
| 16–18 | 0 | 2 |
| 18–20 | 0 | 0 |
| Total | 106 | 106 |

*Table 5.* General scheme of survival analysis.

Read Birr monthly rainfalls and mean monthly temperatures for years 1873 to 1978

↓

Calculate Kilcolgan monthly rainfalls and mean temperatures

↓

Calculate 2-monthly rainfall totals and store if > 200 mm

↓

Calculate changes in water level for months T + 1 and T + 2

↓

Read temperatures for months T + 1 and T + 2

↓

Calculate whether eggs would be laid before water disappeared and store result

↓

If there are 6 failures in succession then store as an extinction

*Table 6.* Results for site A.

| Initial water level (cm) | No. of failures to reach adult | No. of extinctions |
|---|---|---|
| 400 | 0 | 0 |
| 300 | 0 | 0 |
| 250 | 0 | 0 |
| 200 | 0 | 0 |
| 150 | 4 | 0 |
| 100 | 19 | 0 |
| 50 | 40 | 0 |
| 25 | 54 | 0 |
| 20 | 59 | 0 |

Drain rate = 7.6 $cm \cdot d^{-1}$
Rainfall multiplier = 29.8
No. of floodings = 186

*Table 7.* Results for site B.

| Initial water level (cm) | No. of failures to reach adult | No. of extinctions |
|---|---|---|
| 450 | 67 | 0 |
| 400 | 89 | 0 |
| 350 | 109 | 3 |
| 300 | 119 | 4 |
| 250 | 132 | 7 |
| 200 | 146 | 10 |
| 150 | 156 | 13 |
| 100 | 159 | 15 |

Drain rate = 26.9 $cm \cdot d^{-1}$
Rainfall multiplier = 49.3
No. of floodings = 186

Two of the ponds in Site A are now semi-permanent ponds since the farmer has replaced sheep grazing with heavy cattle (Charollais) which appear to have compacted the soil in some ponds. In these the *Tanymastix* have disappeared and more permanent pond predators are appearing. Site B is so large that this is very unlikely to happen.

Populations of *Tanymastix* never appear to become extremely dense in the Irish habitats in which they have been found. It is possible that this is principally due to the limited food supply. This appears to be largely microorganisms many of which are presumably washed into the ponds from surrounding soil and vegetation. It frequently happens, particularly in spring or autumn, that there is significant growth of algae. Usually however the *Tanymastix* have by this time finished laying eggs and their place has been taken by copepods and cladocera – some of which are of great interest (Ali *et al.*, 1987).

It would seem that the simplest way in which to harvest the *Tanymastix* would be to introduce young fish such as trout. However the areas concerned are, when present, inhabited in autumn and winter by large numbers of wading birds, diving and dabbling ducks and wild swans (Whooper & Bewick) which could predate the fish. So this problem remains unsolved.

Lastly I would suggest that the distribution of *Tanymastix* may be the result of the transfer of resting eggs either by birds, people shooting duck, or the transfer of both cattle and sheep from one farm to another (which happens on an extensive scale in Ireland). It is not clear why *Tanymastix* has not become established in more of the similar and, apparently suitable, temporary water bodies in W Ireland which are found in abundance in Galway, Mayo, Roscommon and Clare and are locally known as turloughs (or winter lakes). One is tempted to try some introductions into new areas to try and find out what are the criteria that need to be observed for a population to survive and develop. It is hoped that the present study is a contribution towards this end.

## Acknowledgements

I am grateful to the Director of the Irish Meteorological Office and the Minister for Communications for permission to use the meteorological data analysed in this paper.

## References


Al-Tikrity, M. R., 1979. Factors affecting the egg and general biology of *Tanymastix stagnalis* (L.). Ph.D thesis, University of Dublin.

Al-Tikrity, M. R. & J. N. R. Grainger, 1990. The effect of temperature and other factors on the hatching of the resting eggs of *Tanymastix stagnalis* (L.) (Crustacea, Anostraca). J. therm. Biol. 15: 87–90.

Ali, T. H., J. M. C. Holmes & J. N. R. Grainger, 1987. *Diaptomus cyaneus* Gurney, a freshwater copepod new to Britain and Ireland. Ir. Nat. J. 22: 240–241.

Freiner, D. & O. Gruttner, 1984. Der Eichener Kiemenfuss-krebs. Natur Mus., Frankfurt, 114: 273–286.

Grainger, J. N. R., 1976. Further records for the fairy shrimp *Tanymastix stagnalis* (L.). Ir. Nat. J. 18: 326.

Grainger, J. N. R., 1981. The effect of constant and changing temperatures on the period from egg hatching to egg laying in *Tanymastix stagnalis*. J. therm. Biol. 6: 353–356.

Keane, T. (ed.), 1986. Climate, weather and Irish Agriculture. 332 pp. Agricultural Trust, Dublin.

Loubresse, N. G. de, 1965a. Caracteres les plus apparents de la reproduction de *Tanymastix lacunae* G. en elevage. C. r. hebd. Seanc. Acad. Sci. Paris, 260: 1751–1754.

Loubresse, N. G. de, 1965b. Elevage en Laboratoire du Phyllopode *Tanymastix lacunae* Guerin. Bull. Soc. zool. Fr., 90: 301–341.

Mura, G., 1985. Prelimary report on the sympatric occurrence of two species of fairy shrimps (Crustacea, Anostraca) in some temporary ponds from Italy. Riv. Idrobiol. 24: 73–82.

Pearson, K., 1901. On lines and planes of closest fit to systems of points in space. Phil. Mag. 2 (6th ser.): 559–572.

Pybus, C. & M. J. Pybus, 1980. An ecological study of Lough Murree, a brackish water lake in County Clare. Proc. R. Ir. Acad., 80B: 367–384.

Young, R., 1975. *Tanymastix stagnalis* (L.) new to Britain and Ireland. Ir. Nat. J. 18: 203.

Young, R., 1976. *Tanymastix stagnalis* (Linn.) in County Galway, new to Britain and Ireland. Proc. R. Ir. Acad. 76B: 369–378.

*Hydrobiologia* **212**: 83–86, 1991.
D. Belk, H. J. Dumont & N. Munuswamy (eds), *Studies on Large Branchiopod Biology and Aquaculture*.



# Why only one of two common Central Texas Anostraca atop Enchanted Rock?

Denton Belk
*Biology Department, Our Lady of the Lake University of San Antonio, San Antonio, Texas 78207-4666, USA*

*Key words:* coexistance, competative exclusion, dispersal

## Abstract

*Branchinecta packardi* Pearse, 1912 grows rapidly and begins reproducing at the age of seven days. The first clutch contains only one or two eggs. Females deposit a clutch each day. Reproductive populations occur during all seasons in Central Texas. *Streptocephalus texanus* Packard, 1871 has a slower growth rate and does not begin reproduction until 11 days of age. Females deposit a clutch every second day. The species is a warm season form in Central Texas. These life history differences most likely explain why *B. packardi* inhabits the shallow, exposed weather-pit pools atop Enchanted Rock and *S. texanus* does not.

*Branchinecta packardi* Pearse, 1912 is the only fairy shrimp ever found inhabiting the shallow water-pit pools (Blank, 1951) located at the top of the 130 m high granite dome known as Enchanted Rock. This precambrian granite pluton is the centerpiece of Enchanted Rock State Natural Area, Llano County, Texas, USA (30° 30′ 22″ N, 98° 49′ 06″ W) (Fig. 1). Moore (1950) was the first biologist to report *B. packardi* (missidentified as *Branchinecta coloradensis*, see Lynch, 1964) in the pools on Enchanted Rock. He studied them during the summers of 1948 and 1949. Tribbey (1965) studied succession in three of these pools between October 1964 and April 1965. He reports repeatedly finding *B. packardi* (misidentified as *Branchinecta lindahli*, see Lynch, 1964) in the summit pools. He made casual observations of apparently similar pools on an area of granite at the foot of the dome on its north side. In these pools, he found only the other common Central Texas anostracan *Streptocephalus texanus* Packard, 1871. He never found *S. texanus* in the pools at the summit. During a number of visits from 1965 through 1985, I observed that *B. packardi* was still the only anostracan in the summit pools.

In attempting to explain the absence of *S. texanus* from the pools atop Enchanted Rock, Tribbey (1965) speculated that its exclusion resulted from a high degree of community organization developed over years of competitive interaction under restricting environmental conditions. The fact that he did not know of any cases of the two species coexisting in the same pond reenforced his idea that competitive exclusion was the most probable cause.

However, if community organization within the summit pools is a valid part of the explanation, its operation must be in some way exclusive to these pools as *S. texanus* and *B. packardi* are able to coexist in ponds in a number of other locations. Sublette & Sublette (1967) reported both species from five playa lakes on the Staked Plains of Texas and New Mexico. In a pond near Casper, Wyoming, Lynch (1964) collected these two along

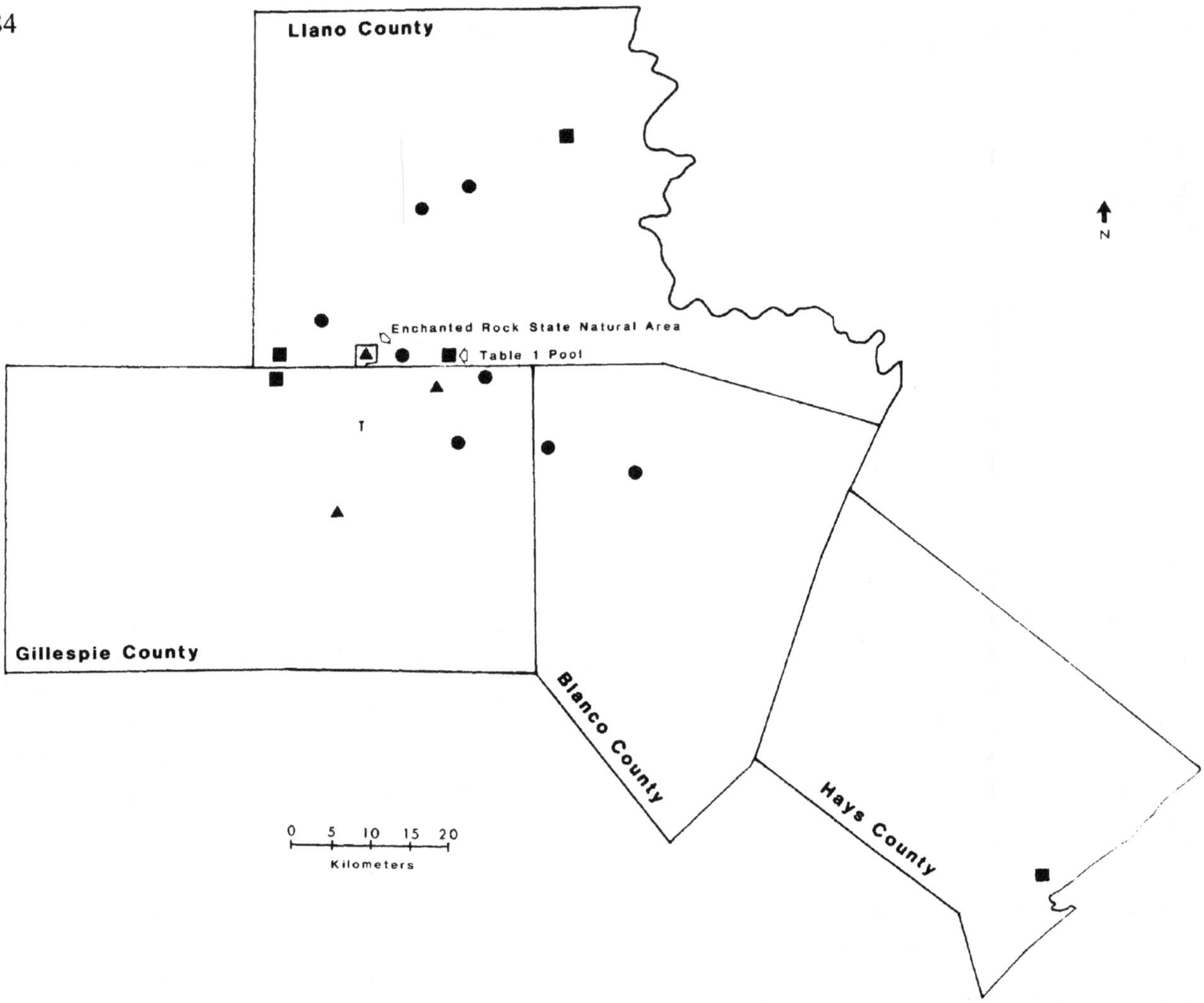


*Fig. 1.* Map showing Enchanted Rock State Natural Area, Table 1 pool, the Teague Ranch weather station (T), Horne's (1971) Rock Pond in Hays County, and pools where Denton Belk collected anostracans during the period 1979–1988: squares = coexisting *B. packardi* and *S. texanus*, circles = only *S. texanus*, and triangles = only *B. packardi*.

with the anostracans *Branchinecta lindahli* and *Thamnocephalus platyurus*. Prophet (1963a) found *S. texanus* and *B. packardi* (misidentified as *Branchinecta lindahli*, see Lynch, 1964) together in several pools near Calwell, Kansas. I collected these two species together in a rock depression pool in northeastern Arizona near Mexican Water (DB 5, catalog number in my collection) and in a deep rock depression pool at Hueco Tanks State Park in west Texas (DB 713). Closer to Enchanted Rock, Horne (1971) studied a pond in a rocky ephemeral creek at San Marcos, Hays County, Texas where these two still coexist (Fig. 1). During numerous visits to the three county area around Enchanted Rock from 1979 to 1988, I located four sites in which both species coexisted (Fig. 1). These four coexisting populations are certainly an underestimate of the true number of such situations in this region of Central Texas as my collection effort was neither extensive nor intensive.

Any suggestion that *S. texanus* is absent from the summit pools because of a failure of dispersal to these pools seems unlikely. The Enchanted Rock pools have undoubtedly existed for hundreds, if not thousands of years. Thus time for dispersal is not a problem. *Streptocephalus texanus* is a common species in the area (Fig. 1).

There are even populations at the base of Enchanted Rock. And a bird known to be capable of dispersing *S. texanus* eggs is also common. Proctor *et al.* (1967) demonstrated that a small shorebird known as the killdeer (*Charadrius vociferus*) can serve as a dispersal agent for *S. texanus*. They fed female *S. texanus* carrying shelled eggs to killdeer. Feces collected 30 min later hatched *S. texanus* in 28 out of 30 trials. They did not study maximum retention time for *S. texanus* eggs. However, they did such test for eggs of another anostracan. In these studies, they found that some *Artemia* cysts were retained for a time in the caeca of the killdeer and passed out of the birds after 24–26 h. These cysts were apparently unaffected by caecal digestive processes and hatched as well as eggs voided after much shorter passage through the digestive system. It seems reasonable to suspect that the same would hold true for eggs of *S. texanus*. I see killdeer, along with other small shorebirds, every time I visit the pools on Enchanted Rock. These same birds are common at all the aquatic habitats in Central Texas. Thus there are no lack of *S. texanus* and dispersal agents around the Enchanted Rock area.

Collections from a pond about 10.8 km east of Enchanted Rock caused me to consider the possibility that *S. texanus* may be excluded from the summit pools by having a growth and reproductive rate that is to slow for it to survive in such shallow, exposed pools (Table 1). The Table 1 pond filled during a 10-day rainy period starting on 20 October 1984. The closest weather station to Table 1 pond is at Teague Ranch 14.6 km to the southwest (30° 26′ N, 98° 49′ W). At Teague Ranch, the heaviest rains fell during the first two days of the rainy period with 40.6 and 58.2 mm being measured. During the next eight days, from 25.6 to 5 mm fell in any one day, and the total for the eight days was 83.1 mm (NOAA, 1984). Maximum hatching of *S. texanus* and *B. packardi* eggs occurs around 24 h after wetting (Prophet, 1963b). Thus the anostracans sampled on 31 October 1984 were probably 9–10 days old. All of the *B. packardi* were fully mature and the females were carrying shelled eggs in their broodpouches. None of the *S. texanus* were mature (Table 1).

*Table 1.* Reproductive condition and minimum/maximum total lengths for ***Branchinecta packardi*** and ***Streptocephalus texanus*** collected 31 October 1984 and 7 November 1984 from a pool formed in a natural drainage off the east side of Texas Highway 16 at 3.9 km south of Farm Road 965 in Llano County, Texas U.S.A. (30°31′47″N, 98°42′39″W). The maturity of males was judged on development of antennal morphology. Immaturity in females was based on development of broodpouch. Collection effort was approximately the same on both dates. Specimens were narcotised with $CO_2$, fixed in 10% formalin for 3-5 h, and stored in 70% isopropyl alcohol.

| | 31 October 1984 | | 7 November 1984 | |
|---|---|---|---|---|
| | No. | mm | No. | mm |
| *B. packardi* | | | | |
| mature males | 161 | 7–11 | 7 | 8–10 |
| females carrying shelled eggs | 47 | 8–12 | 11 | 9–11 |
| *S. texanus* | | | | |
| immature males | 118 | 5–10 | 60 | 8–11 |
| immature females | 134 | 5–10 | 79 | 7–10 |
| mature males | 0 | – | 161 | 9–12 |
| females carrying shelled eggs | 0 | – | 214 | 9–12 |

In laboratory cultures maintained at 20–22 °C, Prophet (1963c) observed that *S. texanus* first mated at 11 days of age. Under the same conditions, reproduction began in *B. packardi* at seven days of age. He noted that the preadult development of *B. packardi* 'was very rapid' and that 'initial clutches rarely contained more than one or two eggs'. Moore (1950) noted that most of the female *B. packardi* in the Enchanted Rock pools during June of 1949 were 7 mm in total length and had only two or three eggs in their broodpouch. Another important difference between these two species is the length of time eggs are held in the broodpouch. *Streptocephalus texanus* usually deposits eggs at two day intervals (Prophet, 1963c), whereas, *B. packardi* typically clears the broodpouch and produces a new clutch each day (Prophet, 1962). From these observations, it is clear that *B. packardi* is much better adapted than *S. texanus* for survival in short duration ephemeral pools. In this regard, Maynard &

Romney (1975) report that *B. packardi* inhabit shallow rock pools in sandstone at Canyonlands National Park, Utah. They point out that these pools typically retain water for a maximum of 12–15 days.

The largest pool Tribbey (1965) studied on Enchanted Rock was approximately 9 m at its longest point by 3 m at its widest point. Its maximum depth was 12 cm with a mean depth of 8.2 cm. He observed that in the absence of additional rain a 30 March 1965 filling to about 35–50% of maximum volume of the summit pools resulted in only a 14 day wet phase. Tribbey told me that the pools at the base of Enchanted Rock held water longer than the summit pools. These lower pools are on the north side of the dome and thus sheltered from the severest drying conditions in this semiarid northern hemisphere environment where the contrast between north and south facing slope vegetation is obvious to even the casual observer.

Of course, not all wet phases in the summit pools are of short duration. Tribbey (1965) studied a 56 day aquatic phase starting on 26 October 1964, and a 34-day wet period starting 4 February 1965. Temperatures around 20 °C during the October filling would have been favorable for *S. texanus*, however, temperatures around 10 °C during the February wet phase would not. Eggs of *S. texanus* do not hatch below 15 °C (Prophet, 1963b; Horne, 1967) and the species is known as a warm season form in Central Texas with eggs usually hatching at 22–30 °C (Horne, 1971). Being a warm season anostracan probably works against *S. texanus* becoming established in the summit pools. This is because partial fillings will more often be followed by rapid drying during warmer periods. Given the rainfall patterns in Central Texas, the pools are likely to receive water anytime during the year.

*Branchinecta packardi* is not limited to any particular season. Reproductive populations have been observed in the pools atop Enchanted Rock during all seasons. Moore (1950) studied them in the summer months of June, July, and August. Tribbey (1965) followed their population changes during the fall, winter, and spring months of October, November, February, and March. In addition to these months, I collected them in April (DB 256,257).

In conclusion, I think differences in growth rate and reproductive biology are the most likely reasons why *B. packardi* inhabits the shallow, exposed weather-pit pools atop Enchanted Rock and *S. texanus* does not.


## Acknowledgements

I thank my wife, Mary Schug Belk, for sharing both the fun of field trips and the work of sorting, counting, and measuring specimens.


## References


Blank, H. R., 1951. Exfoliation and weathering on granite domes in Central Texas. Tex. J. Sci. 3: 376–390.

Horne, F., 1967. Effects of physical-chemical factors on the distribution and occurrence of some southeastern Wyoming phyllopods. Ecology 48: 472–477.

Horne, F. R., 1971. Some effects of temperature and oxygen concentration on phyllopod ecology. Ecology 52: 343–347.

Lynch, J. E., 1964. Packard's and Pearse's species of *Branchinecta*: Analysis of a nomenclatural involvement. Am. Midl. Nat. 71: 466–488.

Maynard, S. S. & S. V. Romney, 1975. The occurrence of four new anostracan (Crustacea) phyllopods in Utah, multispecies records, and some notes on their ecology. S. West Nat. 52: 6–11.

Moore, W. G., 1950. A new locality record for *Branchinecta coloradensis*, with habitat notes on two species of fairy shrimp in Central Texas. Ecology 31: 655–657.

National Oceanic and Atmospheric Administration, 1984. Climatological data Texas August 1984 89(8): 26.

Prophet, C. W., 1962. Ecology and reproduction of five species of Anostraca in Oklahoma. Unpublished Ph.D. dissertation, University of Oklahoma, Norman.

Prophet, C. W., 1963a. Physical-chemical characteristics of habitats and seasonal occurrence of some Anostraca in Oklahoma and Kansas. Ecology 44: 798–801.

Prophet, C. W., 1963b. Some factors influencing the hatching of anostracan eggs. Trans. Kans. Acad. Sci. 66: 150–159.

Prophet, C. W., 1963c. Egg production by laboratory cultured Anostraca. S. West Nat. 8: 32–37.

Proctor, V. W., C. R. Malone & V. L. DeVlaming, 1967. Dispersal of aquatic organisms: viability of disseminules recovered from the intestinal tract of captive killdeer. Ecology 48: 672–676.

Sublette, J. E. & M. S. Sublette, 1967. The limnology of playa lakes on the Llano Estacado, New Mexico and Texas. S. West Nat. 12: 369–406.

Tribbey, B. A., 1965. A field and laboratory study of ecological succession in temporary ponds. Unpublished Ph.D. dissertation University of Texas, Austin.

*Hydrobiologia* **212**: 87–94, 1991.
*D. Belk, H. J. Dumont & N. Munuswamy (eds), Studies on Large Branchiopod Biology and Aquaculture.*



# Natural growth and reproduction in *Triops granarius* (Lucas) (Crustacea: Notostraca)

M.T. Seaman, D.J. Kok, B.J. von Schlichting & A.J. Kruger
*Department of Zoology and Entomology, University of the Orange Free State, Bloemfontein 9300, Republic of South Africa*

*Key words:* temporary water, tadpole shrimps, southern Africa, pans


## Abstract

Field sampling during successive inundations was followed by laboratory studies on growth, egg production and hatching of *Triops granarius* from temporary pans in southern Africa.

The species was common where mean inundation was less than one month. Sexual maturity was reached in as little as six days at a carapace length of 6.5 mm. A biomass of 2.0 g, at carapace length of 25 mm was achieved in two months. Males and females occurred in similar numbers and mating was a prerequisite for sustained fertile egg production. Eggs were in similar numbers in both pouches, while successive clutches were larger. Hatching was light-dependent. Desiccation of eggs was not necessary but a period of up to 20 days was required before undesiccated eggs would hatch. In the event of desiccation, no minimum period thereof was necessary and hatching occurred within as little as a day of rehydration. Hatching success seldom exceeded 50 percent.

As a candidate for aquaculture, the species has failings such as low hatching rate and cannibalism, but it grows rapidly and produces eggs over an extended period.


## Introduction

Both notostracan genera *Triops* and *Lepidurus* have been widely studied (Longhurst, 1956; Fryer, 1988), and this is related to their widespread distribution. Particular attention has been given to morphological and taxonomic aspects. They are ecologically similar but *Triops* is restricted to waters that dry out and prefers warmer regions (Fryer, 1988).

Longhurst (1956) suggested that the eggs of *Triops* spp. probably do not hatch without prior desiccation. The water must have a low osmotic pressure for this to occur, a condition unlikely to be found on the bottom of a pool where the eggs are laid, but most likely when the pool refills after rain and the eggs float to the surface. In this statement he implies that salinity increases so rapidly that even the first batch of eggs laid will not hatch, that all eggs laid sink and that only eggs that float will hatch.

The conditions of short period of inundation and the rapid changes in physical and chemical conditions associated therewith, may be expected to affect patterns of survival, growth, reproduction and hatching.

*T. granarius* is the only species which occurs in southern Africa according to Longhurst (1956) and Rayner & Bowland (1985). Although it is possible that genetic variability is great, which would allow temporarily isolated populations to react differently, it is unlikely that separate races occur because these animals disperse and colonise so easily.

This paper examines the conditions and patterns of survival, growth and reproduction of naturally occurring *T. granarius* populations. In order to do this it is first necessary to look at the climate and habitat and then at the specific characteristics of the animals. Laboratory experiments are used to clarify some of the characteristics observed in the field.

## Climate and habitat

Notostraca inhabit temporary pools of fresh or brackish water in the drier parts of the world (Longhurst, 1956). Our studies (Seaman & Kok, 1987; unpublished data) have shown that *T. granarius* is a common species of the drier western parts of southern Africa, in agreement with the results of Barnard's (1929) survey (Fig. 1). It also occurs in isolated temporary waters in the wetter east (Rayner & Bowland, 1985), where it appears to be considerably less common. Southern Africa is essentially dry, with most of its area having an annual mean precipitation of less than 600 mm (Fig. 1). The area in which *T. granarius* is common by our experience, complies with the fairly broad climatic criteria found by Williams & Busby (pers. comm.) to predict the distribution of *T. australiensis*. The areas on the Australian continent which were predicted by their model to be the most suitable were also those where rainfall was most variable. Essentially, this is where the dominant water body is an episodically-filled temporary water body as opposed to a permanent water body. So the distribution of *Triops* spp. is governed by the availability of a suitable habitat, which in turn depends on particular rainfall and evaporation characteristics. The further west one goes in southern Africa, the lower and more sporadic the rainfall and the greater the evaporation rate (Tyson, 1986). In this area pans are particularly common (Goudie & Thomas, 1985). In the western part of

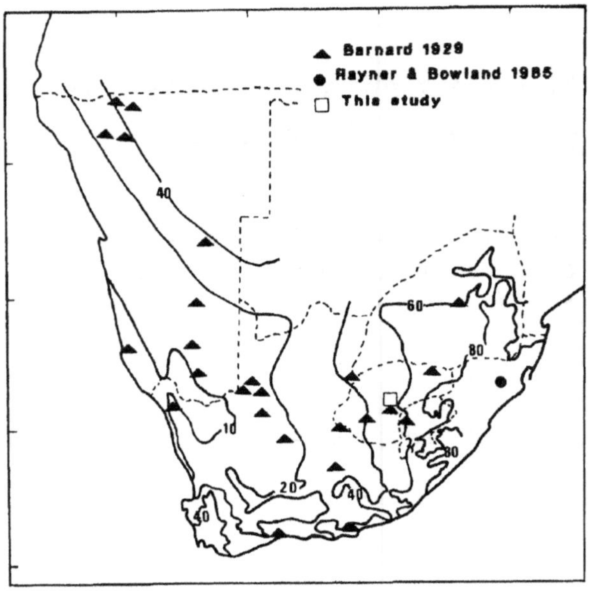


*Fig. 1. Triops granarius* distribution in southern Africa, study area and mean annual rainfall.

the Orange Free State province alone, there are thousands of pans (Le Roux, 1978).

Specific habitat requirements are broad. In this study we have relied on information gathered from two water bodies, namely Bainsvlei Pan and Greyvenstein Pan, both about 10 km west of Bloemfontein (29° 03′ S, 26° 07′ E). Bainsvlei Pan is divided up into a number of closely associated, potentially interlinked basins, each less than one ha, with a maximum depth less than 30 cm. Greyvenstein Pan is equally small and seldom exceeds 50 cm in depth. The former seldom holds rainwater for more than a month. Over one rain season (1986/1987) the average inundation ($n$ = 8) was 9.9 days (S.D. 5.7 days). The latter is regularly inundated for a number of months.

Table 1 summarises the range of conditions found in these pans over a period of approximately two years between 1985 and 1987. The maximum turbidity and conductivity values reflect the situation of the last sampleable water before final drying out. At least in Bainsvlei Pan, this water still contained very active *T. granarius* individuals.

*Table 1.* Size and range of conditions measured in two temporary waters containing *T. granarius* in the western Orange Free State.

| | Bainsvlei | Greyvenstein |
|---|---|---|
| Size (ha) | < 1 ha | < 1 ha |
| Temp. (°C) | 16–36 | 4–28 |
| Conductivity (μS/cm) | 11–911 | 28–668 |
| Turbidity (NTU) | 140–80000 | 96–2500 |
| pH | 6.8–8.9 | 6.4–8.4 |

The temperature range in Bainsvlei Pan reflects the summer condition when the rains occur, while Greyvenstein Pan holds water well into the winter, as is reflected by a minimum of 4 °C. Obviously such shallow water bodies have great diurnal ranges which also fall within the ranges given.

The maximum pH values occur with photosynthetic activity by algae some time after initial filling.

No fish occurred in these or indeed any of the natural temporary pans in the western part of southern Africa.

## Materials and methods

*T. granarius* were collected regularly during inundations. A composite water sample of 5 l was made up of numerous scoops, using a 350 ml cup attached to a 1 m shaft. Samples were concentrated by means of a 112 μm net and preserved in 4 percent formaldehyde.

Laboratory stock cultures and cultures for the determination of growth and survival were kept in standard aerated aquaria 500 × 300 × 150 mm. All cultures were fed on commercial tropical fish flakes and all studies were carried out at 18 to 24 °C.

For the determination of egg production transparent plastic containers 160 × 100 mm filled to 35 mm were used. The floor was covered with a piece of under-gravel aquarium filter which allowed eggs to fall through the slits in its surface while preventing the adults from reaching the eggs to eat or otherwise disturb them.

Hatching was examined by keeping the eggs in white plastic domestic ice trays with hollows, each holding less than 5 mm of dechlorinated tap water.

## Results

### *Survival*

During the short inundations in Bainsvlei Pan *T. granarius* was common and community diversity was low (Fig. 2). The first cohort of *T. granarius* invariably died out through desiccation and not from senescence or other natural

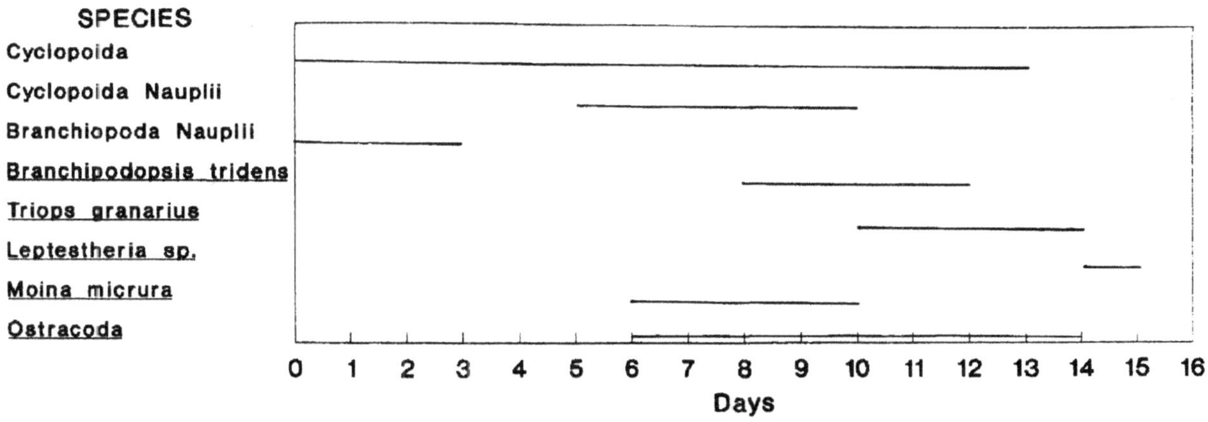


*Fig. 2.* Community composition and succession during a typical short inundation in Bainsvlei Pan.

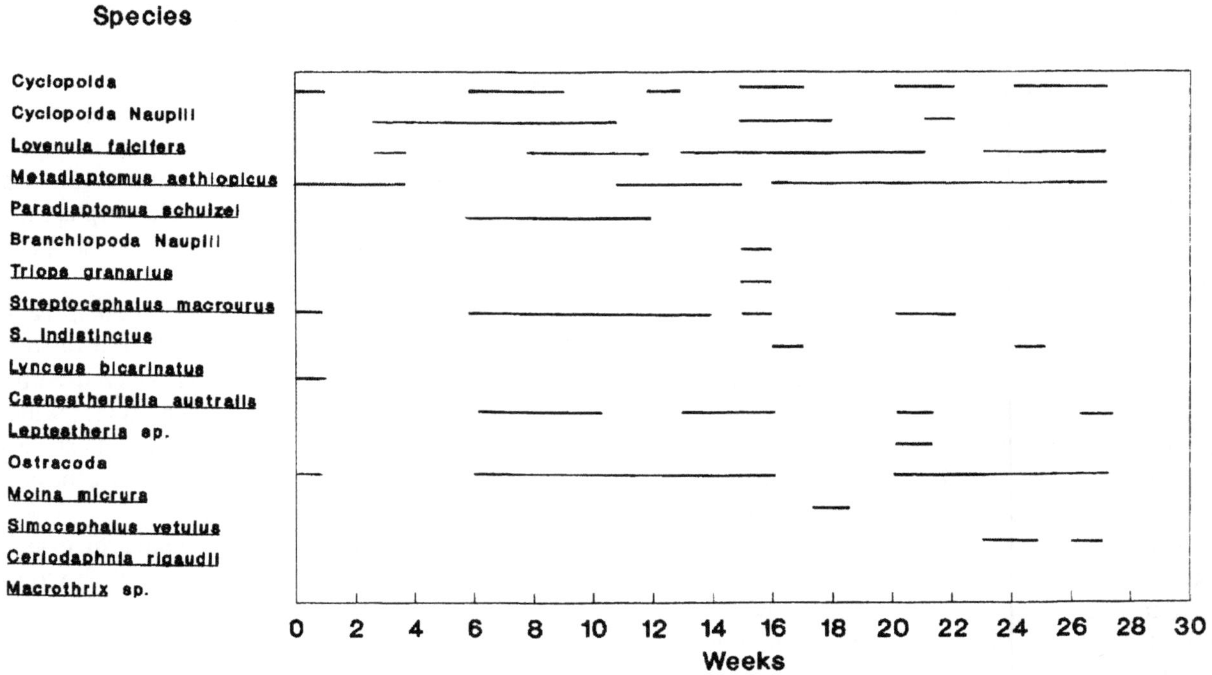


*Fig. 3.* Community composition and succession during a typical long inundation in Greyvenstein Pan.

causes. In the last stages of drying, population density became very high but individual activity at this time was apparently unimpaired.

During the long inundations in Greyvenstein Pan *T. granarius* was less common. Community diversity was high and the chance of its presence in samples was low even though it probably was present in the water body. More than one cohort probably occurred, separated by a period of weeks (Fig. 3). High population densities under extreme conditions do not occur here.

Survival in culture barely reached 30 days under the best conditions (Fig. 4). Density had no apparent influence on survival.

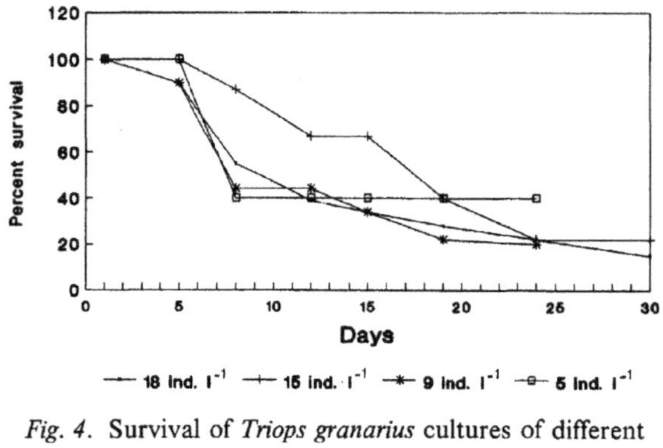


*Fig. 4.* Survival of *Triops granarius* cultures of different density.

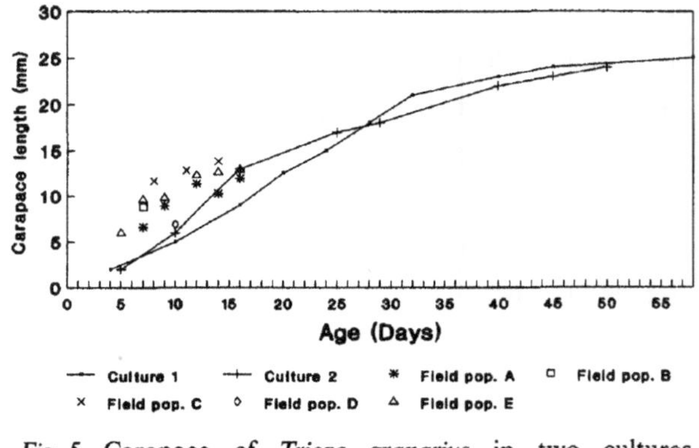


*Fig. 5.* Carapace of *Triops granarius* in two cultures (4ind./20l) and five field populations in Bainsvlei Pan.

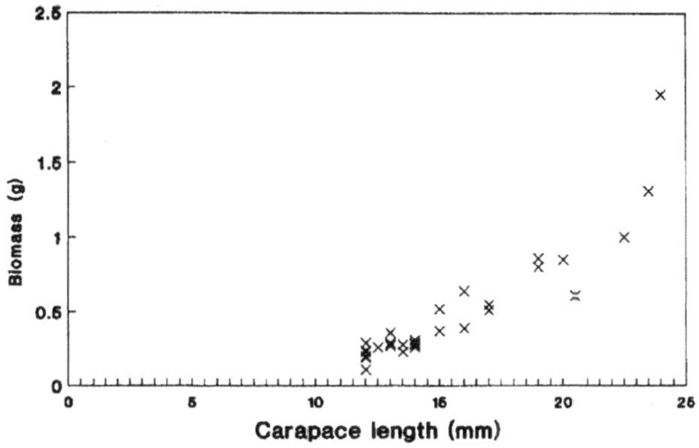


*Fig. 6.* Individual wet biomass relative to carapace length of *Triops granarius* in Bainsvlei Pan.

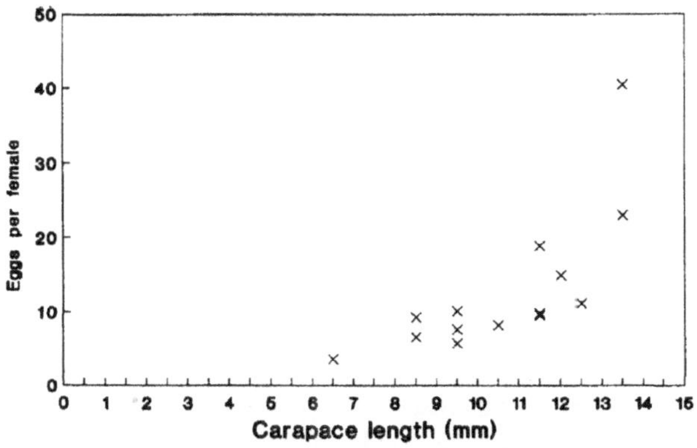


*Fig. 8.* Brood size in relation to carapace length in *Triops granarius* in Bainsvlei Pan.

*Growth*

Growth in the field was more rapid than in the laboratory but maximum size was similar, although seldom reached. No significant difference in male and female growth rates was found. At birth individuals were 650 $\mu$m long, reached 900 $\mu$m by 12 h and 1100 $\mu$m by 16 h. At five days under natural conditions, carapace length reached 6 mm (Fig. 5). In most life spans, 15 mm was not exceeded. The largest individual measured in a field sample had a carapace length of 26 mm, which was similar to the maximum carapace length of 25 mm, measured at 55 days in the laboratory.

Maximum individual biomass exceeded 2.0 g, but was generally between 0.5 and 1.0 g. At a carapace length of 10 mm which is the size at about 10 days of inundation biomass was only about 0.25 g (Fig. 6).

Density affected growth at densities above 5 individuals/l in laboratory cultures (Fig. 7), but field densities did not reach this level except during the last stages of drying.

*Reproduction, fertility and hatching*

The first eggs were borne by females at a carapace length of 6.5 mm (Fig. 8) when about 6 days old. At 10 days (the mean inundation time in

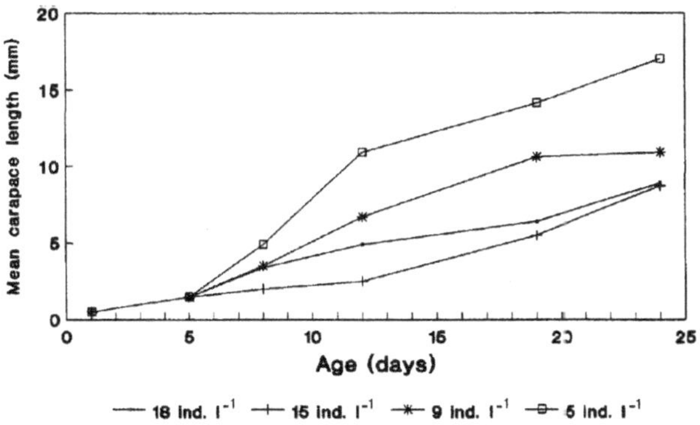


*Fig. 7.* The influence of populations density on growth in *Triops granarius* cultures.

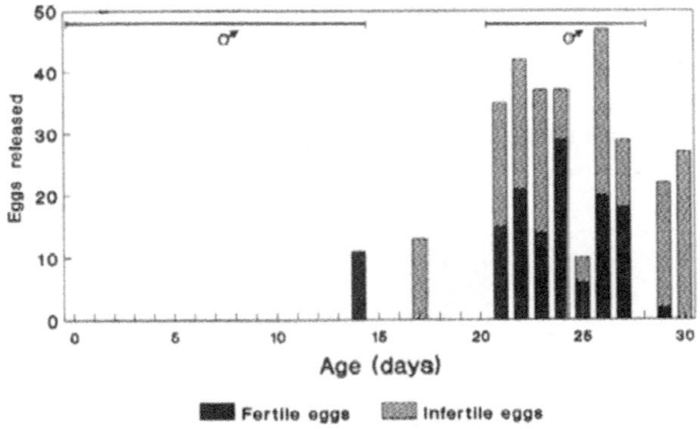


*Fig. 9.* Egg production by a single female *Triops granarius* and the fertility of the eggs, with and without the presence of a male.

Bainsvlei) the brood size was 10 eggs per female, while the maximum brood size at 14 days exceeded 40. The two brood pouches held almost equal numbers of eggs. Eggs were laid randomly and not in clusters although all the eggs in both brood pouches were released simultaneously, whereupon they sank to the bottom and adhered to a surface to which they remained attached. They remained sticky for up to three hours which made them difficult to handle.

Successive broods were produced and released at intervals of 12 to 48 hours, without a fixed pattern. Consequently, one brood and occasionly two were produced daily, as on days twenty-two and twenty-six in Fig. 9, which indicates that eggs were borne in the brood pouches for one day or less. On many days, no eggs were produced.

The presence of a male stimulated egg production and fertile eggs were produced only when a male was present.

The fertility of broods (not subjected to desiccation but incubated separate from the adults) was found to be highly variable even when a male had been present. Brood size or female age did not seem to affect fertility.

Absolutely no hatching occurred in the dark, but even in light hatching success did not exceed 50 percent and the time taken to hatch varied from less than one day to weeks, but usually occurred within three days after rehydration.

Desiccation of eggs was not necessary to ensure hatching, but in some cases they took more than 20 days to hatch if they were not desiccated. Furthermore, period in water prior to desiccation and period desiccated had no influence on the hatching success of the eggs (Table 2).

*Table 2.* Number of eggs out of an original 10 that hatched after a certain number of days in water prior to desiccation and days desiccated. Incubation period in water was 14 days.

| | | Period in water prior to desiccation (days) | | | |
|---|---|---|---|---|---|
| | | 1 | 2 | 7 | 16 |
| Period of desiccation (days) | 1 | 3 | 2 | 2 | 2 |
| | 2 | 2 | 2 | 3 | 0 |
| | 4 | 3 | 2 | 2 | 3 |
| | 8 | 3 | 3 | 3 | 2 |
| | 16 | 1 | 2 | 5 | 2 |

The pre-nauplius, enclosed in an embryonal membrane, hatched from the outer egg shell and remained in this condition for two to five hours, moving continually within the membrane and rotating along its long axis. Shortly before emergence movement increased sharply, the membrane broke and the nauplius swam away.

## Discussion

This study has concentrated more on aspects of the life history than have most others. Although *T. granarius*, the subject of our study, has a life history almost identical to that of most other *Triops* spp. and similar to that of *Lepidurus* spp., anostrocans and conchostracans, there are specific differences in detail at which we must look and upon which we will comment.

According to William's (1985) scheme which is a modification of an earlier one by Wiggins *et al.* (1980), there are four categories of organisms living in temporary waters. Three contain vagile organisms, while *Triops* spp. and similar phyllopods belong to the category in which the organisms are restricted to the temporary habitat where they must survive from one inundation to the next, sometimes for a period of years when rainfall fails. Conversely when rainfall does occur best use must be made of the favourable conditions to reproduce while at the same time not risking losing the entire population if the inundation is too short to allow reproduction. So the animals must breed when there is water, but they must hold some reproductive material in reserve in case that effort is thwarted.

Life history strategies are made up of a number of key complementary traits (Stearns, 1976), namely brood size, size of young, the age distribution of reproductive effort, the interaction of reproductive effort with adult mortality and the variation in these traits among an individual's progeny. The theoretical problem to predict which combination of traits will evolve in organisms living in specified circumstances. The extreme opportunist example of a combination of these traits would, according to the model, be a r-strategist and the opposite, an organism living in very stable conditions at carrying capacity, would be a K-strategist (Pianka, 1970). *T. granarius* uses a different combination of traits and its strategy is less easily defined.

Survival during an inundation depends on the initial absence of predators, which allows the organisms to reach maturity. The greatest mortality is among adults, so reproduction is early. After first reproduction, if conditions remain favourable, the organism grows and continues to reproduce (i.e. iteroparity). Each successive batch of eggs beyond the critical initial number necessary for maintenance of the population is a bonus.

Selection has not favoured survival beyond 30 days probably because there is increasing community diversity and therefore greater predation and a decreasing supply of suitable resources such as the organic detritus available in a newly inundated pool.

*T. granarius* is not the classical opportunist or r-strategist. Although it grows rapidly and reproduces early, it does not have semelparity and its young do not repeat the cycle immediately in order to utilise the windfall conditions while they are still suitable. Rather, it is a conservative organism, laying out a specific investment of a certain proportion of the available egg store which hatches to grow rapidly while aggressively utilising the available food, reproduce and so replenish the store of eggs.

In some respects it might be more like Romanovsky's (1985) 'violents' or 'competitors' which he described as large bodied, rapid growing species characterised by high susceptibility of juveniles to reduced food supply, high resistance of adult instars to starvation and rather high biotic potentials ($r_{max}$). They can dominate or outcompete their competitors in enriched waters.

The species is indeed a pioneer as described by Takahashi & Gohda (1981) but it is not replaced by other species through competition and succession in a Markovian series as might be the case with plants and some other animals (Price *et al.*, 1984).

The strategy is refined in subtle ways, particular to the species. These include the length of time that the eggs can survive desiccation, the extent to which eggs can hatch should the pool remain inundated, the possible presence of more than one type of egg or a range thereof, the precise hatching process, period of survival of the reproductive adult, age at first reproduction, number of eggs produced in a lifetime, optimal density of individuals, optimal temperature during inundation and temperature extremes that the eggs in particular can withstand.

It is possible to comment on some of these in the light of our findings while the rest remain to be answered. We do not know how long the eggs will survive desiccation, although some certainly must survive at least one dry season.

Eggs can certainly hatch without going through the process of desiccation, which is a confirmation of the findings on other populations (Longhurst, 1956). Longhurst also raised an idea of Mathias' that *Triops* has two types of eggs, one drought resistant, the other not, but this seems unlikely because there is no apparent external morphological difference between eggs, unlike *Artemia* eggs for example. There may be merely a variability in physiology of the eggs which makes some more likely to hatch without desiccation and others to be more drought resistant. A variability among eggs is noticed in the very different periods needed for those that will hatch to do so. The proportion that does hatch seldom exceeds 50 percent, which is low compared to the 78.33 percent registered for *T. longicaudatus* by Scott & Grigarick (1979).

The hatching process certainly is strange that allows a hatching individual to be 'half-hatched' and vulnerable for up to five hours in a delicate membrane in which it increases its visibility by moving continually. Aptly, when many of these eggs are hatching, soon after inundation, there are few predators present.

Adult survival extends to about 30 days, after which there is a sharp die-off, much as Takahashi and Gohda (1981) found for *T. cancriformis*.

The age at first reproduction (6 days) in *T. granarius* was earlier than the 12–19 days in

*T. cancriformis* (Takahashi & Gohda, 1981), even though the latter is a smaller species. In a situation such as that in Bainsvlei Pan, with a mean inundation period of 9.9 days, it seems that *T. granarius* would be the more successful competitor.

The number of eggs is not particularly large, there seldom being more than 50 in a brood, produced at about a brood per day over a period from 6 days of age to 30 or more days, which resembles the rate shown for *T. cancriformis* by Takahashi & Gohda (1981).

It is not possible to show optimal density in the field situation because, as the pool dries up, the density changes markedly. In culture a density of less than 5 individuals $l^{-1}$ seemed optimal. Above this density there was impaired growth. However, cultures stabilised at about 2 individuals $l^{-1}$, which may be a better indication of the optimal density. The factor of cannibalism and the avoidance of being eaten is relevant here.

Optimal temperature and the ability of the organism to withstand high temperatures as adults and both high and low temperatures as eggs deserve attention. *T. granarius* in the present study inhabits an area of summer inundations, but it is also found in the Mediterranean part of north Africa and adjacent winter rainfall regions, where the temperature regime will have a different influence on life histories.

Should this organism be considered for aquaculture, the following observations may be relevant. On the negative side, it has a generally poor and unpredictable hatching rate, small brood size, depression of growth at a relatively low density, a high rate of cannibalism which will be aggravated at high densities, and it is not an 'easy' subject for culture notwithstanding its ability to withstand extreme physical and chemical conditions in nature. It also has a high chitin to biomass ratio.

On the positive side it has a large maximum biomass, it grows rapidly to adulthood and it can withstand the broad range of conditions just mentioned.

## Acknowledgements

The Foundation for Research Development of the South African Council for Scientific and Industrial Research and the University of the Orange Free State provided funds and facilities for this study.

## References

Barnard, K. H., 1929. Contributions to the crustacean fauna of South Africa. No. 10. A revision of the South African Branchiopoda (Phyllopoda). Ann. S. Afr. Mus. 29: 181–270.

Fryer, G., 1988. Studies on the functional morphology and biology of the Notostraca (Crustacea: Branchiopoda). Phil. Trans. r. Soc. Lond. B 321: 27–124.

Goudie, A. S. & D. S. G. Thomas, 1985. Pans in southern Africa with special reference to South Africa and Zimbabwe. Z. Geomorph. N.F. 29: 1–19.

Le Roux, J. S., 1978. The origin and distribution of pans in the Orange Free State. S. afr. Geogr. 6: 167–176.

Longhurst, A. R., 1956. A review of the Notostraca. Bull. Br. Mus. nat. Hist. Zool. 3: 1–57.

Pianka, E. R., 1970. On r- and K-selection. Am. Nat. 104: 592–597.

Price, P. W., C. N. Slobodchikoff & W. S. Gaud, 1984. A new ecology. Wiley-Interscience, N.Y.

Rayner, N. A. & A. E. Bowland, 1985. Notes on the taxonomy and ecology of *Triops granarius* (Lucas) (Notostraca: Crustacea) in South Africa. S. afr. J. Sci. 81: 500–505.

Romanovsky, Y. E., 1985. Food limitation and life-history strategies in cladoceran crustaceans. Arch. Hydrobiol. Beih. 21: 363–372.

Scott, S. R. & A. A. Grigarick, 1979. Laboratory studies of factors affecting egg hatch of *Triops longicaudatus* (Leconte) (Notostraca: Triopsidae). Hydrobiologia 63: 145–152.

Seaman, M. T. & D. J. Kok, 1987. Ecological diversity in Orange Free State pans. In R. D. Walmsley & L. Botten (eds), Symposium on the ecology and conservation of wetlands in South Africa, Foundation for Research Development, CSIR, Pretoria, Occasional Report Series 28: 260–273.

Takahashi, F. & M. Gohda, 1981. The type of egg production curve with respect to population preservation: ecological studies of tadpole shrimps (Branchiopoda: Triopsidae) for the biological control of weeds in rice paddies. Verh. Int. Ver. Limnol. 21: 1531–1540.

Tyson, P. D., 1986. Climatic change and variability in South Africa. Oxford University Press, Cape Town.

Williams, W. D., 1985. Biotic adaptations in temporary lentic waters, with special reference to those in arid and semiarid regions. Hydrobiologia 125: 85–110.

*Hydrobiologia* **212**: 95–104, 1991.
*D. Belk, H. J. Dumont & N. Munuswamy (eds), Studies on Large Branchiopod Biology and Aquaculture.*



# Physical and chemical characteristics and phyllopod fauna of temporary pools in north-eastern Natal, Republic of South Africa

M.L. Hamer & C.C. Appleton
*Department of Zoology and Entomology, University of Natal, P.O. Box 375, Pietermaritzburg, Natal, 3200 South Africa*

*Key words:* South Africa, temporary pools, physico-chemistry, phyllopods

## Abstract

Fourteen phyllopod (Branchiopoda: Crustacea) species were collected from numerous temporary pools in north-eastern Natal, South Africa. Physical and chemical data for 10 pools, and results for hourly (over 41 hours) and 3-day interval (throughout periods when phyllopods were present) monitoring of 3 study pools are presented. Comparisons were made with other pools with phyllopods in Africa. The data reveal the broad tolerance to variation and extremes in both physical and chemical conditions in African temporary waterbodies. Species diversity appears to be related to pool size and vegetation. Phyllopod distribution does not follow a pattern associated with physical and chemical conditions.

## Introduction

A variety of temporary freshwater habitats exists in Africa with variation both between and within regions. A general description is therefore impossible. Wiggins *et al.* (1980), however, defined temporary pools, in their typical form as 'an accumulation of surface water in an isolated basin that at no time during the year has either an outlet or an inlet, and [in which] water is entirely absent from the surface for part of the year'. This definition is appropriate to the habitats involved in the present study. Relatively little information is available on the physical features of pools with phyllopods in Africa but it shows nonetheless that these habitats vary considerably in area, depth and vegetation. Thiéry's (1987) study is particularly comprehensive in this regard.

Hutchinson *et al.* (1932) described a series of 'pans' in the southern Transvaal (South Africa) which ranged from 'mud pans', small and lacking vegetation, to lake-sized, heavily vegetated 'pans'. Seaman & Kok (1987) examined pools in the Orange Free State ranging in size from less than one to over 50 ha. As in other phyllopod habitats, these pools usually had an average maximum depth of less than 60 cm. The physical features of the pool (area, depth and substratum) and the climate (rainfall, temperature and evaporation rate) are associated with both the number of inundations occurring within a year and their duration. Up to three aquatic phases may occur per year and each may last for as little as a few days as recorded by Weir (1969) in Wankie (= Hwange), Zimbabwe to as long as 2.5 months in Morocco (Thiéry, 1987) and 6 months in the Orange Free State (Seaman & Kok, 1987). Apart from this physical variation, monitoring of water temperature and chemistry in several African, phyllopod-inhabited, temporary water-bodies (Thiéry, 1987; Seaman & Kok, 1987; Rzoska, 1961; Weir, 1969; Hildrew, 1985) has revealed a wide range in these variables.

The crustacean subclass Branchiopoda includes four orders, three of which are restricted to temporary freshwater habitats. Collectively this group is known as the 'phyllopods', a term regarded as of no taxonomic value but useful when referring to the Notostraca, Anostraca and the Conchostraca (Fryer, 1987). In 1929, Barnard recorded 46 phyllopod species from southern Africa, and of these only two were from Natal: *Streptocephalus cafer* (Loven) was collected near Mahlabatini in Zululand, and *Branchipodopsis natalensis* Barnard from Van Reenen on the border between Natal and the Orange Free State. In 1985, *S. cafer*, the notostracan *Triops granarius* (Lucas), and the conchostracan *Lynceus truncatus* Barnard were collected from an animal wallow in the Umfolozi Game Reserve in Zululand (Rayner & Bowland, 1985). There are no other published records of phyllopods in Natal.

During 1986 and 1987, many ephemeral pools in north-eastern Natal (Zululand) were sampled for phyllopods and certain physical and chemical features of the habitats were recorded to give an indication of conditions tolerated by phyllopods of the region, and to provide a comparison with similar habitats elsewhere in Africa. This paper presents the physical and chemical data and species lists for temporary freshwater habitats within the study area. Data on faunal succession in the pools will be published separately.

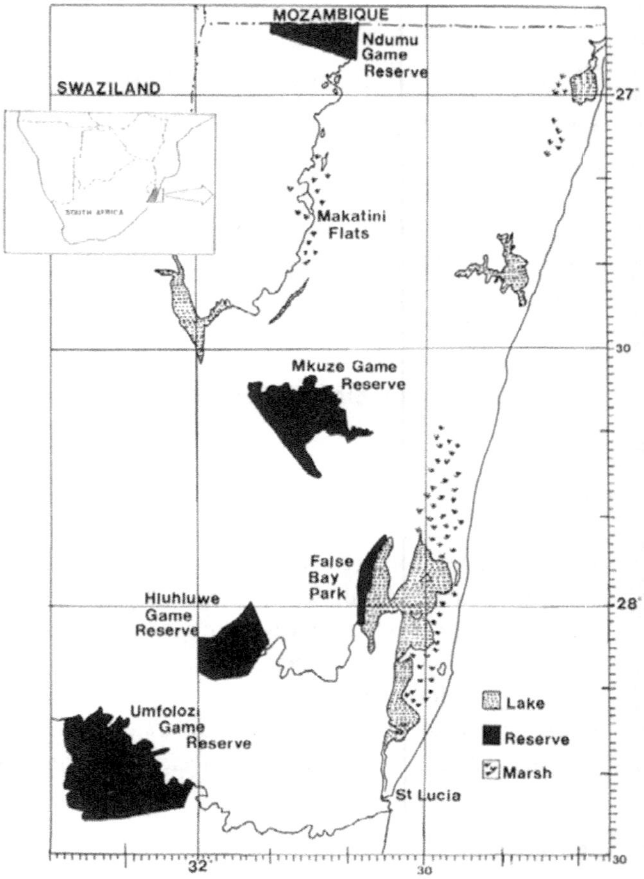


*Fig. 1.* Map of the north-eastern Natal region, its situation in southern Africa (inset) and the localities where phyllopod-harbouring pools were examined.

## Materials and methods

Temporary pools in Ndumu, Umfolozi and Mkuze Game Reserves, False Bay Park and the Makatini Flats (Fig. 1) were sampled in December 1986 and January 1987, two to three weeks after filling. Hand dip-nets (15 × 20 cm and 10 × 12 cm) were used to sweep all areas of the pools. Animals were preserved in 70% ethanol. Barnard's (1929) monograph was used to identify phyllopods and Professor R. Hartland-Rowe (University of Botswana) confirmed the *Streptocephalus* identifications.

Temperature, dissolved oxygen concentration, pH and conductivity were measured in the field. Sediment samples, collected from the dried pools in winter, were analysed by the Department of Agriculture and Water Supply at Cedara. During monitoring of 3 pools on the Makatini Flats at 3-day intervals, measurements were taken at both the edge and centre. Temperature and dissolved oxygen were recorded at the surface (at a depth of less than 10 cm) and approximately 5 cm above the substratum at the centre. Hourly changes in temperature, pH and dissolved oxygen concentration were monitored over a 41-hour period in December 1987. Temperatures were measured at different positions in a Makatini pool using an MCS 120 Multichannel Datalogger and a multipoint digital thermometer.

## Results

### *Phyllopod distribution and diversity in north-eastern Natal*

Figure 1 shows the localities in north-eastern Natal at which phyllopods were collected and Table 1 lists species from pools for which physical-chemical data are given in Table 2. A total of 14 species was recorded, four apparently new for South Africa and not yet named.

### *Description of temporary pools*

Water-bodies with phyllopods ranged from small hoof-print-type depressions to large (24 000 $m^2$) pools. All pools examined in the study area, were temporary, and filled during the summer rainfall season, but the time for which they held water varied from 3 days to 5 months. Vegetation, rooted and floating in the pool basin, was completely absent from pools such as the rhino-wallows common in Umfolozi Game Reserve, while others, such as Makatini 2 (Table 2) had dense masses of floating *Ludwigia stolonifera* as well as various emergent *Cyperus* spp. at the periphery.

### *Rainfall, depth and duration of aquatic phase*

The relationship between rainfall, depth and duration of the aquatic phase was examined in the Makatini pools. These first filled after the winter dry phase on August 29, 1987, following 48 mm of rain. This first inundation was, however, short and pools 1 and 2 had dried up by September 15. They refilled on 28 and 29 September after 95 mm of rain fell over the two days. Thereafter, the main areas of the pools held water until after the end of 12-week sampling period in January 1988 and by this stage all phyllopod populations had disappeared. Intermittent rain (Fig. 2) prevented large changes in area and depth and only the small, hoofprint-type pools and peripheral shallow regions of larger ones dried out after the second inundation.

Sediment analysis of the three Makatini pools showed that while pool 1 was largely clay/silt,

*Table 1.* List of phyllopod species collected from ten different pools in north-eastern Natal (see Fig. 1).

| Locality | Makatini | | | Mkuze | | Ndumu | Umfolozi | | False bay | |
|---|---|---|---|---|---|---|---|---|---|---|
| Pool | 1 | 2 | 3 | 1 | 2 | 1 | 1 | 2 | 1 | 2 |
| Notostraca | | | | | | | | | | |
| Triops granarius | × | × | | × | | × | × | × | | × |
| Anostraca | | | | | | | | | | |
| Streptocephalus cafer | × | × | × | × | × | × | × | × | × | × |
| Streptocephalus macrourus | × | × | × | | × | | | | | |
| Streptocephalus cladophorus | × | × | | × | | | | | | |
| Streptocephalus sp A | × | | | | | | | | | × |
| Streptocephalus sp B | | × | | | | | | | | |
| Branchipodopsis wolfi | | | | | | | × | × | | |
| Branchipodopsis sp. | | × | | | | | | | | |
| Conchostraca | | | | | | | | | | |
| Leptestheria rubidgei | × | × | | × | | × | × | × | × | |
| Eulimnadia africana | × | × | | | | | | | | × |
| Caenestheria australis | | × | × | | | | | | | × |
| Lynceus bicarinatus | | | | | | | | | × | |
| Lynceus truncatus | | | | | | | | × | | |
| Lynceus sp. | × | × | | | | | | | | |

*Table 2(a).* Physical and biological features of 10 ephemeral pools in north-eastern Natal.

| Locality | Approximate area (m) | Average depth (cm) | Dominant vegetation | ± Vegetation cover (%) | Number of phyllopod species |
|---|---|---|---|---|---|
| Makatini 1 | 1600 | 40 | Sedges, *Nymphea* grasses, reeds | 50 | 9 |
| Makatini 2 | 4200 | 50 | Sedges, *Ludwigia* grasses | 80 | 9 |
| Makatini 3 | 600 | 45 | Unvegetated | 0 | 3 |
| Mkuze 1 | 120 | 40 | Small amount grass | 8 | 4 |
| Mkuze 2 | 96 | 20 | Grass at edges | 5 | 2 |
| Ndumu 1 | 48 | 12 | Grass at edges | 10 | 3 |
| Umfolozi 1 | 600 | 25 | Grass and sedges at edges | 10 | 4 |
| Umfolozi 2 | 1200 | 40 | Unvegetated | 0 | 5 |
| False Bay 1 | 100 | 20 | Sedges | 60 | 3 |
| False Bay 2 | 72 | 12 | Grass | 10 | 4 |

*Table 2(b).* Chemical features of 10 ephemeral pools in north-eastern Natal.

| Locality | Sediment Analysis | | | | | Conductivity | pH | D.O. | Turbidity |
|---|---|---|---|---|---|---|---|---|---|
| | % Sand | % Silt | % Clay | Texture | E.D.M. (%)* | μs cm$^{-1}$ | | mg l$^{-1}$ | NTU* |
| Makatini 1 | 34 | 20 | 46 | Clay | 3.35 | 160 | 6.2 | 5.8 | 280 |
| Makatini 1 | 34 | 20 | 46 | Clay | 3.35 | 160 | 6.2 | 5.8 | 280 |
| Makatini 2 | 82 | 5 | 13 | Loamy-sand | 2.15 | 130 | 6.3 | 6.3 | 470 |
| Makatini 3 | 83 | 3 | 14 | Loamy-sand | 0.14 | 205 | 7.4 | 8.0 | 520 |
| Mkuze 1 | 53 | 17 | 30 | Sandy, clay loam | 1.72 | 160 | 6.0 | 6.5 | 260 |
| Mkuze 2 | – | – | – | – | – | 76 | 5.8 | 5.6 | 280 |
| Ndumu 1 | 36 | 20 | 48 | – | – | 80 | 7.4 | – | 230 |
| Umfolozi 1 | 19 | 20 | 61 | Ciay | 1.39 | 360 | 5.8 | – | – |
| Umfolozi 2 | 40 | 22 | 38 | – | – | 280 | 6.0 | – | 320 |
| False Bay 1 | 44 | 14 | 42 | Clay | – | 320 | 6.9 | 8.1 | 140 |
| False Bay 2 | 48 | 7 | 45 | Clay | 6.30 | 325 | 6.4 | 5.6 | – |

Pool 3 had a substratum with a high sand and low clay and organic matter content (Table 2). Since clay and organic matter retain water more effectively than sand, it is strange that Pool 3 held water over the period in which Pools 1 and 2 dried up. Sediments of pools in other areas were generally dominated by clays and silts though the single example from a coastal forest area in False Bay Park yielded a particularly high organic content. In Moroccan pools, Thiéry (1987) found that the sediments were typically fine-grained clays and silts (up to 20 μm diameter) with a smaller fraction of coarser sands in the 20–200 μm range.

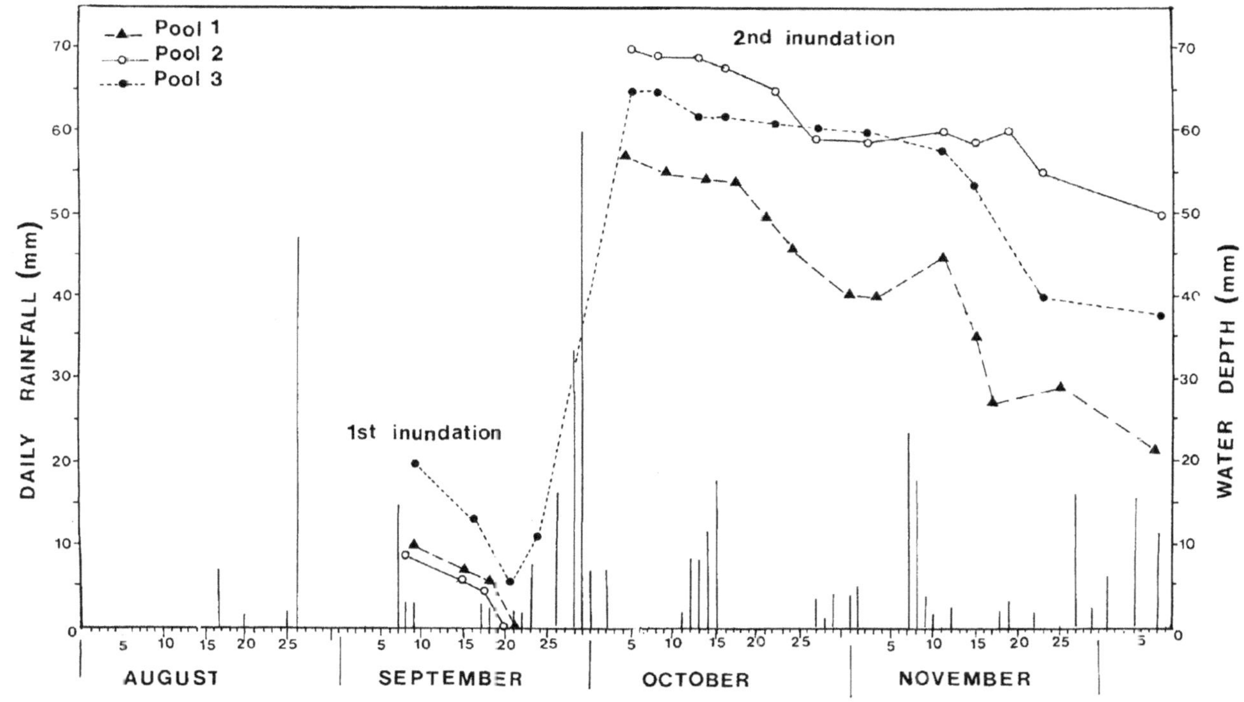


*Fig. 2.* Daily rainfall and water depth in the study pans during the sampling period.

*Physical-chemical monitoring pH*

The pH of the pools in Zululand fell within a narrow range (5.79–7.4) as did values over the study period in the Makatini pools viz; 5.35–7.25 in two pools and 5.95–8.86 in the third (Table 3). No noticeable trends in fluctuations, at either hourly or 3-day intervals, were evident.

*Conductivity*

Conductivity in the Zululand pools was always below 360 $\mu s\ cm^{-1}$ with a minimum value of 76 $\mu s\ cm^{-1}$ in a pool at Mkuze (Table 2). In the Makatini pools, conductivity increased over the study period from between 64 and 102 $\mu s\ cm^{-1}$ to 178 to 302 $\mu s\ cm^{-1}$ (Table 3).

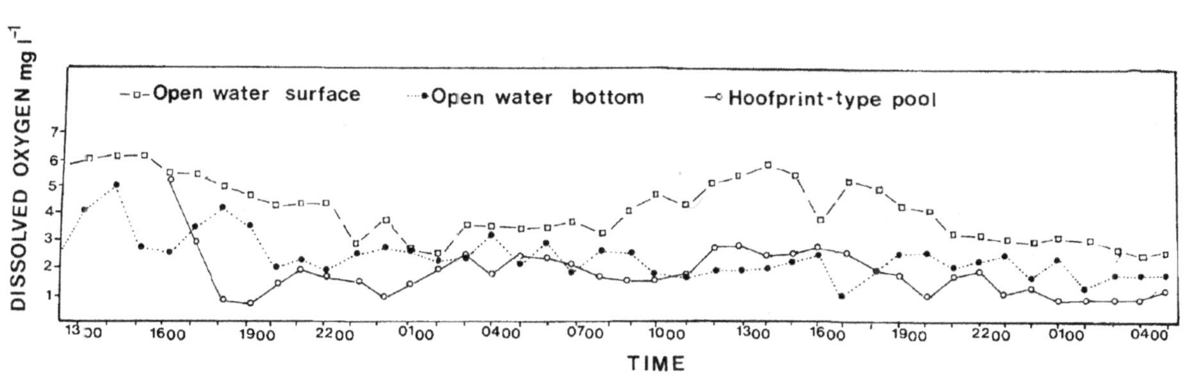


*Fig. 3.* Hourly changes in dissolved oxygen concentration at various positions in Makatini pool 1 over a 41-hour period on 8 and 9 December, 1989.

*Table 3.* Physico-chemical variation in the pools of the Makatini Flats between September 1987 and January 1988.

| | | | Pool 1 | | Pool 2 | | Pool 3 | |
|---|---|---|---|---|---|---|---|---|
| | | | Edge | Center | Edge | Center | Edge | Center |
| Conductivity | | min | 88.7 | 98.0 | 64.5 | 78.0 | 95.0 | 112.0 |
| ($\mu$s/cm$^{-1}$) | | max | 178.0 | 244.0 | 188.0 | 185.0 | 302.0 | 298.0 |
| pH | | min | 5.85 | 5.65 | 5.35 | 5.10 | 5.95 | 6.15 |
| | | max | 6.87 | 7.25 | 7.25 | 7.19 | 8.85 | 8.86 |
| Temperature | S | min | 19.5 | 19.0 | 19.5 | 19.0 | 21.0 | 21.0 |
| (°C) | | max | 37.0 | 35.0 | 39.0 | 39.0 | 35.5 | 35.0 |
| | B | min | – | 19.0 | – | 17.0 | – | 18.0 |
| | | max | – | 29.0 | – | 26.0 | – | 31.0 |
| Dissolved | S | min | 1.5 | 2.8 | 2.2 | 2.1 | 4.0 | 4.2 |
| oxygen | | max | 8.6 | 7.2 | 7.3 | 7.2 | 9.0 | 9.7 |
| (mg l$^{-1}$) | B | min | – | 1.7 | – | 1.3 | – | 4.1 |
| | | max | – | 6.8 | – | 6.8 | – | 8.5 |
| Area | | max | 60 × 80 | | 51 × 68 | | 20 × 34 | |
| (m$^3$) | | min | 48 × 52 | | 42 × 60 | | 18 × 26 | |
| Depth at | | max | 1.0 | | 1.30 | | 1.0 | |
| deepest point (m) | | min | 0.20 | | 0.53 | | 0.40 | |

*Dissolved oxygen concentration*

Levels of dissolved oxygen were generally low (5–6 mg l$^{-1}$). At Makatini, regular monitoring usually showed low levels (4–6 mg l$^{-1}$) but a minimum and maximum of 0.3 and 8.6 mg l$^{-1}$ respectively were reached (Table 3). A feature here was the difference in oxygen levels in the centre of the pools, between the surface water and just above the substratum, and between vegetated and open areas. This was also evident during hourly monitoring, particularly in the daytime when differences of up to 4.00 mg l$^{-1}$ were recorded between surface and bottom and 2.00 mg l$^{-1}$ between vegetated and open areas (Fig. 3). Even during windy periods equilibrium was not reached.

*Water temperature*

Water temperature in the Makatini pools was variable, and related to the time of day, weather and position in the pool. Hourly readings at various levels and positions in the pools showed that temperature differences between surface and substratum were as high as 9–11 °C during the day, but reached equilibrium at night. Temperatures were more extreme in small, shallow hoofprint-type habitats, and in one, a change of 17 °C (a decrease from 41 to 24 °C) occurred over four hours. This represents a cooling rate of 4.3 °C h$^{-1}$ (Fig. 4). Cooling rates of 1.3–3.6 °C h$^{-1}$ were measured in other pools. On one occasion a temperature of 42 °C was maintained for 3 h and 15 min (13h15–15h30) in one of the small pools and this probably resulted in the disappearance of a population of 6–7-day-old anostracans and conchostracans during this period. This was the only instance in which phyllopod mortality appeared to coincide with stressful physico-chemical conditions.

*Water turbidity*

Turbidities of the sample supernatant from a number of pools in north-eastern Natal were all relatively high and ranged from 140 Nephelometric Turbidity Units (NTU) to 520 NTU.

## Discussion

*Physical-chemical factors*

Freshwater temporary pools have been recognised as stressful habitats and recent reviews of such systems (Belk & Cole, 1980; Hartland-Rowe, 1972; Williams, 1985) have emphasised the need for organisms living in them to be adapted to extremes of physical-chemical factors. Indeed, the phyllopods of north-eastern Natal were exposed to wide fluctuations in, and extremes of temperature and low dissolved oxygen concentrations. In spite of this, large populations survived and reproduced.

A phenomenon observed in some temporary pools (Barclay, 1966; Yaron, 1964; Thiéry, 1987) is that of a 'microthermocline' down the water column. This is contrary to expectation, since, as Hartland-Rowe (1972) pointed out, the shallow nature of such pools would cause even a slight breeze to result in holomixis. Although the presence of a distinct thermocline was not investigated in the present study, a steep gradient between surface and bottom temperatures persisted during windy periods and in areas where vegetation was absent. Thiéry (1987) suggested that a thermocline would only exist if the water was turbid, i.e. a Secchi depth < 5 cm, deeper than 20–25 cm and warmer than 20 °C. Since these conditions were common in the Makatini pools, it is possible that the gradient measured was in fact a 'microthermocline'.

Anostracan populations, which were concentrated in the deeper central regions of the pool (Hamer, 1989), could have avoided high water temperatures (above 35 °C) at the surface by moving down the water column but no concentrations of these animals were collected deeper than 20 cm. This may have been because dissolved oxygen levels were usually lower near the substratum (Fig. 3) and food, such as unicellular algae, was likely to be scarce deeper down in the turbid water. In addition, two anostracan species which inhabited the centre of the pools, *S. cafer* and *S. macrourus*, were noted in the laboratory to be positively phototactic. Those species (notostracans and heavy-carapaced conchostracans) which typically inhabited the shallow, peripheral regions of the pools (Hamer, 1989) where temperatures equalled or exceeded those at the surface of the centre, were predominantly benthic in habit (Fryer, 1987; Hamer, 1989) and would be able to use the substratum to gain some relief from temperatures of 30 °C and above.

Phyllopod populations which were exposed to high temperatures survived, and a number of other studies have shown that these animals are able to tolerate extreme temperatures (Mitchell, 1987; Rzoska, 1962; Sluzhevskaya-Drobysheva, 1981). Cloudsley-Thompson (1966) set the lethal temperature over a 24-hour period at 34 °C for *Streptocephalus* sp. and since, in the Makatini pools, temperatures over 35 °C were only maintained for approximately two hours, they were unlikely to cause mortality in adult populations. The effects of high temperatures on larval stages have not been well researched and may result in mortality of early life stages. The rapid heating and cooling rates (Fig. 4) measured in small pools may have caused phyllopods some degree of stress but they survived these conditions during the monitoring period.

Populations were also exposed to low oxygen concentrations (1–3 mg $l^{-1}$). Horne (1971), however, found that oxygen levels below 1 mg $l^{-1}$ were tolerated for short periods by a number of phyllopod species. Swimming at the surface (in the upper 1–2 cm of water) with the phyllopodous limbs exposed to the air–water interface has been suggested as a reaction to low

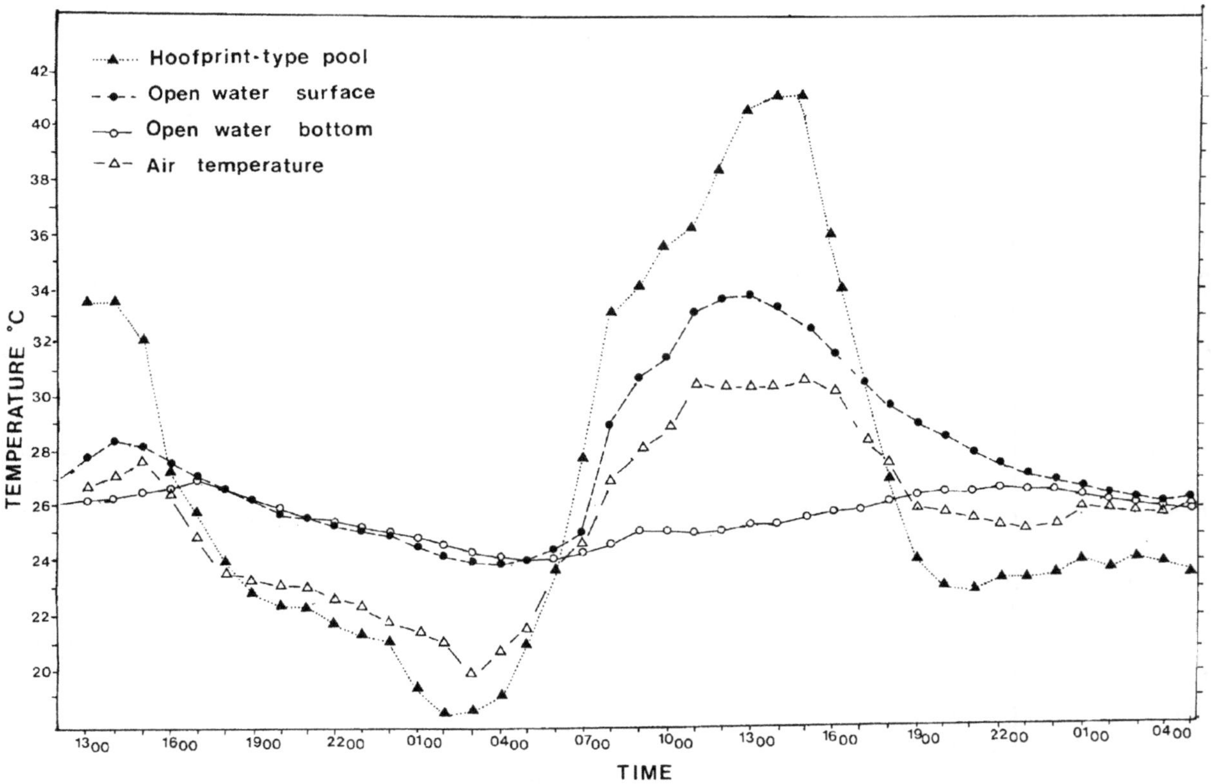


*Fig. 4.* Hourly temperature changes at various positions of Makatini pool 1 over a 41-hour period on 8 and 9 December, 1989.

oxygen concentrations (Hartland-Rowe, 1972; Williams, 1985). This behaviour was observed on a number of occasions. It is quite possible that, as Williams (1985) suggested, successful adaptations have overcome the impact of this factor and that under the circumstances, oxygen depletion may not be a stressor.

The development of respiratory pigments in response to low oxygen concentrations, observed in genera such as *Triops* and *Artemia* (Williams, 1985) was not apparent in the Zululand anostracans. Rather, intraspecific variation in colour was usually associated with water turbidity. Those animals collected in highly turbid waters, for example Makatini 3 (520 NTU), were translucent while those from clearer water (Makatini 1; 280 NTU) frequently had blue-green or violet bodies. Williams (1985) mentioned light intensity as a stress factor in temporary pools, and that pigments might act as 'quenchers' as an adaptation to high light intensities. Possibly, anostracans inhabiting pools with high levels of light penetration have developed pigments to avoid various kinds of light damage. Variation in the colours of *T. granarius* and the conchostracans *Eulimnadia africana* and *Leptestheria rubidgei* was observed but this could not be clearly related to water turbidity and respiratory pigments may have been responsible.

pH ranged between mildly acidic and mildly alkaline. Thiéry (1987) measured similar pH values in Moroccan pools. Dilution caused by intermittent rainfall kept conductivities consistently below 300 $\mu$s cm$^{-1}$ while phyllopod populations were present. However, even under drier conditions conductivities may not increase to intolerable levels. In a survey of temporary pools in Canada Hartland-Rowe (1972) found that in those in which initial conductivities were low, very high concentrations of ions were never reached. The initial measurements in the Makatini pools were less than 100 $\mu$s cm$^{-1}$, and high conductivities were therefore not expected. Conductivity in pools studied by Thiéry (1987) followed a similar trend to that in the Makatini pools, increasing with the length of the aquatic phase from <200 to approximately 800 $\mu$s cm$^{-1}$ but reaching 1400–2000 $\mu$s cm$^{-1}$ in some pools. In other parts of Africa phyllopods have been found in pools with conductivities of 1200 $\mu$s cm$^{-1}$ (Sudan) (Rzoska, 1962) and over 4000 $\mu$s cm$^{-1}$ (Orange Free State) (Seaman & Kok, 1987).

Low water transparency is a common feature of temporary pools. In Morocco for instance, Thiéry (1987) recorded Secchi disc depths of 1–7 cm and turbidity measurements ranging from 100–10000 NTU, the latter being far higher than those occurring in the Makatini pools.

*Phyllopod survival in temporary pools*

It is evident that the physical-chemical conditions prevalent in the Makatini temporary pools do not usually pose a threat to the survival of phyllopod populations and these animals, because various adaptations, are able to inhabit a wide range of ephemeral habitats. Predation, and possibly competition for food and space may present a greater threat to survival of Zululand phyllopods. This was evident in the vegetated, peripheral regions of pools where the disappearance of phyllopod populations coincided with the arrival and emergence of large numbers of adult and larval insects and tadpoles (Hamer, 1989). Habitat durations and predictability also appear to pose a greater threat to phyllopod survival than do water temperature, dissolved oxygen concentration, pH and conductivity.

*Species diversity*

The number of species inhabiting temporary pools in Zululand appears to be related to both pool area and the presence of vegetation. From Tables 2a and b it is evident that Makatini pools 1 and 2 were the largest of the 10 pools examined, had the greatest amount and most diverse vegetation, and the largest number of phyllopod species. Physico-chemical factors do not appear to influence species diversity since the range of conditions measured in the 10 Zululand pools was not wide but the number of phyllopod species inhabiting these pools was between 2 and 9. In addition, the Orange Free State pools, which had conductivities of 668–11170 $\mu$s cm$^{-1}$, high turbidities and pH values of 6.3–9.1 were inhabited by a total of 8 phyllopod species (Seaman & Kok, 1987), a number similar to that of the physico-chemically different Makatini pools 1 and 2. In temporary pools in Zimbabwe, Weir (1969) also found that the greatest faunal diversity was associated with vegetational abundance and diversity.

*Species distribution*

Data for localities other than Makatini are incomplete as they were sampled over a relatively short period and the intermittent occurrence of species and the sudden disappearance of populations (Hamer, 1989) are phenomena commonly recorded for phyllopods. However, even in the small Makatini area (12.5 km$^2$) certain species (*Branchipodopsis* sp., *Streptocephalus* sp. B and *Lynceus bicarinatus*) were restricted to one or two of the 12 pools sampled, while others, such as *S. cafer*, *S. macrourus* and *T. granarius* occurred widely. Physico-chemical factors do not appear to influence phyllopod distribution since the range of values recorded was small and there was much overlap between pools. There is little information concerning phyllopod dispersion and it is not known whether some species have a better dispersion ability than others. The presence of a species in one pool and its absence from others within, for example, a 12.5 km$^2$ area, may be the result of a chance dispersion or could be related to the movement of livestock in the area (Thiéry, 1987).

## Acknowledgements

The authors thank the C.S.I.R. for financial assistance, the Makatini Research Station for rainfall data, and the Natal Parks Board for permission to collect phyllopods in conservation areas and the assistance of its staff in the field.

## References

Barnard, K. H., 1929. A revision of the South African Branchiopoda (Phyllopoda). Ann. s. afr. Mus. 29: 181–271.

Belk, D. & G. A. Cole, 1975. Adaptational biology of desert temporary pond inhabitants. In: N. F. Hadley (ed.), Environmental Physiology of Desert Organisms. Wiley, New York. 207–226.

Cloudsley-Thompson, J. L., 1966. Orientation responses of *Triops granarius* (Lucas) (Branchiopoda: Notostraca) and *Streptocephalus* sp. (Branchiopoda: Anostraca). Hydrobiologia 27: 33–38.

Fryer, G., 1987. A new classification of the branchiopod Crustacea. Zool. J. linn. Soc. 91: 357–383.

Fryer, G., 1988. Studies on the functional morphology and biology of the Notostraca (Crustacea: Branchiopoda). Phil. Trans. r. Soc. Lond. B. 321: 27–124.

Hamer, M. L., 1989. Studies on the phyllopod fauna of ephemeral pools in north-eastern Natal. MSc. thesis, University of Natal, Pietermaritzburg, South Africa. 192 pp.

Hartland-Rowe, R., 1972. The limnology of temporary waters and the ecology of the Euphyllopoda. In R. B. Clarke and R. J. Wooton (eds), Essays in Hydrobiology: 15–31. University of Exeter, UK.

Hildrew, A. G., 1985. A quantitative study of the life history of a fairy shrimp (Branchiopoda: Anostraca) in relation to its habitat, a Kenyan rainpool. J. anim. Ecol. 54: 99–110.

Horne, F., 1967. Effects of physico-chemical factors on the distribution and occurrence of some southeastern Wyoming phyllopods. Ecology 48: 472–477.

Hutchinson, G. E., G. E. Pickford & J. F. M. Schuurman, 1932. A contribution to the hydrobiology of pans and other inland waters of South Africa. Arch. Hydrobiol. 24: 154 pp.

Mitchell, S. A., 1987. Factors affecting the mass culture of *Streptocephalus macrourus* Daday (Anostraca). Ph.D thesis, University of the Orange Free State, Bloemfontein, South Africa. 262 pp.

Rayner, N. A. & A. E. Bowland, 1985. A note on *Triops granarius* Lucas, *Lynceus truncatus* Barnard and *Streptocephalus cafer* Loven from Umfolozi Game Reserve, South Africa. J. Limnol. Soc. sth. Afr. 11: 11–13.

Rzoska, J., 1961. Observations on tropical rainpools and general remarks on temporary waters. Hydrobiologia 17: 265–286.

Seaman, M. T. & D. J. Kok, 1987. Ecological diversity in Orange Free State pans. In R. D. Walmesly & M. L. Botten (eds), Ecology and Conservation of Wetlands in South Africa. national Programmes C.S.I.R., Pretoria.

Sluzhevskaya-Drobysheva, D., 1981. Effect of temperature and feed on growth, maturation and survival rate of *Streptocephalus torvicornis*. Hydrobiol. J. (Washington). 18(5): 95–98.

Thiéry, A., 1987. Les Crustaces Branchiopodes Anostraca, Notostraca et Conchostraca des milieux limniques temporaires (dayas) au Moroc. Taxonomie, biogeographie, ecologie. Ph.D. thesis, Universitaire de Droit Economie et des Sciences D'aix, Marseille, France. 405 pp.

Wiggins, G. B., R. J. Mackay & I. M. Smith, 1980. Evolutionary and ecological strategies of animals in annual temporary pools. Arch. Hydrobiol. Supp. 58: 97–206.

Williams, W. D., 1985. Biotic adaptations in temporary lentic waters with special reference to those in arid and semi-arid regions. Hydrobiologia 125: 85–110.

*Hydrobiologia* **212**: 105–116, 1991.
*D. Belk, H. J. Dumont & N. Munuswamy (eds), Studies on Large Branchiopod Biology and Aquaculture.*



# Life history adaptations of phyllopods in response to predators, vegetation, and habitat duration in north-eastern Natal

M.L. Hamer & C.C. Appleton
*Department of Zoology and Entomology, University of Natal, P.O. Box 375, Pietermaritzburg, 3200, Natal, South Africa*

*Key words:* phyllopods, temporary pools, life history strategies

## Abstract

Phyllopod populations were monitored in three temporary pools differing in the amount of submerged, peripheral vegetation present, surface area and duration. The effects of these factors on the life history strategies employed by phyllopods were investigated. *Triops granarius*, various conchostracan species and the anostracan *Branchipodopsis* sp. inhabited the periphery of two pools where rooted, submerged vegetation was abundant while three *Streptocephalus* species dominated the central, unvegetated regions of the pools and the unvegetated pool. This distribution pattern appeared to be related to the animals' morphology and feeding habits. The peripheral regions of the pools were stressful habitats since they were colonized by large numbers of predators and competitors 30–40 days after inundation and they dried out sooner than the centre. The 'peripheral' species exhibited typical r-selected life history strategies; they grew rapidly, reproduced early and had short lifespans and in this way they overcame the threats presented by their habitat. The 'central' species took advantage of their predator-free, more stable habitat and exhibited life history patterns which tended towards the K-end of the r-K continuum. A degree of intraspecific variation in growth and reproduction was obvious and appeared to be related to differences in habitat duration of the three pools.

## Introduction

The unpredictable nature of temporary pools and the need for their inhabitants to be adapted to survive under such conditions has been noted by many authors. In Williams' (1985) review of the biotic adaptations of animals living in temporary lentic waters of arid and semi-arid regions, he devoted a section to life history adaptations and predicted the importance of r-selection in highly ephemeral pools. Organisms in such habitats should display the following features: intense mortality in the early part of the life cycle, rapid development, high rates of maximal increase ($r_{max}$), early reproduction, few periods of reproduction, short life cycles and density-independent mortality. However, Williams (1985) stressed, as had Pianka (1970) that 'no organism is entirely r-, K- or A-selected, for according to the risks involved, a balance will be struck which maximises the adaptive value of features drawn from each type of selection'. This may apply not only between species but intraspecifically too.

Wiggins *et al.* (1980) suggested that in temporary pools only limited amounts of nutrients are available to planktonic algae in the water column because these nutrients may be trapped in bottom sediments or adsorbed on organic particles and are consequently not returned to the dissolved form during the next wet phase. Such nutrients would, however, be available to the roots of vascular plants and in this way vegetation may serve to recycle sediment-bound nutrients. This system would only be efficient in pools exposed to the sun. Pools which are heavily shaded depend largely on leaf fall for nutrients, a system which appears to be less efficient in the nutrient budget than basins supporting heavy growth of rooted vascular plants. Wiggins *et al.* (1980) therefore considered resource levels (the type of vegetation) and habitat duration to be the most important factors determining the type of habitat and thus the community structure and faunal succession in temporary pools. They also suggested that these factors may affect the life history strategies of the fauna but noted that at a detailed level, little was known about the influences that habitat duration and resource levels have on individual species.

Literature comparing life history patterns of 'phyllopods' (used here in the sense of Fryer, 1987), from different types of temporary pools is scant. Although numerous studies have provided data on the biology of different species, features such as growth, reproduction and population densities have seldom included comparative information from habitats with different periods of inundation, degrees of predictability and types of vegetation.

Studies such as those of Belk (1977) and Hildrew (1985) showed that different reproductive and growth patterns, respectively, could be related to different habitat durations, predictability and in Belk's (1977) study, resource levels and predators as well. Takahashi (1976) however, suggested that temperature, population density and food sources were the causes of variation in growth and reproduction in *Triops granarius* (Lucas) populations in rice paddies. Laboratory studies of other species (Lake, 1968; Mitchell, 1987; Sluzhevskaya-Drobysheva, 1981) have provided similar results. In a study on the aquatic snail *Lymnea elodes* Say, Brown *et al.* concluded that although environmental factors had a greater influence on life history, genetic-based adaptations to a particular habitat (whether temporary or permanent) did cause measurable variation.

Physical and chemical factors undoubtedly influence growth, reproduction and longevity but the effect of the habitat on these aspects of life history has received little attention. The aim of this study was to monitor the phyllopod populations in habitats differing in duration, stability and the amount of vegetation present. Wiggins *et al.* (1980) and Williams' (1985) suggestions as to the importance of r-selected strategies to the inhabitants of ephemeral pools and the possibility of variation in these strategies being related to the habitat, where also examined.

## Description of sites studied

Three pools in the Makathini area of north-eastern Natal, South Africa (Fig. 1) were selected. Pool 1 covered an area of about 15000 $m^2$ when full and consisted of one large deep ($\pm$ 60 cm) basin, numerous smaller isolated or semi-isolated pools and shallow reed, sedge (*Cyperus distans* L.f., *Kyllinga erecta* Schumach. and *Cyperus fastigiatus* Rottb.) or grass-covered marshy areas. Pool 2 was approximately 2400 $m^2$ in area at maximum inundation. Initially this pool was vegetated only at the edges, mainly by sedges and reeds, but after about three weeks mats of floating *Ludwigia stolonifera* became established in the central part. Pool 3 was the smallest of the pools, attaining a maximum area of 600 $m^2$. Apart from being sparsely grassed around the perimeter, it remained devoid of any floating or submerged vegetation throughout the study.

The pools first filled after about 48 mm of rain fell in early August 1987. Pools 1 and 2 dried up after two weeks which resulted in the death of the initial phyllopod populations. Pool 3, however, held water until early January, i.e. for approximately 20 weeks. Heavy rains at the end of September refilled pools 1 and 2 and the main areas of these held water until January, by which time all phyllopod populations had disappeared. Only the small hoofprint-type pools and the peripheral, shallow areas of the larger habitats dried up during the study period, after the second inundation.

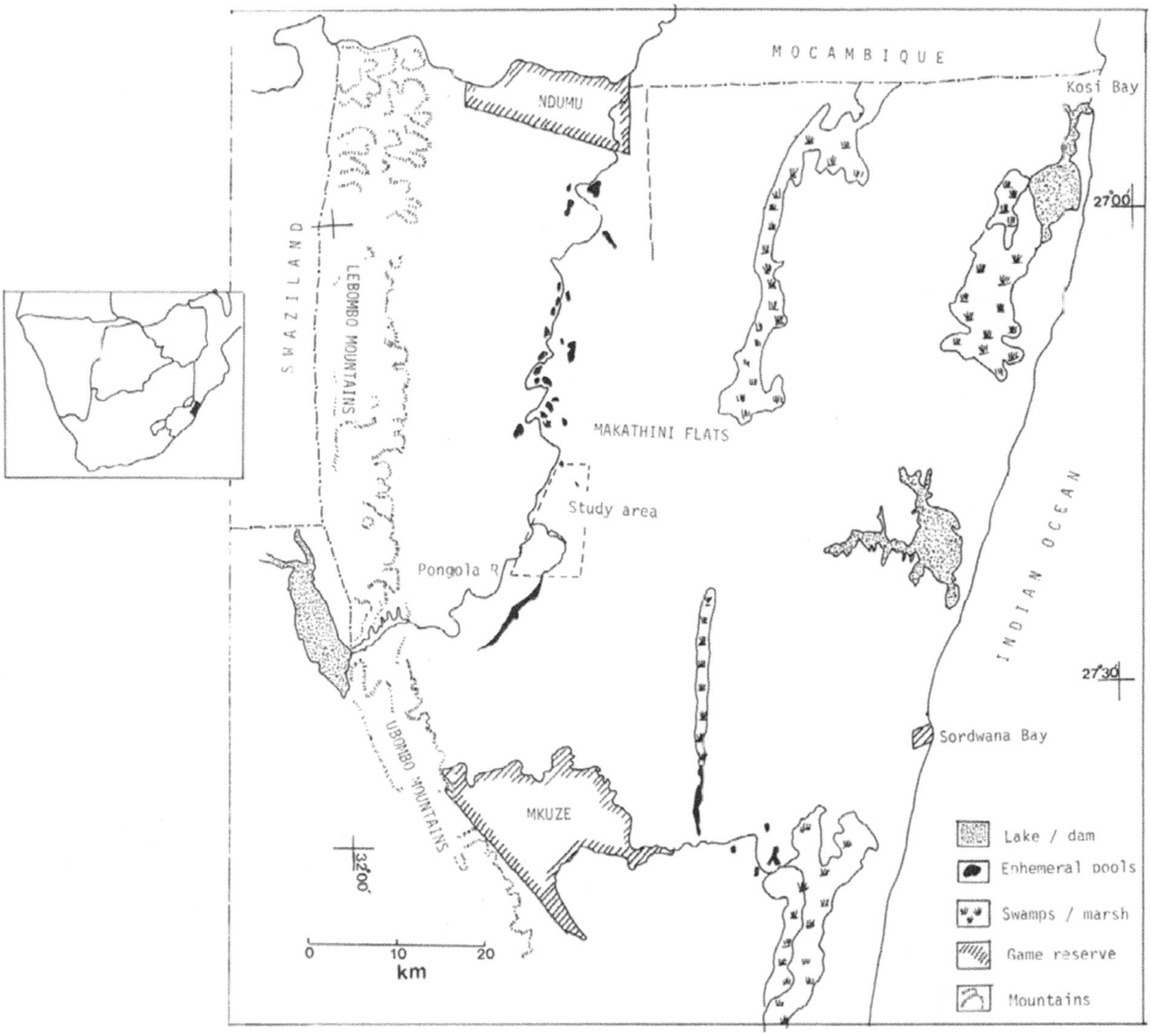


*Fig. 1.* Map of southern Africa showing the position of north-eastern Natal and the study area.

## Materials and methods

### *Population abundance*

A hand-held net with a mouth of 15 × 20 cm and a mesh size of 2 mm was used for taking quantitative samples. Transverse sweeps 80 cm in length were made and the depth to which the net had been submerged during the sweep was recorded. Trials in a clear pool showed that, provided disturbance of the water was kept to a minimum, and movement of the net during the sweep was swift, few animals appeared to escape. In shallow or heavily vegetated areas of the pans 80 cm sweeps were impossible and shorter 20 or 40 cm sweeps were made.

A total of 12 samples was taken at three to four-day intervals from each pool starting a week after the initial filling and continuing until all phyllopod populations had died off. Six of these samples were 'edge' samples taken not more than 2 m from the perimeter of the pool, and six were 'centre' samples, taken from the central, deeper basin. Phyllopods caught in each sample were identified, sexed and measured (median carapace length for Notostraca, total body length excluding the caudal furcae for anostracans and the length of the conchostracan carapace). Reproductive condition was also noted before they were returned to the pool. Samples too large to analyse in the field were preserved in 70% ethanol and examined in the laboratory.

*Analysis of samples*

Abundance estimates were expressed as the number of animals $m^{-3}$ of water. The mean of the six samples $\pm$ standard errors (SE) of shrimps per volume of water that passed through the net were calculated for the numerically dominant species in each pool. The $\chi^2$ (variance to mean ratio) test for agreement with a Poisson series for small samples (Elliot, 1985) was used to determine the distribution of individuals in the population. This was necessary since it affected the analysis of the samples. The distribution of the phyllopods varied from day to day which made previously used equations such as Taylor's Power Law (Hildrew, 1985) and other transformations inapplicable to the data. The abundance data were therefore presented in a straightforward manner. The mean number of each species $m^{-3}$ of water was plotted, together with 95% confidence limits for each sampling day. Edge and centre samples were compared using the non-parametric Man-Whitney-U test (Elliot, 1985), and where these did not differ significantly, (ie. *Streptocephalus macrourus* Daday and *Streptocephalus cafer* (Lovén) in pool 3) the two sets of data were combined.

The growth of the numerically dominant phyllopod species populations in the pans was monitored by measuring all individuals caught during sampling. The mean sizes for males and females, together with standard error were calculated using the equation for small sample sizes where the number of animals measured was less than 30 (Elliot, 1985).

**Results**

*Population density and distribution*

Considerable variation, both temporal and spatial was observed in phyllopod populations. Initially the *Streptocephalus* spp. immatures were aggregated but as populations declined, they became randomly distributed although on some days they were found to be regularly spaced. The conchostracan and *T. granarius* populations were either randomly or regularly distributed. Different species, however, appeared to prefer different regions of the pools. The vegetated, peripheral regions of pools 1 and 2 were inhabited by up to eight species which included the notostracan *T. granarius*, the conchostracans, *Eulimnadia africana* (Brauer), *Leptestheria rubidgei* (Baird), *Caenestheriella australis* (Loven) and two *Lynceus* spp. and the anostracan *Branchipodopsis* sp. The deeper central regions of the pools were dominated numerically by three anostracan species, *S. cafer*, *S. macrourus* and *Streptocephalus* sp.

Populations of the latter group showed similar density trends. Large numbers (100–2600 $m^{-3}$) were present initially but these decreased rapidly about two weeks after inundation of the pool, with a smaller population persisting for a further 30 to 40 days before disappearing (Fig. 2). In pool 3 small populations of *S. macrourus* and *S. cafer* were present for a total of 125 and 107 days respectively (Figs. 3, 4). The fluctuations shown in these figures are probably the result of sampling errors and changes in distribution, since it was clear that only a single cohort was present and that the population did not increase (Hamer, 1989).

The presence of anostracan populations at the edges of pools 1 and 2 appears somewhat contradictory to the separation made here of species into 'edge' and 'centre' groups and the placement of the *Streptocephalus* spp. in the latter group. However, these populations were usually more dense in the centre (Fig. 2) and peaks in the densities of populations at the edge usually coincided with decreases in their numbers in the centre. It is therefore possible that these fluctuations were

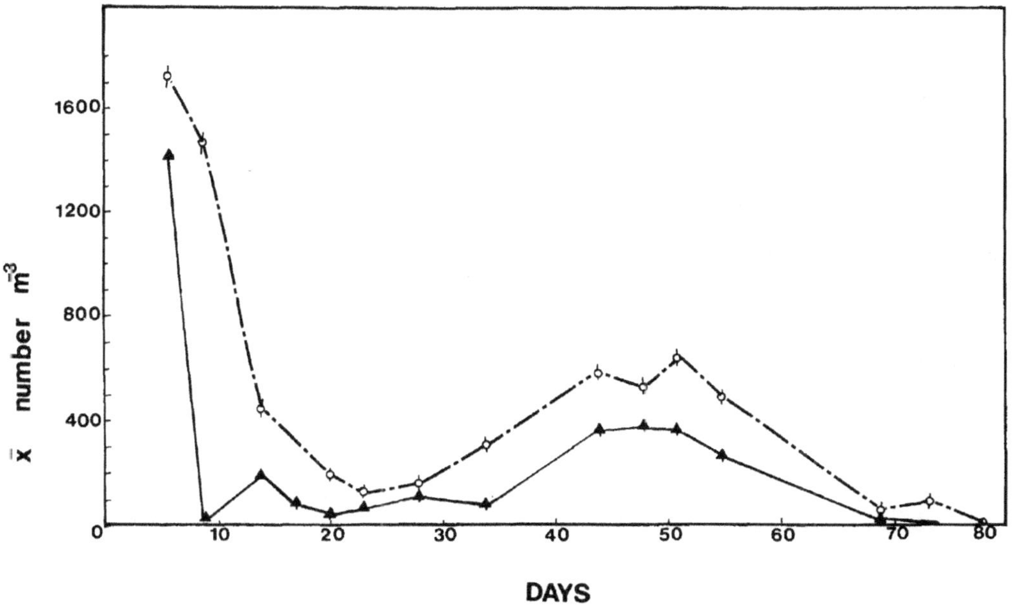


*Fig. 2.* Density of *Streptocephalus macrourus* in pool 2; mean number of animals $m^{-3}$ water $\pm 95\%$ confidence limits ($p = 0.95$). Dotted line = centre; solid line = edge.

associated with changes in the distributions of the populations. *Streptocephalus cladophorus* Barnard was present in the peripheral regions of pool 1 but nearer the centre of the basin in pool 2. This species could, therefore, represent an intermediate between 'edge' and 'centre' species. Growth and reproduction data presented in Table 1 support this possibility.

The abundance of the other phyllopod species was difficult to represent graphically because

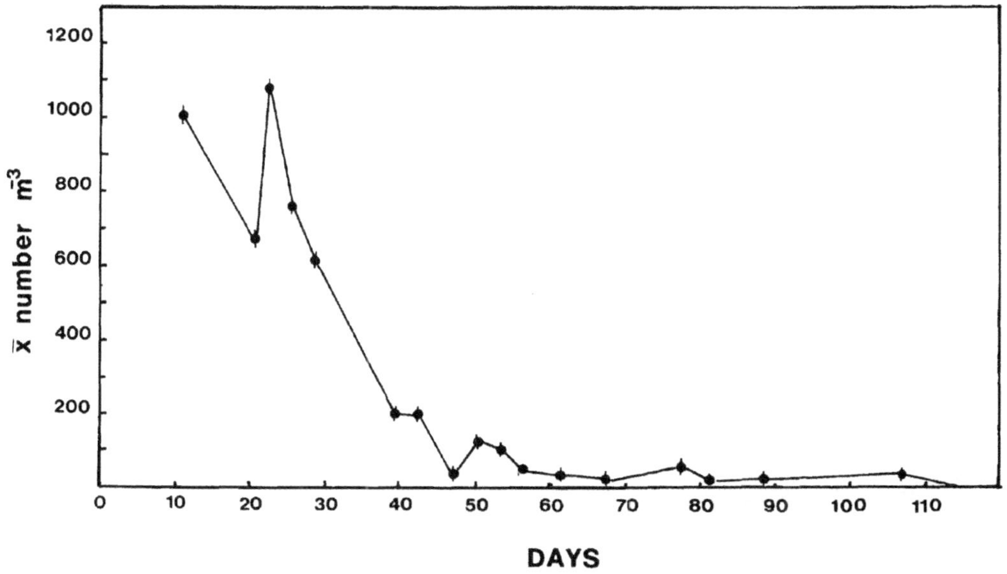


*Fig. 3.* Density of *Streptocephalus macrourus* in pool 3; mean number of animals $m^{-3}$ water $\pm 95\%$ confidence limits ($p = 0.95$). Where confidence limits are not indicated, these were too small to illustrate clearly. Edge and centre samples were combined.

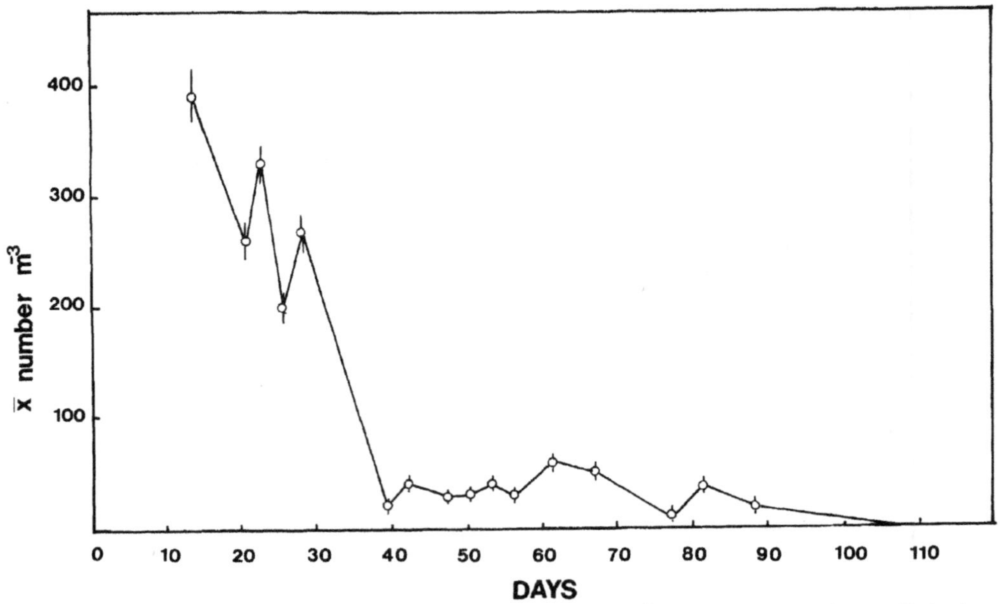


*Fig. 4.* Density of *Streptocephalus cafer* in pool 3; mean number of animals $m^{-3}$ water $\pm 95\%$ confidence limits ($p = 0.95$). Edge and centre samples were combined.

most of these were present in the pools for only short periods and some occurred in small numbers (eg. the mean number of *Lynceus* spp. caught on any sampling day was 47.5 $m^{-3}$ or 6 in a total of 12 subsamples). However, the population density trends of *E. africana* and *Branchipodopsis* sp. appeared to conform to a pattern shown by the other species occurring at the edges of the pools. These populations did not undergo the sudden early decrease in numbers that was exhibited by

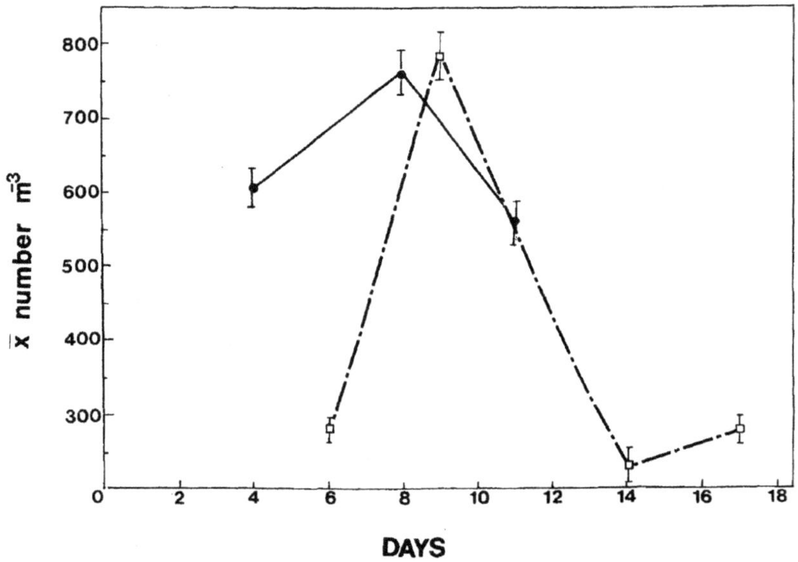


*Fig. 5.* Density of *Branchipodopsis* sp. in edge region of pool 1 (solid line) and of *Eulimnadia africana* (broken line) in the edge region of pool 2 (broken line). Mean number of animals $m^{-3}$ water $\pm 95\%$ confidence limits ($p = 0.95$).

*Table 1.* Summary of growth and reproductive data for the dominant phyllopod species of the Makatini pools.

| Species and locality | Average growth $day^{-1} \pm$ S.E. (mm) | % of maximum size increase $day^{-1}$ | $\pm$ age at 1st reproduction (days) | $\overline{X} \pm$ S.E. size 1st reproduction | $\overline{X}$ + S.E. maximum size (mm) | Maximum lifespan (days) | No. eggs/clutch (range) |
|---|---|---|---|---|---|---|---|
| Notostraca | | | | | | | |
| *Triops granarius* Pool 1 | 0.48 ± 0.18 | 3.81 | 7 | 5.65 ± 0.61 | 12.59 ± 0.95 | 26 | 4 – 26 |
| Conchostraca | | | | | | | |
| *Eulimnadia africana* Pool 1 | 0.28 ± 0.17 | 5.21 | 10 | 4.64 ± 0.44 | 5.37 ± 0.73 | 19 | 225 – 296 |
| *Eulimnadia africana* Pool 2 | 0.39 ± 0.17 | 5.50 | 8 | 6.48 ± 0.48 | 7.09 ± 0.74 | 18 | – |
| Anostraca | | | | | | | |
| *Branchipodopsis* sp. Pool 1 | 0.77 ± 0.36 | 8.37 | 7 | 8.13 ± 1.00 | 9.20 ± 1.20 | 12 | 42 – 89 |
| *Streptocephalus cladophorus* Pool 1 | 0.48 ± 0.22 | 5.00 | 10 | 7.28 ± 0.28 | 9.60 ± 0.98 | 20 | 38 – 66 |
| *Streptocephalus cladophorus* Pool 2 | 0.50 ± 0.22 | 3.06 | 14 | 13.08 ± 1.62 | 16.35 ± 1.28 | 33 | – |
| *Streptocephalus cafer* Pool 1 | 0.44 ± 0.23 | 2.83 | 10 | 13.00 ± 0.74 | 15.55 ± 1.39 | 35 | 64 – 252 |
| *Streptocephalus cafer* Pool 3 | 0.28 ± 0.22 | 1.09 | 21 | 11.13 ± 0.96 | 25.77 ± 2.07 | 91 | 81 – 347 |
| *Streptocephalus macrourus* Pool 1 | 0.25 ± 0.16 | 1.80 | 18 | 9.38 ± 0.41 | 13.68 ± 0.68 | 54 | 34 – 106 |
| *Streptocephalus macrourus* Pool 2 | 0.21 ± 0.18 | 1.43 | 14 | 9.62 ± 0.33 | 14.70 ± 0.62 | 70 | – |
| *Streptocephalus macrourus* Pool 3 | 0.22 ± 0.20 | 0.93 | 26 | 9.87 ± 0.94 | 23.73 ± 2.02 | 110 | – |
| *Streptocephalus* sp. Pool 2 | 0.27 ± 0.18 | 1.54 | 16 | 10.90 ± 0.60 | 17.75 ± 1.82 | 65 | 18 – 68 |

the 'central' species. Instead, in relation to their life-span, they either maintained a relatively stable number of animals (e.g. *Branchipodopsis* sp., Fig. 5) or numbers declined gradually over a two to three week period, after which the entire population disappeared within three days. This disappearance coincided with the invasion of the vegetated edge regions of the pools by a diverse community of insect, amphibian and arachnid predators and/or competitors for space and possibly also food.

*Growth and reproduction*

Growth was rapid in all cases and data for seven species from all three pans are summarized in Table 1. Two types of growth pattern were evident: one for the anostracans inhabiting the central areas of the pools (*S. macrourus*, *S. cafer*) another for species inhabiting the peripheral areas. In the 'central' species, initial growth was rapid and almost exponential. This was followed by a period of little or no growth and then by a

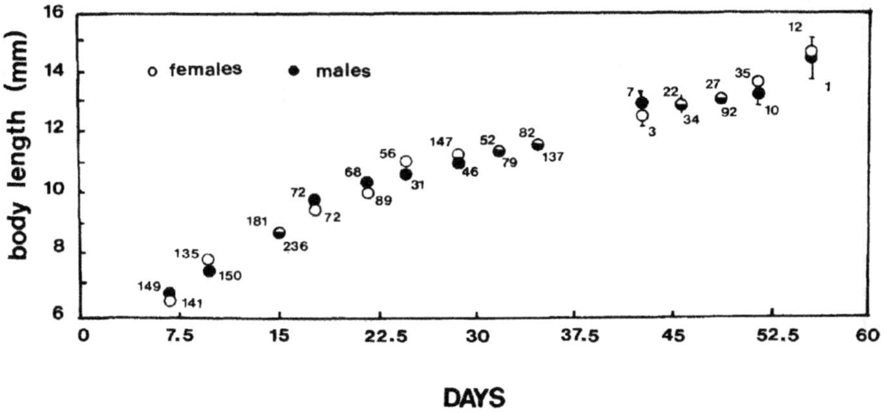


*Fig. 6.* Growth rate of *Streptocephalus macrourus* in pool 1. Mean total length, excluding the caudal furcae (edge and centre samples combined) ± S.E. Where S.E. are not represented, these were too small to illustrate. Numbers written above and below the symbols indicate the number of females and males measured (*n*) respectively.

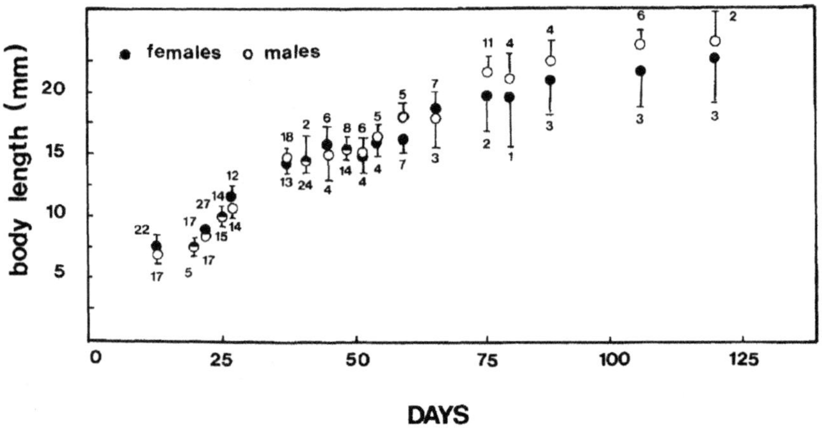


*Fig. 7.* Growth rate of *Streptocephalus macrourus* in pool 3. Mean total length, excluding the caudal furcae ± S.E. Numbers above and below the symbols indicate the number of males and females measured (*n*) respectively.

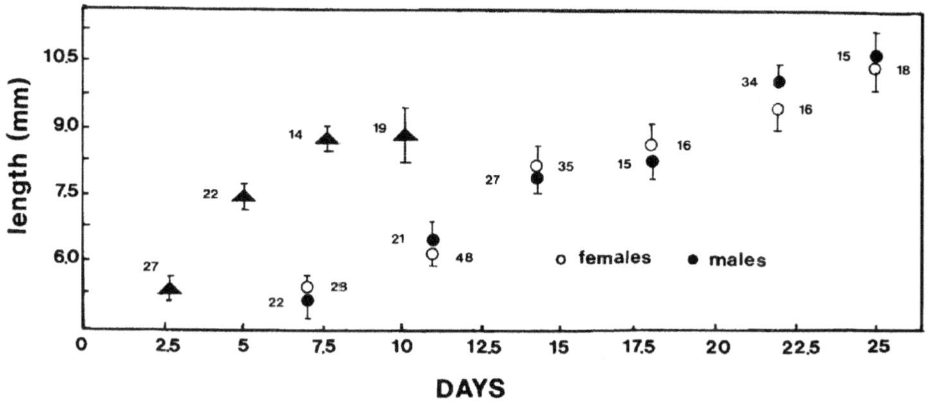


*Fig. 8.* Growth rate of *Triops granarius* in the edge region of pool 1 (open circles = females, shaded circles = males). Mean median carapace length ± S.E. Triangles represent the growth rate of *Branchipodopsis* sp. Mean total body length (excluding caudal furcae) ± S.E. (sexes do not separated). Numbers adjacent to symbols indicate the number of animals measured (*n*).

second period of growth until the population disappeared (Figs. 6, 7). Variation in the timing of these three growth phases, of the sizes reached and in size increases per day, was evident between species and intraspecifically in the different habitats (Table 1).

*T. granarius* and *Branchipodopsis* sp. in pool 1 provided examples of growth patterns typical of the species which inhabited the peripheral areas of pool 1 and the margins of pool 2. Although they experienced rapid initial growth, they then reached a period of slower growth with little subsequent increase in size (Fig. 8). These species disappeared from the habitat soon after this plateau had been reached and in both 'edge' and 'centre' species the decline in the growth phase appeared to follow closely the onset of egg production by the females (Table 1).

In addition to the different patterns of growth shown by 'edge' and 'centre' species, the former group had a greater size increase per day in terms of both length and in the maximum size reached (Table 1; columns 1 and 2). Reproductive trends, too, varied according to species and habitat. Those species living in the peripheral areas started reproducing sooner than the anostracans of the central regions (Table 1). However, because of the peripheral species' relatively brief lifespans, they had a shorter reproductive period than the late-reproducing, but longer-living 'central' species. Laboratory cultures of *S. macrourus*, *S. cafer* and *T. granarius* at 23 °C showed that egg broods are produced at two to three day intervals and that the number of eggs per brood increased with body size. Females of each of the 'central' species would, therefore, be expected to have a larger total egg-output than the females of the 'edge' species. This was true, particularly for species such as *T. granarius* and *Branchipodopsis* sp. which produced small clutches and had short lifespans (Table 1).

## Discussion

This study has shown that the phyllopods of north-eastern Natal employ life history strategies typical of r-selection; they grow rapidly, reproduce early, have short life cycles with few periods of reproduction. However, there were distinct differences in their life history patterns, and also variation in the degree to which they were r-selected.

Distribution of the species within the pools appeared to be related to the presence of rooted submerged and emergent vegetation (and thus detritus) and water depth in the different regions of the habitats. Anostracans are generally filter-feeders, abstracting particles from suspension in the open water (Fryer, 1987) and they would thus have difficulty in feeding amongst the dense vegetation common in the shallow, peripheral regions of the pools. Relatively small species such as *Branchipodopsis* sp. ($\bar{x}$ body length = 7.5 mm) may, however, have been able to feed successfully in these areas at low densities. *T. granarius* and the conchostracans are omnivorous, feeding on detritus (Fryer, 1987) and these species were often seen swimming on the substratum and in the laboratory at least, only swam freely for short periods. Fryer (1988) also noted this behaviour in *Triops* and stated that the morphology of the Notostraca is not adapted to surface swimming but rather to a benthic habit. For species with such living and feeding habits the peripheral habitat, where rooted vegetation and thus detritus was abundant and the water shallow, would probably be more suitable than the deep and vegetation-scarce central regions. In addition to the lack of suitable food in the centre, the bottom waters of the central basin were usually oxygen-depleted relative to the surface and marginal waters (Hamer, 1989). Over a 24-hour period for instance, the range of dissolved oxygen concentrations was 6.3–3.3 mg $l^{-1}$ while at the bottom of the central region it was 5.0–1.5 mg $l^{-1}$. Species which spend much time on the substratum would have to move closer to the surface when the availability of oxygen became stressful and for heavy-shelled animals such as the conchostracans and *T. granarius*, this would be energetically expensive.

However, the peripheral regions of the pools appeared to be stressful in other ways: they became colonized by a diverse non-phyllopod fauna many of which were predators and others possible competitors, and they dried up sooner than the central basin. The inability of phyllopods to avoid predators has been mentioned by a number of authors (Bishop, 1967; Hartland-Rowe, 1972; Wiggins *et al.*, 1980) and their lack of mobility means that they cannot leave a habitat which is drying up. The 'edge' species have therefore developed life history strategies which ensure their continued existence, while the 'central' species were able to take advantage of the relative stability (in terms of longer duration and an absence of predators) of this part of the habitat.

Although essentially r-selected, the life history patterns of *S. macrourus*, *S. cafer* and *Streptocephalus* sp., when compared with those of the 'peripheral' species tended towards the K-end of the r-K continuum. Growth was, on average, slower (0.21–0.28 mm $d^{-1}$), the time and size at which egg production started varied (Table 1) but even though the females were still relatively small (9.38–13.0 mm), they were between 4 and 26 days older than gravid females of species occurring at the edge. As in K-selected organisms, they attained a large body size (particularly in pool 3), and outlived those species living at the edge. The long lifespan meant too, that numerous egg broods were produced per female. Population density and survivorship patterns of the 'central' species were, however, atypical of K-selected animals inasmuch as a high population density which declines rapidly early in the life history is a trait usually associated with r-selection. The reason for these early population decreases, which occurred in all the *Streptocephalus* spp. sampled, is uncertain but may be related to resource availability. Wiggins *et al.* (1980) suggested that only limited amounts of nutrients are available to planktonic organisms in the water column possibly because nutrients may be trapped in bottom sediments. The initial food available might have been insufficient to maintain a large anostracan populations, but no evidence was collected to support this hypothesis.

While the 'central' species avoided predation by inhabiting, predominantly, the predator-scarce unvegetated, deeper regions of the pool, the 'peripheral' species achieved this by completing their life cycle in less than 40 days, i.e. before predators became abundant in their part of the habitat. By compacting their life history into a short period, they maximised the chances of reaching maturity and reproducing before their unstable, short-duration habitat dried out. The life history strategies shown by the 'peripheral' species are, mainly, typical of r-selection. Growth was rapid (0.28–0.77 mm $d^{-1}$) and in most cases the leveling off shown by the *Streptocephalus* spp. inhabiting the centre, occurred while the animals were still small, and growth did not resume afterwards. Reproduction was early (7–10 days post-hatching) in all the 'peripheral' species, but since lifespans were only 12 to 33 days, probably only a few egg broods per female were produced. Population density was not typical of r-selected organisms; populations were relatively stable or declined gradually and the entire population disappeared as a result of a sudden crash. The disappearance of the 'peripheral' species occurred even if the habitat did not dry up, but whether this happened before the arrival of predators or as a result of them was not clear.

Since the life cycles of 'edge' species were short, and few broods of eggs per female were produced, there may have been some advantage in maintaining the initial population densities. the larger the population of egg-producing females, the greater the total egg output. The detritus available in these areas would probably have contained sufficient nutrients to support these populations. This scenario is in contrast to that documented for the 'central' species in which, although their populations rapidly became depleted, each surviving female produced many clutches of eggs. Since larger females produced more eggs, they achieved an even greater total egg output and this may account for their denser initial populations than those of the 'edge' species (Figs. 2, 3, 4).

Intraspecific variation in life history was also recorded in the three study pools. This was particularly noticeable in respect of growth, reproduction and lifespan of *S. macrourus* and *S. cafer* in pool 3 when compared with pools 1 and 2 and it would appear that this variation is correlated with habitat duration. Pool 3 did not dry up after the first inundation in August 1987, while the other two did, despite the fact that the former had a substratum with a higher sand and lower clay and organic matter content than the latter two (Hamer, 1989). The prolonged duration of pool 3 was probably a result of underground seepage and meant that the anostracan populations there were able to reach sizes at which many eggs per clutch per female could be produced (up to approximately 347), and since they lived for about 100 days, many clutches were produced. However, this and the other records of intraspecific variation observed may also have been related to food quality and availability.

## Conclusions

The results of this study support the predictions of Wiggins *et al.* (1980) that vegetation (which influences the quantity and source of detritus) and habitat duration determine community structure, and that these factors determine the type of habitat and the life history strategies of its inhabitants. Williams' (1985) predictions that r-selected strategies are dominant in the life histories of ephemeral pool inhabitants as well as the presence of both inter- and intraspecific variation in this respect are relevant too. However, as has been shown by other studies (Lake, 1968; Mitchell, 1987; Takahashi, 1976), population density, food, temperature and water chemistry also influence growth, reproduction and longevity of phyllopods. Although Brown *et al*'s (1985) study of the snail *L. elodes* showed that food availability influenced growth and reproduction to a greater extent than the origin of the snails, those snails from temporary pools always grew more slowly, matured at a smaller shell size and had higher fecundities than individuals from permanent pools. This lends support to the possibility that adaptations to specific habitat types are genetically based. Belk's (1977) evidence for variation in the reproduction trends of an anostracan species being related to the predictability of the habitat gives further strength to this hypothesis. Hildrew's (1985) study showed that different growth and reproduction rates occurred within the same population in response to different inundation patterns.

Obviously further research is necessary to conclusively identify the effects of the various factors on life history strategies in temporary waterbodies. Translocation experiments, laboratory culture of animals from different habitat types, and reactions of phyllopods to food type and predators would provide interesting data on this subject.

**Acknowledgements**

The senior author was in receipt of a CSIR (FRD) bursary for the duration of this study.

**References**

Belk, D., 1977. Evolution of egg size strategies in fairy shrimps. S. West Nat. 22: 99–105.

Bishop, J. A., 1967. Some adaptations of *Limnadia stanleyana* King (Crustacea: Branchiopoda: Conchostraca) to a temporary freshwater environment. J. anim. Ecol. 36: 599–609.

Brown, L. R., D. R. Devries & B. K. Leathers, 1985. Causes of life history variation in the freshwater snail *Lymnea elodes*. Malacologica 26: 191–200.

Elliot, J. M., 1985. Some Methods for the Statistical Analysis of Samples of Benthic Invertebrates. 2nd edn. Freshwater Biological Association Scientific Publications, London.

Hamer, M. L., 1989. Studies on the Phyllopod Fauna of Ephemeral Pools in North-Eastern Natal. M.Sc. thesis. University of Natal, Pietermaritzburg, South Africa.

Hartland-Rowe, R., 1972. The limnology of temporary waters and the ecology of Euphyllopoda. In R. B. Clarke & R. J. Wooton (eds), Essays in Hydrobiology. University of Exeter (U.K.): 15–31.

Hildrew, A. G., 1985. A quantitative study of the life of a fairy shrimp (Branchiopoda: Anostraca) in relation to the temporary nature of its habitat, a Kenyan rainpool. J. anim. Ecol. 54: 99–110.

Lake, P. S., 1968. Effects of temperature on growth, longevity and egg production in *Chirocephalus bundyi* Prevost (Crustacea: Anostraca). Hydrobiologia 33: 342–351.

Mitchell, S. A., 1987. Factors Affecting the Mass Culture of *Streptocephalus macrourus* Daday (Anostraca). Ph.D. thesis. University of the Orange Free State, Bloemfontein, South Africa.

Pianka, E. R., 1970. On r- and K-selection. Am. Nat. 104: 592–597.

Sluzhevskaya-Drobysheva, E. B., Effect of temperature and feed on growth, maturation and survival rate of *Streptocephalus torvicornis*. Hydrobiol. J. 18: 95–98.

Takahashi, F., 1976. Pioneer life of the tadpole shrimp, *Triops* spp. (Notostraca: Triopsidae). Appl. ent. Zool. 12: 104–117.

Wiggins, G. B., R. J. Mackay & I. M. Smith, 1980. Evolutionary and ecological strategies of animals in annual temporary pools. Arch. Hydrobiol. Suppl. 58: 97–206.

Williams, W. D., 1985. Biotic adaptations in temporary lentic waters with special reference to those in arid and semi-arid regions. Hydrobiologia 125: 85–110.

*Hydrobiologia* **212**: 117–136, 1991.
*D. Belk, H. J. Dumont & N. Munuswamy (eds), Studies on Large Branchiopod Biology and Aquaculture.*



# Multispecies coexistence of branchiopods (Anostraca, Notostraca & Spinicaudata) in temporary ponds of Chaouia plain (western Morocco): sympatry or syntopy between usually allopatric species

Alain Thiéry
*Laboratoire de Biologie Animale-Hydrobiologie, Faculté des Sciences, 33 rue Louis Pasteur, F-84000 Avignon, France*

*Key words:* Anostraca, Notostraca, Spinicaudata, temporary ponds, sympatry, coexistence

## Abstract

The study of the structure of communities of Crustacean branchiopods Anostraca, Notostraca and Spinicaudata of temporary ponds in Morocco revealed the existence of a geographical area with a particularly diversified number of species. Over a few hectares, 11 species, among the 18 known in Morocco, are present. While most of them are usually allopatric and belong to distinct associations, indicative of particular ecological and climatic conditions, they can coexist in Chaouia plain, either in adjoining ponds (sympatric species) or in the same pond (syntopic species). In one pond we could find up to 10 phyllopods: 6 Anostraca, 2 Notostraca and 2 Spinicaudata.

This faunistic diversity may result from abiotic features (the location of the ponds at the boundary of two climatic areas) and from a heterogeneous geological substratum, creating a patchwork of temporary ponds with various mesologic conditions.

Secondarily, this diversity may result from different life history characteristics of each coexisting species. As all species hatch synchronically, coexistence is only possible if their growth rates and life span are different. This implies that horizontal and vertical distribution, and exploitation of food supplies in the habitat differ for each ecophase of the syntopic species (spatial and trophic exclusion).

## Introduction

The pioneering work of Simon (1886), Sars (1896) and Daday (1910, 1914, 1923) addressed the systematics of the Anostraca, Notostraca and Conchostraca ( = Spinicaudata and Laevicaudata sensu Fryer, 1987) living in temporary ponds, but their community structure was neglected. Most authors found that rarely more than 2 or 3 anostracans and 1 or 2 conchostracans live together in one biotope (Gauthier, 1928 in North Africa; Dexter, 1946, 1953; Dexter & Kuenhle, 1951; Prophet, 1963; Moore, 1963; Maynard & Romney, 1975; Huggins, 1974; Belk, 1977 and Daborn, 1978 in North America; Alonso, 1981 in Spain; Geddes, 1981 in Australia; Mura, 1985, 1987 and Mura & Cottarelli, 1984 in Italy). The factors determining the composition of these communities are still not well understood. At first, Gauthier (1928, 1930), Rzóska (1961), and Belk (1977), studied populations by a species versus abiotic factors approach. They concluded that a major factor controling the communities was climate, although Horne (1967) stressed the importance of physical and chemical factors. Only Donald (1971), in Canada, stressed that coexistence of congeneric species may result from differences in their reproductive cycle and their food niche. The importance of these biotic factors was recently shown by Hamer & Appleton (1991) in temporary ponds of Natal.

The Notostraca *Triops* and *Lepidurus* are usually allopatric (Williams, 1968: Australia; Thiéry, 1988b: France). Data concerning the coexistence of conchostracan species are limited (Mattox, 1957; Wooton & Mattox, 1958), but in Anostraca, Donald (1983) reported the coexistence of five species. He adds that the percentage of occurrence of three or five anostracan species per pond in the U.S.A. makes up only 1% of the cases, while more than 80% of the ponds are inhabited by one species and 15% by two species. These conclusions confirm Daborn's data (1978) in nearctic regions and Dimentman's (1981) on Israel.

In Morocco, most temporary ponds have communities of one, two or three Anostraca, and often one or two Spinicaudata, and the notostracan *Triops* or *Lepidurus* (Thiéry, 1987). However, in the area of eastern Chaouia, temporary ponds occur which have many coexisting species, particulary in one pond we regularly collected up to

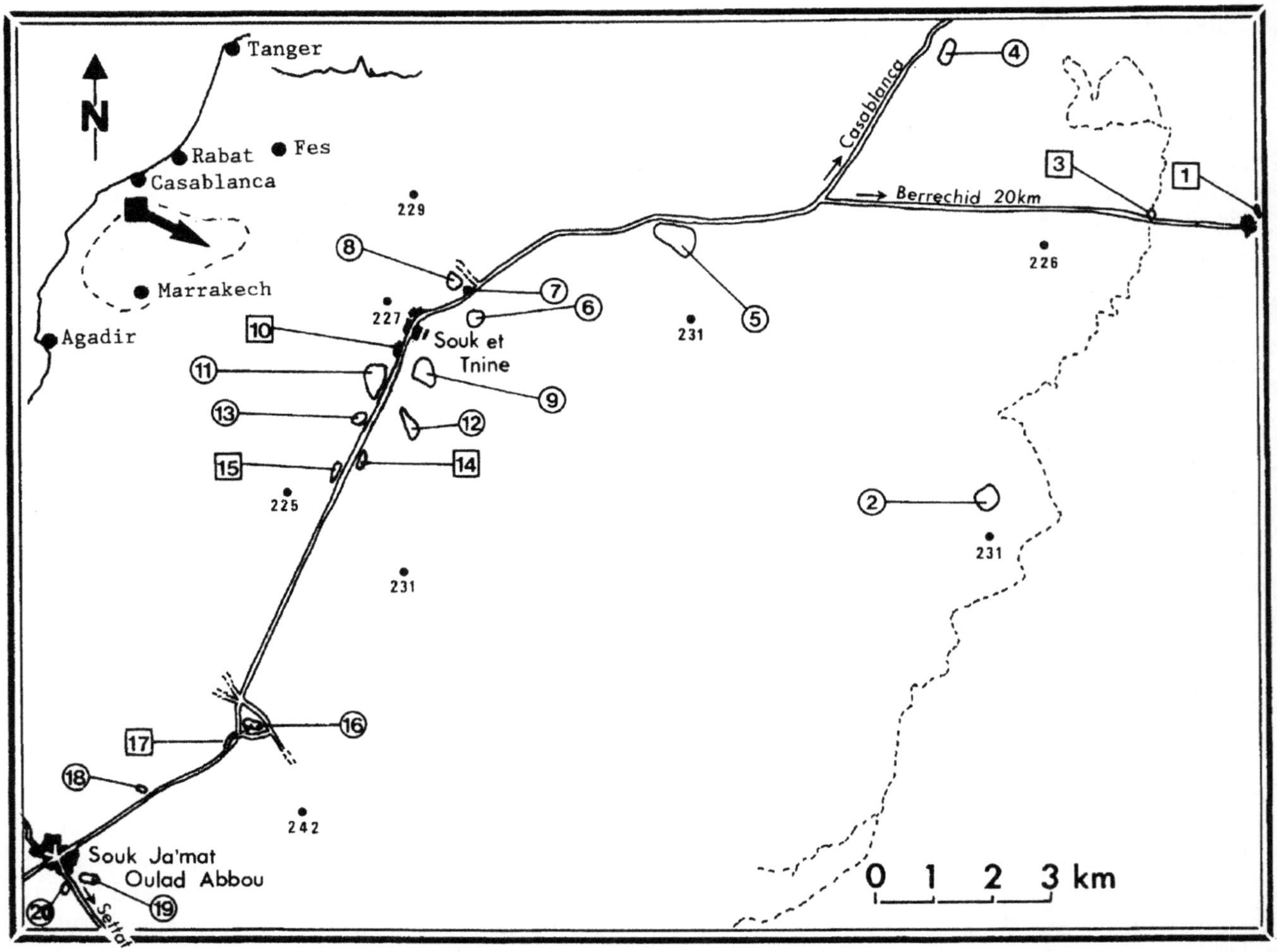


*Fig. 1.* Map of study area. Insert shows a map of occidental Morocco with location of the semi-arid/arid zone boundary (arid zone inside dotted line). Black points indicate altitude (meters). Numbers encircled are typical temporary ponds (dayas) and numbers in squares are road-ditches.

six anostracans. This situation was never found in the literature (but see Maeda-Martinez, this volume). In the present work, these populations are studied from a biological point of view (body size and feeding ecology), beside the synecological one (species versus abiotic factors), with an aim to determine the importance of specific factors in structuring communities and to explain why Chaouia plain communities differ from others.

## Description of study area and habitats

### *Localisation*

The study area is situated in west Morocco, from Souk Jemmaa Oulad Abbou to Berrechid (33° 06′–33° 13′ N; 7° 56′–7° 57′ W) in the eastern Chaouia plain (south of Casablanca), at 240 m above sea level (Fig. 1).

### *Geology*

The temporary ponds of the Chaouia ponds of the Chaouia plain are located on Cretaceous calcareous marl and clay in a synclinal basin South of Berrechid. The Quaternary period was affected by a succession of wet and dry periods that caused bad draining characteristics for surface water, thus favouring the presence of many temporary ponds (= dayas, Arab term) and road-ditches (Ruhard, 1975). In some cases, a capillary contact with the oligohaline groundwater ($Cl^-$ 1 to 2 $g\,l^{-1}$) salinizes the surface.

### *Climatic features*

The region experiences a thermo-mediterranean climate, with approximately 5–6 dry months (a month is biologicaly dry when precipitation is lower than twice the thermic mean (Fig. 2)). Rain, about 400 mm per year, falls from October to March, and floods the temporary ponds until they dry out between spring and late summer (Fig. 3). Annual average temperature is c. 18 °C. The differences between summer and winter ($\Delta t°$ = 15.9 °C) is important in spite of the proximity of the ocean.

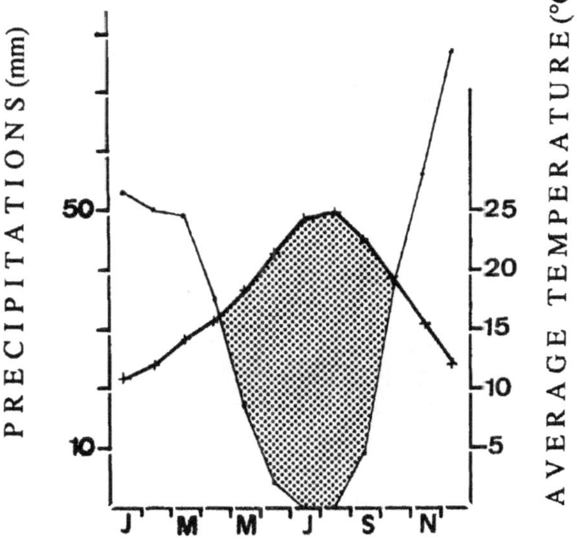


*Fig. 2.* Ombrothermic diagram at Berrechid-Averroes station (near Casablanca-Nouasser airport). Dry months, in grey on the graph, are characterized by precipitations ≤2 T °C.

Continentality increases rapidly with distance from the ocean (Kutsch, 1978). As shown on the climagram modified after Sauvage (1963) (Fig. 4), the study area forms the boundary of two climatic zones; the oceanic semi-arid climate of the oceanic coast, and the arid zone of the Jbilets mountains of Marrakech (with 7–8 dry months).

### *Physical properties and water chemistry*

When flooded, the ponds cover between many hectares (stations 2, 5, 9, 11) and 250 to 300 $m^2$, or less (road ditch, station 10, 14, 15, 17, rock-pools, station 18). Their depth does not exceed 1.50 m, and mostly averages 0.30–0.40 m. (Fig. 5). Depth and surface area vary along and across years, in relation with rainfall variability. As in most temporary ponds (Rzóska, 1961, 1984; Prophet, 1962; Moore, 1970; Daborn, 1976b; Williams, 1985), conductivity also varies

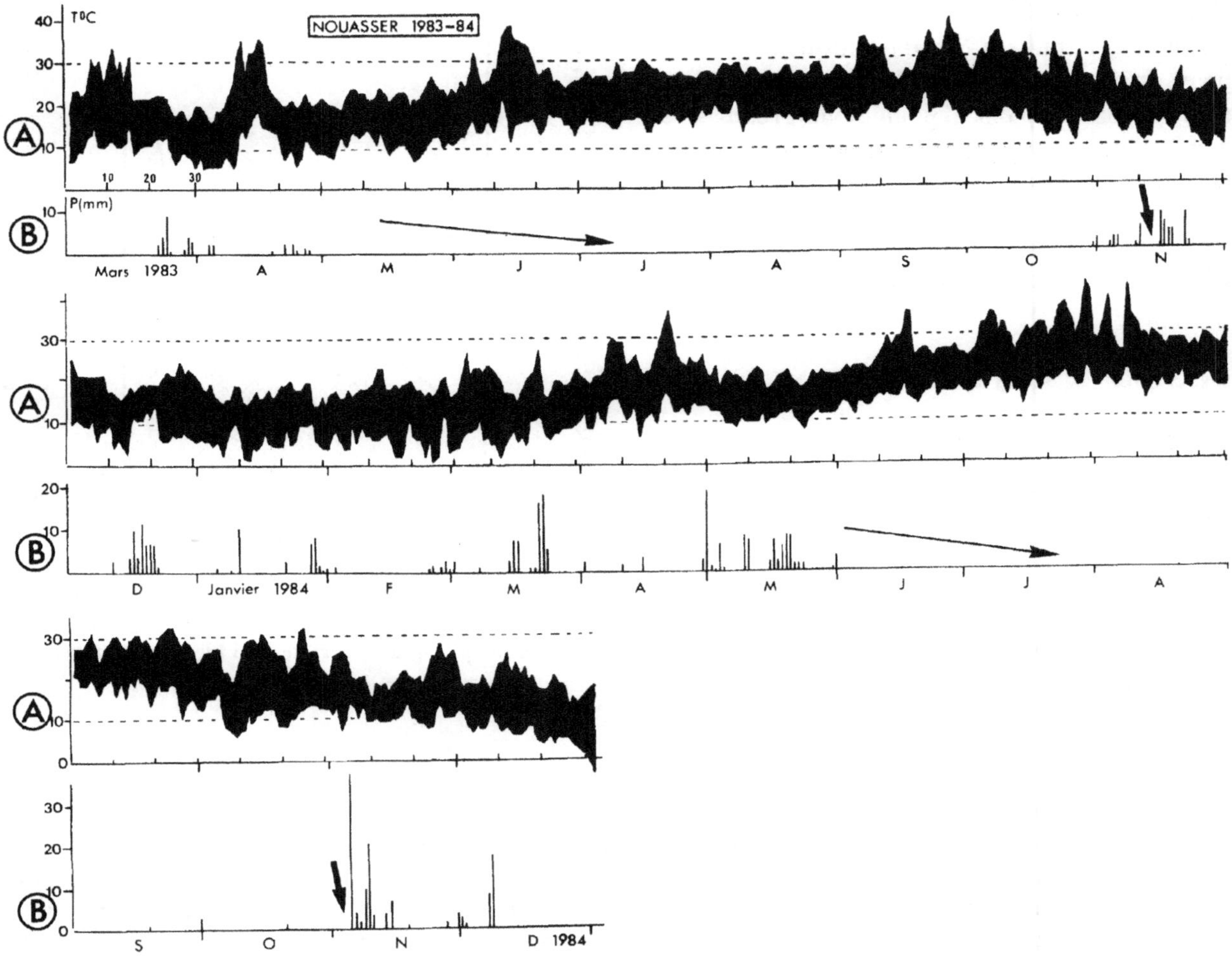


*Fig. 3.* Climatic features (daily data between march 1983 and december 1984) at Nouasser airport station. A. (Upper): temperatures min and max ( °C), b. (lower): rains (mm). Horizontal arrows from May to July indicate the progressive drying of ponds. Vertical arrows in November indicate the flooding of ponds.

through the hydrologic cycle. It is low (30–60 $\mu$S cm$^{-1}$) at flooding and increases (500–800 $\mu$S cm$^{-1}$) as drying up proceeds (Fig. 6). Conductivity is also related to dry residue (Fig. 9), or to transparency, as measured by Secchi disc (Fig. 8). Two distinct salinity types occur: fresh-water ones, not in contact with the groundwater (conductivity 80–600 $\mu$S cm$^{-1}$) and brackish ponds connected to the oligohaline groundwater (conductivity 400–2000 $\mu$S cm$^{-1}$) (Dayas 7, 8 for example). In addition, two types of ponds are distinguishable by substrate type (marl, clay...). Most have a pH between 6.8 and 8.2, but some reach 8.4 to 9.8 (Dayas 2, 5, 8 and 11). In all, much organic matter and clay are in stable suspension (Secchi disc often <5 cm) (Fig. 7a, b, c, d). The suspended colloidal or particulate matter is linked to pH and temperature (Thiéry, 1987). X-ray analysis of the clays shows an almost exclusive presence of Kaolinites and Illites. These clays are the last components of a decaying substratum and result from successive phases of dissolution and erosion, which are characteristic of the genesis of temporary ponds (dayas) in dry areas of North Africa (Thiéry, 1987). They are a suitable substratum for bacterial

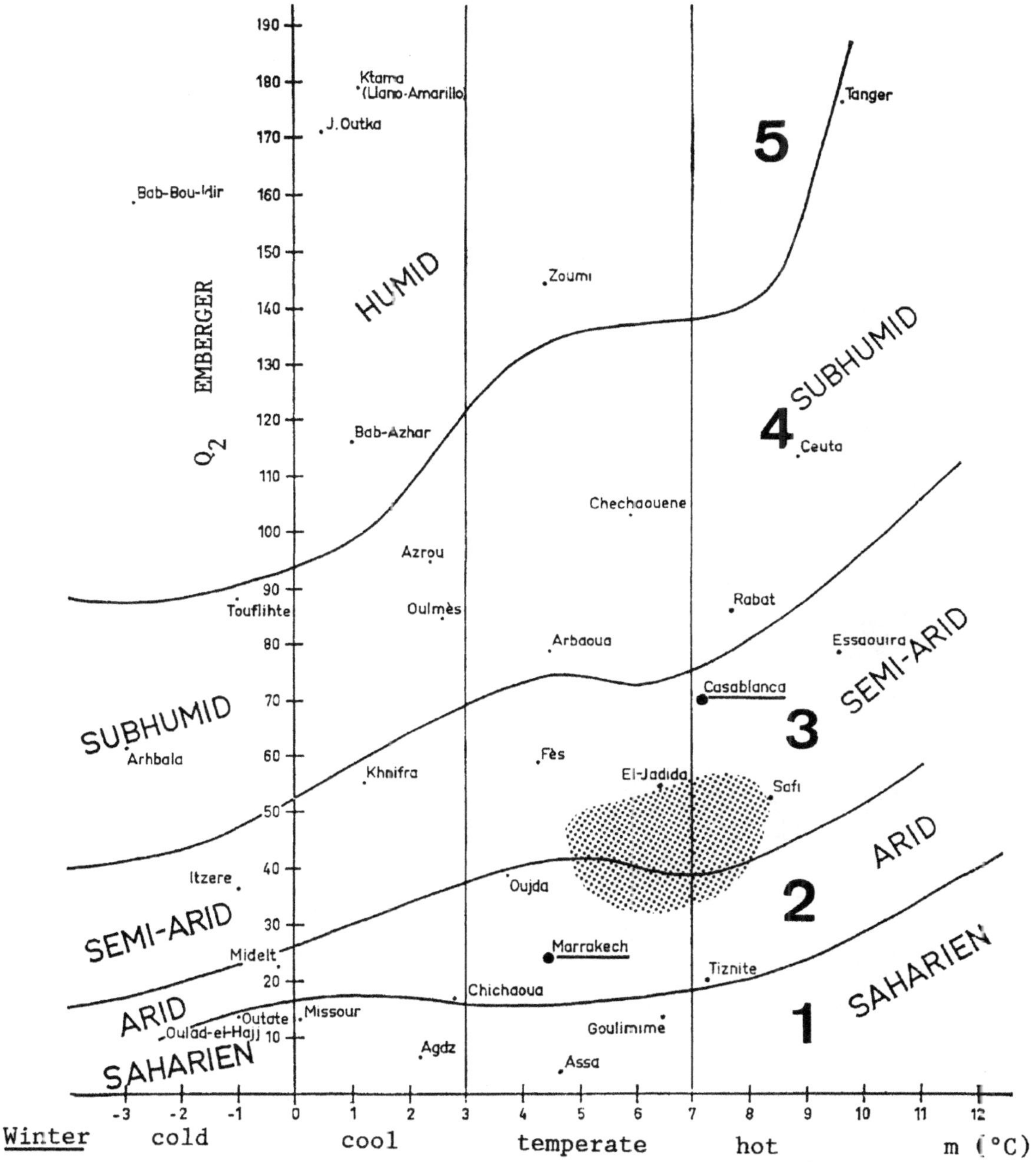


*Fig. 4.* Situation of the Chaouia plain (grey area) on the climagram of Sauvage (1963), at the interface between semi-arid and arid climates. 1, 2, 3, 4 & 5 are bioclimatic stages.

populations (Felton *et al.*, 1967; Pedrós-Alió & Brock, 1983; Arruda *et al.*, 1983; Melack, 1985). A close relationship exists between total dissolved and particulate organic matter and dry residue ($r^2 = 0.904$) (Fig. 10). It is low when flooding occurs and increases when ponds dry up. The amount of dry residue evolves inversely to transparency. Waters are well oxygenated with dissolved oxygen at midday from 80% saturation in open water to 180–200% in beds of *Ranunculus*

*Fig. 5.* Height of twenty habitats in which associations of co-occurring species of Branchiopods occur. a) daya 6. b) daya 7. c) road side ditch 14. d) daya 13. e) daya 16. f) rock-pool 18. g) daya 19. h) daya 20. All photographs taken during flood period.

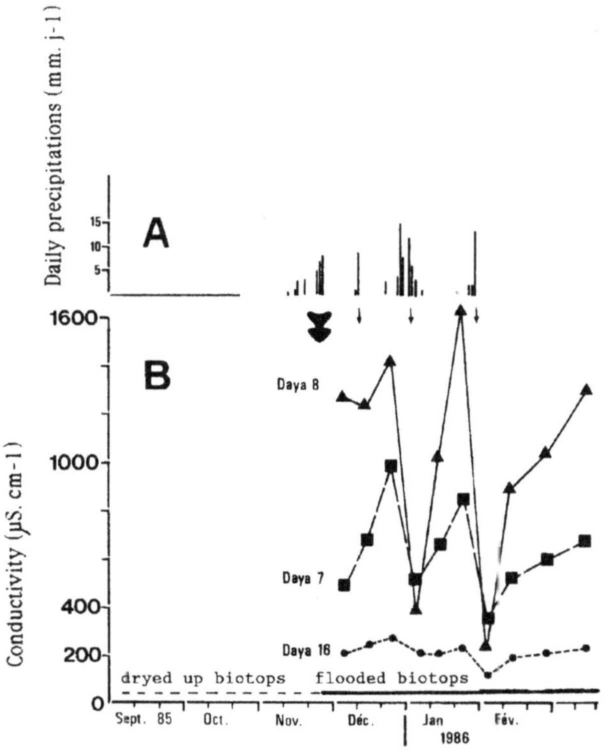


*Fig. 6.* Evolution of the conductivity -B- in 3 temporary ponds of Chaouia plain during the first 4 months of an hydrologic cycle. Black arrow indicate flooding. Dilutions are linked to rainfalls (daily data upper)-A-. In large daya 16, we noted slighter conductivity changes than in little daya 7, where differences were more important. Daya 8 is sometimes fed by oligohaline groundwater, so its conductivity is high.

and *Spirogyra*. However, during the night, low oxygen concentrations (30–40%) occur in the deepest zones of the ponds.

*Vegetation*

Many species of macrophytes and algae are present in the dayas. Dense beds of *Ranunculus aquatilis baudotii*, *Eleocharis palustris*, *Glyceria fluitans* occur in the deepest parts, and *Damasonium alisma*, *Isoetes velata* and *Marsilea pubescens* near the shores, entangled with *Spirogyra* spp. and *Chara* spp. (Fig. 7a, b, c). The distribution of *Ranunculus* and *Glyceria* is not homogeneous: zones with dense vegetation and zones with free water are found (Fig. 5e; 7b).

*Associated vertebrate fauna*

Aquatic vertebrates are mainly composed of the Anoura *Bufo mauritanicus*, *Bufo bufo*, *Hyla meridionalis*, *Rana esculenta* and the Urodele *Pleurodeles waltlii*. Among birds, *Ciconia ciconia*, *Himantopus himantopus* and *Bubulcus ibis* are present but not abundant. Some of these vertebrates (particularly *Pleurodeles*) have a predatory impact on the euphyllopods (Thiéry, 1987). They also dispense eggs among the ponds of one area (amphibians) (Mathias, 1937) or among ponds of distant areas (birds) (Balfour-Browne, 1909; Decksbach, 1924; Proctor *et al.*, 1967; Thiéry & Pont, 1987). The gut contents of *Pleurodeles waltlii* were rich in Anostraca, Notostraca and Spinicaudata. One female from Daya 8, 11 November 1985, contained 1 *Triops cancriformis mauritanicus* (juvenile) and 97 *Tanymastigites brteki*, with numerous ovigerous females and viable eggs in the rectum.

In Daya 8, 24 January 1986: 151 *Tanymastigites brteki* with ovigerous females and viable eggs were found in the digestive tract of one *Pleurodeles* female.

In Daya 9, on December 1985 one female contained 1 *Triops cancriformis mauritanicus* juvenile, 3 *Branchipus schaefferi* of which 2 ovigerous females, 17 *Cyzicus bucheti*, with some Ostracods (*Eucypris virens*) and insect larvae, and in the rectum, many eggs of *Branchipus* and valves of *Cyzicus*.

However, birds and amphibians are not alone involved in cyst dispersal. The transport of eggs between ponds can also be caused by sheep; when they come to drink eggs stick on their feet (Fig. 7e). We scraped feet of five sheep leaving the daya and we counted 7 eggs of *Triops*, 3 eggs of *Branchipus* and 1 egg of *Tanymastigites*.

## Methods

Twenty ponds were sampled approximately monthly from March 1983 to June 1986, between 11 a.m. and 2 p.m., except during summer when biotopes were dry. Qualitative samples were col-

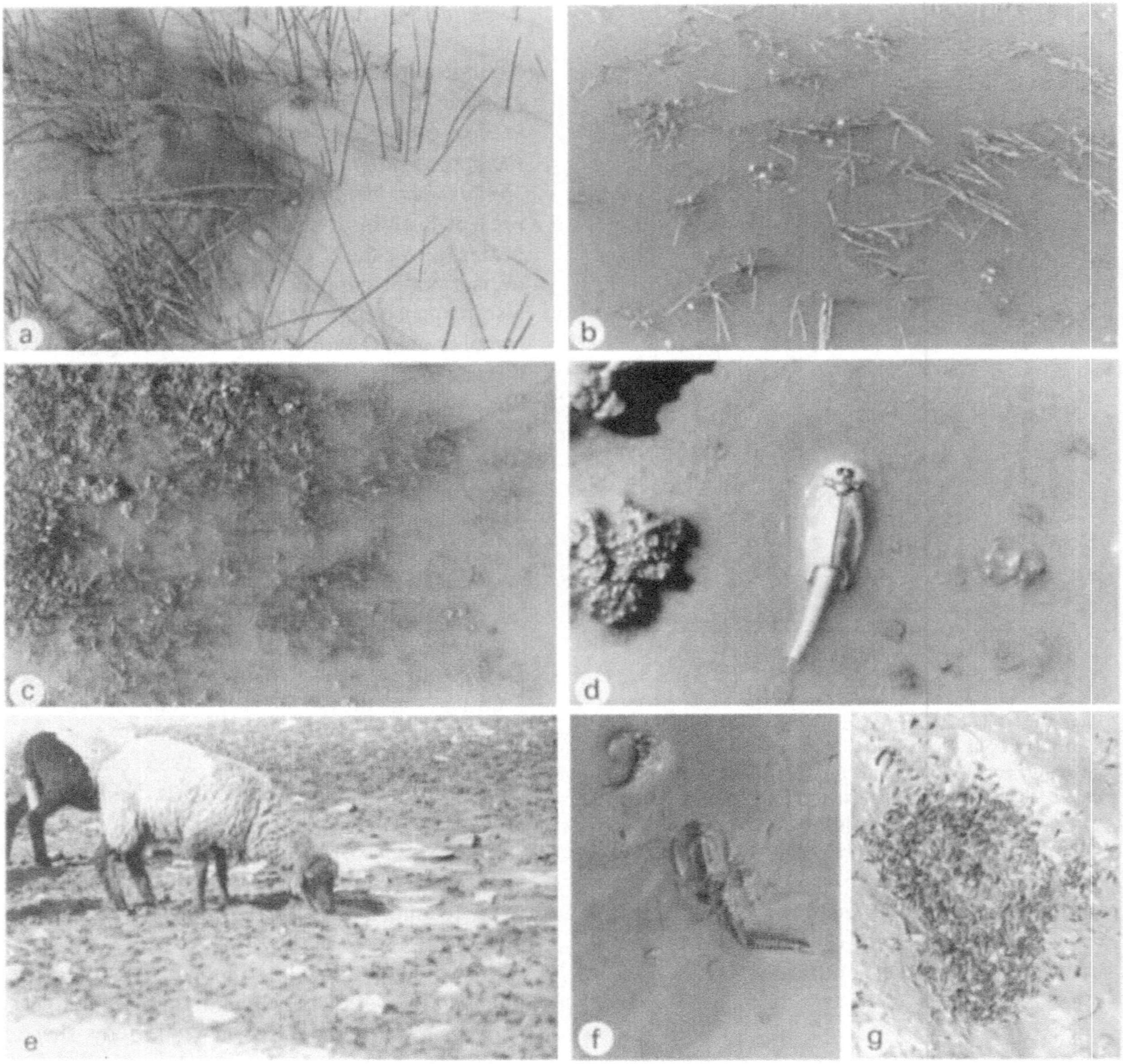


*Fig. 7.* a) shore of road-ditch 14 with *Spirogyra* entangled in *Eleocharis palustris*. b) *Glyceria fluitans* and *Ranunculus aquatilis* in daya 16. c) *Marsilea pubescens* and *Isoetes velata* at shore of daya 16. d) *Triops granarius* in turbid water during drying period (Secchi disc < 1 cm). e) sheep drinking in a daya with their legs in the mud. f) dead *Triops granarius* in dry daya. g) *Leptestheria mayeti* gathered in deep zone of dry daya.

lected from bottom to surface with a 200 $\mu$m net, towed for about 10 minutes across each pond, in and out of the aquatic vegetation. Samples were pooled and preserved in 10% neutralised formalin. At every visit water for analysis of physical and chemical variables of the habitat was collected. Specific Conductance, pH, dissolved Oxygen, temperature, depth and turbidity were measured in the field. Dry residue was evaluated after dehydration at 105 °C for 24 hours. Total organic matter, particulate and in solution, was determined after calcination of dry residue at

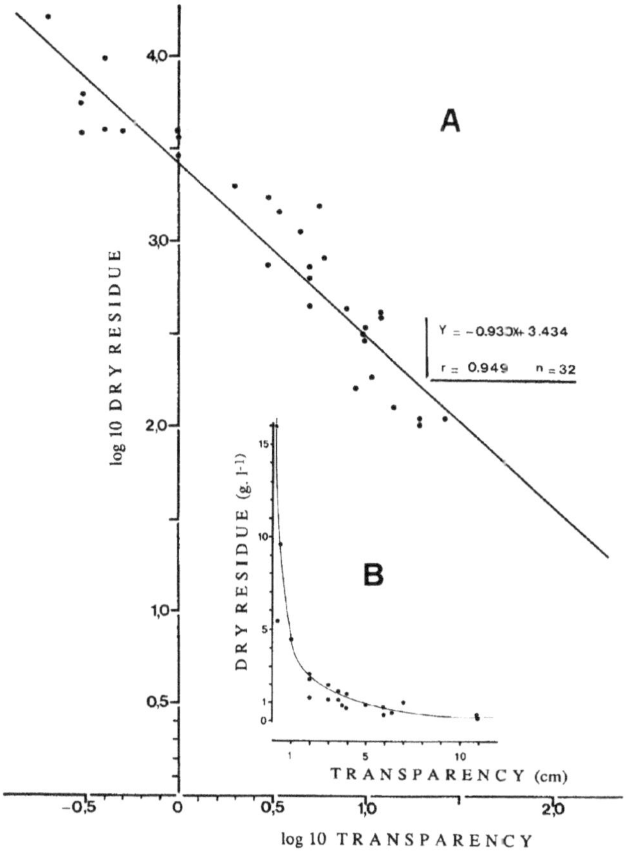


*Fig. 8.* A. Relation between dry residue and Transparency (log scale): $Y = -0.930\ X + 3.434$ ($r^2 = 0.900$, $n = 32$) B. (Lower), same relation in untransformated data.

525 °C during 6 hours. X-Ray analysis was made with a 600 W powder diffractometer.

Species were determined with the help of Daday (1910, 1914, 1923), Linder (1941), Longhurst (1955), Straskraba (1966), Flössner (1972), and Thiéry (1987, 1988a). Anostraca were measured under a binocular microscope with an eye micrometer. Total length was taken from the front to extremity of the cercopods. Notostraca were measured along the dorsal carina of the cephalothoracic carapace (Longhurst, 1955; Linder, 1952, and Thiéry, 1987). For Spinicaudata the length of the valves was measured. For constructing growth curves, the length of individuals of the three orders were grouped into one millimeter size-classes. Data on rainfall and temperature at Berrechid and Casablanca-Nouasser Airport were provided by the National Meteorologic Department in Casablanca.

For bacterial analysis, samples of water and sediment were stored in ice and brought to the laboratory where they were studied. Enumerations were made after plating out on nutritive gelose (Plate count Agar). All cultures were incubated during 7 days at room temperature (20–25 °C). Bacterial identifications were made by classic staining techniques and using culture medium and identification galeries 'Api System R' 20 E. ref. 2010 & 20 N.E. ref. 2005.

## Results

Twelve species were collected: *Chirocephalus diaphanus* Prévost, *Streptocephalus torvicornis bucheti* Daday, *Branchinecta ferox* (Milne-Edw.), *Tanymastigites jbiletica* Thiéry, *Tanymastigites brteki* Thiéry, *Branchipus schaefferi* Fischer, *Tanymastix affinis* Daday, *Triops granarius* (Lucas), *Triops cancriformis mauritanicus* Ghigi, *Cyzicus bucheti* (Daday), *Leptestheria mayeti* Simon, and *Maghrebestheria maroccana* Thiéry (Table 1). Thus, 66.6% of the moroccan euphyllopod fauna is present in the Chaouia plain, much more than in any other geographical and climatic area in Morocco studied (Fig. 13).

The species assemblage of Chaouia ponds is composed of species from two of the bioclimatic areas of Thiéry (1986):

– species from arid areas: *Triops granarius*, *Tanymastigites jbiletica*, *Leptestheria mayeti* and even *Branchipus schaefferi*, although it prefers high continental areas.

– species from semi-arid areas with oceanic influence: *Cyzicus bucheti*, *Maghrebestheria maroccana*, *Streptocephalus torvicornis bucheti*, and *Triops cancriformis mauritanicus*.

The relationship between the two groups of species and the climatic areas is shown on Fig. 11, along a transect between the coastal plain and the Sahara, through the High Atlas. The limits between the various associations are distinct and the transitions are rapid, so their respective species can be considered as allopatric. Only few

*Table 1.*

| Station number | 1 (a) | | | 2 | | | 3 (a) | | | 4 | | | 5 | | | 6 | | | 7 | | | 8 | | | 9 | | | 10 (a) | | |
|---|---|---|---|---|---|---|---|---|---|---|---|---|---|---|---|---|---|---|---|---|---|---|---|---|---|---|---|---|---|---|
| Maximum depth (m) | 0.4 | | | 0.3 | | | 1.5 | | | 0.4 | | | 0.4 | | | 0.6 | | | 1.5 | | | 0.4 | | | 0.3 | | | 0.3 | | |
| Maximum volume ($m^3$) | 20 | | | 3.000 | | | 6 | | | 1.5 | | | 10 | | | 750 | | | 1.980 | | | 1.500 | | | 1.125 | | | 1 | | |
| | A | B | C | A | B | C | A | B | C | A | B | C | A | B | C | A | B | C | A | B | C | A | B | C | A | B | C | A | B | C |
| | | | | | | (b) | | | | | | (b) | | | | | | | | | | (b) | | | | | | | | |
| *Branchipus schaefferi* | • | • | • | | | | • | • | • | | | | • | • | • | | • | • | • | • | • | | | | • | • | • | • | | • |
| *Tanymastix affinis* | | | | | | | | | | | | | • | | | | | | | | | | | | | | | • | • | • |
| *Streptocephalus torvicornis bucheti* | | | | | | | | | | | | | • | | | | | | • | • | | | | | | | | | | |
| *Chirocephalus diaphanus* | | | | | | | | | | | | | • | | | | | | • | • | • | | | | | | | | | |
| *Tanymastigites brteki* | | | | • | • | | | | | | | | | • | • | • | | | | | | | • | • | | | | | | |
| *Tanymastigites jbiletica* | | | | | | | | | | | | | | | | | | | | | | | | | | | | | | |
| *Branchinecta ferox* | | | | | | | | | | | | | | | | | | | | | | | • | • | | | | | | |
| *Triops cancriformis mauritanicus* | | | | | | | | | | | | | • | • | | | • | • | • | • | • | | | | | | | | | |
| *Triops grnarius* | | | | • | • | | | | | • | • | | • | • | • | • | | • | • | • | • | | • | • | • | • | • | | | |
| *Leptestheria mayeti* | | | | | | | • | • | • | • | • | | | | | • | | | • | • | | | • | • | • | • | • | • | | |
| *Cyzicus bucheti* | | | | | | | • | | | | | | | | | | | | • | • | | | | | | | | | | |
| *Maghrebestheria maroccana* | | | | | | | | | | | | | | | | | | | | | | | | | | | | | | |
| Richness (S) | 1 | | | 2 | | | 3 | | | 2 | | | 7 | | | 5 | | | 7 | | | 4 | | | 3 | | | 3 | | |

| Station number | 11 | | | 12 | | | 13 | | | 14 (a) | | | 15 (a) | | | 16 | | | 17 (a) | | | 18 | | | 19 | | | 20 | | |
|---|---|---|---|---|---|---|---|---|---|---|---|---|---|---|---|---|---|---|---|---|---|---|---|---|---|---|---|---|---|---|
| Maximum depth (m) | 0.3 | | | 0.3 | | | 1.2 | | | 1.2 | | | 0.5 | | | 1.5 | | | 0.2 | | | 0.1 | | | 0.3 | | | 0.4 | | |
| Maximum volume ($m^3$) | 7.500 | | | 550 | | | 45 | | | 36 | | | 4 | | | 5.625 | | | 1.5 | | | 0.25 | | | 190 | | | 380 | | |
| | A | B | C | A | B | C | A | B | C | A | | C | A | B | C | A | B | C | A | B | C | A | B | C | A | B | C | A | B | C |
| | | | | | | | | | | | | | (b) | | | | | | (b) | | | | | | | (b) | (b) | | | |
| *Branchipus schaefferi* | | | | | | | • | • | • | | • | • | | • | • | • | • | • | | | | | | | • | | | | | |
| *Tanymastix affinis* | | | | | | | | | | | • | • | | | | | • | • | | • | • | • | • | • | | | | | | |
| *Streptocephalus torvicornis bucheti* | | | | • | • | • | | | | • | • | • | | • | • | • | • | • | | | | • | | | • | | | | | |
| *Chirocephalus diaphanus* | | | | | | | | | | | | | | | | • | • | • | | | | | | | | | | | | |
| *Tanymastigites brteki* | • | • | • | | | | | | | | | | | | | • | • | • | | | | | | | | | | | | |
| *Tanymastigites jbiletica* | | | | | | | | | | | | | | | | | • | • | | | | | | | | | | | | |
| *Branchinecta ferox* | | | | | | | | | | | | | | | | | | | | | | | | | | | | | | |
| *Triops cancriformis mauritanicus* | | | | | | | | | | • | • | • | | • | • | • | • | • | | | | | | | • | | | • | • | • |
| *Triops grnarius* | • | • | | • | • | • | • | • | • | | | | | | | • | • | • | | | | | | | | | | | | |
| *Leptestheria mayeti* | | | | | • | | • | • | • | | | | | | | | • | • | | | | | | | | | | | | |
| *Cyzicus bucheti* | | | | | | | | | | • | • | • | | | | • | • | • | | | | | | | | | | • | • | • |
| *Maghrebestheria maroccana* | | | | | | | | | | | • | • | | • | • | | | | | | | | | | | | | • | | • |
| Richness (S) | 2 | | | 3 | | | 3 | | | 6 | | | 4 | | | 10 | | | 1 | | | 2 | | | 3 | | | 3 | | |

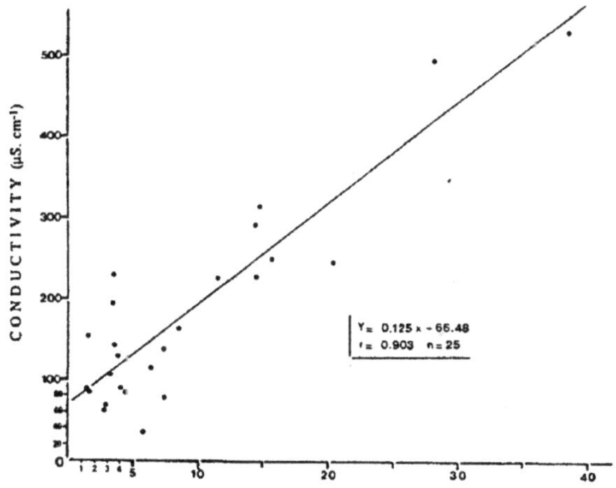


*Fig. 9.* Relation between conductivity and dry residue: $Y = 0.125\ X + 66.48$ ($r^2 = 0.810$, $n = 25$).

ponds are inhabited by species belonging to arid and semi-arid groups. So, it is clear that the Chaouia region is a particularity. It represents an ecotonic zone wider than most other moroccan transitory zones.

In addition to these two groups we noted the presence of species that have no particular bioclimatic requirement, but are stenotopic, such as *Tanymastix affinis* which is restricted to cupulary ponds, rock-pools, and weedy ditches. The presence of *Tanymastigites brteki* and *Branchinecta ferox* is determined by mineralization of the water (Thiéry & Garcin, 1986).

Table 1 shows the presence of a real patchwork of temporary biotopes inhabited by more or less rich and diversified communities. The richness of the communities is related to biotope surface but also to depth. Figure 12 shows that rich communities always occur in deeper ponds (open circles, maximum depth above 0.6 meter). Here (station 16), up to six Anostraca species, two Notostraca and two Spinicaudata may live together and hatch year after year. This regularity of occurrence was not observed in the species studied by Donald (1983). In the temporary ponds of Chaouia, as in all dayas of Morocco, all species hatch in the few hours or days following flooding (Figs. 14, 15). The study of their respective growth curve (Figs. 14, 15) shows a great diversity in the length of their biological cycle. Species living in arid zones (*T. jbiletica*, *T. granarius*, *L. mayeti* and *B. schaefferi*) grow more rapidly but die earlier than species living in semi-arid zones, even if their feeding mode, habitat and way of moving are homologous (*C. bucheti*, *S. torvicornis bucheti*, *T. cancriformis mauritanicus*, *M. maroccana*). This difference in intrinsic growth between arid and semi-arid zone species is illustrated by the growth curves of *Triops granarius* and *Triops cancriformis cancriformis* in one and the same daya (station 7) (Fig. 14). The growth curves, fitted by the Von Bertalanffy model ($L_t = L\infty\ (1 - e^{-K(t - t0)})$), show that their $K$ coefficients are close ($K_{Triops\ granarius} = 0.039$ and $K_{Triops\ cancriformis\ mauritanicus} = 0.034$), but maximum length (theoritical value) is different ($L\infty_{Triops\ granarius} = 41.71$ mm and $L\infty_{Triops\ cancriformis\ mauritanicus} = 30.83$ mm). The Von Bertalanffy equations are: $L_{t \cdot Triops\ granarius}$ (mm) $= 41.71\ (1 - e^{-0.039\ T})$ and $L_{t\ Triops\ cancriformis\ mauritanicus}$ (mm) $= 30.83\ (1 - e^{-0.034\ T})$.

In Daya 16, at the community level, the

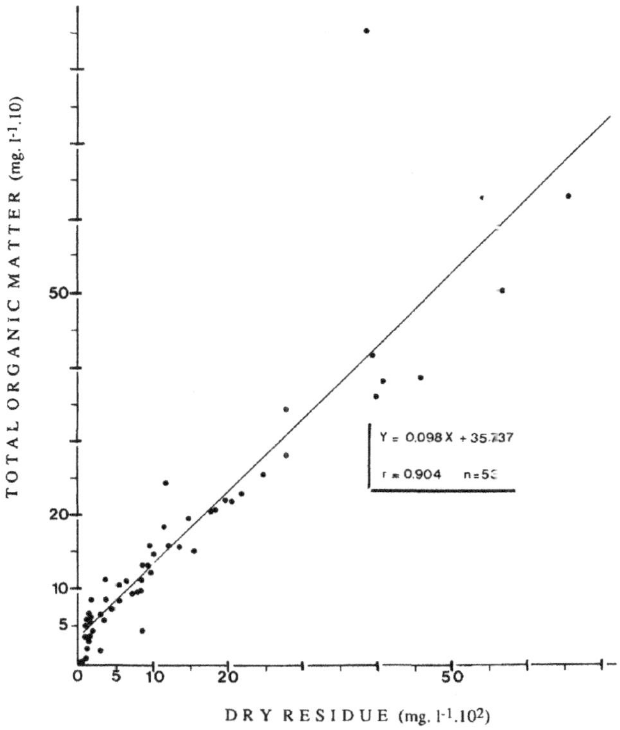


*Fig. 10.* Relation between total organic matter, and dry residue: $Y = 0.098\ X + 35.74$ ($r^2 = 0.811$, $n = 53$).

sequence for the various species was as follows (Fig. 15): *T. affinis* is adult after a period of six–seven days and disappears around the 16th day; *Tanymastigites jbiletica* completes its biological cycle in 26–30 days. The cycle of *Branchipus schaefferi* and *Streptocephalus torvicornis bucheti* takes 66–75 days and their optimal sizes are similar to each other. The growth curve of *Chirocephalus diaphanus*, in which a gynandromorph has been discovered and described by Thiéry (1983), is not presented because the population was not abundant. *Triops granarius* grows rapidly but disappears after about 56 days, while *Triops cancriformis mauritanicus* grows slowlier but survives for 160–180 days. Similarly, among Spinicaudata *Leptestheria mayeti* grows twice as fast as *Cyzicus bucheti* but disappears after about 66–70 days, while *C. bucheti* only disappears 180 days after of flooding. We found that in Chaouia ponds, species of the arid zone had the same growth as in their typical climatic area.

Predation: as shown by the digestive tract analysis, adults of *Pleurodeles* have an impact on the populations of phyllopods, young feeding on

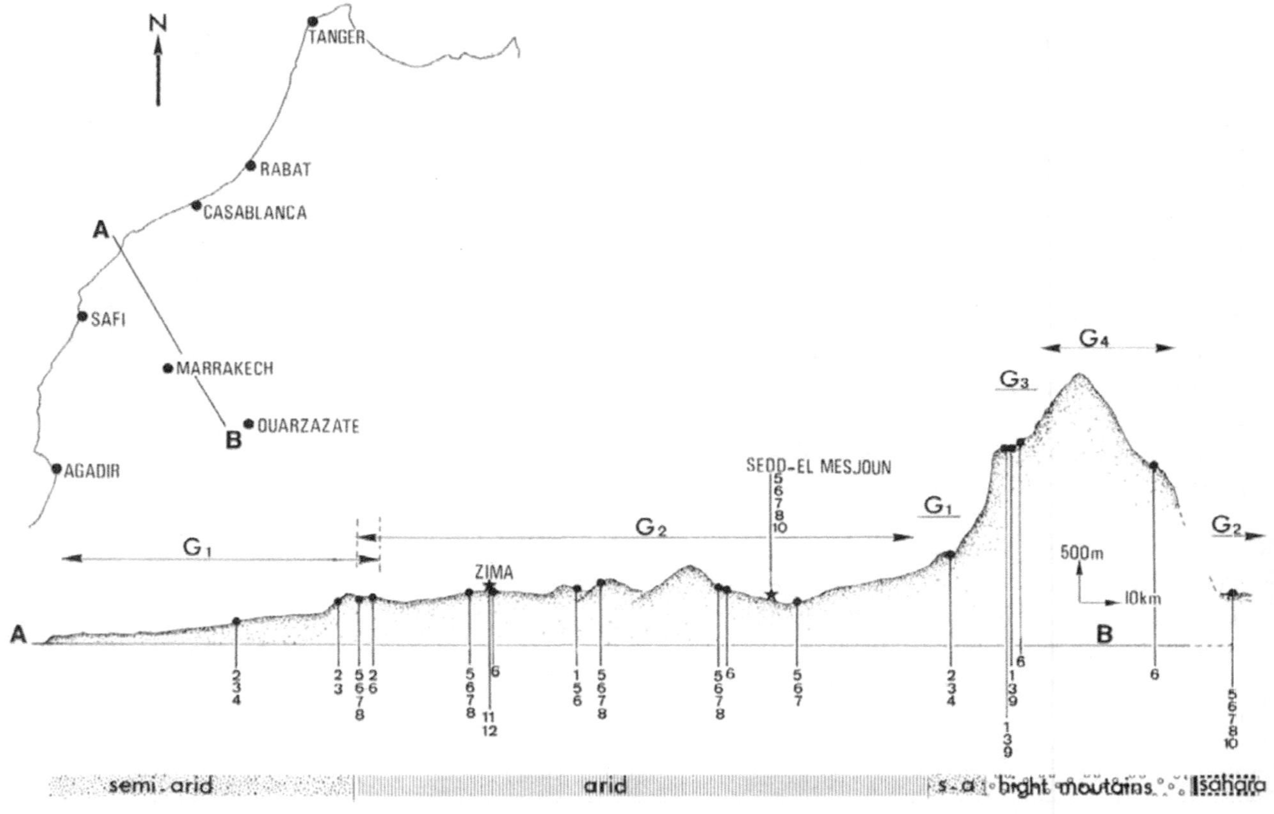


*Fig. 11.* Altitudinal and bioclimatic transect N. N.W.-S. S.E. showing the zonation of the species groups. 1. *Tanymastix affinis.* (stenotopic of ephemeral ponds), 2. *Streptocephalus torvicornis bucheti*, 3. *Triops cancriformis mauritanicus*, 4. *Cyzicus bucheti* (2–4 species of semi-arid zone), 5. *Tanymastigites jbiletica*, 6. *Branchipus schaefferi*, 7. *Triops granarius*, 8. *Leptestheria mayeti* (5–8 species of arid zone), 9. *Chirocephalus diaphanus*, 10. *Eocyzicus saharicus*, (stenotopic of oligohaline and hot waters), 11. *Artemia* sp, 12. *Branchinella spinosa* (11 & 12 salt waters species).

Black points are temporary ponds along the transect A–B; stars (Zima and Sedd El Mesjoun) are sebkhra's (inland salt waters). Group 1 (G.1.): semi-arid zone; Group 2 (G.2): arid zone; Group 3 (G.3): mountain zone; Group 4 (G.4): high mountain above 2500 m.

Below the transect the semi-arid, arid, high mountain and sub desertic zones are shown. Their borders are according to Sauvage's map (1963). The transition between the different groups, corresponding to bioclimatic stages, are distinct. Only one interpenetration between group 1 and Group 2 was recorded.

benthic Ostracoda and Cladocera. However, as *Pleurodeles* appears only in spring, when phyllopods are in the second half of their growth curve, and that *Pleurodeles waltlii* are low in densities ($<0.2\ m^{2-1}$), their predatory impact on phyllopods is not important. Other amphibians, as *Bufo*, eat terrestrial fauna (insects, annelids...) and birds appear only in low densities at the end of hydrological cycle.

Predation by insects (Dytiscidae and Odonata) was not estimated.

## Discussion

At any point in time all co-occurring species present in a pond showed differences in body-length, resulting from a different neonatal length and a different growth rate. Several authors (Mozley, 1932; Rzóska, 1961; Moore, 1963; Geddes, 1981, 1986; Donald, 1971 and Mura, 1985) observed the same variation in size among sympatric species resulting from different growth speeds. The growth curves obtained from Retallack & Clifford's (1980) data for the pairs *Eubranchipus bundyi* – *E. intricatus*, and for *Lynceus brachyurus* – *L. mucronatus*, also combine different sizes with different growth rates. As well as the anostracans *Eubranchipus bundyi* (life cycle of 2 months) and *E. intricatus* (life cycle of 3–4 months), the life cycle of the Laevicaudata was of two types: *L. mucronatus* grows and lives for 60 days, while *L. brachyurus* lasts 118 days. A third coexisting species, *Cyzicus mexicanus*, persists throughout summer. These results are close to those of Hamer & Appleton (1991) which show the importance of differences between growth rate of phyllopods coexisting in the same temporary pool.

According to Daborn (1976a, 1979) the size range of the individuals belonging to the same or to different species seems to be of major importance. Size variation implies a diversity in feeding behaviour, related to the interspace between filtering setae on the endites of the foliaceous legs. The size of the filtered particles is different for the coexisting species, and filtering segregation may be a major factor permitting coexistence of ento-

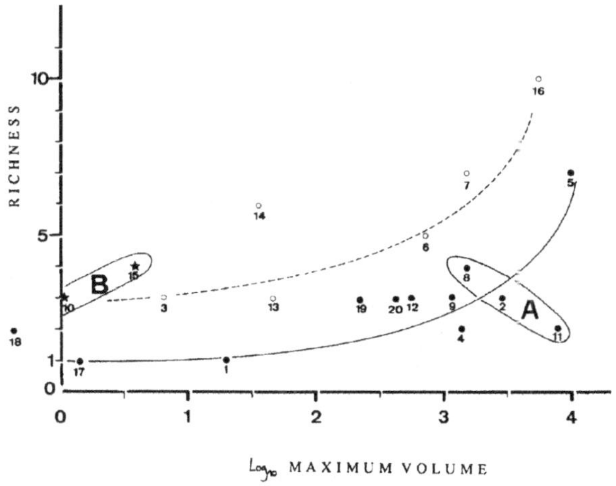


*Fig. 12.* Relation between species richness and volume of each pond (log transformation). Black points correspond to depths < 0.50 m, white circles to depths > 0.60 m. A is a group of ponds with brackish water, and B are road-ditches.

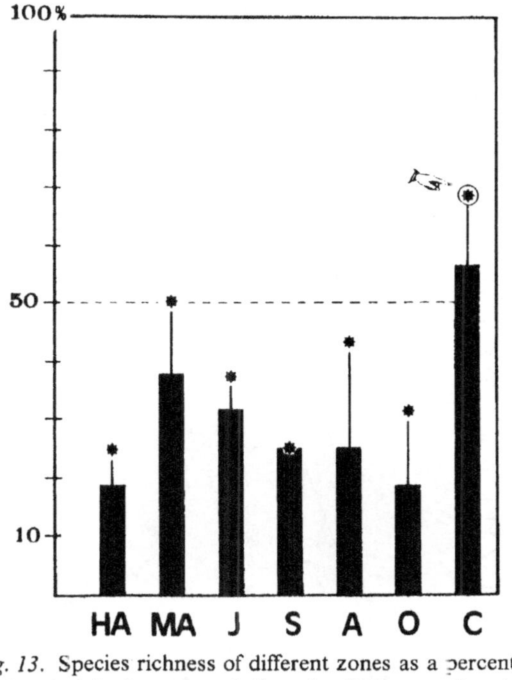


*Fig. 13.* Species richness of different zones as a percentage of the total freshwater euphyllopods of Morocco (brackish-water species, *Artemia* and *Branchinella* excluded). Black bars: maximum species richness in one pond in %; Stars: maximum richness of each zone, all biotopes together, in %. HA: High Atlas; MA: Middle Atlas; J: Jbilet mountains and Marrakech area; S: Sahara (east of High-Atlas; A: Atlantic coast; O: oriental plateau (east of Middle-Atlas); C: study area, Chaouia plain between Berrechid and Souk Jemma Oulad Abbou, the richest zone.

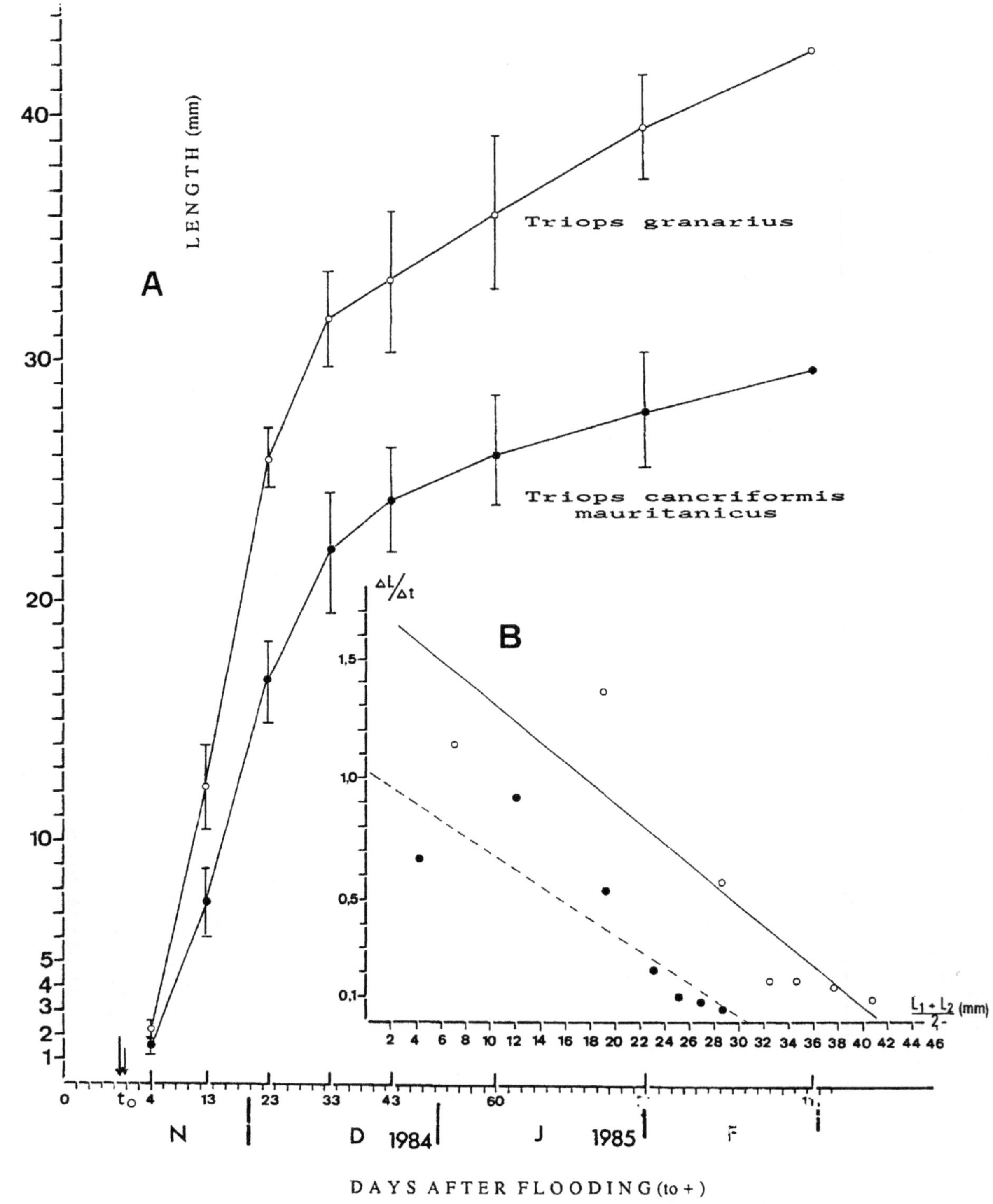


*Fig. 14.* A: growth-curves of *Triops granarius* (open dots) and *Triops cancriformis mauritanicus* (full points) in Daya 7 between flooding (black arrows) in November 1984 and February 1985. Each average on the growth curves with standard error. *B* (below): graphic determination of $L\infty$ (intersection of with x-axis) and *K* (slope of regression line).

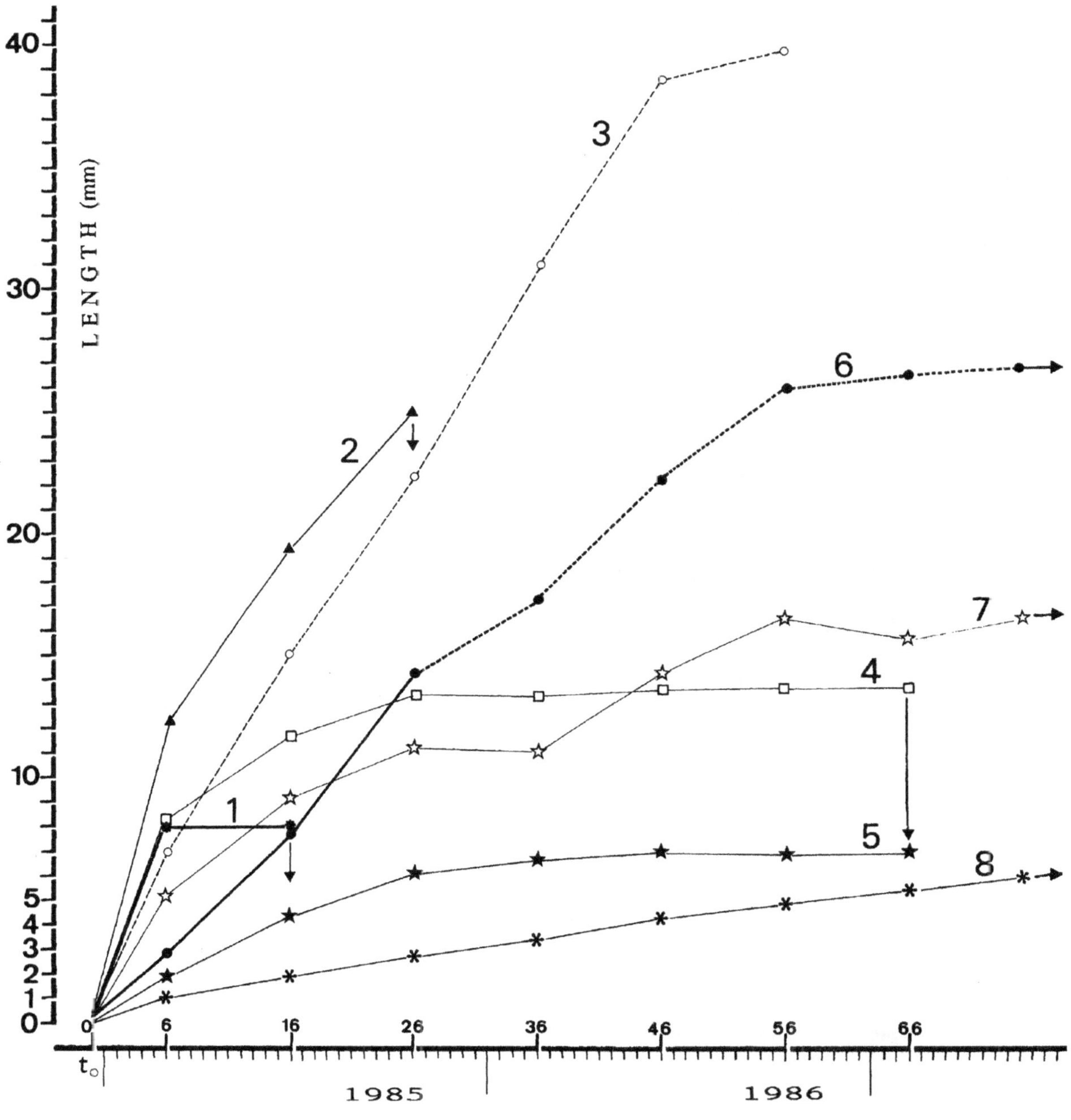


*Fig. 15.* Growth curves of all co-occurring species of daya 16 from December 1985 (flooding) to March 1986. 1) *Tanymastix affinis* 2) *Tanymastigites jbiletica* 3) *Triops granarius* 4) *Branchipus schaefferi* 5) *Leptestheria mayeti* 6) *Triops cancriformis mauritanicus* 7) *Streptocephalus torvicornis bucheti* 8) *Cyzicus bucheti.* Growth curve of *Chirocephalus diaphanus* not given. Vertical arrows indicate natural extinction of populations (species 1, 2, 3, 4, 5); horizontal arrows indicate that the biological cycle lasts between 2 or 3 months. (species 6, 7, 8).

mostracan species (Walter, 1970; Makarevicz & Likens, 1975; Hebert, 1982; Mertens *et al.*, this volume). In our case, this phenomenon is closely associated with the nature of food supplies. In the turbid water of temporary ponds the production and biomass of phytoplankton is limited (Marzolf & Osborne, 1972; for a review see Stern & Stickle, 1978). In the dayas studied, few phytoplankton cells occur (range 1.2. $10^4$ cell $l^{-1}$ – 1.5. $10^6$ cell $l^{-1}$; sometimes 4. $10^8$ cells $l^{-1}$). Moreover these algae have poor nutrient value because of a rapid intestinal transit (less than half an hour) and a low degradation rate. So, bacterial and protist biomass are the main food source for filtering invertebrates (Saunders *et al.*, 1980). As shown by Tezuka (1974) and Berk *et al.* (1976) for Cladocera and Copepods, there is no growth if bacterial concentration is below $10^7$ cell $ml^{-1}$. In turbid dayas the concentration of free-living bacteria (*Micrococcus* represents the dominant biomass) is always lower than $10^7$ $ml^{-1}$ (Thiéry, 1987) and therefore the growth of euphyllopods must result from bacteria attached to clay. Many authors (Felton *et al.*, 1967, and Pedrós-Alió & Brock, 1983) have shown that an increase in attachment area stimulates bacterial activities and organic matter fixation (Murphy, 1936). Unlike Montmorillonite clays, Kaolinite and Illite are not toxic to planktonic and benthic animals, even at saturation (Robinson, 1957).

By the continuous rhythmic beat of their thoracic legs branchiopods (Cannon, 1933; Barlow & Asleigh, 1980) create a current which concentrates particles (clays, organic flocks). The ingested bacterial biomass may then be very high.

The distribution of *Triops granarius* inside one pond shows that the individuals do not uniformly occupy all the area. Young individuals are present near the edges of a pond, while older individuals tend to gather in the center (Fig. 16). *Tanymastigites jbiletica*, *Triops cancriformis mauritanicus*, and the Spinicaudata *Leptestheria mayeti* and *Cyzicus bucheti* had a similar distribution inside their ponds. Similar observations are made by Hamer & Appleton (1991) in temporary ponds of Natal. They distinguish two main groups, the 'edge' species and the 'central' species.

Therefore, when two species belonging to various bioclimatic groups live together in the same pond, both differences in growth-speed and spatial spreading may allow these species to occupy distinct micro-habitats.

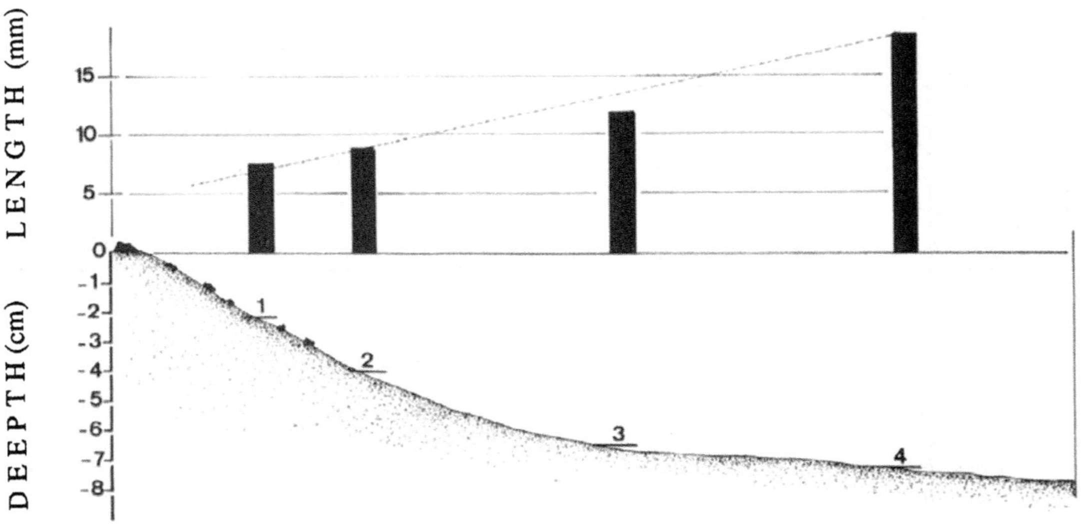


*Fig. 16.* Differential distribution of one cohort of *Triops granarius* as related to depth in a single pond. Black bars represent average length of cephalothoracic carapace (mm) and numbers (1, 2, 3 and 4) are four stations with different depths (cm) in the pond. The largest *Triops* gather in the deepest zones.

## Conclusion

Explanations of the coexistence of otherwise allopatric species of euphyllopods are of four types:

1) Climatic and water characteristics favour the co-occurrence of naturally allopatric species. The concept of a climatic ecotone containing a patchwork of temporary ponds (with or without link with the oligohaline groundwater), some being more, others less temporary, and shallow or deep, may explain the presence of every species in the Chaouia plain. Large and deep ponds allow the biological cycle of more species to get completed than small, ephemeral ponds where only species with short cycles can reach maturity. The former may also develop heterogeneity across the water column. Consequently, their theorical richness should be greater (Table 1), shown also by Bengtsson (1989), and Fryer (1985). Co-occurrence is facilitated by importation of resting eggs by amphibians, birds and sheep (zoochory).

2) In turbid environment a diversity of food types, greatly paired to organic matter attached to clay, may allow a resource partitioning. Anostraca, Notostraca and Spinicaudata, although filtering species, do not exploit the same resources. While Anostraca are principaly microphagous filter-feeders, abstracting particles from suspension in the open water (Fryer, 1987), Notostraca are non-filtering, benthic omnivores, feeding on detritus (Fryer, 1987, 1988). Sometimes Notostraca can also be diggers and sifters of organic bottom material and feed on small animals like Tubificidae (Thiéry, 1987). Spinicaudata are also omnivorous (Royan, 1976) but sometimes obtain food by scraping the periphyton attached to vegetation. As the source of food is diversified (benthic bacterial flocks, planktonic ones, detritus, periphyton, small benthic prey...), an obvious result is that the resources exploited by co-occurring species of these different orders are distinct enough to permit this coexistence. As the nutritional resources based on argillotrophic system are most of the time in excess, this typical characteristic of temporary ponds of arid and semi-arid zones (Yaron, 1964; Brown, 1968; Daborn, 1975; Rzóska, 1984; Thiéry, 1987) seems a major factor regulating the life history of euphyllopods. All branchiopods of semi-arid and arid zones of Morocco could be considered as 'argillotrophobacterivorous' animals.

3) In coexisting Anostraca, Notostraca and Spinicaudata in Chaouia dayas, significant size differences exist between species. These data are close to these of Hutchinson (1951), Cole (1961), Hammer & Sawchyn (1968), Anderson & Fabris (1970) and Champeau (1970) on Copepod populations. As these authors suggest, size differences may result in a difference in food niche. Presumably, larger species will ingest larger food particles than smaller species.

4) Horizontal or vertical distribution of the species of the different orders within the pond seems to be of great importance. Our data on *Triops granarius* show that Notostraca of different length (but also, Anostraca and Spinicaudata) may be separated horizontally and vertically). The vertical and horizontal heterogeneity of the pond, mainly vegetation structure or any other environmental factor correlated with turbidity, like vertical thermic microstratification (Thiéry, 1987), favours the optimal occupation of microhabitats by all the syntopic species.

In consequence, following Mac Arthur & Wilson's (1967) theory, according to which interspecific competition increases local extinction rates and limits the regional richness of species in islands or restricted areas, we could predict that the lack or the low level of interspecies competition between Anostraca, Notostraca and Spinicaudata is an important factor favoring richness of species in pond communities. The results of Bengtsson (1989) on three allotopic *Daphnia* species reinforce this interpretation: while these species, using common food resources and of similar population dynamics, increase their extinction rate when they coexist in experimental conditions, they never occur together in the same habitat (natural rockpools or ponds in Sweden or Finland (trophical exclusion).

Our study suggests that the spatial and temporal distribution of co-occurring species in a temporary pond results, at each time, from differences in body size. Neonatal length and mass, length of juvenile growth period, feeding behaviour, intrinsic growth rates, are major factors in separating ecophases of all species. Occupying optimally all the available microhabitats, these sympatric species, (to the Chaouia plain geographical scale), must be considered as syntopic species in the larger dayas, or allotopic in smaller ones (in the sense of Rivas, 1964).

Although our results, based on natural observations, do not fully explain the coexistence of branchiopod species, they are a first step towards further research which should consider laboratory culture of animals, food types in relation to length, and behaviour under different conditions (vegetation, open water, depth, turbidity, temperature, light....).

## Acknowledgements

I am grateful to Dr. H.J. Dumont for useful comments and advice.

## References


Alonso, M., 1981. Las comunidades de entomostraceos de las lagunas de Villafafila (Zamora). Actas del Primer Congreso Espanol de Limnologia, Barcelona: 61–67.

Anderson, R. S. & G. L. Fabris, 1970. A new species of diaptomid copepod from Saskatchewan with notes on the crustacean community of the pond. Can. J. Zool. 48: 49–54.

Arruda, J. A., G. R. Marzolf & R. T. Faulk, 1983. The role of suspended sediments in the nutrition of zooplankton in turbid reservoirs. Ecology 64: 1225–1235.

Balfour-Browne, F., 1909. Note on the rediscovery of *Apus cancriformis* in Britain. Ann. Scot. nat. Hist. 1909: 118.

Barlow, D. I. & M. A. Sleigh, 1980. The propulsion and use of water currents for swimming and feeding in larval and adult *Artemia*. In G. Persoone, P. Sorgeloos, O. Roels & E. Jaspers (eds), The Brine shrimp *Artemia*. Vol. I. Morphology, Genetics, Radiobiology, Toxicology. Universa Press, Wetteren, Belgium: 61–73.

Belk, D., 1977. Zoogeography of the Arizona Fairy Shrimps (Crustacea: Anostraca). Arizona Academy of Science 12: 70–78.

Bengtsson, J., 1989. Interspecific competition increases local extinction rate in a metapopulation system. Nature 340: 713–715.

Berk, S. G., R. R. Colwell & E. B. Small, 1976. A study of feeding responses to bacterial prey by estuarine ciliates. Trans. Am. Micros. Soc. 95: 514–520.

Brown, G. W., 1968. Desert biology; Special topics on the physical and biological aspects of arid regions. Academic Press ed., 1: 635 pp.

Cannon, H., 1933. On the feeding mechanism of the Branchiopoda. With an appendix on the mouth parts of the Branchiopoda by Cannon H. & F. M. C. Leak. Phil. Trans. r. Soc. London B 222: 267–352.

Champeau, A., 1970. Recherches sur l'écologie et l'adaptation à la vie latente des Copépodes des eaux temporaires provençales et corses. Thèse Doctorat es Sciences, Université Aix-Marseille I, 360 pp.

Cole, G. A., 1961. Some calanoid copepods from Arizona with notes on congeneric occurrences of *Diaptomus* species. Limnol. Oceanogr. 6: 432–442.

Daborn, G. R., 1975. The argillotrophic lake system. Verh. Int. Ver. Limnol. 19: 580–588.

Daborn, G. R., 1976a. The life history of *Branchinecta mackini* Dexter (Crustacea, Anostraca) in an argillotrophic lake of Alberta. Can. J. Zool. 55: 161–168.

Daborn, G. R., 1976b. Physical and chemical features of a vernal temporary pond in Western Canada. Hydrobiologia 51: 33–38.

Daborn, G. R., 1978. Distribution and biology of some nearctic tundra pool phyllopods. Verh. Int. Ver. Limnol. 20: 2442–2451.

Daborn, G. R., 1979. Limb structure and sexual dimorphism in the Anostraca (Crustacea). Can. J. Zool. 57: 894–900.

Daday, E., 1910. Monographie systématique des Phyllopodes Anostracés. Ann. Sc. Nat. Zool. sér. 9, 9: 91–489.

Daday, E., 1914. Monographie systématique des phyllopodes Conchostracés. Ann. Sc. Nat. Zool. série 9, 20: 39–330.

Daday, E., 1923. Monographie systématique des phyllopodes Conchostracés. Ann. Sc. Nat. Zool. série 10, 6: 255–390.

Decksbach, N. K., 1924. Zür Verbreitung und Biologie der Apusidae in Russland. Russ. Hydrobiol. Zeits. 3: 143–155.

Dexter, R. W., 1946. Further studies on the life history and distribution of *Eubranchipus vernalis* (Verrill). Ohio J. Sci. 46: 31–44.

Dexter, R. W., 1953. Studies of North American fairy shrimps with the description of two new species. Am. Midl. Nat. 49: 751–771.

Dexter, R. W. & C. H. Kuehnle, 1951. Further studies on the fairy shrimp populations of North eastern Ohio. Ohio J. Sci. 51: 73–86.

Dimentman, C., 1981. The rainpool ecosystems of Israel: geographical distribution of freshwater Anostraca (Crustacea). Israel J. Zool. 30: 1–15.

Donald, D. B., 1971. Limnology and plankton of temporary ponds. Thesis, University of Calgary, Alberta, Canada. 135 pp.

Donald, D. B., 1983. Erratic occurrence of Anostracans in a temporary pond: colonization and extinction or adaptation to variations in annual weather? Can. J. Zool. 61: 1492–1498.

Felton, M., J. J. Cooney & W. F. Moore, 1967. A quantitative study of the bacteria of a temporary pond. J. gen. Microbiol. 47: 25–31.

Flössner, D., 1972. Krebstiere, Crustacea, Kiemen- und Blattfusser, Branchiopoda Fischlause, Branchiura. Tierwelt Deutschlands, Veb Gustav Fischer Verlag Jena Ed. 60: 1–501.

Fryer, G., 1985. Crustacean diversity in relation to the size of waterbodies: some facts and problems. Freshwat. Biol. 15: 347–361.

Fryer, G., 1987. A new classification of the branchiopod Crustacea. Zool. J. Linn. Soc. 91: 357–383.

Fryer, G., 1988. Studies on the functional morphology and biology of the Notostraca (Crustacea: Branchiopoda). Phil. Trans. r. Soc., London B, 321: 27–124.

Gauthier, H., 1928. Recherches sur la faune des eaux continentales de l'Algérie et de la Tunisie. Thèse, Lechevallier Ed., Paris, 419 pp. + 4 pl. h. t.

Gauthier, H., 1930. Sur la répartition en Algérie et en Tunisie de certains Entomostracés d'eau douce Européens (Branchipes et *Apus*). C. r. Assoc. fr. Avanc. Sci.: 572–575.

Geddes, M. C., 1981. The Brine shrimps *Artemia* and *Parartemia*. Comparative physiology and distribution in Australia. Hydrobiologia 81: 169–179.

Geddes, M. C., 1986. Understanding Zooplankton communities in Farm Dams: the importance of predation. In P. De Deckker & W. D. Williams (eds), Limnology in Australia, Monographiae Biologicae Vol. 61, Dr W. Junk, Dordrecht: 387–401

Hamer, M. L. & C. C. Appleton, 1991. Life history adaptations of phyllopods in response to predators, vegetation and habitat duration in north-eastern Natal. In Belk, D., H.J. Dumont & N. Munuswamy (eds), Studies on Large Branchiopod Biology and Aquaculture. Developments in Hydrobiology 64. Kluwer Academic Publishers, Dordrecht: 105–116. Reprinted from Hydrobiologia 212.

Hammer, U. T. & W. W. Sawchyn, 1968. Seasonal succession and congeneric associations of *Diaptomus* spp. (Copepoda) in some Saskatchewan ponds. Limnol. Oceanogr. 13: 476–484.

Hebert, P. D. N., 1982. Competition in zooplankton communities. Ann. Zool. Fenn. 19: 349–356.

Horne, F., 1967. Effects of physical-chemical factors on the distribution and occurrence of some southeastern Wyoming Phyllopods. Ecology 48: 472–477.

Huggins, D. G., 1974. The sympatric occurrence of three species of Eubranchiopoda in Douglas County, Kansas. Southw. Nat. 20: 577–578.

Hutchinson, G. E., 1951. Copepodology for the Ornithologist. Ecology 32: 571–577.

Kutsch, H. G., 1978. Le pouvoir d'évaporation du climat marocain. Ministère de l'Agriculture et de la Réforme Agraire. Direction de la Recherche Agronomique ed., Rabat 378 pp.

Linder, F., 1941. Contributions to the morphology and taxonomy of the Branchiopoda Anostraca. Zoo. Bidr. Uppsala 20: 101–302 + 1 pl. h. t.

Linder, F., 1952. Contributions to the morphology and taxonomy of the Branchiopoda Notostraca, with special reference to the North American species. Proc. U. S. Nat. Mus. 102: 1–69.

Longhurst, A. R., 1955. A review of the Notostraca Bull. Brit. Mus. (Nat. Hist.) Zool. London 3: 1–57.

Mac Arthur, R. H. & E. O. Wilson, 1967. The theory of island biogeography. Princeton University Press ed.

Maeda-Martinez, A., 1991. Distribution of species of Anostraca, Notostraca, Spinicaudata, and Laevicaudata in Mexico. In Belk, D., H.J. Dumont & N. Munuswamy (eds), Studies on Large Branchiopod Biology and Aquaculture. Developments in Hydrobiology 64. Kluwer Academic Publishers, Dordrecht: 209–219. Reprinted from Hydrobiologia 212.

Makarewicz, J. C. & G. E. Likens, 1975. Niche analysis of a zooplankton community. Science 190: 1000–1003.

Marzolf, G. R. & J. A. Osborne, 1972. Primary production in a Great Plains reservoir. Verh. int. Ver. Limnol.: 18: 126–133.

Mathias, P., 1937. Biologie des Crustacés Phyllopodes. Act. Sci. Ind. 447, Hermann Paris, 107 pp.

Mattox, N. T., 1957. A new estheriid conchostracan with a review of other North American forms. Am. Midl. Nat. 58: 367–377.

Maynard, S. S. & S. V. Romney, 1975. The occurrence of four new Anostracan (Crustacea) Phyllopod in Utah, multispecies records, and some notes on their ecology. Utah Acad. Proc. 52: 6–11.

Melack, J. M., 1985. Interactions of detrital particulates and plankton. Hydrobiologia 125: 209–220.

Mertens, J., N. Munuswamy & H. J. Dumont, 1991. The filtration apparatus of Anostraca (Crustacea): species specific setulation in the genus *Streptocephalus*. In Belk, D., H.J. Dumont & N. Munuswamy (eds), Studies on Large Branchiopod Biology and Aquaculture. Developments in Hydrobiology 64. Kluwer Academic Publishers, Dordrecht: 187–193. Reprinted from Hydrobiologia 212.

Moore, W. G., 1963. Some interspecies relationships in Anostraca populations of certain Louisiana ponds. Ecology 44: 131–139.

Moore, W. G., 1970. Limnological studies of temporary ponds in Southeastern Louisiana. Southwest. Nat. 15: 83–110.

Mozley, A., 1932. A biological study of a temporary pond in western Canada. Am. Nat. 66: 235–249.

Mura, G., 1985. Preliminary report on the sympatric occurrence of two species of Fairy shrimps (Crustacea, Anostraca) in some temporary ponds from Italy. Riv. Idrobiol. 24: 73–82.

Mura, G., 1987. Occurrence of *Artemia* in solar saltworks and coastal brine ponds in Sardinia, Italy. J. Crust. Biol. 7: 697–703.

Mura, G. & V. Cottarelli, 1984. Gli Anostraci delle acque temporanee di Castel Porziano (Roma). Riv. Idrobiol. 23: 165–171.

Murphy, H. F., 1936. The role of Kaolinite in phosphate fixation. Hilgardie 12: 343–382.

Pedrós-Alió, C. & T. D. Brock, 1983. The importance of attachment to particles for planktonic bacteria. Arch. Hydrobiol. 98: 354–379.

Proctor, V. W., E. R. Malone & V. L. De Vlaming, 1967. Dispersal of aquatic organisms: viability of disseminules recovered from the intestinal tract of captive killdeer. Ecology 48: 672–676.

Prophet, C. W., 1962. Ecology and reproduction of five species of Anostraca in Oklahoma. Dissert. Abs. USA 23: 1129.

Prophet, C. W., 1963. Physical-chemical characteristics of habitats and seasonal occurrence of some Anostraca in Oklahoma and Kansas. Ecology 44: 798–801.

Retallack, J. T. & H. F. Clifford, 1980. Periodicity of Crustaceans in a saline Prairie Stream of Alberta, Canada. Am. Midl. Nat. 103: 123–132.

Rivas, L. R., 1964. A reinterpretation of the concepts 'Sympatric' and 'Allopatric' with proposal of the additional terms 'Syntopic' and 'Allotopic'. Syst. Zool. 13: 42–44.

Robinson, M., 1957. The effects of suspended materials on the reproductive rate of *Daphnia magna*. Pub. Inst. mar. Sci. 4: 265–277.

Royan, J. P., 1976. Studies on the gut contents of *Leptestheriella maduraiensis* (Conchostraca: Branchiopoda) Nayar and Nair. Hydrobiologia 51: 209–212.

Ruhard, J. P., 1975. Chaouia et plaine de Berrechid. In Ressources en eau du Maroc. Tome 2. Plaines et bassin du Maroc atlantique. Notes et Mémoires du service géologique n° 231. Service Géologique du Maroc, Rabat, pp. 185–221.

Rzóska, J., 1961. Observations on tropical rainpools and general remarks on temporary waters. Hydrobiologia 17: 265–286.

Rzóska, J., 1984. Temporary and other waters. In L. L. Cloudsley-Thompson (ed.), Key environments: Sahara desert, Pergamon Press, pp. 105–114.

Sars, G. O., 1896. Phyllocarida and Phyllopoda. In Fauna Norvegiae Vol. I. 140 pp.

Saunders, G. W., K. W. Cummins, D. Z. Gak, E. Pieczynska, V. Straskrabova & R. G. Wetzel, 1980. Organic matter and decomposers. In E. D. Le Cren & R. H. Lowe-McConnell (eds), The functioning of freshwater ecosystems. I. B. P. 22, Cambridge University Press pp. 341–392.

Sauvage, C., 1963. Etages bioclimatiques. In Atlas du Maroc. Comité National de Géographie du Maroc, Section II. Physique du Globe et Météorologie: 44 pp.

Simon, E., 1886. Etude sur les Crustacés du sous Ordre des Phyllopodes. Ann. Soc. ent. Fr., S6, 6: 393–460.

Stern, E. M. & W. B. Stickle, 1978. Effects of turbidity and suspended material in aquatic environments litterature review. Technical Report D-78-21, United States Army Corps of Engineers Waterways Experiment Station, Vicksburg, Mississippi, USA.

Straskraba, M., 1966. Taxonomical studies on Czechoslovak Conchostraca III. Family Leptestheriidae; with some remarks on the variability and distribution of Conchostraca and a key to the Middle-European species. Hydrobiologia 27: 571–589.

Tezuka, Y., 1974. An experimental study on the food chain among Bacteria, *Paramecium* and *Daphnia*. Int. Revue ges. Hydrobiol. 59: 31–37.

Thiéry, A., 1983. Un cas de gynandromorphisme naturel chez *Chirocephalus diaphanus* Prévost (Crustacea, Anostraca). Riv. Idrobiol. 22: 84–87.

Thiéry, A., 1986. Les Crustacés Branchiopodes Anostracés, Notostracés et Conchostracés du Maroc occidental: I. Inventaire et répartition. Bull. Soc. Hist. nat. Toulouse 122: 145–155.

Thiéry, A. & C. Garcin, 1986. Les Crustacés Branchiopodes (Anostraca, Notostraca, Conchostraca) du Maroc occidental. II: caractérisation de la structure du peuplement par l'analyse factorielle des correspondances. 1st Colloque interuniversitaire Maroco-français, Fes, 5 pp. (unpublished).

Thiéry, A., 1987. Les crustacés Branchiopodes Anostraca, Notostraca & Conchostraca des milieux limniques temporaires (Dayas) au Maroc. Taxonomie, Biogéographie, Ecologie. Thèse Doctorat es Sciences, Université Aix-Marseille III, 405 pp.

Thiéry, A., 1988a. *Maghrebestheria maroccana* n. Gen., n. sp., nouveau représentant des Leptestheriidae au Maroc (Conchostraca). Crustaceana 54: 43–56.

Thiéry, A., 1988b. *Triops* Schrank, 1903 et *Lepidurus* Leach, 1816, Crustaces Branchiopodes Notostracés. In Documents pour un Atlas Zoogéographique du Languedoc-Roussillon, Univeristé Paul Valéry, Montpellier ed., n° 33 pp. 1–4.

Thiéry, A. & D. Pont, 1987. *Eoleptestheria ticinensis* (Balsamo-Crivelli, 1859). Conchostracé nouveau pour la faune de France (Crustacea, Branchiopoda). Vie Milieu 37: 115–121.

Walter, G. M., 1970. Limnological studies of temporary ponds in southeastern Louisiana. Southwest. Nat. 15: 83–110.

Williams, W. D., 1968. The distribution of *Triops* and *Lepidurus* (Branchiopoda) in Australia. Crustaceana 14: 119–126.

Williams, W. D., 1985. Biotic adaptations in temporary lentic waters, with special reference to those in semi-arid and arid regions. Hydrobiologia 125: 85–110.

Wootton, D. M. & N. T. Mattox, 1958. Notes on the distribution of conchostracans in California. Bull. South. Calif. Acad. Sci. 57: 122–124.

Yaron, Z., 1964. Notes on the ecology and Entomostracan fauna of temporary rainpools of Israel. Hydrobiologia 24: 489–513.

*Hydrobiologia* **212**: 137–143, 1991.
D. Belk, H. J. Dumont & N. Munuswamy (eds), *Studies on Large Branchiopod Biology and Aquaculture.* 


# Factors affecting contributions of the tadpole shrimp, *Lepidurus packardi*, to its oversummering egg reserves

J.S.B. Ahl
*Department of Biological Sciences, California State University, Chico, Ca; present address: Department of Biology, St. Ambrose University, Davenport, IA 52803-2989, USA*

*Key words: Lepidurus packardi*, tadpole shrimp, parasitic castration, phenology, reproductive biology

## Abstract

The phenology and reproductive biology of the tadpole shrimp, *Lepidurus packardi*, which inhabits temporary annual pools in northern California, U.S.A., were investigated to identify factors affecting its contributions to its egg reserves that oversummer in the dried pond sediments. Field observations throughout 2 seasons revealed the nearly continuous presence of juveniles, indicating multiple generations, and a predominance of males during the mid to late stages of the pond. Observation of the first oviposition, and dissections revealed that sexual maturation occurred in shrimp 10–12 mm carapace length. Fecundity increased with size, but was drastically reduced in parasitized individuals. Metacercariae of an echinostome fluke caused parasitic castration in 53% of the population by mid-season. In the laboratory, oversummered eggs collected from dried pond sediments hatched within 17 days, when incubated at 10 °C. Eggs laid in the laboratory hatched within 25 days at 10 °C, without prior dehydration, but hatching was reduced at higher temperatures.

## Introduction

Temporary annual pools (Alexander, 1976) located throughout the state of California, USA, are referred to as ecological islands, because they support recently evolved endemic plant species (Stebbins, 1976), that populate the pool margins and basins as the seasonal waters recede. Less well known, however, are the faunal populations indigenous to these ephemeral waters, particularly benthic crustaceans, such as, tadpole shrimp.

*Lepidurus packardi* Simon 1886 is a tadpole shrimp that inhabits temporary annual pools in northern California exclusively (Linder, 1952; Longhurst, 1955), and virtually nothing is known about its biology. Since its initial collection by Packard in 1883, only the environmental conditions affecting hatching of oversummered eggs have been investigated (Lanway, 1974).

Information in the biology of *Lepidurus*, in general, is limited and incomplete. Observations have been made on the seasonal occurrence of *Lepidurus arcticus* Pallas in lakes and ponds of Greenland (Poulsen, 1940; Arnold, 1966) and Norway (Sømme, 1934), and of *Lepidurus lemmoni* Holmes from a large alkali lake in California (Lynch, 1966). Initially it was thought that *Lepidurus* only inhabited permanent bodies of water, and that its eggs could not tolerate desiccation (Brauer cf. Fox, 1949). However, this notion has since been dispelled by reports of *Lepidurus* species occurring in ponds that dry regularly (Desportes & Andrieux, 1944; Fox, 1959), and from laboratory studies of the effects of desiccation and temperature on egg hatching (Chaigneau, 1959; Braswell, 1967). Finally, there is one brief account of the biology of *L. apus*, that includes molting and copulation (Desportes & Andrieux, 1944).

In this study, the phenology and reproductive biology of *L. packardi* is described, and factors affecting contributions to the egg reserves that oversummer in the dried pond sediments are discussed.

## Habitat description

The tadpole shrimp populations studied were from three closely associated temporary ponds located on the Vina Plains in Tehama County (R1W, T24N, Sec 20) 25.6 km north of Chico, California. They are assumed to be of the same system because water from Pond I drains into II when filled to overflowing, and III is connected to I by a culvert. The ponds are shallow (< 1 m) depressions in an open treeless range land frequented by livestock and migrating water fowl. Their margins vary from cobbley hardpan to soft clay mud, and some areas receive strong wave action from the prevailing winds. In the first season of observation, flooding occurred in November, when the mean ambient minimum and maximum temperatures were 3.0 °C and 18.0 °C (U.S. Climatological Data, 1982). Standing water lasted until the first week in May. The following year, flooding occurred earlier, in late October, during warmer weather with 10.5 °C minimum, and 23.0 °C maximum temperatures. The ponds were dry by late May.

## Methods and materials

The ponds were sampled at weekly intervals between 10 am and 2 pm. Shrimp were collected with a dip net (28.5 cm × 23.5 cm opening) utilizing stirring motions at mid-depth. The turbulence created pulled the shrimp off the bottom of the pond allowing them to be caught while swirling in the water column. This technique minimized injury that would have occurred had the net been raked along the bottom, picking up stones with the shrimp.

Carapace lengths (CL) of the shrimp were measured with a vernier caliper mid-dorsally from the anterior edge to the median carinal spine on the posterior sulcus. Small shrimp were measured with an ocular micrometer in a dissecting microscope. Sexes were determined according to descriptions of *L. lemmoni* by Lynch (1966).

The shrimp were kept in glass fingerbowls (17 cm in diam.) containing pond soil and aerated, demineralized, distilled water as the culture medium (Moore, 1957). The cultures were maintained in environmental chambers at 10 °C, and 10/14 L/D cycle. They were fed fresh pond detritus at weekly intervals (Desportes & Andrieux, 1944). Reproductive shrimp were maintained in pairs, and the females were inspected daily for the presence of eggs in their foot capsules. When eggs appeared, the female was isolated in a 400 mL beaker until oviposition. In the beaker, the shrimp rested on a plastic mesh (1 mm × 1 mm) that was wedged 3 cm below the surface of the water, which allowed the eggs to fall to the bottom where they could not be consumed. After oviposition, the female was measured and returned to the male, because successive clutches of eggs will not descend into the foot capsules unless copulation has occurred (Desportes & Andrieux, 1944). The eggs were collected with a pipet and counted. A record was kept for each female that included the time the eggs spent in the foot capsules, the number of eggs deposited (clutch size), and the time between oviposition and the descent of eggs into the foot capsules.

Newly oviposited eggs (904) were incubated in fresh hatching medium at 10 °C and 10/14 LD to determine of they were diapause or non-diapause. The hatching medium consisted of demineralized, distilled water that had been thoroughly mixed with pond soil, then filtered. When hatching was observed, it prompted an investigation of the influence of temperature on this kind of egg. So two groups of 100 newly deposited eggs were incubated at 15 °C and 20 °C.

Oversummered diapause eggs collected from

dried pond sediments were rehydrated to observe hatching synchrony. Prior to rehydration, the eggs were subjected to freezing temperatures for 24 h to break the diapause (Braswell, 1967). Incubation was at 10 °C and 10/14 L/D (Lanway, 1974).

Shrimp ranging in size from 6 to 12 mm CL were dissected to ascertain the onset of sexual maturation as evidenced by the development of the gonads. Eggs in the oviducts were counted to assess the relationship of fecundity to size and age.

## Results

The distribution of size ranges in the population as the season progressed is shown in Fig. 1.

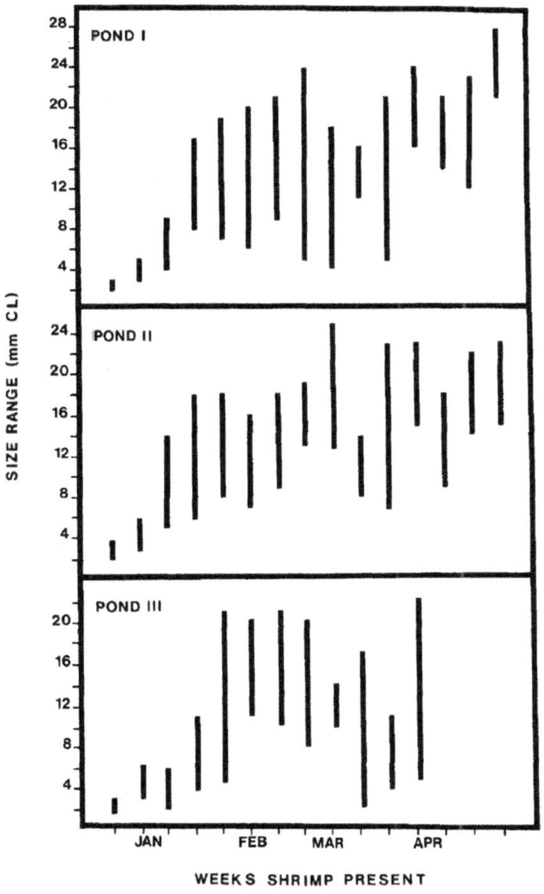


*Fig. 1.* Range in size (mm CL) of shrimp collected throughout the first season in each of the study ponds.

Juvenile *L. packardi* appeared approximately 3 weeks after the ponds flooded in early December, and adults appeared 3–4 weeks later in mid-January. Juveniles were collected intermittently in all 3 ponds throughout the season. In pond I they were present through the end of February, then reappeared for one week in mid-March. In pond II, they were collected through mid-February, and again at the beginning of March, then disappeared entirely. In pond III juveniles were present through the end of January, and one week in mid-February. Then after a brief absence they reappeared at the beginning of March, and were present when the pond dried 3 weeks later. Adult shrimp were present continuously from their initial appearance until the ponds dried.

When the shrimp first appeared, they were few in number, but became abundant by mid-January. At the end of the season, fewer shrimp were found; mostly small males (10–20 mm CL) and a few large females (>20 mm CL).

The sex ratio fluctuated during both seasons of observation especially in ponds I and II (Fig. 2). In the first year, the sexes occurred in equal frequency in all three ponds except during March. In pond I, males significantly outnumbered females, but the opposite occurred in pond II. The following year, males were more abundant in all three ponds, except during the mid to late season in ponds I and II. During February the number of males decreased significantly ($p \leq 0.05$) in pond II, then increased again to greatly outnumber the females. In pond I, there was a decline in the number of males during March and April; but it did not affect the expected sex ratio, and it did not increase significantly before the pond dried.

The dissections revealed that the gonads were not well developed in either sex until 10–12 mm CL. Within this size range, 20% of the females were reproductive (deposited eggs) at 10 mm CL, 80% at 11 mm CL, and 100% at 12 mm CL. Fecundity increased with size based upon the number of eggs present in the oviducts, foot capsules, and the frequency of egg deposition between molts. Small shrimp (11 mm CL) had as many as 23 eggs in the oviducts, and deposited

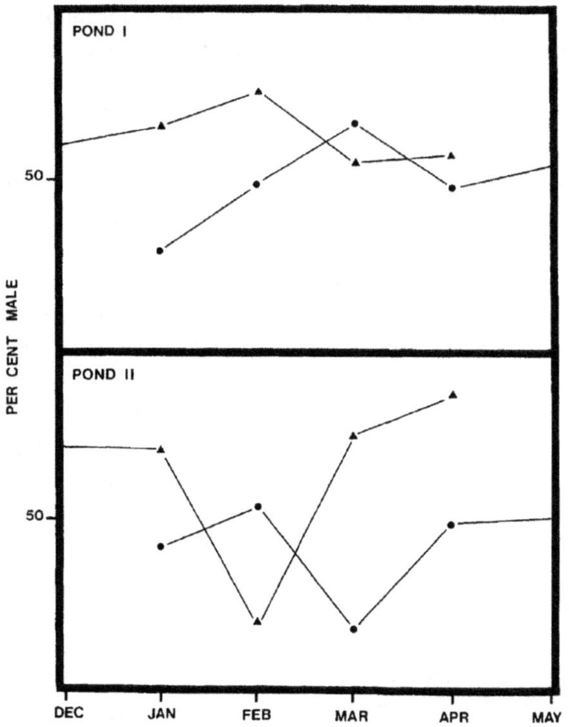


*Fig. 2.* Seasonal variation in the sex ratio in ponds I and II. ● first year; ▲ second year.

1–2 clutches averaging 8 eggs each. Larger shrimp (20 + mm CL) had up to 537 eggs in the oviduct, and deposited as many as 6 clutches averaging 32–61 eggs. It was estimated that a shrimp 25 mm CL could deposit approximately 861 eggs during her lifetime, and half of those eggs would be deposited after she was 20 mm CL in size.

Egg production appeared to be continuous as evidence by numerous oocytes found in various developmental stages. The oviducts were not always full indicating only partial evacuation when the eggs descend into the foot capsules. Some oviducts were completely empty, but the ovaries had many large ova on tetrads or in the follicle ducts. Thus, filling and evacuation of the oviduct was sporadic.

A large percentage of shrimp collected in February and March were afflicted with parasitic castration (Fig. 3). The gonads of both sexes were greatly reduced in size or absent, and their body cavities were filled, instead, with many metacercariae. Snails collected from the ponds released cercariae that were identified as belonging to the Family Echinostomatidae.

Observations of hatching of individual larvae revealed that the neonate was a metanauplius similar to that of *Triops* (Scott & Grigarick, 1978). This stage lasted 1.5–2 hours, then molted into the larval form resembling a miniature (1 mm CL) adult.

Hatching responses varied in newly oviposited eggs (Fig. 4). At 10 °C, 77% hatched within 16–47 ($\bar{x}$ = 25) days. Of the remaining unhatched eggs, 12.5% succombed to mold, and 10.8% remained unchanged. When subjected to higher temperatures hatching decreased, and developmental times were shortened. At 15 °C, 46% hatched within 13–19 ($\bar{x}$ = 18) days, and 13% hatched within 9–33 ($\bar{x}$ = 19) days at 20 °C.

Eggs that had oversummered in the pond sediments hatched within 11–26 ($\bar{x}$ = 17) days when rehydrated at 10 °C (Fig. 3). Initially 67% began development, as evidence by the shell cracking and extrusion of the embryo, but some became moldy and died. Of the 33% that did not initiate development, all but 3% were attacked by mold. These latter eggs remained unchanged after 50 days of incubation.

Hatching was synchronous in all treatment groups, except for newly oviposited eggs incubated at 20 °C, in which hatching continued intermittently for 33 days (Fig. 4). The oversummered eggs were closely synchronized, in that 64% hatched within 4 days. Of the non-diapause newly

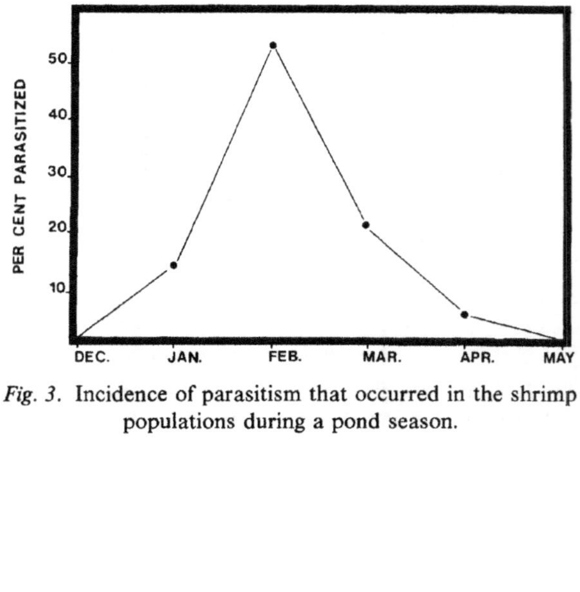


*Fig. 3.* Incidence of parasitism that occurred in the shrimp populations during a pond season.

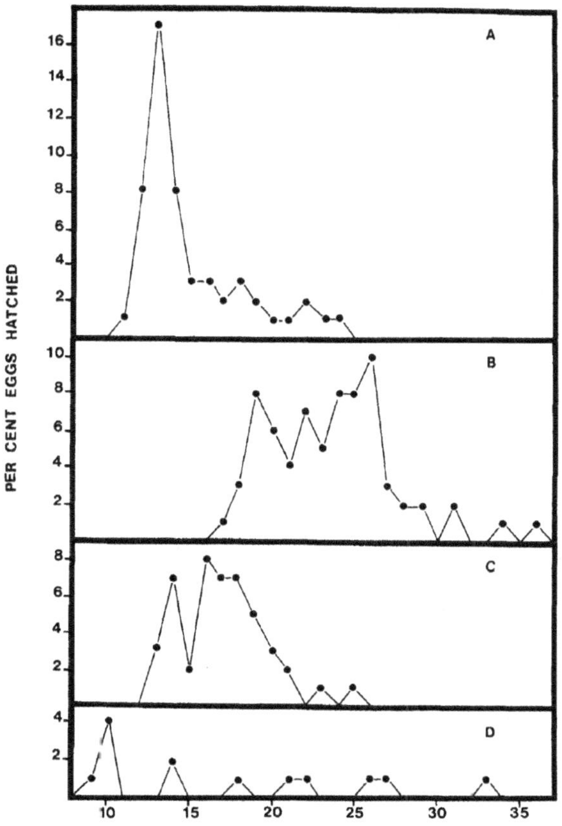


*Fig. 4.* Hatching profiles of shrimp eggs at various temperatures. A) Rehydrated oversummered eggs incubated at 10 °C. B) Newly deposited eggs incubated at 10 °C; C) 15 °C; and D) 20 °C.

oviposited eggs, hatching at 15 °C was more synchronous than at 10 °C. In the former treatment group, 96% hatched within 8 days compared to 82% hatching within 10 days in the latter.

Egg deposition observed in field collected shrimp revealed that oviposition in *L. packardi* was not coupled with molting. One to 6 clutches were deposited between molts, but ecdysis occurred only after the release of the eggs, sometimes as long as several days later, even in those shrimp that deposited only one clutch. Generally, the number of eggs per clutch, and the number of clutches deposited between molts increased with the size of the shrimp. Eggs remained in the foot capsules 1–8 days. After oviposition, the length of time the foot capsules were empty was quite variable. In some cases eggs descended within minutes of copulation, in others, the foot capsules remained empty as long as 11 days.

## Discussion

Successful adaptation of the tadpole shrimp to the temporary annual pool habitat is due to high fecundity, and the production of drought resistant diapause eggs. These eggs accumulate in the pond sediments, where they remain, sometimes more than a year, until favorable conditions return. This egg reserve is maintained by limited hatching when flood conditions return, and by replacement contributions from each ensuing population. Deficits to this reserve could occur with the production of non-diapause eggs, or decreased fecundity. The occurrence of both of these factors was investigated in *L. packardi* by combined field and laboratory studies of phenology and reproductive biology.

*L. packardi* were present 15 of the 20 weeks the ponds were flooded, and larvae were observed almost continuously. For juveniles to be present with the adults suggests that hatching from oversummered eggs is either heterogeneous, or that more than one generation occurs in a season. Although it is not unlikely that some of the larvae were from a latent hatch, it is more probable that they hatched from eggs deposited that season.

Rehydrated oversummered eggs hatched over a 14 day period, but most (64%) hatched within the first 4 days. Similarly, many (82%) of the newly deposited eggs hatched within 10 days of each other, when incubated at 10 °C, and 96% hatched within a period of 8 days at 15 °C. The synchronous manner in which the eggs hatched suggests that the larvae present in the early part of the pond flood phase were from oversummered eggs, and those appearing later were a second generation that developed from eggs deposited by shrimp from the initial hatch.

The production of non-diapause eggs can be viewed as maximizing fitness (Taylor, 1980), if there is enough time for another generation to mature before the season ends. However, it could also be considered detrimental with respect to unexpected changes in environmental conditions. For instance, unseasonable evaporation of the pond during dry windy weather could result in the loss of the second generation before sexual maturation.

Nevertheless, recruitment within the same pond season was particularly important for *L. packardi* from the Vina Plains pools, because there was much predation by migrating water fowl, and fecundity was drastically reduced by parasitic castration. The incidence of parasitism was greatest (53%) in mid-season when the shrimp were first reproductive. Recruitment may offset parasitism by replacing parasitized individuals, and by providing additional (or preferred) hosts for the cercariae. Although it was not observed, it is possible that the parasite entered the shrimp by burrowing into the thin cuticle of the flabella (gills) on the appendages, or any part of the body surface shortly after ecdysis when the cuticle was still soft. In the laboratory, shrimp which had just recently molted were pale colored, and easily sustained injuries when handled. Since larval tadpole shrimp molt frequently, their cuticles would be in a soft post-molt condition more often than the adults. Consequently, they would be more susceptible to penetration by the cercariae.

The population appeared to be predominated by males in the mid to late stages of the pond, which has been reported previously by others. Jezequel (1921) observed 83% occurrence of males in a bisexual population of *L. apus* in May. Desportes & Andrieux (1944) reported more males (55%) in another population of the same species during March. Sømme (1934) collected males from a hermaphroditic population of *L. arcticus* during September.

Although the factors affecting sex determinism were not investigated, high temperatures occurring in early fall flooding, or during the mid to late stages of the pond may have influenced the development of males. The labile nature of sex determinism, and the influence of temperature on the conversion of potential females into males has been well documented in amphipods (Ginsburger-Vogel, 1974; Ginsburger-Vogel & Magniette-Mergault, 1981), and in copepods (Egloff, 1967) (but see Sassaman, this volume).

The predominance of males could be considered a deficit in terms of productivity, because fewer shrimp would be available to contribute eggs to the reserve. However, it could also be regarded as differential mortality compensation (Egloff, 1967). The decline in the males in the mid to late season could have been due to mortality, then the subsequent increase in their numbers could have been from recruitment. Females presumably live longer, as evidenced by the presence of large individuals late in the pond season. A few large males were found, but not as frequently as small ones. Since the population begins to dwindle towards the end of the season, an increase in males would optimize reproductive success. Copulation with the large, fecund females would ensure the deposition of hundreds of fertilized eggs before the pond dried.

In conclusion, parasitic castration was the major limiting factor affecting contributions to the diapause egg reserve of *L. packardi*. Hatching of eggs deposited during the season produced a second generation, and provided replacements for losses due to predation and parasitism. The surplus of males created by this hatch optimized reproductive success, by increasing the chances of copulation with the large fecund females present at the end of the season.

**Acknowledgements**

I thank Dr. D.G. Alexander for his enthusiastic encouragement and support.

**References**

Alexander, D. G., 1976. Ecological aspects of the temporary annual pool fauna. In S. Jain (ed.), Vernal Pools. Their Ecology and Conservation. Instit. Ecol. Publ. 9: 32–36.

Arnold, G. P., 1966. Observations on *Lepidurus arcticus* (Pallas) (Crustacea, Notostraca) in east Greenland. Ann. & Mag. nat. Hist. 9: 599–617.

Braswell, A. L., 1967. Preincubation treatment required for hatching of *Lepidurus apus*. Utah Acad. Proc. 44: 344–352.

Chaigneau, J., 1959. Action de la desiccation et de la température sur l'éclosion de l'œuf de *Lepidurus apus* (Leach) (Crustacé, Phyllopode). Bull. Soc. zool. Fr. 84: 398–407.

Desportes, C. & L. H. Andrieux, 1944. Sur la biologie de *Lepidurus apus*. Bull. Soc. zool. Fr. 69: 61–68.

Egloff, D. A., 1967. Ecological aspects of sex ratio and reproduction in experimental and field populations of the marine copepod *Tigriopus californicus*. Ph.D. Thesis. Stanford University, 150 pp.

Fox, M. H., 1949. On *Apus*: its rediscovery in Britain, nomenclature and habits. J. Zool. 119: 693–702.

Ginsburger-Vogel, T., 1974. Temperature-sensitive intersexuality and its determinism in *Orchestia gamarella* Pallas. In R. Reinboth (ed.), Intersexuality In The Animal Kingdom. Springer-Verlag, Berlin: 106–120.

Ginsburger-Vogel, T. & F. Magniette-Mergault, 1981. The effects of temperature on sexual differentiation in the temperature sensitive thelygynic-intersexual offspring of *Orchestia gammarella* (Pallas) (Amphipoda: Crustacea) II. Effects of temperature during embryonic and post-embryonic development. Int. J. Invert. Reprod. Dev. 4: 51–65.

Jezequel, J., 1921. Prédominance abnormale de mâles dans une population d'*Apus* (*Lepidurus productus* Bosc). Bull. Soc. zool. Fr. 46: 99–100.

Lanway, C. S., 1974. Environmental factors affecting crustacean hatching in five temporary ponds. Masters Thesis. Calif. St. Univ., Chico. 89 pp.

Linder, F., 1952. Contributions to the morphology and taxonomy of the branchiopod Notostraca, with special reference to the North American species. Proceed. U.S. nat. Mus. 102: 1–69.

Longhurst, A. R., 1955. A review of the Notostraca. Bull. b. Mus. Zool. (nat. Hist.) 3: 1–57.

Lynch, J. E., 1966. *Lepidurus lemmoni* Holmes: A redescription with notes on variation and distribution. Trans. Am. micro. Soc., 85: 181–192.

Moore, W. G., 1957. Studies in the laboratory culture of Anostraca. Trans. Am. micro. Soc. 76: 159–173.

Packard, A. S., 1883. A monograph of the phyllopod crustacea of North America. U.S. Geol. Geogr. Surv. Terr. Rep. 12: 295–516.

Poulsen, E. M., 1940. Biological remarks on *Lepidurus arcticus* Pallas, *Daphnia pulex* deGreer and *Chydorus sphaericus* O.F.M. in East Greenland. Medd. Grønland. 131: 1–50.

Scott, S. R. & A. A. Grigarick, 1978. Observations in the biology and laboratory rearing of the tadpole shrimp *Triops longicaudatus* (LeConte) (Notostraca: Triopsidae). Wasmann J. Biol. 36: 116–126.

Sømme, S., 1934. Contributions to the biology of Norwegian fish food animals. I *Lepidurus arcticus*. Avh. Norsk. Vidensk. Akad. Oslo. 6: 1–36.

Stebbins, G. L., 1976. Ecological islands and vernal pools of California. In S. Jain (ed), Vernal Pools. Their Ecology and Conservation. Instit. Ecol. Publ. 9: 1–4.

Taylor, F., 1980. Timing in the life histories of insects. Theor. Pop. Biol. 18: 112–124.

U.S. Climatological Data. California. 1982. vol. 86.

*Hydrobiologia* **212**: 145–149, 1991.
*D. Belk, H. J. Dumont & N. Munuswamy (eds), Studies on Large Branchiopod Biology and Aquaculture.*



# Uptake of lead, cadmium and zinc by the fairy shrimp, *Branchinecta longiantenna* (Crustacea: Anostraca)[1]

Alan Mizutani, Eric Ifune, Andrew Zanella & Clyde Eriksen
*Joint Science Department, The Claremont Colleges, Claremont, CA 91711, USA*

*Key words:* lead, cadmium, zinc, metal uptake, fairy shrimp, *Branchinecta longiantenna*

## Abstract

Individuals of the fairy shrimp, *Branchinecta longiantenna*, were subjected to 5 concentrations (0.1 to 15 mg $l^{-1}$) of Pb in diluted habitat water at 13 °C. Lead concentrations (mg $kg^{-1}$ wet weight) in the animals were determined at 2-day intervals by digestion in nitric acid followed by atomic absorption analysis. The shrimp were also subjected to 0.1 mg $l^{-1}$ media of Cd and Zn, separately.

Uptake rates by the fairy shrimp for the three metal ions at 0.1 mg $l^{-1}$ were: 0.111, 0.0885, and 0.0460 mg $kg^{-1}$ $day^{-1}$ for Zn, Pb, and Cd, respectively. After 2 days in 1.0 mg $l^{-1}$ Cd or Zn, the animals expired; but they surviced for 8 days in a 10 mg $l^{-1}$ Pb medium and for 2 days in 25 mg $l^{-1}$ Pb. Lead uptake demonstrated a linear dependence on the Pb concentration in the media.

Shrimp survived at much higher tissue accumulations of Pb compared to Zn and Cd. Estimated lethal doses were 20, 1.2–2.4, and 0.4–1.4 mg $kg^{-1}$ wet weight for Pb, Zn, and Cd, respectively. Pb was found to be at much lower concentration than Cd or Zn in the natural pond water but between Cd and Zn levels in the sediment. Thus Cd and Zn probably present a greater threat to *B. longiantenna* than Pb, although Pb may be in higher concentration in the environment.

## Introduction

Temporary rain pools are important resources which are being rapidly eliminated by urban and agricultural development. Although filled with water perhaps but a few weeks a year, temporary waters are sites of great species diversity, often containing a number of endemic and, increasingly, threatened and endangered species (Holland, 1978, 1988; Holland & Jain, 1977; Jain, 1976; Jain & Moyle, 1984; Eng *et al.*, 1990). Perhaps most significant among the animals of temporary waters are fairy shrimp, primitive crustaceans of the Order Anostraca. Fairy shrimp hatch from embryonated eggs, which lie dormant in the soil, soon after water fills the pool basins and suitable temperatures predominate. Depending on species and temperatures, life cycles may be completed in 2 to 12 weeks. Fairy shrimp filter-feed phytoplankton, fine particulate organic matter, and nutrient-rich colloidal particles, thus being major secondary producers. As such, they are a significant food source for migratory birds which use the pools along their flight paths.

Because such a large proportion of remaining temporary waters are either surrounded by agriculture, urban development, and roads, or are directly impacted by off-road recreational vehicles (Eriksen *et al.*, 1986), it is unfortunate that so

[1]Contribution 47, Laboratory of Ecology, The Claremont Colleges, Claremont, CA 91711, USA. Send reprint requests to Clyde Eriksen.

little information exists concerning the effects of agricultural chemicals, heavy metals, or other pollutants on the flora and fauna of these waters (Bagshaw *et al.*, 1986). Research interests both in rain pools and heavy metals led us to examine the effects of Pb (auto exhaust), and Cd and Zn (tire wear) on the newly described California endemic fairy shrimp, *Branchinecta longiantenna* (Eng *et al.*, 1990).

We tested two hypotheses: (1) the rate of uptake of Pb, Cd, and Zn would be proportional to their concentrations in the media and duration of exposure, and (2) relative toxicities would be similar to those known for other organisms (e.g., Cd and Zn much greater than Pb). Secondary goals were to: (1) enlarge the data base concerning the biology of this possibly endangered species, and (2) determine if fairy shrimp, as exemplified by *B. longiantenna*, might be sensitive 'indicator organisms' of the health of temporary waters and their surroundings.

## Materials and methods

Specimens of *Branchinecta longiantenna* were collected from a turbid rain pool located between Soda Lake and California Valley, San Luis Obispo County, CA (35° 18′ N, 119° 57′ W). The site was selected because it appeared to have experienced little human impact. Shrimp were maintained in the laboratory in 10 l aquaria containing diluted pool water (1 : 5 pool water to deionized water) at 13 °C. (Straight pool water was not used due to our inability to transport the volume needed for the entire series of experiments). TDS of the diluted water was about 65 mg $l^{-1}$ with bicarbonate being the principle anion (Eriksen, *unpubl.*). Since *B. longiantenna* has been collected in a reasonably wide range of TDS (95–1450 mg $l^{-1}$, Eng *et al.*, 1990), we assumed that the diluted medium would not have an effect on survival. However, dilution does decrease the food particle concentration, so 3.0 ml of yeast suspension (0.1 g yeast per ml of water) was added to each aquarium every 3 days.

Stock solutions (1000 mg $l^{-1}$) of the metals were prepared by dissolving anhydrous salts, $Pb(NO_3)_2$, $Cd(NO_3)_2$, or $ZnCl_2$, in deionized water. Using pool water to dilute aliquots of stock solutions, the following series of experimental media were prepared for each metal: 0.1, 1.0, 5.0, 10.0, 15.0, 25.0, 50.0, and 100.0 mg $l^{-1}$.

In an initial series of experiments designed to estimate toxicity levels, groups of 20 shrimp were exposed to metal concentrations of 25, 50, and 100 mg $l^{-1}$. Experiments were carried out using 4-liter Nalgene tanks lined with nylon netting to facilitate removal of the animals for analysis. Since all shrimp died within 2 days, media containing 0.1, 1.0, 5.0, 10.0, and 15.0 mg $l^{-1}$ metal ion were selected for the uptake studies. The shrimp survived for 8 days or more in the Pb media, but in the Cd and Zn media they survived for greater than 2 days only at the 0.1 mg $l^{-1}$ concentration.

Lead uptake experiments began with 100 fairy shrimp exposed to each of the five concentrations (0.1–15 mg $l^{-1}$) at 13 ± 1 °C. To insure that metal ion concentrations remained constant, solutions were renewed every 3 days. Twenty shrimp were removed for analysis from each solution every 2 days. They were placed in deionized water for 30 min to purge digestive tract contents, including any heavy metals (Patten, 1980). The shrimp were damp-dried with Kimwipes, weighed to 0.1 mg, and dissolved by heating in equal portions by volume of concentrated $HNO_3$ and water. The resulting solution was concentrated by evaporation to achieve a 10.0 ml sample for analysis in a Varian Model 1100 atomic obsorption flame spectrophotometer with a reproducibility of ± 10%.

Habitat water was treated with concentrated $HNO_3$ (200 ml water: 25 ml acid), digested, and passed through Whatman 451 filter paper to remove particulates. The filtrate was then concentrated to 25.0 ml by heating and analyzed by A.A. Sediment samples (12 g each) were dried at 110 °C, weighed, treated with 25 ml water and 25 ml concentrated $HNO_3$, digested, filtered, diluted to 100 ml, and analyzed by A.A.

## Results and discussion

The relatively higher values for Zn in habitat water and sediment (Table 1) presumably reflect its greater natural abundance. The very low level of Pb in the water probably results from the insolubility of lead salts, such as $PbCO_3$, since Pb is not so low in the sediment. The fact that ambient Zn levels are initially fivefold higher than Pb or Cd in the fairy shrimp may simply reflect the fivefold or higher concentration of Zn in the water and sediment of the pool compared to the other two metals. However, with Zn being a necessary trace element in the shrimp's metabolism, it may be selectively concentrated by the shrimp (Rainbow & White, 1989).

Animals subjected to $> 25\ mg\,l^{-1}$ of Pb expired in 2 days or less, as did those in any Cd or Zn concentration $> 0.1\ mg\,l^{-1}$. Animals treated with 15 mg $l^{-1}$ Pb media expired after 6 days and showed 20 mg $kg^{-1}$ Pb in their tissues at that time. In contrast, shrimp in 0.1 mg $l^{-1}$ Pb showed about 0.5 mg $kg^{-1}$ Pb in tissues after 6 days and survived for 8 days when sampling ceased. Shrimp in 0.1 mg $l^{-1}$ Cd or Zn showed tissue levels of about 0.4 and 1.2 mg $kg^{-1}$ Cd and Zn, respectively, after 6 days shortly before death (i.e., they did not survive for a sampling on the eighth day).

Since shrimp died relatively quickly at high levels of Cd and Zn ions, we studied their rates of accumulation only at the lowest concentration,

*Table 1. Branchinecta longiantenna:* Environmental and tissue metal levels[1]

| Environmental (background) | | | | Laboratory Lethal tissue level |
|---|---|---|---|---|
| Metal | Water | Sediment | Fairy shrimp | |
| Pb | ≤0.005 | 0.012 | 0.10 | 20.0[2] |
| Cd | 0.060 | 0.005 | 0.10 | 0.40[3]–1.4[4] |
| Zn | 0.300 | 0.055 | 0.53 | 1.20[3]–2.4[4] |

[1] mg $l^{-1}$ or mg $kg^{-1}$ wet wt; standard deviations are approx. 10%
[2] at 15 mg $l^{-1}$ Pb
[3] at 0.1 mg $l^{-1}$ Cd or Zn
[4] at 25 mg $l^{-1}$ Cd or Zn

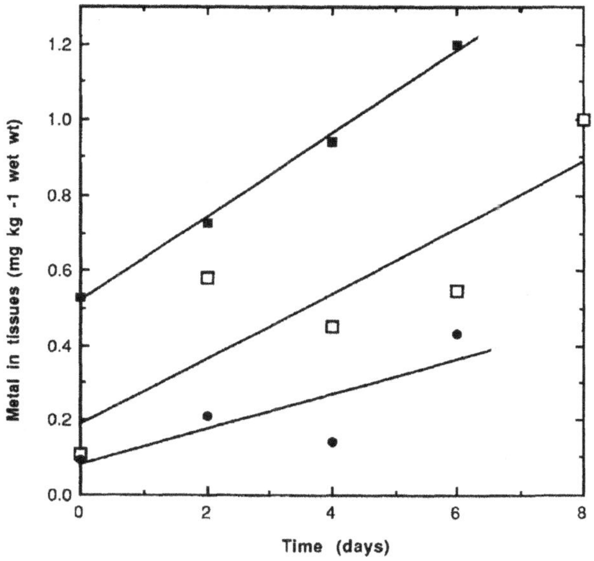


*Fig. 1. Branchinecta longiantenna:* metal uptake at 0.1 mg $l^{-1}$ (Zn, filled squares; Pb, open squares; Cd, circles). Least squares fits of data yield slopes of: 0.111, 0.0885, and 0.0460 for Zn, Pb, and Cd, respectively, with corresponding R-values of 0.998, 0.869, and 0.807.

0.1 mg $l^{-1}$. Analyses of the tissues at 2-day intervals showed a linear increase in metal concentration with time. The line slopes, for Zn (0.111) and for Cd (0.0460), indicate that Zn was taken up over twice as fast (2.41 ×) as Cd (Fig. 1). Given its lower toxicity, Pb uptake could be determined at 4 different concentrations higher than 0.1 mg $l^{-1}$. At all 5, tissue Pb increased linearly with time (e.g., Fig. 1, 0.1 mg $l^{-1}$ and Fig. 2, 1.0 mg $l^{-1}$). The Pb uptake slope at 0.1 mg $l^{-1}$ (0.0885) was only slightly less (0.8 ×) than Zn but nearly twice (1.92 ×) that of Cd at 0.1 mg $l^{-1}$ (Fig. 1). By comparing uptake rates at all Pb ion concentrations, a positive linear correlation is demonstrated (Fig. 3).

Thus *B. longiantenna* concentrates heavy metals at varying rates depending upon the metal and linearly with time for at least one week. The positive linear relationship between the rate of Pb uptake and its medium concentration parallels that for Pb uptake in the bay mussel, *Mytilus edulis*, also a filter feeding organism (Schulz-Baldes, 1974). Since similar results were found for

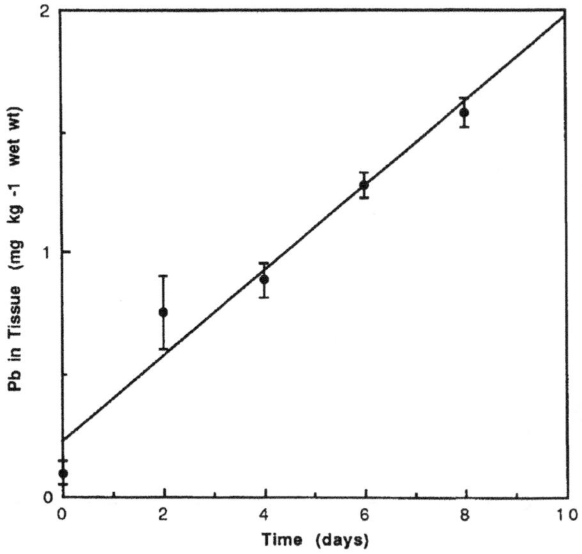


*Fig. 2. Branchinecta longiantenna:* Pb uptake at 1.0 mg $l^{-1}$. The equation of the regression line fitted by least squares is y = 0.174 + 0.225 with $R = 0.979$. Error bars represent $\pm 1$ standard deviation.

mussels (Duckwiler, 1979) and several crustaceans (Rainbow & White, 1989) in both Cd and Zn and for the anostracan crustacean, *Artemia salina*, in Cd (Jennings & Rainbow, 1979), we would also expect fairy shrimp to accumulate Cd and Zn linearly with increasing concentration. However, to test this hypothesis, studies need to be done at much lower ($<0.1$ mg $l^{-1}$) levels because of the toxicity of Cd and Zn to at least this fairy shrimp and probably all softer water species. We say this because *A. salina*, also an anostracan, but one inhabiting brines in which most ions must also be highly concentrated, survives in at least 10 mg $l^{-1}$ Cd for at least 5 days and ultimately accumulates approximately 100 mg Cd per kg wet weight of animal (estim. from Jennings & Rainbow, 1979).

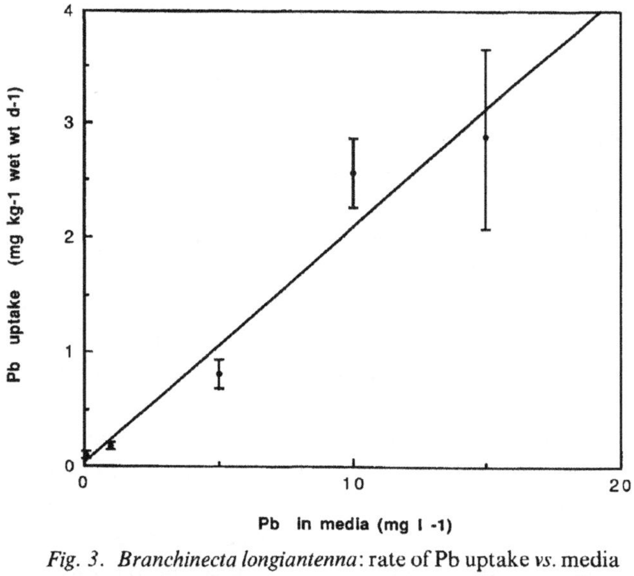


*Fig. 3. Branchinecta longiantenna*: rate of Pb uptake *vs.* media Pb concentration. The equation of the regression line fitted by least squares is 0.206 x + 0.002 with $R = 0.975$. Error bars represent $\pm 1$ standard deviation of the line slopes for uptake at each concentration of Pb in media.

*Branchinecta longiantenna* survived at relatively high levels of Pb, both in media and in tissues, compared to Cd and Zn (Table 1). Since Zn is apparently an essential trace element for crustaceans (Rainbow & White, 1989), we found it interesting that Zn appears to be almost as toxic as Cd. However, lethal doses in mammals for these metals, though much higher, follow a similar trend as found here: Pb, 1490; Cd, 54; and Zn, 36 mg $kg^{-1}$ body weight (estimated from % metal in chloride salts from Tox. Sub. List 1972).

Based on estimates of the lethal concentrations for the metals in *Branchinecta longiantenna* (Table 1), the animals appear to be at risk in their natural waters since the lethal levels are only twice as high as the ambient levels for Zn (1.2 vs 0.53) and four times as high for Cd (0.40 vs 0.10). In contrast, the background lead concentration (0.10 mg $kg^{-1}$) in the tissues is 200 times lower than the lethal concentration (20 mg $kg^{-1}$ Pb). The fact that the natural tissue levels of Pb and Cd are the same despite the much lower concentration of Pb in the water may, we suspect, be due to the faster rate of uptake for Pb. In addition, Pb in sediment is about twice that of Cd, and the shrimp probably ingest metals on sediment particles commonly dispersed into the water by wind mixing. The latter pathway could also contribute to the high level of Zn found in the shrimp, since Zn is over fourfold higher than Pb in the sediment.

*Branchinecta longiantenna* appears to be a sensitive indicator organism for the presence of Pb,

Cd, and Zn. These metals can pose a danger to the survival of this species and probably to other species of fairy shrimp, with Cd, then Zn, clearly being more toxic than Pb. A recent study showed that soil along heavily travelled southern California roads contains over 100 times more Pb than Cd (Singler, 1989), so that the shrimp may face potentially detrimental exposure to Pb as well. However, extrapolating the rates of metal uptake under laboratory conditions to the natural environment is difficult because the metal concentrations in a pool depend upon myriad factors: amount and frequency of precipitation, evaporation rate, leaching rate, types of other ions in solution, and turbidity. The fairy shrimp studied here came from a pool which has endured relatively little human impact, but situations certainly exist where the organisms are exposed to much more intrusion by vehicles and agriculture, as well as to metal-rich local geologies. Studies on systems such as desert playas which are heavily used for recreational purposes should tell us more about the effects of heavy metals on fairy shrimp.

## References


Bagshaw, J. C., P. Rafiee, C. O. Matthews & T. H. MacRae, 1986. Cadmium and zinc reversibly arrest development of *Artemia* larvae. Bull. envir. Contam. Toxicol. 37: 289–296.

Duckwiler, G., 1979. Parameters affecting the heavy metal uptake by the common mussel, *Mytilus edulis*. Senior thesis, Claremont McKenna College, Claremont, CA, 72 pp.

Eng, L. L, D. Belk & C. H. Eriksen, 1990. California anostraca: distribution, habitat, and status. J. Crust. Biol. 10: 247–277.

Eriksen, C. H., G. E. Prettyman & J. E. Mœur, 1986. The effects of soil disturbance by off road vehicles on the eggs and habitat of playa lake crustaceans. In R. G. Zahary (ed.), Desert ecology 1986: a research symposium. So. Calif. Acad. Sci. and So. Calif. Desert Studies Consortium: 50–65.

Holland, R. F., 1978. The geographic and edaphic distribution of vernal pools in the Great Central Valley, California. Calif. Native Plant Soc., Sp. Publ. No. 4: 1–12.

Holland, R. F., 1988. Vernal pools. In M. E. Barbour & J. Major (eds), Supplement to Terrestrial vegetation of California. Calif. Native Plant Soc., Sp. Publ. No. 9: 1012–1014.

Holland, R. F. & S. Jain, 1977. Vernal pools. In M. E. Barbour & J. Major (eds), Terrestrial vegetation of California. Wiley & Sons, NY: 515–533.

Jain, S. (ed.), 1976. Vernal pools, their ecology and conservation. Institute of Ecol., Univ. Calif., Davis, CA. Publ. 9: 1–93.

Jain, S. & P. B. Moyle (eds), 1984. Vernal pools and intermittent streams. Institute of Ecol., Univ. Calif., Davis, CA. Publ. 28: 1–280.

Jennings, J. R. & P. S. Rainbow, 1979. Accumulation of cadmium by *Artemia salina*. Mar. Biol. 51: 47–53.

Patten, M., 1980. A comparison of the feeding behaviors of two anostracan crustaceans, *Artemia salina* Linnaeus and *Branchinecta longiantenna* in different particle concentrations. Senior thesis, Claremont McKenna College, Claremont, CA, 35 pp.

Rainbow, P. S. & S. L. White, 1989. Comparative strategies of heavy metal accumulation by crustaceans: zinc, copper, and cadmium in a decapod, an amphipod and a barnacle. Hydrobiol. 174: 245–262.

Schulz-Baldes, M., 1974. Lead uptake from sea water and food, and lead loss in the common mussel, *Mytilus edulis*. Mar. Biol. 25: 177–193.

Singler, N., 1989. An analysis of lead and cadmium in soil adjacent to heavily travelled roads. Senior thesis, Claremont McKenna College, Claremont, CA, 53 pp.

Toxic Substances List, 1972. National Institute for Occupational Safety and Health, U.S. Department of Health, Education, and Welfare, Washington, D.C.: 114, 311, 547.

Hydrobiologia **212**: 151–168, 1991.
D. Belk, H. J. Dumont & N. Munuswamy (eds), Studies on Large Branchiopod Biology and Aquaculture.



# Genetic characterization and intra-generic relationships of *Artemia monica* Verrill and *A. urmiana* Günther

F. Alberto Abreu-Grobois[1] & J. A. Beardmore[2]
[1]*Estación Mazatlán, Instituto de Ciencias del Mar y Limnologia (U.N.A.M.), Apartado Postal 811, Mazatlán, Sinaloa 82000, Mexico*; [2]*Sch. of Biol. Sciences, University College of Swansea, Singleton Park, Swansea SA2 8PP, Wales, UK*

*Key words:* *Artemia*, genetic variation, genetic distances, parthenogenesis, reproductive isolation, phylogeny

## Abstract

Allozyme variation encoded for by 22 enzyme loci analyzed with starch-gel electrophoresis from samples of gonochoristic *Artemia monica* and *A. urmiana* collected from Lake Urmia, Iran and from Mono Lake, CA, USA, respectively, are compared and contrasted with data from representative gonochoristic populations of *A. franciscana*, *A. persimilis* and *A. salina*, as well as diploid, triploid, tetraploid, and pentaploid parthenogenetic forms. Values for parameters measuring degree of genetic variability (number of alleles per locus, proportion of polymorphic loci, and expected heterozygosity) in *A. monica* and *A. urmiana* were among the highest in the genus (1.76, 0.46, 0.19, and 2.19, 0.55, and 0.14, respectively). For *A. monica*, possible factors promoting its high genetic variability are discussed. Mean values for Nei's *D* between *A. monica* and *A. franciscana* were very low (0.09) and consistent with reported cytological and morphological similarities, thus the physiological differentiation which distinguishes the former must involve differences at very few loci. *D* values from *A. urmiana* to rest of the gonochoristic species ($\geq 0.95$) are consistent with its taxonomic status and distinctive morphology, while its low genetic distance to the monophyletic *Artemia* parthenogenetic lineage (mean $D = 0.48$) suggests a recent common ancestry.

## Introduction

Populations of the anostracan branhiopod genus *Artemia* are found in more than 350 habitats dispersed across the world where natural or man-made salterns are found (Vanhaecke *et al.*, 1987). Their distribution reflects the flight paths of some migratory bird species and recent deliberate inocculations for commercial purposes by man (Persoone & Sorgeloos, 1980).

Of the species in this genus, the least studied remains *A. urmiana* Günther, a gonochoristic brine shrimp reportedly confined to Lake Urmia, Azărbăijān, Iran (Browne & MacDonald, 1982; Vanhaecke *et al.*, 1987). Clark & Bowen (1976) demonstrated the reproductive isolation of this species through laboratory crosses. More recently, however, Barigozzi *et al.* (1987) described a population from Lake Urmia as being composed exclusively of parthenogenetic (di-, tetra- and pentaploid) organisms. Though it is premature to ascertain from the available reports whether nor not this site is currently dominated by an exotic asexual brine shrimp species, the ease by which *Artemia* cysts are transported and inoculated in novel environments without control or documentation permits the indiscriminant and irreversible displacement of local strains which often exhibit unique genetic adaptations (Beardmore, 1987). Because of the apparent risk for the loss of bisexual *A. urmiana*, the urgency for its characterization and the preservation of its gene pool is greatly increased.

The habitat of *A. monica* Verrill, on the other hand, has been much researched ecologically (Dana & Herbst, 1977; Lenz, 1980; Lenz, 1987). This gonochoristic species is reproductively isolated in nature due to ecological barriers: the high-carbonate concentrations of its native waters make it toxic to *Artemia* nauplii of low-carbonate habitats and, in turn, *A. monica* cannot survive the high-chloride concentrations commonly found in most of the other localities where brine shrimp are found (Bowen, *et al.*, 1985).

The present report provides primary genetic data derived from electrophoretic analysis of the products of 22 enzyme loci from *A. monica* and *A. urmiana*. In order to compare and contrast their levels of genetic variation to those of their closest relatives and to analyze their evolutionary and taxonomic relationships to the rest of the genus, original data from representatives of the various taxa in the genus are also included. Electrophoresis of protein loci has been amply demonstrated to be suited for both systematic (Avise, 1974; Ayala, 1983; Thorpe, 1983) and evolutionary studies (cf. Oxford & Rollinson, 1983).

## Materials and methods

### *Origin of populations*

*Artemia* samples were obtained as cysts, processed and transported from the following sites (code designating the populations and date of collection, if known, inside brackets). See Vanhaecke *et al.* (1987) for geographical settings.

Gonochoristic species: *A. franciscana* Kellog – Great Salt Lake, UT, USA (GS77 or GSL77, 1977); San Francisco Bay, Alviso, CA (SF76); Pekelmeer, Bonaire, Caribbean Sea (PEK, 1977); *A. monica* Verrill – Mono Lake, CA, USA (MON); *A. salina* Leach (used in this work as equivalent to *A. tunisiana* Bowen & Sterling)- San Lucar de Barrameda, Cadiz, Spain (BAR); *A. persimilis* Piccinelli & Prosdocimi, 1968- Salinas de Buenos Aires, Buenos Aires, Argentina (BA); *A. urmiana* Günther – Lake Urmia, Azärbãijãn, Iran (URM, December 1981).

Parthenogenetic forms: Diploid- Burgas-Pomerije, Bulgaria (BP(2)); Calpe, Alicante, Spain (CAL); Lake Techirgiol, Constanta, Rumania (LT); Triploid- Eilat, Israel (EIL); Izmir, Smyrna, Turkey (IZM(3)); Tuticorin, India (TUT); Tetraploid- Delta del Ebro, Tarragona, Spain (DE); Pentaploid-Burgas-Pomerije, Bulgaria (BP(5)). Among parthenogenetic populations, there is often a coexistance of more than one ploidy and more than one genotype. In order to avoid confusion, ploidy designation is given by Arabic numerals inside parenthesis, following the results of Abreu-Grobois & Bearmore (1989), while the genotype designation (Roman numerals) is given according to results of a broad electrophoretic survey of the genus (Abreu-Grobois, 1983). Criteria used for gonochoristic species identification follows Abreu-Grobois (1987).

### *Cyst hatching and culture methods*

All cyst samples were either decapsulated (Sorgeloos *et al.*, 1986) or disinfected with a 7% solution of 'antiformin' (Finamore & Clegg, 1969) before hatching using standard methods (Sorgeloos *et al.*, 1986) in 35 p.p.t. artificial sea water (ASW, Instant Ocean$^{TM}$). Nauplii were separated from empty cysts and transferred to 1 l culture containers kept at 28 °C in a constant-temperature cabinet. Developing *Artemia* were fed *ad libitum* on exponentially-growing *Dunaliella tertiolecta* Butcher cells. Culture of *Artemia* from MON required a special medium ('artificial Mono Lake water', AML) devised by Dr. Sarane Bowen (pers. comm.) made up by mixing equal volumes of a carbonates solution (40 g $Na_2CO_3$, 32 g $NaHCO_3$, 8 g $K_2SO_4$ in 1 litre distilled water) and ASW with the resulting precipitate removed by filtration. These nauplii were hatched in a 1 : 1 solution of AML with distilled water then reared in full strength AML and fed on micronized rice bran (see Sorgeloos *et al.*, 1986).

### *Electrophoresis*

Horizontal starch gel electrophoresis (Smith, 1976) using 12.5% Connaught hydrolysed starch was used for analyses of genetic variation. Individual *Artemia* were rinsed with distilled water, excess liquid absorbed and the whole organism homogenized, after a freeze-thaw cycle, with 25 $\mu$l of 0.02 M Tris-HCL pH 8.0 buffer containing 2 mg $ml^{-1}$ phenyl methyl sulphonyl fluoride and a trace of bromophenol blue tracer as well as 2 mg $ml^{-1}$ of NADP. Homogenates were absorbed into Whatman No. 3, 4 × 5 mm filter papers and placed onto the gel matrix, 5 cm from the cathodal end. Separations were performed at 4 °C. Enzymes from the homogenates were separated using one of three electrophoresis running buffers- Tris-EDTA-Borate pH 8.6 (Wilson *et al.*, 1973) for: glutamate-oxaloacetate transaminase (GOT), phosphoclucomutase (PGM), mannose-phosphate isomerase (MPI), leucine amino-peptidase (LAP). Tris-EDTA-Citrate pH 7.1 (F.J. Ayala, pers. comm.) for: esterases (EST), dimeric esterase (EST-D), 6-phospho-gluconate dehydrogenase (6PGDH), isocitrate dehydrogenase (IDH), malate dehydrogenase (MDH), lactate dehydrogenase (LDH). Poulik pH 8.7/8.2 (Poulik, 1957) for: catalase (CAT), superoxide dismutase (SOD), phophoglucose isomerase (PGI), malic enzyme (ME), amylase (AMY), peptidases (PEP). For enzyme stain recipes see Harris & Hopkinson (1976).

### *Electromorph designation*

Allozyme bands were identified according to their mobilities relative to that of the most common allelomorph in *A. franciscana* from SF76 which was chosen as the reference standard. The most common SF76 band was arbitrarily assigned a mobility of '100' according to standard practice. Other bands were designated a code corresponding to 100 plus (or minus, if slower) the difference, in millimeters, between their electrophoretic mobilities and that of the SF76 100 band.

### *Analysis of data*

Standard methods for the analysis of genetic variation within populations were employed providing values of observed ($H_e$) and estimates of unbiased expected heterozygosity ($H_e$) and their standard errors over all loci (Nei, 1975; Nei & Roychoudhury, 1974). In the case of parthenogenetic forms, the calculated $H_e$ values do not relate directly to the frequency of observed heterozygotes as, in them, the unit of segregation is the composite genotype and not single genes as in panmictic populations. Nevertheless, an analogous measure is calculated in the same way and is a true reflection of genetic diversity (Nei, 1975). Deviations from Castle-Hardy-Weinberg equilibrium in the genotype frequencies were tested using a normal $\chi^2$ test, except for samples where one or more of the expected genotype classes contained less than 5 entries, in which case Elston & Forthofer's (1977) suggested methods were followed. For the determination of the average number of alleles, these were counted if their frequency was $\geq 0.01$. A locus was considered polymorphic if the frequency of its most common allele was $\leq .95$.

The allele frequency data from the populations was utilized to calculate genetic distances ($D$) and their standard errors using Nei's methods (Nei, 1972; Nei & Roychoudhury, 1974).

For the phylogenetic reconstruction of the relationships between the populations studied, Swofford's (1982) WAGPROC computer program was employed using the Farris (1970) algorithm for Wagner Trees 'simple' addition sequences and global branch swapping to search for multiple minimum-length trees. The network was rooted at the midpoint of the greatest patristic distance, and the optimization of the HTU character states was by 'minimum homoplasy' criteria.

*Table 1.* Variation within allozyme loci in gonochoristic and parthenogenetic *Artemia*

| Locus | Alleles | GONOCHORISTIC *Artemia* *franciscana* | | | *monica* | *persimil* | *salina* | *urmiana* | PARTHENOGENETIC *Artemia* *diploid* | | | | *triploid* | | *tetra-* | *penta-* |
|---|---|---|---|---|---|---|---|---|---|---|---|---|---|---|---|---|
| | | GS77 | SF76 | PEK | MON | BA | BAR | URM | BP(2) | CAL-I | LT-VI | EIL | IZM(3) | TOT | DE | BP-I(5) |
| Cat | N | 48 | 48 | 47 | 54 | 48 | 50 | 25 | 3 | 30 | 5 | 32 | 22 | 28 | 40 | 7 |
| | 87 | | | | | 1.00 | | | | | | | | | | |
| | 89 | | 0.16 | | | | 1.00 | | 0.50 | 1.00 | 0.50 | 0.33 | 0.33 | 0.33 | 0.50* | 0.50* |
| | 98 | | | | | | | 0.12 | | | | | | | | |
| | 99 | | | | | | | | | | | | | | 0.50* | 0.50* |
| | 100 | 1.00 | 0.84 | 1.00 | 1.00 | | | | | | | | | | | |
| | 106 | | | | | | | | | | 0.50 | 0.67 | 0.67 | 0.67 | | |
| | 109 | | | | | | | 0.88 | | | | | | | | |
| | 112 | | | | | | | | | | | | | | | |
| | 113 | | | | | | | | 0.50 | | | | | | | |
| | H obs | | 0.25 | | | | | 0.24 | | | | | | | | |
| | H exp | | 0.28 | | | | | 0.22 | | | | | | | | |
| | chi$^2$ | 0.25 | | | | | 0.23 | | | | | | | | | |
| Est-1 | N | 79 | 48 | 43 | 54 | 42 | 48 | 25 | 3 | 30 | 5 | 32 | 22 | 28 | 40 | 7 |
| | 100 | 1.00 | 1.00 | 1.00 | 1.00 | 1.00 | | | | | | | | | | |
| | 102 | | | | | | 1.00 | 1.00 | 1.00 | 1.00 | 1.00 | 1.00 | 1.00 | 1.00 | 1.00 | 1.00 |
| Est-4 | N | 79 | 48 | 43 | 54 | 42 | 48 | 25 | 3 | 30 | 5 | 32 | 22 | 28 | | |
| | 100 | 1.00 | 1.00 | 1.00 | 1.00 | 1.00 | | | | | | | | | | |
| | 105 | | | | | | 1.00 | 1.00 | 1.00 | 1.00 | 1.00 | 1.00 | 1.00 | 1.00 | 1.00 | 1.00 |
| Est-D | N | 79 | 48 | 43 | 54 | 58 | 48 | 25 | 3 | 30 | 5 | 32 | 22 | 28 | 40 | 7 |
| | 85 | | | | | 0.10 | | | | | | | | | | |
| | 88 | | | | 0.04 | | | | | | | | | | | |
| | 94 | | | 0.13 | | | | 0.06 | | | | | | | | |
| | 98 | | | | | 0.90 | 1.00 | | | | | | | | | |
| | 100 | 0.48 | 1.00 | 0.74 | 0.64 | | | 0.94 | 1.00 | 1.00 | 1.00 | 1.00 | 1.00 | 1.00 | 0.75 | 0.80 |
| | 107 | | | 0.13 | | | | | | | | | | | | |
| | 110 | 0.42 | | | 0.21 | | | | | | | | | | | |
| | 120 | 0.06 | | | 0.11 | | | | | | | | | | | |
| | 129 | 0.03 | | | | | | | | | | | | | 0.25 | 0.20 |
| | H obs | 0.58 | | 0.37 | 0.54 | 0.21 | | 0.12 | | | | | | | | |
| | H exp | 0.59 | | 0.42 | 0.54 | 0.19 | | 0.12 | | | | | | | | |
| | chi$^2$ | 0.08 | | 0.06 | 0.04 | 0.37 | | 1.09 | | | | | | | | |

*Table 1.* (Continued)

| Locus | Alleles | GONOCHORISTIC *Artemia* *franciscana* | | | *monica* | *persimil* | *salina* | *urmiana* | PARTHENOGENETIC *Artemia* *diploid* | | | | *triploid* | | *tetra-* | *penta-* |
|---|---|---|---|---|---|---|---|---|---|---|---|---|---|---|---|---|
| | | GS77 | SF76 | PEK | MON | BA | BAR | URM | BP(2) | CAL-I | LT-VI | EIL | IZM(3) | TOT | DE | BP-I(5) |
| Idh-2 | N | 156 | 48 | 48 | 51 | 48 | 48 | 25 | 3 | 30 | 5 | 32 | 22 | 28 | 40 | 7 |
| | 86 | | | | | 0.16 | | | | | | | | | | |
| | 89 | | | | | | | | | | | | | 0.33 | 0.25 | 0.20 |
| | 93 | 0.01 | | 0.92 | | 0.07 | | | | | | | | | | |
| | 95 | | | | | | | | 1.00 | 1.00 | 1.00 | 1.00 | | | | |
| | 97 | | | | | | 0.35 | 0.08 | | | | | | | | |
| | 98 | | | | | 0.70 | | | | | | | | 0.67 | 0.75 | 0.80 |
| | 100 | 0.90 | 0.98 | 0.08 | 0.54 | | | 0.90 | | | | | | 0.67 | 0.75 | 0.80 |
| | 104 | 0.09 | | | 0.10 | | 0.65 | 0.02 | | | | | 1.00 | | | |
| | 110 | | 0.02 | | 0.36 | | | | | | | | | | | |
| | H obs | 0.19 | 0.24 | 0.17 | 0.45 | 0.42 | 0.42 | 0.20 | | | | | | | | |
| | H exp | 0.18 | 0.24 | 0.15 | 0.57 | 0.38 | 0.46 | 0.19 | | | | | | | | |
| | $chi^2$ | 1.23 | 0.01 | 0.22 | 2.86 | 0.06 | 0.38 | 0.28 | | | | | | | | |
| Lap-2 | N | 77 | 48 | 52 | 51 | 46 | 42 | 25 | 3 | 30 | 5 | 32 | 22 | 28 | 40 | 7 |
| | 97 | | | | | | 1.00 | 1.00 | | | | | | | | |
| | 99 | 1.00 | 1.00 | 1.00 | 1.00 | | | | 1.00 | 1.00 | 1.00 | 1.00 | 1.00 | 1.00 | 1.00 | 1.00 |
| | 100 | | | | | 1.00 | | | | | | | | | | |
| Lap-3 | N | 77 | 48 | 52 | 51 | 46 | 42 | 25 | 3 | 30 | 5 | 32 | 22 | 28 | 40 | 7 |
| | 95 | | | | | | | | | | | | | | | |
| | 96 | | | | | 1.00 | | | | | | | | | | |
| | 98 | | | | 0.46 | | 1.00 | 0.92 | 0.50 | 1.00 | 1.00 | 1.00 | 1.00 | 1.00 | 1.00 | |
| | 100 | 1.00 | 1.00 | 1.00 | 0.54 | | | | | | | | | | | |
| | 101 | | | | | | | 0.08 | 0.50 | | 1.00 | | | | | |
| | H obs | | | | 0.37 | | | 0.08 | | | | | | | | |
| | H exp | | | | 0.50 | | | 0.15 | | | | | | | | |
| | $chi^2$ | | | | 3.19 | | | 0.85 | | | | | | | | |
| Ldh | N | 176 | 48 | 50 | 54 | 45 | 42 | 25 | 3 | 30 | 5 | 32 | 22 | 28 | 40 | 7 |
| | 100 | 1.00 | 1.00 | 1.00 | 1.00 | 1.00 | | | | | | | | | | |
| | 103 | | | | | | 1.00 | 1.00 | 1.00 | 1.00 | 1.00 | 1.00 | 1.00 | 1.00 | 1.00 | 1.00 |

*Table 1.* (Continued)

| Locus | Alleles | GONOCHORISTIC *Artemia* *franciscana* | | | *monica* | *persimil* | *salina* | *urmiana* | PARTHENOGENETIC *Artemia* *diploid* | | | | *triploid* | | *tetra-* | *penta-* |
|---|---|---|---|---|---|---|---|---|---|---|---|---|---|---|---|---|
| | | GS77 | SF76 | PEK | MON | BA | BAR | URM | BP(2) | CAL-I | LT-VI | EIL | IZM(3) | TOT | DE | BP-I(5) |
| Got-1 | N | 22 | 48 | 46 | 54 | 48 | 48 | 25 | 3 | 30 | 5 | 32 | 22 | 28 | 40 | 7 |
| | 93 | | | | | | 1.00 | | | | | | | | | |
| | 95 | | | 0.10 | | | | | | | | | | | | |
| | 100 | 0.95 | 1.00 | 0.90 | 1.00 | 0.95 | | | | | | | | | | |
| | 104 | | | | | 0.05 | | 1.00 | | | | | | | | |
| | 110 | 0.05 | | | | | | | 1.00 | 1.00 | 1.00 | 1.00 | 1.00 | 1.00 | 1.00 | 1.00 |
| | H obs | 0.09 | | 0.20 | | 0.10 | | | | | | | | | | |
| | H exp | 0.09 | | 0.18 | | 0.10 | | | | | | | | | | |
| | chi$^2$ | 0.02 | | 0.24 | | 0.64 | | | | | | | | | | |
| Got-2 | N | 22 | 48 | 46 | 54 | 48 | 48 | 25 | 3 | 30 | 5 | 32 | 22 | 28 | 40 | 7 |
| | 91 | | | | | | | | | | | | | | 0.25 | 0.20 |
| | 93 | 0.05 | | | 0.06 | | | | 1.00 | | | | | | | |
| | 100 | 0.93 | 1.00 | 1.00 | 0.61 | 0.97 | | 0.96 | | 1.00 | 1.00 | 1.00 | 1.00 | 1.00 | 0.75 | 0.80 |
| | 101 | | | | | | 0.14 | | | | | | | | | |
| | 106 | 0.02 | | | 0.33 | 0.03 | | 0.04 | | | | | | | | |
| | 107 | | | | | | 0.86 | | | | | | | | | |
| | H obs | 0.14 | | | 0.43 | 0.06 | 0.27 | 0.08 | | | | | | | | |
| | H exp | 0.13 | | | 0.52 | 0.06 | 0.24 | 0.08 | | | | | | | | |
| | chi$^2$ | 0.93 | | | 0.92 | 2.35 | 0.65 | 0.02 | | | | | | | | |
| Idh-1 | N | 157 | 48 | 48 | 53 | 48 | 43 | 25 | 3 | 30 | 5 | 32 | 22 | 28 | 40 | 7 |
| | 87 | 0.01 | | | | 0.10 | | | | | | | | | | |
| | 93 | 0.89 | 0.10 | 0.08 | 0.42 | 0.90 | | 0.02 | | | | | | | | |
| | 98 | | | | | | | 0.04 | | | | | | | 0.25 | 0.20 |
| | 100 | 0.08 | 0.90 | 0.84 | 0.58 | | | | | | | | | | | |
| | 102 | | | 0.07 | | | | | | | | | | | | |
| | 103 | | | | | | 1.00 | | | | | | | | | |
| | 105 | | | | | | | 0.90 | 1.00 | 1.00 | 1.00 | 1.00 | 1.00 | 1.00 | 0.75 | 0.80 |
| | 106 | 0.01 | | | | | | | | | | | | | | |
| | 108 | | | | | | | 0.02 | | | | | | | | |
| | 110 | | | | | | | | | | | | | | | |
| | 112 | | | | | | | 0.02 | | | | | | | | |
| | H obs | 0.21 | 0.21 | 0.31 | 0.47 | 0.21 | | 0.16 | | | | | | | | |
| | H exp | 0.19 | 0.19 | 0.28 | 0.49 | 0.19 | | 0.19 | | | | | | | | |
| | chi$^2$ | 1.60 | 0.29 | 1.03 | 0.06 | 0.29 | | 2.00 | | | | | | | | |

*Table 1.* (Continued)

| Locus | Alleles | GONOCHORISTIC *Artemia* *franciscana* | | | *monica* | *persimil* | *salina* | *urmiana* | PARTHENOGENETIC *Artemia* *diploid* | | | | *triploid* | | *tetra-* | *penta-* |
|---|---|---|---|---|---|---|---|---|---|---|---|---|---|---|---|---|
| | | GS77 | SF76 | PEK | MON | BA | BAR | URM | BP(2) | CAL-I | LT-VI | EIL | IZM(3) | TOT | DE | BP-I(5) |
| Mdh-1 | N | 176 | 48 | 50 | 54 | 46 | 48 | 25 | 3 | 30 | 5 | 32 | 22 | 28 | 40 | 7 |
| | 85 | | | | | 0.02 | | | | | | | | | | |
| | 86 | | | | | | 1.00 | | | | | | | | | |
| | 90 | | | | | 0.97 | | 0.06 | | | | | | | 0.25 | |
| | 92 | | | | | | | | | | | | 0.67 | | | |
| | 93 | | | 0.11 | | | | | | | | | | | | |
| | 94 | | | | | | | 0.70 | | | | 0.67 | | 0.67 | | |
| | 97 | | | | | 0.01 | | | | | | | | | | |
| | 99 | | | | | | | 0.24 | 1.00 | 1.00 | 1.00 | | 0.33 | | 0.75 | 1.00 |
| | 100 | 1.00 | 1.00 | 0.89 | 0.99 | | | | | | | | | | | |
| | 105 | | | | 0.01 | | | | | | | 0.33 | | 0.33 | | |
| | H obs | | | 0.18 | 0.02 | 0.07 | | 0.36 | | | | | | | | |
| | H exp | | | 0.20 | 0.02 | 0.06 | | 0.46 | | | | | | | | |
| | $chi^2$ | | | 0.23 | | 2.24 | | 0.37 | | | | | | | | |
| Mdh-2 | N | 176 | 48 | 50 | 54 | 46 | 48 | 25 | 3 | 30 | 5 | 32 | 22 | 28 | 40 | 7 |
| | 89 | | | | | 1.00 | | 0.02 | | | | | | | | |
| | 92 | | | | | | 1.00 | | | | | 0.33 | | | | |
| | 97 | | | | | | | | | | | | 1.00 | | | |
| | 99 | | | | | | | 0.98 | 0.50 | 1.00 | 1.00 | | | | 1.00 | 1.00 |
| | 100 | 1.00 | 1.00 | 1.00 | 1.00 | | | | | | | | | | | |
| | 101 | | | | | | | | | | | | | 1.00 | | |
| | 108 | | | | | | | | 0.50 | | | 0.67 | | | | |
| | H obs | | | | | | | 0.04 | | | | | | | | |
| | H exp | | | | | | | 0.04 | | | | | | | | |
| | $chi^2$ | | | | | | | 0.00 | | | | | | | | |
| ME | N | 138 | 48 | 48 | 54 | 86 | 50 | 25 | 3 | 30 | 5 | 32 | 22 | 28 | 40 | 7 |
| | 96 | | | | | 1.00 | 1.00 | 1.00 | | | | | | | | |
| | 98 | | | | | | | | 1.00 | 1.00 | 1.00 | 1.00 | 1.00 | 1.00 | 1.00 | 1.00 |
| | 100 | 1.00 | 1.00 | 1.00 | 1.00 | | | | | | | | | | | |

*Table 1.* (Continued)

| Locus | Alleles | GONOCHORISTIC *Artemia* | | | | | | | PARTHENOGENETIC *Artemia* | | | | | | | |
|---|---|---|---|---|---|---|---|---|---|---|---|---|---|---|---|---|
| | | *franciscana* | | | *monica* | *persimil* | *salina* | *urmiana* | *diploid* | | | | *triploid* | | *tetra-* | *penta-* |
| | | GS77 | SF76 | PEK | MON | BA | BAR | URM | BP(2) | CAL-I | LT-VI | EIL | IZM(3) | TOT | DE | BP-I(5) |
| Mpi | N | 42 | 48 | 48 | 54 | 48 | 48 | 23 | 3 | 30 | N.A. | 32 | 22 | 28 | 40 | 7 |
| | 78 | | | | | 0.73 | | | | | | | | | | |
| | 84 | | | | | 0.20 | | | | | | | | | | |
| | 90 | | | | | 0.07 | | | | | | | | | | |
| | 91 | | | | | | 0.01 | | | | | | | | | |
| | 95 | 0.01 | | | | | | | | | | | | | | |
| | 97 | | | | | | 0.37 | | | | | | | | | |
| | 100 | 0.96 | 1.00 | 0.85 | 1.00 | | 0.31 | | | | | | | | | |
| | 102 | | | | | | 0.30 | 0.02 | | | | | | 0.67 | 0.50 | 0.60 |
| | 105 | | | | | | | 0.09 | 1.00 | | | 1.00 | 1.00 | | | |
| | 106 | 0.02 | | 0.15 | | | | | | | | | | | | |
| | 110 | | | | | | | 0.63 | | | | | | | | |
| | 113 | | | | | | | 0.26 | | 1.00 | | | | | | |
| | 116 | | | | | | | | | | | | | 0.33 | 0.50 | 0.40 |
| | H obs | 0.07 | | 0.29 | | 0.46 | 0.46 | 0.30 | | | | | | | | |
| | H exp | 0.07 | | 0.25 | | 0.43 | 0.68 | 0.54 | | | | | | | | |
| | chi² | 2.02 | | 0.83 | | 0.82 | 1.65 | 2.32 | | | | | | | | |
| Pep-1 | N | 138 | 48 | 48 | 53 | 46 | 48 | 25 | 3 | 30 | 5 | 32 | 22 | 28 | 40 | 7 |
| | 92 | | | | | 1.00 | 1.00 | | | | | | | | | |
| | 95 | | | | 0.08 | | | | | | | | | | | |
| | 100 | 1.00 | 1.00 | 1.00 | 0.91 | | | | | | | | | | | |
| | 101 | | | | | | | 1.00 | 1.00 | 1.00 | 1.00 | 1.00 | 1.00 | 1.00 | 1.00 | 1.00 |
| | 104 | | | | 0.01 | | | | | | | | | | | |
| | H obs | | | | 0.19 | | | | | | | | | | | |
| | H exp | | | | 0.17 | | | | | | | | | | | |
| | chi² | | | | 0.26 | | | | | | | | | | | |
| Pep-4 | N | 138 | 48 | 45 | 54 | 46 | 46 | 25 | 3 | 30 | 5 | 32 | 22 | 28 | 40 | 7 |
| | 90 | | | 0.05 | | | | | | | | | | | | |
| | 93 | | | | | | 0.34 | | | | | | | | | |
| | 97 | | | | | | | 0.04 | | | | | | | | |
| | 98 | | | | | | 0.66 | | | | | | | | | |
| | 100 | 1.00 | 1.00 | 0.95 | 0.91 | 1.00 | | 0.94 | 1.00 | 1.00 | 1.00 | 1.00 | 1.00 | 1.00 | 1.00 | 1.00 |
| | 103 | | | | 0.09 | | | | | | | | | | | |
| | 105 | | | | | | | 0.02 | | | | | | | | |
| | H obs | | | 0.10 | 0.15 | | 0.40 | 0.12 | | | | | | | | |
| | H exp | | | 0.10 | 0.17 | | 0.46 | 0.12 | | | | | | | | |
| | chi² | | | 0.64 | 0.48 | | 0.72 | 1.09 | | | | | | | | |

*Table 1.* (Continued)

| Locus | Alleles | GONOCHORISTIC *Artemia* | | | | | | | PARTHENOGENETIC *Artemia* | | | | | | | |
|---|---|---|---|---|---|---|---|---|---|---|---|---|---|---|---|---|
| | | *franciscana* | | | *monica* | *persimil* | *salina* | *urmiana* | *diploid* | | | | *triploid* | | *tetra-* | *penta-* |
| | | GS77 | SF76 | PEK | MON | BA | BAR | URM | BP(2) | CAL I | LT VI | EIL | IZM(3) | TOT | DE | BP-I(5) |
| 6Pgdh | N | 176 | 44 | 47 | 53 | 47 | 47 | 25 | 3 | 30 | 5 | 32 | 22 | 28 | 40 | 7 |
| | 82 | | | | | | | | 0.50 | | | | 0.50* | | | |
| | 89 | | | | | | 0.27 | 0.08 | 0.50 | | 0.50 | 0.50* | 0.50* | 0.50* | 0.50* | 0.50* |
| | 93 | | | | | | 0.02 | | | | | | | | | |
| | 94 | | | | | | | 0.78 | | 0.50 | 0.50 | 0.50* | | 0.50* | 0.50* | 0.50* |
| | 99 | | | | | | 0.05 | 0.14 | | 0.50 | | | | | | |
| | 100 | 0.08 | 0.52 | 0.37 | 0.72 | 0.79 | | | | | | | | | | |
| | 103 | | | | | | 0.46 | | | | | | | | | |
| | 105 | 0.91 | 0.48 | 0.62 | 0.26 | 0.21 | | | | | | | | | | |
| | 108 | | | | | | 0.20 | | | | | | | | | |
| | 109 | 0.01 | | | 0.02 | | | | | | | | | | | |
| | H obs | 0.19 | 0.46 | 0.54 | 0.40 | 0.37 | 0.74 | 0.44 | | | | | | | | |
| | H exp | 0.17 | 0.50 | 0.47 | 0.42 | 0.33 | 0.68 | 0.37 | | | | | | | | |
| | $chi^2$ | 1.34 | 0.32 | 1.16 | 0.18 | 0.52 | 0.03 | 1.23 | | | | | | | | |
| Pgi | N | 138 | 48 | 47 | 54 | 86 | 50 | 25 | 3 | 30 | 5 | 32 | 22 | 28 | 40 | 7 |
| | 59 | | | | | 0.03 | | | | | | | | | | |
| | 69 | | | | | 0.28 | | | | | | | | | | |
| | 70 | | | | | | | | | | | | | | | 0.40 |
| | 75 | | | | | | 0.03 | 0.94 | | | | | | 1.00 | 0.50 | |
| | 77 | | | | | | | | 1.00 | | 1.00 | 1.00 | 1.00 | | | |
| | 80 | 0.10 | | | 0.41 | | | | | | | | | | | |
| | 82 | | | | | 0.68 | | | | | | | | | | |
| | 83 | | | | | | 0.93 | 0.06 | | | | | | | | |
| | 85 | | | | | | | | | 1.00 | | | | | 0.50 | 0.60 |
| | 90 | 0.43 | 0.31 | 0.17 | 0.30 | | | | | | | | | | | |
| | 92 | | | | | | 0.04 | | | | | | | | | |
| | 100 | 0.47 | 0.69 | 0.82 | 0.29 | | | | | | | | | | | |
| | 109 | 0.01 | | 0.01 | | | | | | | | | | | | |
| | 118 | | | | 0.01 | | | | | | | | | | | |
| | H obs | 0.57 | 0.42 | 0.30 | 0.52 | 0.50 | 0.14 | 0.12 | | | | | | | | |
| | H exp | 0.59 | 0.43 | 0.30 | 0.67 | 0.46 | 0.13 | 0.12 | | | | | | | | |
| | $chi^2$ | 0.06 | 0.05 | 0.15 | 0.26 | 1.53 | 0.27 | 1.09 | | | | | | | | |
| Pgi-P | N | N.A. | N.A. | N.A. | N.A. | 86 | N.A. | N.A. | N.A. | N.A. | N.A. | N.A. | N.A. | N.A. | N.A. | N.A. |
| | 69 | | | | | 0.01 | | | | | | | | | | |
| | 82 | | | | | 0.62 | | | | | | | | | | |
| | 95 | | | | | 0.37 | | | | | | | | | | |
| | H obs | | | | | 0.51 | | | | | | | | | | |
| | H exp | | | | | 0.48 | | | | | | | | | | |
| | $chi^2$ | | | | | 0.03 | | | | | | | | | | |

*Table 1.* (Continued)

| Locus | Alleles | GONOCHORISTIC *Artemia* | | | | | | | PARTHENOGENETIC *Artemia* | | | | | | | |
|---|---|---|---|---|---|---|---|---|---|---|---|---|---|---|---|---|
| | | *franciscana* | | | *monica* | *persimil* | *salina* | *urmiana* | *diploid* | | | | *triploid* | | *tetra-* | *penta-* |
| | | GS77 | SF76 | PEK | MON | BA | BAR | URM | BP(2) | CAL-I | LT-VI | EIL | IZM(3) | TOT | DE | BP-I(5) |
| Pgm | N | 63 | 48 | 52 | 54 | 46 | 48 | 24 | 3 | 30 | 5 | 32 | 22 | 28 | 40 | 7 |
| | 83 | | | | | | | 0.02 | | | | | | | | |
| | 86 | | | | | | | | 0.50 | 0.50 | 0.50 | 0.33 | 0.33 | 0.33 | 0.50 | 0.40 |
| | 89 | | | 0.11 | | | | | | | | | | | | |
| | 91 | | | | | | 0.94 | 0.35 | 0.50 | 0.50 | | 0.67 | 0.67 | 0.67 | 0.25 | 0.40 |
| | 92 | | | 0.14 | | | | | | | | | | | | |
| | 94 | | | | | | 0.06 | | | | | | | | | |
| | 95 | 0.01 | | | | | | | | | | | | | | |
| | 96 | | | | | | | 0.62 | | | | | | | | |
| | 97 | 0.68 | 0.12 | | | | | | | | 0.50 | | | | | |
| | 99 | | | 0.75 | | | | | | | | | | | | |
| | 100 | 0.31 | 0.87 | | 1.00 | 0.18 | | | | | | | | | | |
| | 102 | | | | | | | | | | | | | | 0.25 | 0.20 |
| | 104 | | | | | 0.82 | | | | | | | | | | |
| | H obs | 0.44 | 0.17 | 0.38 | | 0.28 | 0.12 | 0.50 | | | | | | | | |
| | H exp | 0.44 | 0.22 | 0.41 | | 0.30 | 0.12 | 0.49 | | | | | | | | |
| | $chi^2$ | 0.01 | 2.06 | 0.21 | | 0.13 | 0.38 | 0.03 | | | | | | | | |
| Sod | N | 138 | 48 | 47 | 54 | 46 | 50 | 25 | 3 | 30 | 5 | 32 | 22 | 28 | 40 | 7 |
| | 79 | 0.01 | | | | | | | | | | | | | | |
| | 83 | | | | | | 1.00 | 0.02 | | | | | | | 0.25 | 0.20 |
| | 87 | | | | | 1.00 | | | | | | | | | | |
| | 97 | | | | | | | 0.98 | 1.00 | 1.00 | 1.00 | 1.00 | 1.00 | 1.00 | 0.75 | 0.80 |
| | 100 | 0.99 | 1.00 | 1.00 | 1.00 | | | | | | | | | | | |
| | H obs | 0.01 | | | | | | 0.04 | | | | | | | | |
| | H exp | 0.01 | | | | | | 0.04 | | | | | | | | |
| | $chi^2$ | 0.00 | | | | | | 0.00 | | | | | | | | |

* Possible allele dosage effect seen in heterozygote banding patterns, however as this was too complicated to score accurately, simple symmetric heterozygotes with two alleles at equal dosages were scored.

N.A. no activity observed at this locus for this population/clone.

## Results

The number of genomes sampled, the relative mobilities of the electromorphs, the allele frequencies, the observed and the unbiased expected heterozygosities (Nei & Roychoudhury, 1974) found in the *Artemia* samples are listed in Table 1. Out of a total of 22 loci which were assayed, intraspecific and/or interspecific variation was found in all but one (Amy). One locus (Pgi) was duplicated in one of the species included for comparative purposes (*A. persimilis*). $\chi^2$ values for deviations from Castle-Hardy-Weinberg equilibrium were non-significant for all results from the gonochoristic populations.

A summary of the genetic variation over all loci scored is given in Table 2. The average number of alleles per locus ($N_a$) found both in *A. monica* (1.76) and in *A. urmiana* (2.19) are among the highest in the genus. This is also true for the proportion of polymorphic loci (P), 0.46 and 0.55 respectively, and the mean overall values for expected heterozygosity ($H_a$), 0.19 and 0.14, respectively. Because of the characteristically high standard errors of these parameters, however, no significant differences were found for pair-wise comparisons of the means for these parameters (Games and Howell's method; in Sokal & Rohlf, 1981).

Nei standard genetic distances (*D*) calculated for all loci between the species studied range from 0.05 for one *A. monica*-*A. franciscana* comparison (MON–SF76), to 2.98 between a diploid parthenogenetic clone (BP(2)) and *A. persimilis* (Table 3). Highest *D* value from *A. monica* was between it and *A. salina* (BAR, 2.43); that from *A. urmiana* was between it and *A. franciscana* (PEK, 1.65). The lowest *D* from *A. urmiana* was obtained in a comparison with a triploid pathenogenetic clone (TUT, 0.40). Notwithstanding the very low *D* values between *A. monica* and the *A. franciscana* populations studied, significant differences are found between their allelic frequencies at polymorphic loci.

The levels of $N_a$, P and $H_a$ were not equal among the loci studied. However, loci with high levels, of variation in *A. urmiana* and *A. monica* tended to be high in other species as well, with a tendency towards highest levels of variation in loci which code for large enzyme subunits (e.g. Pgm, Est-D, Pgi, 6Pgdh) as has been observed in other studies (Koehn & Eanes, 1978).

The genetic distances measured between *A. urmiana* and *A. monica* and representative examples of the genus (Table 2), situate these two species in clusters corresponding to the geographic distribution of the species and their populations. While *A. urmiana* is clearly differentiated from the remaining amphigonic species (mean $D = 1.21$), it is, from the present data, a close relative to the parthenogenetic forms (mean $D = 0.48$). Compared to Nei's *D* measurements among other organisms (Ayala, 1983), these *D* values are, respectively, equivalent to comparisons between full species and intermediate between incipient species and full species. *A. monica* seems very closely related to the remaining *A. franciscana* populations in the list, with a mean *D* of 0.09, analogous to distances intermediate between local populations and subspecies (Ayala, *loc. cit.*). Most of the genetic differentiation between *A. monica* and *A. franciscana* resides in four of the loci studied: Est-D, Got-2, Lap-3, and Pgi. While no locus in *A. monica* and *A. franciscana* could be considered 'diagnostic' (capable of assigning an individual to the correct taxon with a probability ≥ 99%; see Ayala, 1983), between 52 and 77% of loci were diagnostic in comparing between *A. monica* and any of the other species. For *A. urmiana*, only 36% of loci studied were diagnostic between it and *A. salina*, and 65 to 77% were diagnostic in comparisons between this species and the remaining bisexuals. Only 18% of the loci were diagnostic in comparing *A. urmiana* with the parthenogenetic forms as a whole.

For the Wagner tree derivation of the phylogenetic relationships between the populations, allele frequencies were assigned a binary code for presence (1) or absence (0) of alleles. Six trees of equivalent minimum lengths were obtained (only one is shown in Fig. 1). The qualitative differences between the equivalent trees involve minor changes affecting the root position of branches of

*Table 2.* Summary of genetic variation parameters over all loci

| Locus Alleles | GONOCHORISTIC *Artemia* | | | | | | | PARTHENOGENETIC *Artemia* | | | | | | | |
|---|---|---|---|---|---|---|---|---|---|---|---|---|---|---|---|
| | *franciscana* | | | *monica* | *persimil* | *salina* | *urmiana* | *diploid* | | | | *triploid* | | *tetra-* | *penta-* |
| | GS77 | SF76 | PEK | MON | BA | BAR | URM | BP(2) | CAL-I | LT-VI | EIL | IZM(3) | TOT | DE | BP-I(5) |
| Number of loci scored | 22 | 22 | 22 | 22 | 23 | 22 | 22 | 22 | 22 | 21 | 22 | 22 | 22 | 22 | 22 |
| Average number of genomes samples per locus | 110.7 | 47.8 | 47.6 | 53.5 | 52.2 | 47.1 | 24.9 | 3 | 30 | 5 | 32 | 22 | 28 | 40 | 7 |
| (s.e.) | (54.5) | (0.87) | (2.77) | (1.06) | (14.1) | (2.6) | (0.48) | | | | | | | | |
| Mean number of alleles per locus* | 1.71 | 1.29 | 1.62 | 1.76 | 1.71 | 1.62 | 2.19 | 1.24 | 1.10 | 1.15 | 1.24 | 1.19 | 1.29 | 1.57 | 1.52 |
| (s.e. of mean) | (0.21) | (0.10) | (0.16) | (0.22) | (0.19) | (0.24) | (0.24) | (0.09) | (0.06) | (0.08) | (0.09) | (0.09) | (0.10) | (0.13) | (0.13) |
| Proportion of polymorphic loci[b] | 0.36 | 0.27 | 0.50 | 0.46 | 0.39 | 0.36 | 0.55 | 0.24 | 0.10 | 0.15 | 0.24 | 0.19 | 0.29 | 0.52 | 0.48 |
| Average heterozygosity per population (or per clone in asexual forms) expected | 0.11 | 0.08 | 0.13 | 0.19 | 0.13 | 0.13 | 0.14 | 0.11 | 0.04 | 0.07 | 0.10 | 0.08 | 0.12 | 0.21 | 0.18 |
| pbserved | 0.11 | 0.07 | 0.13 | 0.16 | 0.14 | 0.12 | 0.13 | | | | | | | | |

[a] An allele was counted if its frequency $\geq 0.01$.
[b] Locus was considered polymorphic if the frequency of its most common allele $\leq 0.95$.

*Table 3*. Matrix of Nei's std. genetic distances (above diagonal) and their s.e. (below diagonal)

| Locus Alleles | GONOCHORISTIC *Artemia* | | | | | | | PARTHENOGENETIC *Artemia* | | | | | | | |
|---|---|---|---|---|---|---|---|---|---|---|---|---|---|---|---|
| | *franciscana* | | | *monica* | *persimil* | *salina* | *urmiana* | *diploid* | | | | *triploid* | | *tetra-* | *penta-* |
| | GS77 | SF76 | PEK | MON | BA | BAR | URM | BP(2) | CAL-I | LT-VI | EIL | IZM(3) | TOT | DE | BP-I(5) |
| GS77 | – | 0.09 | 0.13 | 0.09 | 1.03 | 2.81 | 1.54 | 2.52 | 2.11 | 1.94 | 2.08 | 2.05 | 2.06 | 2.16 | 2.15 |
| SF76 | 0.03 | – | 0.09 | 0.05 | 1.03 | 2.43 | 1.42 | 2.56 | 1.88 | 1.85 | 1.88 | 1.89 | 1.87 | 1.98 | 1.96 |
| PEK | 0.07 | 0.05 | – | 0.13 | 1.10 | 2.87 | 1.65 | 2.41 | 1.99 | 1.96 | 1.96 | 1.97 | 1.95 | 2.06 | 2.05 |
| MON | 0.04 | 0.02 | 0.06 | – | 1.03 | 2.43 | 1.53 | 2.27 | 1.97 | 2.12 | 1.94 | 1.91 | 1.92 | 1.99 | 1.98 |
| BA | 0.29 | 0.29 | 0.31 | 0.28 | – | 1.91 | 1.38 | 2.98 | 2.34 | 2.30 | 2.31 | 2.32 | 2,06 | 2.00 | 2.00 |
| BAR | 0.85 | 0.79 | 0.91 | 0.66 | 0.54 | – | 0.95 | 0.98 | 0.91 | 1.13 | 0.94 | 0.87 | 0.89 | 0.82 | 0.83 |
| URM | 0.41 | 0.40 | 0.45 | 0.39 | 0.38 | 0.27 | – | 0.62 | 0.44 | 0.51 | 0.46 | 0.56 | 0.40 | 0.41 | 0.42 |
| BP(2) | 0.69 | 0.66 | 0.70 | 0.61 | 0.99 | 0.27 | 0.19 | – | 0.24 | 0.15 | 0.13 | 0.17 | 0.34 | 0.24 | 0.28 |
| CAL-I | 0.56 | 0.52 | 0.55 | 0.49 | 0.69 | 0.26 | 0.16 | 0.11 | – | 0.17 | 0.21 | 0.28 | 0.29 | 0.15 | 0.12 |
| LT-VI | 0.50 | 0.51 | 0.55 | 0.57 | 0.68 | 0.32 | 0.17 | 0.07 | 0.09 | – | 0.19 | 0.26 | 0.33 | 0.20 | 0.21 |
| EIL | 0.56 | 0.53 | 0.55 | 0.54 | 0.68 | 0.27 | 0.16 | 0.07 | 0.10 | 0.09 | – | 0.14 | 0.19 | 0.21 | 0.27 |
| IZM(3) | 0.54 | 0.53 | 0.55 | 0.47 | 0.68 | 0.25 | 0.18 | 0.09 | 0.12 | 0.11 | 0.08 | – | 0.25 | 0.28 | 0.27 |
| TUT | 0.56 | 0.53 | 0.55 | 0.49 | 0.57 | 0.25 | 0.15 | 0.13 | 0.12 | 0.13 | 0.10 | 0.11 | – | 0.16 | 0.21 |
| DE | 0.50 | 0.54 | 0.57 | 0.51 | 0.53 | 0.24 | 0.15 | 0.09 | 0.07 | 0.09 | 0.09 | 0.10 | 0.08 | – | 0.04 |
| BP(5) | 0.58 | 0.54 | 0.56 | 0.51 | 0.53 | 0.24 | 0.15 | 0.11 | 0.06 | 0.10 | 0.11 | 0.11 | 0.10 | 0.02 | – |

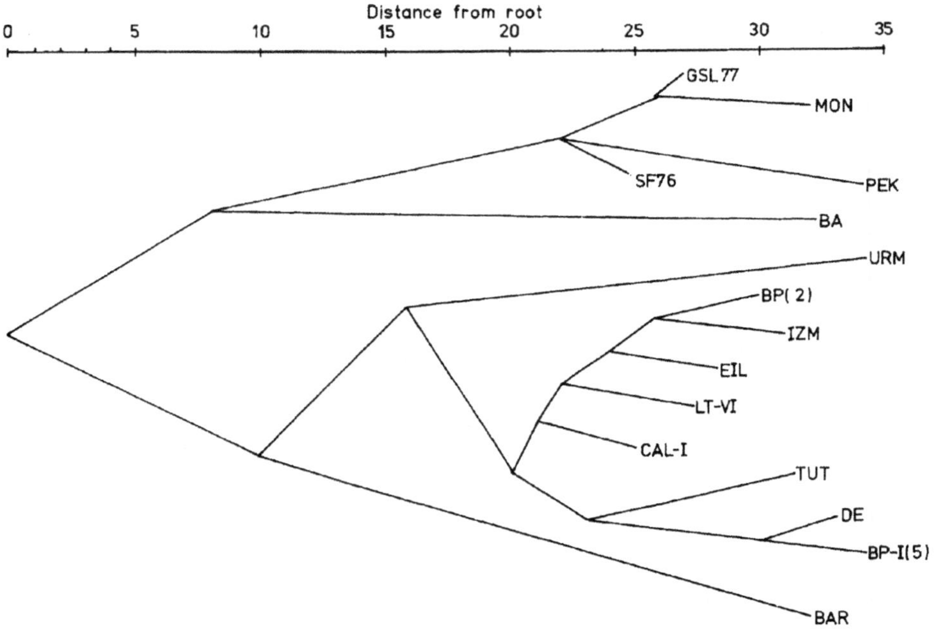


*Fig. 1.* Phylogenetic tree of *A. monica*, *A. urmiana* and representative species of the genus *Artemia*. Rooting at midpoint of longest path. Total length = 185.0. Index of consistency (Kluge & Farris, 1969) = 0.703. Cophenetic correlation (Sneath & Sokal, 1973) = 0.992.

SF76 and CAL-I, and do not affect the conclusions of this paper. In this representation, the lengths of the tree branches are proportional to the number of different alleles between the nodal populations or 'taxonomic units'. This representation of the relationships between *A. monica*, *A. urmiana* and representative populations of the species in the genus gives clusters congruent with the genetic distances measured and suggests *A. urmiana* in the Old World branch is closely related to the parthenogenetic line. *A. monica*, on the other hand, has diverged little from the *A. franciscana* populations.

## Discussion

The value of electrophoresis to detect genetic variation has been extensively demonstrated in numerous organisms (Ferguson, 1980) and in the case of parthenogenetic *Artemia*, at least, has finer resolution than alternative mtDNA methods (Browne & Bowen, 1989). Studies on other crustacean species suggest the presence of widespread polymorphisms at protein loci, although the degree of variability is not evenly distributed among the taxonomic subgroups (Hedgecock *et al.*, 1982). While the levels of genetic variability found in *Artemia*, as a whole, are among the highest in Crustacea (Abreu-Grobois, 1987) those for *A. urmiana* and *A. monica* (see Table 2) are the most prominent in the genus. The results for these two species rank among those for polyploid parthenogenetic *Artemia* clones which, in common with other organisms where asexual polyploidy is found, tend to accumulate heterozygosity (Abreu-Grobois, 1987; Lokki, 1976).

Levels of genetic variation found in these species may be related to the history of their population structures. The geologic evidence for climatic and hydrological changes in the last 12000 years affecting regions where populations of the *A. franciscana* and *A. monica* species are found in western U.S.A. suggest extinction/recolonization events, particularly in the smaller lakes, caused by excessive dilution of salts or desiccation followed by reinoculation at times even by individuals from populations with different medium ionic compositions (Bowen *et al.*, 1988). Dispersal of *Artemia* can be assisted by waterfowl which carry cysts attached to their bodies or in their gut after feeding on brine shrimp (MacDonald, 1980). The sequence of shrinking and expansion of *Artemia* habitats has generated a pattern of scattered and relatively isolated biotopes. Thus, conditions for maximal interpopulational differentiation have occurred in this area where a high degree of genetic interpopulational differentiation is, in fact, typical (Abreu-Grobois, 1987).

High values of $N_e$ and rapid population growth are also characteristic of *Artemia* populations. Both *A. monica* and *A. urmiana* inhabit large and relatively stable lakes which promote for very large values of $N_e$ and buffer against bottleneck effects of population crashes. In Mono Lake, population sizes increase rapidly in June, reaching peaks in excess of 60000 *Artemia* $m^2$, or nine billion individuals for the whole 150 $km^2$ lake (Lenz & Dana, 1987). No data are available for Lake Urmia.

Selection may also play an important rôle in the processes affecting the levels of genetic variability in local populations. While the relationship between observed electrophoretically-detectable variation and phenotypical variation involved in the adaptation processes is open to debate, work on parthenogenetic *Artemia* clones in both field and laboratory-simulated conditions (Browne, 1988; Browne & Hoopes, 1989) provides evidence for correlations between overall P values, $H_e$, and spatial and temporal variations in salinity. If the enzyme loci *themselves* are not the focus of selection, the detectable variation is useful as a tracer for evolution at gene loci linked to them or as a measure of general levels of genetic variability and its correlations with selection processes.

Immigration from differentiated populations may also occur, enriching the number of alleles at polymorphic loci. It is unlikely that *A. monica* genes will be exported to other habitats for, as its cysts sink (Lenz, 1980) and do not accumulate on the surface waters or on the shore, they will not be extracted by migrating birds. Nevertheless, extensive migration of birds such as gulls and grebes, both of which feed heavily on *Artemia* (Cooper *et al.*, 1984; Winkler, 1977), flying from other habitats in western North America into Mono Lake has been reported (Lenz, 1987). Though the chemical composition of Mono Lake waters will tend to eliminate immigration, gene flow into the lake could exist from populations with adaptations to similar media and a capacity to interbreed with *A. monica*. *Artemia* from Fallon Pond, Nevada represents one such extant population (Lenz, 1980; Bowen *et al.*, 1985). The high levels of genetic variability in *A. monica* could be a remnant of higher levels of migration into the lake from differentiated populations adapted to carbonate-rich waters, much like Fallon, but which have become extinct because of changes in the hydrology of the region. How much gene flow into the Mono *Artemia* population from Fallon does occur cannot be determined from the present work but could be further investigated by electrophoresis. Such work would complement the inferences of reproductive isolation derived from physiological data and aid in the taxonomy of the North American *Artemia*.

The isolation of *A. urmiana* may not be so stringent as no ecological barriers have been reported but no report exists detailing the presence and abundance of this species in other nearby or distant habitats (see Vanhaecke *et al.*, 1987).

Of major importance is the analysis of the genetic relationships between *A. monica*, *A. urmiana* and the other members of the genus. The very low *D* value between *A. monica* and its closest relative, *A. franciscana*, although correlating well with their cytological (Abreu-Grobois & Beardmore, 1989) and anatomical similarities (Abreu-Grobois, 1983; Abreu-Grobois, 1987) implies that the physiological adaptations may be dependent upon a relatively small number of genes. These loci do not affect mating behavior or reproductive compatibility as *A. monica* can interbreed with chloride water *Artemia* in carefully contrived media (Bowen *et al.*, 1985).

The genetic distances between *A. urmiana* and other gonochoristic species is concordant with its current specific status and is congruent with its morphological (Abreu-Grobois, 1987, 1989) and cytological characteristics (Abreu-Grobois & Beardmore, 1989) Its inferred phylogenetic relationship with present-day parthenogenetic forms is suggestive of a recent single common ancestral lineage branching to both *A. urmiana* and the asexual forms, probably about 5–6 Mybp (Beardmore & Abreu-Grobois, 1983). While a monophyletic sexual ancestry to parthenogenesis has long been inferred (Artom, 1931; Bowen *et al.*, 1980), former methods of analysis could not identify *A. urmiana* as the most likely bisexual source (Barigozzi, 1974).

The electrophoretic characterization of *A. urmiana* presented here will be of importance in comparing samples of gonochoristic *Artemia* from biotopes in the Middle East and Asia which are only just coming under study. More so if this gonochoristic species has been displaced from its original habitat or now only exists in specific areas of the lake (Barigozzi *et al.*, 1987). A recent study of a proposed new species, *A. sinica* (Yaneng, 1989), is a case in point. This species has great morphological similaries with *A. urmiana* (Abreu-Grobois, 1989) and is reproductively isolated from *A. franciscana*, *A. persimilis* and *A. salina*. Unavailability of samples from *A. urmiana* has prevented is being test-mated with the proposed new species (Yaneng, pers. comm.). Should gonochoristic *Artemia* be extinct in Lake Urmia, the only means by which to identify new populations with strong morphological similarities to *A. urmiana* will be by measurements of genetic distances employing the data presented in this work. The capacity to verify the systematic identity of gonochoristic *Artemia* in these new areas is most important, for accurate knowledge of the extent of the distribution range of *A. urmiana* may help to understand the relative colonizing capacity that the different Old World gonochoristic and parthenogenetic species have in the little studied *Artemia* biotopes east of the Mediterranean.

**Acknowledgements**

We thank the following persons and institutions who kindly donated cysts: Artemia Reference Center (State University of Ghent, Belgium)-GS77, SF76, PEK, BA, BP, EIL, IZM, TUT; Sarane Bowen- MON; Robert Ward- LT; Ali Valamanesch- URM; Francisco Amat Domenecq- BAR, CAL, DE. This work was sponsored by a grant from the Mexican National Council for Science and Technology (CONACyT).

## References

Abreu-Grobois, F. A., 1983. Population genetics of *Artemia*. Ph. D. Thesis, Univ. Coll. Swansea, Great Britain. 438 pp.

Abreu-Grobois, F. A., 1989. Observations and questions with regard to *Artemia* taxonomy. *Artemia* Newsletter No. 12 (April) 5–8.

Abreu-Grobois, F. A., 1987. A review of the genetics of *Artemia*. In P. Sorgeloos, D. A. Bengtson, W. Decleir & E. Jaspers (eds), *Artemia* Research and its Applications. Universa Press, Wetteren, Belgium 1: 61–99.

Abreu-Grobois, F. A. & J. A. Beardmore, 1989 (in press). Chromosomes and chromocenters in the genus *Artemia*. An. Inst. Cienc. del Mar y Limnol. Univ. Nal. Auton. Mexico. 16(2)

Artom, C., 1931. L'origine e l'evoluzione della partenogenesi attraverso i differenti biotipi di una specie collettiva (*Artemia salina* L.) con speciale riferimento al biotipo partenogenetico di Séte. Mem. r. Accad. Ital. 2 = 1–57.

Avise, J. C., 1974. Systematic value of electrophoretic data. Syst. Zool. 23: 465–481.

Ayala, F. J., 1983. Enzymes as taxonomic characters. In G. S. Oxford & D. Rollinson (eds), Protein Polymorphism: Adaptive and Taxonomic Significance. Systematics Association/ Academic Press, Lond.: 3–26.

Barigozzi, C., 1980. Genus *Artemia*: problems of systematics. In G. Persoone, P. Sorgeloos, O. Roels & E. Jaspers. (eds), The Brine Shrimp *Artemia*. Universa Press, Wetteren, Belgium 1: 147–153.

Barigozzi, C., V. Varotto, L. Baratelli & R. Giarrizzo, 1987. The *Artemia* of Urmia Lake (Iran): mode of reproduction and chromosome numbers. Atti Acc. Lincei Rend. fis. 81: 87–90.

Beardmore, J. A. & F. A. Abreu-Grobois, 1983. Taxonomy and evolution in the brine shrimp *Artemia*. In G. S. Oxford & D. Rollinson (eds), Protein Polymorphism: Adaptive and Taxonomic Significance. Systematics Association/ Academic Press, London.: 153–164.

Beardmore, J. A., 1987. Concluding remarks for Symposium

Session I: Morphology, Ecotoxicoloy, Radiobiology, Genetics and Strain characterization. In P. Sorgeloos, D. A. Bengtson, W. Decleir & E. Jaspers (eds), *Artemia* Research and its Applications. Universa Press, Wetteren, Belgium. 1: 345–346.

Bowen, S. I. & G. Sterling, 1978. Esterase and malate dehydrogenase isozyme polymorphisms in 15 *Artemia* populations. Comp. Biochem. Physiol. 61B: 593–595.

Bowen, S. T., M. L. Davis, S. R. Fenster & G. A. Lindwall, 1980. Sibling species of *Artemia*. In G. Persoone, P. Sorgeloos, O. Roels & E. Jaspers (eds), The Brine Shrimp *Artemia*. Universa Press, Wetteren, Belgium 1: 155–167.

Bowen, S. T., E. A. Fogarino, K. N. Hitchner, G. L. Dana, V. H. S. Chow, M. R. Buoncristiani & J. R. Carl., 1985. Ecological isolation in *Artemia*: Population differences in tolerance of anion concentrations. J. Crustacean Biol. 5: 106–129.

Browne, R. A., 1988. Genetic and ecological divergence of sexual and asexual brine shrimp (*Artemia*) populations in the Mediterranean Basin. Nat. Geogr. Res. 4: 510.

Browne, R A. & S. T. Bowen 1989 (in press). Population genetics of *Artemia*. In R. A. Browne, P. Sorgeloos, C. N. Trotman (ed.), CRC Handbook of *Artemia* Biology. CRC Press Inc., Boca Raton, Fla., USA.

Browne, R. A. & C. W. Hoopes, 1989 (in press). Asexuality, microspeciation and genotype selection in *Artemia*. Evolution.

Browne, R. A. & G. H. MacDonald, 1982. Biogeography of the brine shrimp, *Artemia*: Distribution of pathenogenetic and sexual populations. J. Biogeogr. 9: 331–338.

Clark, L. S. & S. T. Bowen, 1976. The genetics of *Artemia salina* VII. Reproductive isolation. J. Hered. 67: 385–388.

Cooper, S. D., D. W. Winkler & P. H. Lenz, 1984. The effect of grebe predation on a brine shrimp population. J. anim. Ecol. 53: 51–64.

Dana, G. & D. B. Herbst, 1977. Secondary producer level. In Winkler, D. W. (ed.) An ecological study of Mono Lake, California. Institute of Ecology Publication No. 12. University of California, Davis.: 57–62.

Elston, R. C. & R. Forthofer, 1977. Testing for Hardy-Weinberg equilibrium in small samples. Biometrics. 33: 536–542

Farris, J. S., 1970. Methods for computing Wagner trees. Syst. Zool. 19: 83–92.

Ferguson, A., 1980. Biochemical Systematics and Evolution. Blackie, Glasgow. 194 pp.

Finamore, F. J. & J. S. Clegg, 1969. Biochemical aspects of morphogenesis in the brine shrimp *Artemia salina*. In G. M. Padilla, G. L. Whitson & I. L. Cameron (eds). The Cell Cycle. Academic Press, N.Y.: 249–278.

Günther, R. T., 1900. Contributions to the natural history of Lake Urmi, N.W. Persia, and its neighbourhoud. J. Linn. Soc. Zool. 27: 345–452.

Harris, H. & D. A. Hopkinson, 1976. Handbook of enzyme electrophoresis in human genetics. North-Holland, Amsterdam. 389 pp.

Hedgecock, D., M. L. Tracey & K. Nelson, 1982. Genetics. In D. E. Bliss (ed.), The Biology of Crustacea. Academic Press, N.Y. 2: 283–400.

Kluge, A. G. & J. S. Farris, 1969. Quantitative phyletics and the evolution of anurans. Syst. Zool. 18: 1–32.

Koehn, R. K. & W. F. Eanes, 1978. Molecular structure and protein variation within and among populations. Evol. Biol. 11: 39–103.

Lenz, P. H., 1987. Ecological studies on *Artemia*: a review. In P. Sorgeloos, D. A. Bengtson, W. Decleir & E. Jaspers (eds), *Artemia* Research and its Applications. Universa Press, Wetteren, Belgium 3: 5–18.

Lenz, P. H., 1980. Ecology of an alkali-adapted variety of *Artemia* from Mono Lake, California, USA. In G. Persoone, P. Sorgeloos, O. Roels, & E. Jaspers (eds), The Brine Shrimp *Artemia*. Universa Press, Wetteren, Belgium. 3: 79–96.

Lenz, P. H. & G. L. Dana, 1987. Life-cycle studies in *Artemia*: a comparison between a sub-tropical and a temperate population. In P. Sorgeloos, D. A. Bengtson, W. Decleir & E. Jaspers (eds), *Artemia* Research and its Applications. Universa Press, Wetteren, Belgium 3: 89–100.

McDonald, G. H., 1980. The use of cysts as food by the flamingo (*Phoenicopterus ruber roseus*) and the shellduck (*Tadorna*). In G. Persoone, P. Sorgeloos, O. Roels & E. Jaspers (eds), The Brine Shrimp *Artemia*. Universa Press, Wetteren, Belgium. 3: 97–104.

Nei, M., 1972. Genetic distance between populations. Am. Nat. 106: 283–292.

Nei, M., 1975. Molecular Population Genetics and Evolution. North-Holland Publishing Co., Amsterdam. 378 pp.

Nei, M. & A. K. Roychoudhury, 1974. Sampling variances of heterozygosity and genetic distance. Genetics 76: 379–390.

Oxford G. S. & D. Rollinson, 1983. Protein Polymorphism: Adaptive and Taxonomic Significance. Systematics Association/ Academic Press, London. 405 pp.

Persoone, G. & P. Sorgeloos., 1980. General aspects of the ecology and biogeography of *Artemia*. In the Brine Shrimp *Artemia*. G. Persoone, P. Sorgeloos, O. Roels & E. Jaspers (eds), Universa Press, Wetteren, Belgium. 3: 3–24.

Piccinelli, M. & T. Prosdocimi, 1968. Descrizione tassonomica delle due specie *Artemia salina* L. e *Artemia persimilis* n.sp. Ist. Lomb. (Rend. Sci.) B 102: 113–118.

Poulik, M. D., 1957. Starch gel electrophoresis in a discontinuous system of buffers. Nature 180: 1477–1479.

Smith, I., 1976. Chromatographic and Electrophoretic Techniques. Heinemann, London.

Sneath, P. H. A. & R. R. Sokal, 1973. Numerical Taxonomy-The Principles and Practices of Numerical Classification. W. H. Freeman & Co., San Francisco, USA.

Sokal, R. R. & F. J. Rohlf, 1981. Biometry. W. H. Freeman & Co., San Francisco, USA 859 pp.

Sorgeloos, P., P. Lavens, Ph. Léger, W. Tackaert & D. Versichele, 1986. Manual for the Culture and Use of Brine Shrimp *Artemia* in Aquaculture. F.A.O./State University of Ghent, Ghent, Belgium.

Swofford, D. L., 1982. Wagner procedure computer program (WAGPROC). Version 3. Dept. Genetics and Dev., Univ. Illinois, Urbana, Illinois, USA.

Thorpe, J. P., 1983. Enzyme variation, genetic distance and evolutionary divergence in relation to levels of taxonomic separation. In G. S. Oxford & D. Rollinson (eds), Protein Polymorphism: Adaptive and Taxonomic Significance. Systematics Association/ Academic Press, London.: 131–152.

Vanhaecke, P., W. Tackaert & P. Sorgeloos, 1987. The biogeography of *Artemia*: an updated review. In P. Sorgeloos, D. A. Bengtson, W. Decleir & E. Jaspers (eds), *Artemia* Research and its Applications. Universa Press, Wetteren, Belgium. 1: 129–155.

Wilson, F. R., G. S. Whitt & C. L. Prosser, 1973. Lactate dehydrogenase and malate dehydrogenase isozyme patterns in tissues of temperature acclimated goldfish (*Carasscus auratus* L.). Comp. Biochem. Physiol. 46B: 105–116.

Winkler, D. W., 1977. An Ecological Study of Mono Lake, California. Inst. Ecol. Univ. California, Davis, California, USA. Publ. No. 12.

Yaneng, C., 1989. New *Artemia* sibling species from People's Republic of China. *Artemia* Newsletter No. 11 (Jan.): 40–41.

Hydrobiologia 212: 169–179, 1991.
D. Belk, H. J. Dumont & N. Munuswamy (eds), Studies on Large Branchiopod Biology and Aquaculture.



# Sex ratio variation in female-biased populations of Notostracans

Clay Sassaman
*Department of Biology, University of California, Riverside, CA 92521, USA*

*Key words:* Triops, sex ratio, genetics, selfing


## Abstract

Females from female-biased populations of the notostracan *Triops newberryi* produce viable eggs when reared in isolation. Clutches produced under such conditions exhibit sex-ratio polymorphism. One category of females (monogenics) produces only female offspring; the second category (amphigenics) produces clutches with a sex ratio of 3 females:1 male. The relative proportions of the two categories of females varies significantly between populations and is correlated with population sex ratio. This correlation, and the pattern of offspring distribution in amphigenics, suggests that sex is determined by an autosomal Mendelian gene locus for which the male-determining allele is recessive. Limited pedigree analysis of lineages under selfing support the genetic model.


## Introduction

Notostracans exhibit an extreme ambivalence to the normal conventions of sexual reproduction. Although males are known for all of the eleven or so nominal species of tadpole shrimps, they vary greatly in their local frequencies. Populations range from complete unisexuality (absence of males) to bisexual states in which male frequencies range from less than 1% to over 70% (Vandel, 1924; Mathias, 1937). This variation has motivated a number of studies on the reproductive biology of notostracans because it is so large as to imply that populations at the extremes of the range must differ in fundamental attributes. Populations in which males are completely lacking are unlikely to have the same mode of reproduction as those in which males are in the majority.

An approach common to much of this literature has been the comparison of completely unisexual populations with those in which males and females are in approximately equal frequency. However, in at least several species there are regular geographic gradients of varying male frequencies, and intermediate populations are characterized by highly female-biased sex ratios. These female-biased populations have often been regarded as somewhat abnormal or irregular and attempts to explain their occurrence have usually explicitly (or implicitly) viewed them as admixtures of unisexual and bisexual populations in different proportions (Vandel, 1931; Longhurst, 1955c; Zaffagnini & Trentini, 1980). Several versions of such explanations, invoking differential range expansion of unisexual and bisexual populations, have been advanced; they differ primarily in their assumptions on the reproductive mode of unisexual populations.

My primary purpose in this paper is to focus attention on these female-biased populations. A brief review of the literature on population sex-ratio variation suggests that female-bias is a widespread phenomenon and that it occurs in some species in which males are never common. A review of studies on the reproductive mode of various notostracans indicates that females from female-biased populations have not been studied extensively, but that their reproductive biology is inferred primarily from studies on bisexual populations.

I will describe the results of experiments on self-compatibility and clutch sex ratio in female-biased populations of North American *Triops* that suggest a mechanism which allows a continuum of population sex ratios from unisexuality to male-female equality. I will close by speculating on the role of this mechanism in the evolution of notostracan sex ratios.

*A taxonomic note*

Any attempt to review the notostracan literature must be tempered by a qualification: the primary taxonomy of the order is unsettled. The modern era of classification dates to the major revisions of Linder (1952) and Longhurst (1955a). Linder's analysis of North American *Lepidurus* has been largely substantiated (e.g., Lynch 1970, Saunders 1980a) although *Lepidurus lynchi* Linder has become a synonym of *Lepidurus lemmoni* Holmes (Lynch, 1966). Longhurst's (1955a) views of *Lepidurus* are somewhat problematic, especially his synonymy of *Lepidurus couesii* Packard with *Lepidurus apus* (Linn.) (see, e.g., Brtek *et al.*, 1984; Thiery, 1986; Fryer, 1988). For purposes of this review I will use Linder's classification for North American *Lepidurus* but I cannot distinguish between *L. couseii* and *L. apus* in Old World records.

In the case of *Triops*, Linder (1952) and Longhurst (1955a) were in essential agreement, and there have been no substantive modifications since their reviews. There are, nevertheless, cracks in the taxonomic edifice. Fryer (1988) has suggested the possibility of cryptic speciation within *Triops cancriformis* (Bosc.) and I and my colleagues have found clear evidence of at least two species within North American populations of *Triops longicaudatus* (LeConte). These differences will be detailed elsewhere (ms. in prep.). For the present I will distinguish between *T. longicaudatus* (a group of populations distributed primarily to the east of the Continental Divide) and *Triops newberryi* (Packard) generally to the west. In reviewing data for populations that are not from North America I follow Longhurst (1955a).

*Sex ratio variation in Notostraca*

I have categorized the literature on sex ratio variation in Notostraca by defining four qualitative patterns of sex ratio as follows: Type I (male bias) – a statistically significant excess of males in the population sample; Type II (equality) – males and females present in frequencies not significantly differing from 1:1; Type III (female bias) – a statistically significant excess of females in the population sample; and Type IV (unisexual) – a complete absence of males in a large sample ($>50$). Table 1 summarizes the distributions of these patterns among the various species of notostracans and indicates the pertinent literature references used in the summary. I have indicated the presence/absence of a particular pattern for each species rather than the frequency of occurrence of reports for each pattern to avoid the potential problem of sampling bias of different types of populations.

Two nominal species of *Lepidurus* are not included in the summary. Too few specimens (all male) of *L. batesoni* Longhurst are known (Longhurst, 1955a), and the validity of the species is itself in question (Longhurst, 1958). There are scant records in the literature concerning *L. packardi* Simon, but the specimens examined by Linder (1952) indicate approximate equality of the sexes. Sex ratios of the remaining North American species of *Lepidurus* are characterized by single values.

Although there are a number of potential sampling biases that may compromise the interpretation of sex ratio distributions based on field samples, several aspects of this summary of notostracan sex ratios are noteworthy. First, the single-most widespread pattern across species is female-bias. Second, unisexuality has been documented in about half of the species and, in every case, species with unisexual populations also have female-biased populations. Finally, several species with female-biased populations are not represented by populations in which males and

*Table 1.* Summary of population sex ratios for various species of Notostraca

| Species | I male bias | II equality | III female bias | IV uni-sexual | References |
|---|---|---|---|---|---|
| *Lepidurus apus* | + | + | + | + | Jezequel 1921, Lundblad 1921, Vandel 1924, 1931, Mathias 1937, Baudoin *et al.* 1937, Desportes & Andrieux 1944 |
| *L. arcticus* | – | – | + | + | Lundblad 1921, Johansen 1922, Somme 1934, Arnold 1966 |
| *L. bilobatus* | – | – | + | – | Saunders 1980a |
| *L. couesii* | – | + | – | – | Packard 1875 |
| *L. lemmoni* | – | + | – | – | Lynch 1966[1] |
| *Triops australiensis* | – | – | + | – | Spencer & Hall 1895, Main 1953 |
| *T. granarius* | + | + | + | – | Sars 1901, Gurney 1925, Peres 1939, Tiwari 1954, Akita 1966, Raynor & Bowland 1985 |
| *T. cancriformis* | – | + | + | + | Walton & Kemp 1911, Vandel 1931, Mathias 1937, Peres 1939, Tiwari 1954, Akita 1966, Hempel-Zawitkowska 1967, Zaffagnini & Trentini 1980 |
| *T. longicaudatus* | + | + | + | + | Linder 1932, Akita 1966, Scott & Grigarick 1978, unpub. obs. |
| *T. newberryi* | – | – | + | + | unpub. obser. |

[1] Without specifics, the aggregate ratio for the species is about 1 : 1.

females are co-equal in frequency. These three points will be important in evaluating hypotheses on the significance of female-bias in the evolution of notostracan sex ratio. Before doing so, however, it is necessary to comment on the literature dealing with mode of reproduction.

*Reproductive mode in Notostraca*

Two general approaches have been used to deduce the reproductive mode of various species (and species populations) of notostracans: (1) histological examination of the female gonadal tissues, and (2) the ability of females to produce viable eggs when reared in isolation. Each of these methods has both advantages and disadvantages and each, in and of itself, may provide equivocal results.

The histological approach as been, by far, the most commonly employed, primarily because this approach can be taken with specimens collected directly from the field; it does not require that animals be reared under controlled laboratory conditions. The principle problem with the technique is that there is no clear consensus concerning the interpretation of the results. The presence of testicular lobes embedded within the ovary of female is usually interpreted to indicate an hermaphroditic mode of reproduction in unisexual populations (e.g., Bernard, 1891a, b; Longhurst, 1955a, b; Wingstrand, 1978) but some workers do not accept that conclusion (e.g., Zaffagnini & Trentini, 1980). Although several studies have indicated that the oocyte is haploid (Longhurst, 1955a, b; Zaffagnini & Trentini, 1980) there is no direct cytogenetic evidence on the mechanism of restoration of diploidy (Longhurst, 1955a, b; Zaffagnini & Trentini, 1980).

The apparent absence of testicular lobes in unisexuals (or their rudimentary nature) is generally taken to imply parthenogenetic reproduction (Hempel-Zawitkowska, 1967; Akita, 1971), and the absence of such lobes in females from sexual populations is a primary criterion of obligatory sexual outcrossing (Longhurst, 1955a, b). The interpretation of these absences is also, however, subject to interpretation. The ambiguity about negative histological data resolves around the fact that we have no information on how small a testicular lobe can be while still retaining adequate male function.

A second approach to the study of reproductive mode is the analysis of female self-compatibility. This approach involves rearing individuals in isolation and determining the viability of the eggs they produce. It has been used primarily to characterize differences between females from sexual species versus those from primarily unisexual ones (Longhurst, 1955a, b). Clearly, females from unisexual populations can be expected to produce viable offspring if reared correctly but such an observation provides no pertinent information on the mechanism by which the reproduction occurred. However, females from bisexual or female-biased populations may or may not be self-compatible.

The literature on reproductive mode in various species of Notostraca is summarized in Table 2. Several pertinent points can be made from this compilation. First, the vast majority of histological studies have concluded that unisexual populations of notostracans are hermaphrodites. In only one species, *Triops cancriformis*, is there negative evidence of an ovotestis and studies on different populations of it have yielded contradictory results. Second, the two species of *Triops* for which unisexual populations are unknown, *Triops australiensis* (Spencer & Hall) and *Triops granarius* (Lucas) (Table 1), appear to be, on both histological and self-compatibility criteria, obligately outcrossing, sexual species. Third, there has been only one intraspecific comparison between unisexual and sexual populations (Longhurst, 1955a, b), in *T. cancriformis*, and in that case the females were from populations with 1 : 1 sex ratio rather than from female-biased populations.

These two bodies of literature together suggest the need for biological information on female-biased populations of Notostraca. These populations occur frequently (Table 1) and in species in which they cannot be explained as recent admixtures of unisexuals and bisexuals because the latter populations are not known to exist. Furthermore, the reproductive biology of female-biased populations has not been characterized. Do these populations represent mixtures of self-incompatible sexuals and self-compatible unisexuals, or do they uniformly resemble one or the other mode of reproduction?

## Experimental analysis of populations of *Triops newberryi*

### *Populations sampled*

The present study is based on five geographically separated populations of *Triops newberryi* from Utah, Nevada, and California. The Utah population is from a livestock watering tank located approximately 27 km southeast of Hurricane (Washington County). The Nevada population is from a small roadside pool about 65 km south of Boulder City (Clark County). The three California populations are: Eureka Valley Dry Lake, a large playa at the southern end of Eureka Valley, approximately 72 km southeast of Bishop (Inyo County), Sheephole Mountain Dry Lake, a large playa 35 km east of Twentynine Palms (San Bernardino County), and Joshua Tree (Squaw Tank), a small water impoundment in the Joshua

*Table 2.* Summary of reproductive modes of various species of Notostraca

| Species | Population type | Mode of reproduction | Method of inference | References |
|---|---|---|---|---|
| *Lepidurus apus* | unisexual | hermaphroditic | histology | Bernard 1891a<br>Wingstrand 1978 |
| *L. arcticus* | unisexual | hermaphroditic | histology | Bernard 1891b<br>Longhurst 1955b<br>Wingstrand 1978 |
| *Triops granarius* | sexual | outcrossing | histology | Longhurst 1955b |
| | | outcrossing | self-incomp | Longhurst 1955b |
| *T. australiensis* | sexual | outcrossing | histology | Spencer 1896<br>Longhurst 1955b |
| | | outcrossing | self-incomp | Longhurst 1955b |
| *T. cancriformis* | unisexual | hermaphroditic | histology | Bernard 1891a<br>Longhurst 1955b<br>Wingstrand 1978 |
| | unisexual | parthenogenetic | histology | Zaffagnini &<br>Trentini 1980<br>Akita, 1966<br>Hempel–<br>Zawitkowska 1967 |
| | sexual | outcrossing | histology | Longhurst 1955b |
| *T. longicaudatus* | unisexual | hermaphroditic | histology | Longhurst 1955<br>Akita 1966 |

Tree National Monument, 24 km south of Twentynine Palms (Riverside County).

Populations were sampled by collecting dried mud samples containing the desiccated eggs (actually encysted embryos). For my purposes, an individual reared directly from such a mud sample is viewed as 'field collected' because it existed as an embryo when collected.

### *Experimental methods*

#### *Rearing*

Animals were reared from field collected mud samples either by hydrating an aliquot of about 100–200 ml of sediment in a 38 l aquarium of demineralized water or by washing the sample through a 250 $\mu$m screen and incubating the retained eggs in small finger bowls of demineralized water. In aquarium rearings, hatchlings were typically allowed to grow in the tank for about three days at ambient temperature (20–25 °C) with continuous aeration and fluorescent illumination. Larvae hatched from screened mud were generally transferred to individual culture as metanauplii.

Individuals were isolated in plastic cups containing 75 ml of culture water. The water was obtained by hydration of mud from the Sheephole Mountain site (in demineralized water) and screened through a 60 $\mu$m sieve. Individuals were initially fed a pinch of finely ground goldfish food (Wardley's). Beginning on the third day after isolation they were fed daily with freshly killed *Artemia* until they were large enough to feed directly on live *Artemia*. Cultures were maintained at approximately 25–30 °C under tungsten illumination or in incubator cabinets in complete darkness.

#### *Sex determination*

Sexes of animals were determined at about 6–7 days of age. Females were identified by the early developmental stages of the egg sac on the 11th thoracic appendage, males by their absence. Confirmation of the sex of males was made by two additional criteria: the greater number of apodous terminal body rings and the relative lengths of endites 5 and 6 of the second thoracic leg (Saunders, 1980b). Sexes initially determined by these criteria in animals subsequently reared to sexual maturity were invariable correlated with the appropriate gonadal development and sexual behavior.

If females were to be kept alive for egg collection, the culture water was changed daily after they reached a body length of about 1 cm. Eggs were collected from unmated females by periodically caging them in a small nytex mesh cage prior to their molts. The eggs were pipetted into clean water in 1 dram shell vials (in lots of 25 to 50 eggs) and were allowed to dry over a period of several days. After air-drying, the vials of eggs were maintained under ambient conditions for periods ranging from several weeks to a year prior to being hatched. Hatching was accomplished by adding distilled water to the storage vial and placing the vial about 20 cm from a 100 watt light. Larvae generally hatched in about 12 hours and were then transferred to individual culture. Several successive clutches of eggs from many of the females were collected, stored, and analyzed independently of each other. All results for clutch sex ratios reported here are based only on those clutches for which a minimum of 20 offspring were reared and sexed.

## Results

*Population sex ratios*

Population sex ratios in *T. newberryi* vary geographically from about 20% males to a complete absence of males. The proportions of males (and the sample sizes on which these proportions are based) for the five populations sampled here are: Joshua Tree, 0.00 ($N = 65$); Eureka Valley, 0.01 ($N = 88$); Hurricane, 0.03 ($N = 65$); Boulder City, 0.06 ($N = 104$); and Sheephole Mountain, 0.16 ($N = 165$). Populations do not distribute themselves into discrete categories of sex ratio, the variation is somewhat continuous. Indeed, there is no objective criterion for classifying a population as unisexual since males can occur at exceedingly low frequencies.

*Viability of eggs from isolated females*

Over 90% (76 of 84) of the females produced viable eggs when reared in isolation and the overall fraction of eggs that hatched was 57.7% ($N = 4257$ eggs). These estimates are minimum values; some eggs may have been viable but did not hatch under this specific protocol. Similarly, the eight females from which no eggs were hatched cannot be regarded necessarily as self-incompatible. In related experiments I have found that eggs produced by females under conditions of food shortage and poor water quality are often sterile. Culture conditions were not optimized for viability, but were merely randomized across populations.

There were no consistent differences between populations in either the fraction of females that produced viable eggs or the fraction of eggs that hatched. Females were equally likely to produce viable eggs irrespective of the sex ratio of the population from which they were isolated. These results indicate, therefore, that self-compatibility is not an attribute that distinguishes unisexual populations from bisexual populations or that differs among bisexual populations of differing sex ratio.

*Clutch sex-ratio polymorphism*

Clutch sex ratios were obtained for 50 females isolated from the five populations. In contrast to all expectations, female were of two distinctly different types in terms of the distributions of progenies that they produced when reared in isolation. The predominant type yielded families that were exclusively female. I term this type 'monogenic' to refer to the uniformity of offspring phenotypes produced. The second category of female, 'amphigenic', produced mixed clutches that included both male and female offspring.

The one unisexual population, Squaw Tank, and the bisexual population with exceedingly rare males, Eureka Valley, were both uniformly composed of monogenics in my samples. The remain-

*Table 3.* Clutch sex ratios from females of *Triops newberryi* reared in isolation.

| Population | Isofemale line No. | No. of males | No. of females | Total | % males |
|---|---|---|---|---|---|
| Joshua Tree, CA Squaw Tank | #21 | 0 | 40 | 40 | 0.0 |
| | #22 | 0 | 44 | 44 | 0.0 |
| | #24 | 0 | 22 | 22 | 0.0 |
| | #25 | 0 | 28 | 28 | 0.0 |
| | #26 | 0 | 61 | 61 | 0.0 |
| | #27 | 0 | 28 | 28 | 0.0 |
| | #28 | 0 | 30 | 30 | 0.0 |
| | #29 | 0 | 28 | 28 | 0.0 |
| | #33 | 0 | 30 | 30 | 0.0 |
| Eureka Valley, CA | #2 | 0 | 34 | 34 | 0.0 |
| | #3 | 0 | 42 | 42 | 0.0 |
| | #12 | 0 | 24 | 24 | 0.0 |
| | #13 | 0 | 35 | 35 | 0.0 |
| | #14 | 0 | 29 | 29 | 0.0 |
| | #17 | 0 | 38 | 38 | 0.0 |
| | #18 | 0 | 39 | 39 | 0.0 |
| | #19 | 0 | 35 | 35 | 0.0 |
| | #25 | 0 | 45 | 45 | 0.0 |
| | #26 | 0 | 27 | 27 | 0.0 |
| Hurricane, UT | #5 | 0 | 43 | 43 | 0.0 |
| | #6 | 0 | 30 | 30 | 0.0 |
| | #7 | 0 | 24 | 24 | 0.0 |
| | #8 | 0 | 24 | 24 | 0.0 |
| | #9 | 11 | 23 | 34 | 32.4 |
| | #10 | 15 | 30 | 45 | 33.3 |
| | #11 | 0 | 28 | 28 | 0.0 |
| | #13 | 0 | 24 | 24 | 0.0 |
| | #16 | 0 | 42 | 42 | 0.0 |
| | #19 | 0 | 36 | 36 | 0.0 |
| | #20 | 0 | 23 | 23 | 0.0 |
| Boulder City, NV | #6 | 5 | 31 | 36 | 13.9 |
| | #32 | 7 | 31 | 38 | 18.4 |
| | #33 | 0 | 22 | 22 | 0.0 |
| | #35 | 0 | 30 | 30 | 0.0 |
| | #36 | 0 | 24 | 24 | 0.0 |
| | #37 | 12 | 33 | 45 | 26.7 |
| | #38 | 9 | 19 | 28 | 32.1 |
| | #39 | 0 | 29 | 29 | 0.0 |
| | #40 | 0 | 53 | 53 | 0.0 |
| | #41 | 0 | 34 | 34 | 0.0 |
| Sheephole MT., CA | #12 | 0 | 41 | 41 | 0.0 |
| | #15 | 9 | 20 | 29 | 31.0 |
| | #18 | 7 | 42 | 49 | 14.3 |
| | #30 | 15 | 32 | 47 | 31.9 |
| | #31 | 5 | 20 | 25 | 20.0 |
| | #32 | 0 | 41 | 41 | 0.0 |
| | #34 | 10 | 44 | 54 | 18.5 |
| | #35 | 0 | 30 | 30 | 0.0 |
| | #52 | 0 | 31 | 31 | 0.0 |
| | #75 | 18 | 46 | 64 | 28.1 |

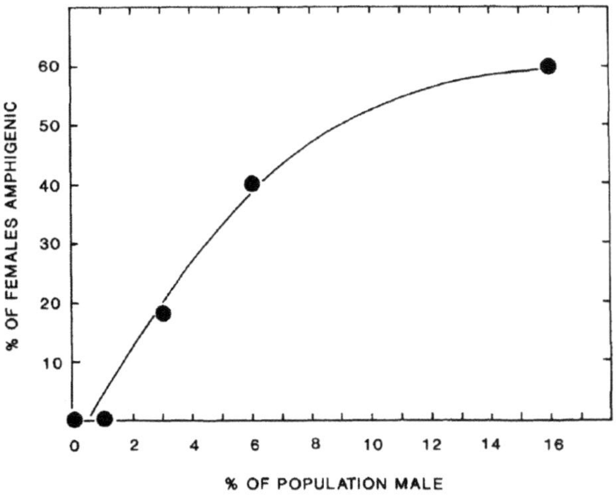


*Fig. 1.* The correlation between the proportion of females amphigenic in a population sample and the proportion of males in the sample, for five population of *Triops newberryi*. Line is eye-fitted.

ing three populations, in contrast, all were polymorphic in having both amphigenics and monogenics in appreciable frequencies (Table 3). Although the number of females tested from each population is small, statistical analysis using the log-likelihood ratio (Sokal & Rohlf, 1981) indicates significant heterogeneity between populations in the relative frequencies of monogenic and amphigenic females ($G = 17.756$, 4 d.f., $P = 0.001$). This variation is furthermore highly correlated with population sex ratio (Fig. 1).

Amphigenic females produce clutches with a characteristic sex ratio. Pooling over all amphigenics within populations, and over all populations, yields the aggregate amphigenic sex ratio of exactly 3 females to 1 male (Table 4). None of the 12 individual clutches from amphigenics deviated from this ratio.

*Genetic analysis of the sex-ratio polymorphism*

Formal genetic studies on the inheritance of clutch sex ratio in *Triops newberryi* are currently in progress; I will briefly summarize the preliminary results. Monogenic females are pure-breeding. I have isolated and progeny-tested 44 daughters from three different monogenic females (from the bisexual Sheephole Mountain population) – all were themselves monogenic. Amphigenic females are not pure-breeding; in addition to producing 25% male offspring under selfing hermaphroditism, they produce both monogenic and amphigenic daughters. I have, to date, reared the clutches of twelve daughters from a single amphigenic female (also from Sheephole Mountain). Nine of these were themselves amphigenic; the remaining three were monogenic, establishing that monogenics can arise from amphigenics.

## Discussion

The existence of clutch sex-ratio polymorphism in *T. newberryi* (Table 3) is not without precedent – we have previously found identical polymorphisms in two species of conchostracans which also exhibit female-biased population sex ratios, *Eulimnadia antlei* Mackin and *Eulimnadia texana* (Packard) (Sassaman, 1988; Sassaman & Weeks, in prep.). Furthermore, the preliminary genetic analysis of the inheritance of amphigenics and monogenics in *T. newberryi* is completely con-

*Table 4.* Amphigenic Clutch sex ratios

| Population | Number of clutches | Sex ratio among amphigenics | | Test of Ho: 3 females: 1 male | |
|---|---|---|---|---|---|
| | | Female | Male | X2 | P |
| Sheephole MT. | 6 | 204 | 64 | 0.18 | 0.67 |
| Boulder City | 4 | 114 | 33 | 0.51 | 0.48 |
| Hurricane | 2 | 53 | 26 | 2.64 | 0.10 |
| Sum | 12 | 371 | 123 | 0.002 | 0.96 |

sistent with studies on the inheritance of these two categories in *E. texana* (Sassaman & Weeks, in prep.). Finally, the relationship between population sex ratio and amphigenic frequency in *T. newberryi* (Fig. 1) is reminiscent of the relationship between population sex ratio and allozyme heterozygosity in *Eulimnadia texana* (Sassaman, 1989) and likely arises from similar factors.

The most parsimonious model for explaining the clutch sex ratio polymorphism of *T. newberryi*, and the relationship of this polymorphism to sex ratio, is to postulate a simple mechanism of sex determination embedded within a mixed mating system. I suggest two alleles (Z and W) at a single locus yielding three genotypes (and phenotypes) as follows: W/W (monogenic self-compatible hermaphrodite), Z/W (amphigenic self-compatible hermaphrodite), and Z/Z (male). This model explains the existence of two types of clutches (Table 3) and the general correlation between the frequencies of amphigenics and males (Fig. 1). If we further assume the Z allele to be recessive to the W allele, then the production of 25% male offspring by isolated amphigenics (Table 4) follows directly. Selfing of the Z/W would also produce monogenics according to this model, a prediction consistent with the preliminary genetic analysis.

This model of the inheritance of sex, while necessary to my view of the genetics of female-biased *T. newberryi* populations, is not sufficient to explain their existence. If females reproduce by selfing, then the Z allele is effectively a recessive lethal that will be rapidly driven to low frequency or eliminated, and the population will be unisexual and entirely monogenic. If, on the other hand, reproduction is exclusively by sexual outcrossing, then monogenic genotypes will not exist in the population and the sex ratio will rapidly equilibrate at 1 : 1. Thus, some intermediate level of selfing mixed with outcrossing is necessary to generate female-bias, and fitness differences between genotypes may also be necessary.

The immediate significance of this model, if it is verified, is in providing an alternative mechanism for the evolutionary dynamics of sex ratio in Notostraca. The viewpoint that female-biased populations are abnormal and secondary mixtures of reproductively isolated unisexuals and bisexuals provides no mechanism for the origin of these differences. In two provocative papers, Akita (1966, 1971) purported to have found evidence of a step gradient of reproductive characteristics in Japanese populations of *Triops longicaudatus*. He found three kinds of populations: bisexuals with males outnumbering females, unisexuals but with well-developed testicular lobes in the ovary, and unisexuals with rudimentary (or absent) testicular lobes in the ovary. Males of the bisexual populations mated with females of the unisexual populations, but such matings were not fertile. These results have been taken to present a case of rapid evolution of unisexuality (Akita, 1971) and as evidence that unisexuals are reproductively isolated from males (Zaffagnini & Trentini, 1980). However, in a retrospective taxonomic analysis of the Japanese populations, Akita (1972, 1976) found that the three kinds of populations were, respectively: bisexual *Triops granarius*, unisexual (hermaphroditic) *T. longicaudatus*, and unisexual (parthenogenetic) *T. cancriformis*.

The alternative, suggested by the present work, is that female-biased populations may represent direct transitional states between sexual and unisexual populations. The advent of self-compatibility within a sexual population with female-heterozygote sex determination would initiate the condition of clutch sex ratio polymorphism. Further spread of that self-compatibility would generate a transient state of female-bias which could be stabilized by a variety of evolutionary forces. Some female-biased populations, unconstrained by any advantages of sexuality, would inbreed to the state of unisexuality, thus generating a continuum of populations ranging from bisexual to unisexual.

**Acknowledgements**

I thank G. Pratt, D. Juliani, and A. Fawcett for assistance in collecting samples used in the present work and M. Fugate for bibliographic assistance, help in rearing animals, and helpful discussions. Collections in Joshua Tree National Monument were made under a permit from the National Park Service. This research has been partially supported by Statewide Critical Applied Research Funds from the Agricultural Experiment Station of the University of California.

## References


Akita, M., 1966. Life history of Triops longicaudatus (LeConte) in Japan. I. Sex ratio and mode of reproduction. Zool. Mag. 75: 178–182.

Akita, M., 1971. On the reproduction of Triops longicaudatus (LeConte). Zool. Mag. 80: 242–250.

Akita, M., 1972. On Triops Longicaudatus (LeConte) from Arizona. Zool. Mag. 81: 217–22.

Akita, M., 1976. Classification of Japanese tadpole shrimps. Zool. Mag. 85: 237–247.

Arnold, G. P., 1966. Observations in Lepidurus arcticus (Pallas) (Crustacea, Notostraca) in east Greenland. Ann. Mag. nat Hist., series 13, 9: 599–617.

Baudoin, R., F. Campan & M. de Larambergue, 1937. Variations dans la proportion relative des sexes chez les phyllopodes. Bull. Soc. zool. Fr. 62: 423–426.

Bernard, H. M., 1891a. Hermaphroditismus bei Phyllopoden. Jena. Z. Naturwiss. 25: 337–338.

Bernard, H. M., 1891b. Hermaphroditism of the Apodidae. Nature 43: 343–344.

Bernard, H. M., 1892. The Apodidae. A morphological study. MacMillan & Co., London, 316 pp.

Bernard, H. M., 1896. Hermaphroditism among the Apodidae. Ann. Mag. nat. Hist., series 6. 17: 296–309.

Brtek, J., L. Forro & J. E. Ponyi, 1984. Contributions to the knowledge of Branchiopoda (Crustacea) fauna of Mongolia. Ann. hist.-nat. Mus. Nat. Hung. 76: 91–99.

Desportes, C. & L. H. Andrieux, 1944. Sur la biologie de Lepidurus apus (L. 1761). Bull. Soc. zool. Fr. 69: 61–68.

Fryer, G., 1988. Studies on the functional morphology and biology of the Notostraca (Crustacea: Branchiopoda). Phil. Trans. r. Soc., Lond., B. 321: 27–124.

Gurney, R., 1925. Some Asiatic specimens of Apus. Rec. ind. Mus. 27: 439–442.

Hempel-Zawitkowska, J., 1967. Natural history of Triops cancriformis (Bosc). Zool. pol. 17: 173–239.

Jezequel, J., 1921. Predominance anormale de males dans une population d'Apus (Lepidurus productus Bosc.). Bull. Soc. zool. Fr. 46: 99–100.

Johansen, F., 1922. Euphyllopod Crustacea of the American Arctic. Rep. Canad. Arctic Exped. 1913–1918 7 Part G; 1–34.

Linder, F., 1952. Contributions to the morphology and taxonomy of the Branchiopoda Notostraca, with special reference to the North American species. Proc. U.S. nat. Mus. 102: 1–69.

Longhurst, A. R., 1955a. A review of the Notostraca. Bull. brit. Mus. nat. Hist. (Zool.) 3: 1–57.

Longhurst, A. R., 1955b. The reproduction and cytology of the Notostraca (Crustacea, Phyllopoda). Proc. zool. Soc. Lond. 125: 671–680.

Longhurst, A. R., 1955c. Evolution in the Notostraca. Evolution 9: 84–86.

Longhurst, A. R., 1958. Abnormal variation in the Notostraca. Syst. Zool. 7: 84–88.

Lundblad, O., 1921. Vergleichende studien uber die nahrungsaufnahme einiger schwedischen phyllopoden, nebst synonymischen, morphologischen und biologischen bemerkungen. Ark. Zool. 13(16): 1–114.

Lynch, J. E., 1966. Lepidurus lemmoni Holmes: a redescription with notes on variation and distribution. Trans. am. micros. Soc. 85: 181–192.

Lynch, J. E., 1970. Lepidurus couesii Packard (Notostraca) redescribed with a discussion of specific characters in the genus. Crustaceana 23: 43–49.

Main, A., 1953. Sex ratio and variation in Apus australiensis Spencer and Hall. West. austr. Nat. 4: 34–39.

Mathias, P., 1937. Biologie des Crustaces Phyllopodes. Hermann and C., Paris. 107 pp.

Packard, A. S., Jr., 1875. New phyllopod crustaceans. Am. Nat. 9: 311–312.

Peres, J. M., 1939. Contribution a la conaissance des Notostraces du Maroc. Bull. Soc. Sci. nat. Maroc 19: 23–32.

Rayner, N. A. & A. E. Bowland, 1985. Notes on the taxonomy and ecology of Triops granarius (Lucas) (Notostraca: Crustacea) in South Africa. S. Afr. J. Sci. 81: 500–505.

Sars, G. O., 1901. On the Crustacean fauna of central Asia. Ann. Mus. Zool. Acad. imp. St. Petersbourg 6: 130–164.

Sassaman, C., 1988. Clutch sex-ratio polymorphism in the clam-shrimp Eulimnadia antlei. Am. Zool. 28: 135A.

Sassaman, C., 1989. Inbreeding and sex ratio variation in female-biased populations of a clam shrimp, Eulimnadia texana. Bull. mar. Sci. 45: 425–432.

Saunders, J. F., III, 1980a. A redescription of Lepidurus bilobatus Packard (Crustacea: Notostraca). Trans. am. micros. Soc. 99: 179–186.

Saunders, J. F., III, 1980b. Eubranchiopoda of Colorado. Part 1. Introduction and Notostraca. Nat. Hist. Invent. Colorado 5: 11–24.

Scott, S. R. & A. A. Grigarick, 1978. Observations on the biology and rearing of the tadpole shrimp Triops longicaudatus (LeConte) (Notostraca: Triopsidae). Wasmann J. Biol. 36: 116–126.

Somme, S., 1934. Contributions to the biology of Norwegian fish food animals. 1. Lepidurus arcticus Pallas 1793 syn L. glacialis Kroyer 1847. Avh. ut. Det Norske Vidensk-Akad. Oslo I. Matem.-Naturvid. Klasse 1934, No. 6: 1–36.

Sokal, R. R. & F. J. Rohlf, 1981. Biometry. Freeman, San Francisco, 859 pp.

Spencer, B., 1896. Report on the work of the Horn Scientific Expedition to Central Australia, II. Zoology. Dulau & Co., London, 431 pp.

Spencer, B. & T. S. Hall, 1895. Preliminary description of a new species of Apus. Vict. Nat. 11: 161–162.

Thiery, A., 1986. Les Crustaces Branchiopodes (Anostraca, Notostraca et Conchostraca) du Maroc occidental. I. Inventaire et repartition. Bull. Soc. Hist. nat. Toulouse 122: 145–155.

Tiwari, K. K., 1954. Sex ratio and variability of apodous segments in Apus (Phyllopoda: Crustacea). J. Bombay nat. Hist. Soc. 52: 641–644.

Vandel, A., 1924. La spanandrie (disette de males) geographique chez un Crustace Branchiopode: Lepidurus apus (L.). C. r. Acad. Sci., Paris 178: 1759–1761.

Vandel, A., 1931. La parthenogenese. Doin & C., Paris, 412 pp.

Walton, H. J. & S. Kemp, 1911. Notes on the occurrence of *Apus* in eastern Asia. Rec. ind. Mus. 6: 351–357.

Wingstrand, K. G., 1978. Comparative spermatology of the Crustacea Entomostraca. 1. Subclass Branchiopoda. Biol. Skr. 22: 1–66.

Zaffagnini, F. & M. Trentini, 1980. The distribution and reproduction of Triops cancriformis (Bosc) in Europe (Crustacea Notostraca). Monit. zool. ital. (N.S.) 14: 1–8.

*Hydrobiologia* **212**: 181–185, 1991.
*D. Belk, H. J. Dumont & N. Munuswamy (eds), Studies on Large Branchiopod Biology and Aquaculture.*



# Characterizing bisexual *Artemia* populations by isoelectric focusing

J.C. Navarro, F. Hontoria, I. Varó & F. Amat
*Instituto de Acuicultura de Torre de la Sal (C.S.I.C.), 12595 Ribera de Cabanes, Castellón, Spain*

*Key words:* isoelectric focusing, *Artemia* populations, isozymes

## Abstract

Isoelectric focusing (IEF) of protein extracts, α-Esterase and Phosphoglucose isomerase, from groups of *Artemia* adults from different bisexual populations were studied. Both gave clear separation between the Old and New World species. The second was more polymorphic, allowing a discrimination among Mediterranean populations.

## Introduction

The anostracan genus *Artemia* contains sibling species, reproductively isolated in nature but identical or similar in outward appearance (Bowen *et al.*, 1978, 1980; Abreu-Grobois & Beardmore, 1980).

Isoelectric focusing (IEF) is a powerful tool usable to differentiate such morphologically similar organisms, (Lundström, 1977).

Several studies have been carried out in order to 'finger-print' *Artemia* species and populations using IEF. Seidel *et al.* (1980) and Seidel & Simpson (1984) studied the general protein pattern of pooled nauplii from different strains. Requintina & Simpson (1987) used specific stains to demonstrate the isozymic banding patterns obtained after focusing protein extracts from pooled nauplii. Requintina (1988) succeeded in isoelectric-focusing the isozymes of individual *Artemia* adults.

The aim of the present work was to assay protein extracts from homogenized adults by means of IEF and staining the gels for the presence of phosphoglucose isomerase (PGI) and α-esterase (α-EST) isozymes. These isozymes, in combination with two pH gradients (3–10 and 4–6.5 respectively), were shown to give the best resolution in differentiating *Artemia* populations by Requintina & Simpson (1987), Requintina (1988), and Navarro *et al.* (1987).

## Materials & methods

Cysts from the different populations were hatched in sea water (38 g $l^{-1}$) at 28 °C. Nauplii were transferred to 5 l plastic tanks, fed the microalga *Tetraselmis suecica*, and kept at 28 °C until reaching adulthood. Prior to preparation, animals were put in vials with clean filtered sea water for 4 hours and were later thoroughly rinsed with distilled water.

Table 1 shows the populations analyzed, their origin, and the number of individuals utilized.

Figure 1 shows the location of the Mediterranean populations. Quemado is located in New Mexico (USA) and GSL is Great Salt Lake, Utah (USA).

The animals, blotted with a paper towel, weighed, transferred to a glass homogeneizer, were crushed in 1% glycine buffer (pH 7) in a proportion of three parts of buffer per one of animals. After 30 min of centrifugation at 4 °C

*Table 1.* Populations analyzed, origin, and number of individuals utilized.

| Code | Name | Origin | Males | Females |
|---|---|---|---|---|
| 1 | Ibiza | Spain | 16 | 16 |
| 2 | Mallorca | Spain | 11 | 11 |
| 3 | Megrine | Tunisia | 15 | 15 |
| 4 | Mala | Spain | 26 | 4 |
| 5 | Bonmatí | Spain | 15 | 15 |
| 6 | S. Antioco | Italy | 15 | 15 |
| 7 | Sfax | Tunisia | 14 | 16 |
| 8 | Cádiz | Spain | 15 | 15 |
| 9 | Quemado* | USA | 10 | 19 |
| 10 | GSL* | USA | 15 | 15 |

* Used as internal standard for comparison purposes.

(Minifuge GL, Heraeus CHRIST 4400) and 4500 rpm, the resulting supernatant was separated from the pellet with a pasteur pipette and stored at − 30 °C without further manipulation.

Acrylamide gels were used. Isoelectric focusing was carried on a LKB Multiphor system (LKB Instruments), at 4 °C. Up to 24 samples were run simultaneously. Strips soaked in 1 M NaOH (cathode) and 1 M $H_3PO_4$ (anode) were applied to the edges of the gel to provide electrical contact with the platinum wire electrodes connected to the power supply. After pre-focusing for 10 min at 20 mA of constant current, 25 W constant power was given to the gel until the proteins attained equilibrium (1.5–2 h). Then the gels were transferred to plastic cuvettes were specific staining was carried out.

The following staining solutions were used: Phosphoglucose isomerase: Fructose 6 phosphate, 80 mg; NAD P, 16 mg; $MgCl_2$ 6 $H_2O$, 80 mg; Glucose 6P dehyd., 0.3 mg; PMS, 5 mg; MTT, 28 mg TRIS-HCl 0.1 M pH 8, 120 ml. α-Esterase: α-Naphthyl acetate, 37.5 mg; Fast Blue B, 125 mg; Phosphate buffer 0.1 M pH 7.4, 125 ml.

The grouping of the populations was carried

*Fig. 1.* Location of the Mediterranean populations. 1: Ibiza, 2: Mallorca, 3: Megrine, 4: Mala, 5: Bonmati, 6: S. Antioco, 7: Sfax, 8: Cadiz.

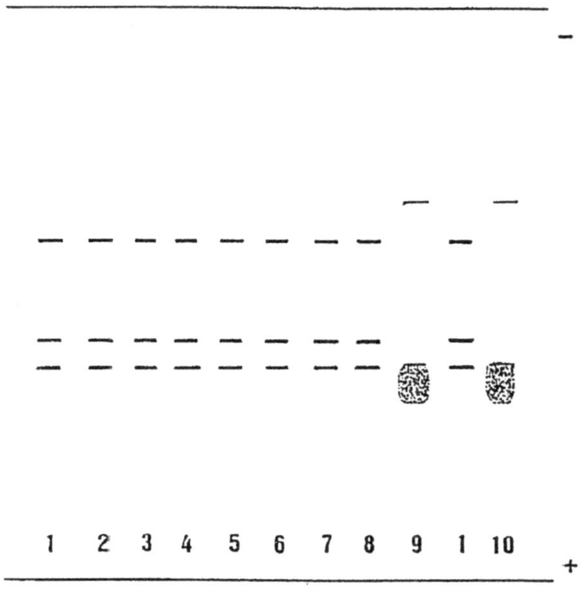


*Fig. 2.* Diagram showing the banding pattern obtained after staining for α-EST. pH gradient: 4–6.5.

out by means of single linkage clustering (Sneath, 1957) applied after calculation of the Marczewski & Steinhaus statistic (Angus *et al.*, 1988):
$S = w/a + b - w$; $w$ = number of bands of common mobility; $a$ = number of bands in unit a; $b$ = number of bands in unit b.

## Results

Figures 2 to 5 show the diagrams for the PGI and α-Est isozyme patterns obtained for the populations analyzed and the dendrograms of the Marczewski & Steinhaus statistic. PGI shows a more polymorphic pattern than α-Est, allowing a discrimination among the populations from the Balearic Islands (Mallorca and Ibiza) and the other locations.

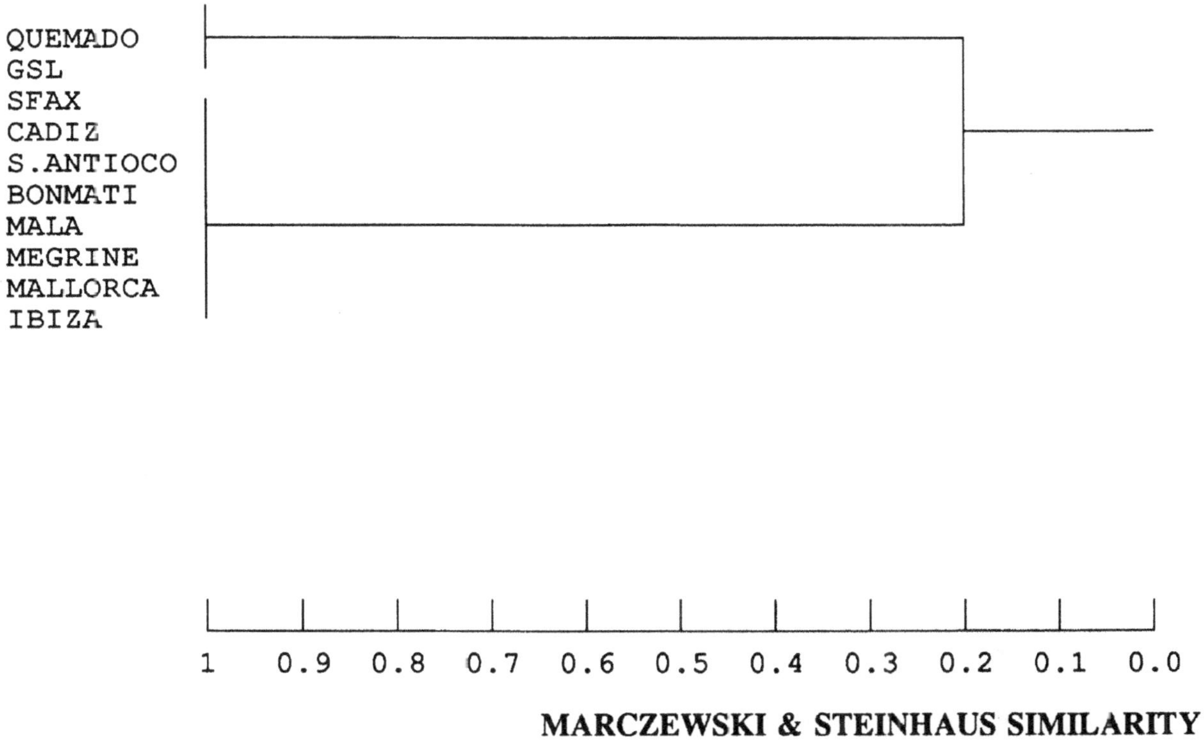


*Fig. 3.* Dendrogram drawn from the results of the Marczewski & Steinhaus statistic applied to the banding pattern obtained for α-EST.

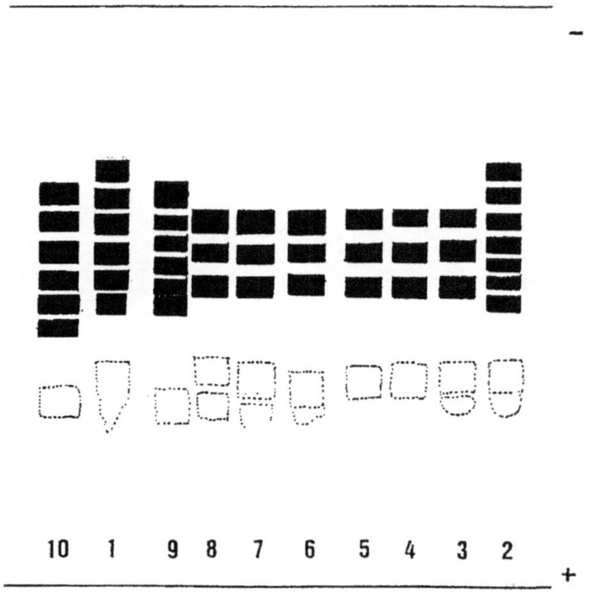


*Fig. 4.* Diagram showing the banding pattern obtained after staining for PGI. pH: 3–10.

## Discussion & conclusions

The analysis of samples pooled from groups of individuals masks individual heterozygosity in such a way that the banding patterns obtained are a reflection of the identity of the population as a whole (Seidel, 1983). For this purpose, the use of adults instead of nauplii is better since with adults the complications introduced by developmental isozymes are not present (Requintina, 1988). A further advantage is the possibility of including in the analyses those populations either not producing oviparous offspring or those from which only biomass is available.

Both enzymatic systems used gave clear separation at the species level between Old and New World bisexuals, as formerly demonstrated by other workers using different methods (Abreu-Grobois & Beardmore, 1980; Barigozzi, 1980). Additionally, our results show a higher polymor-

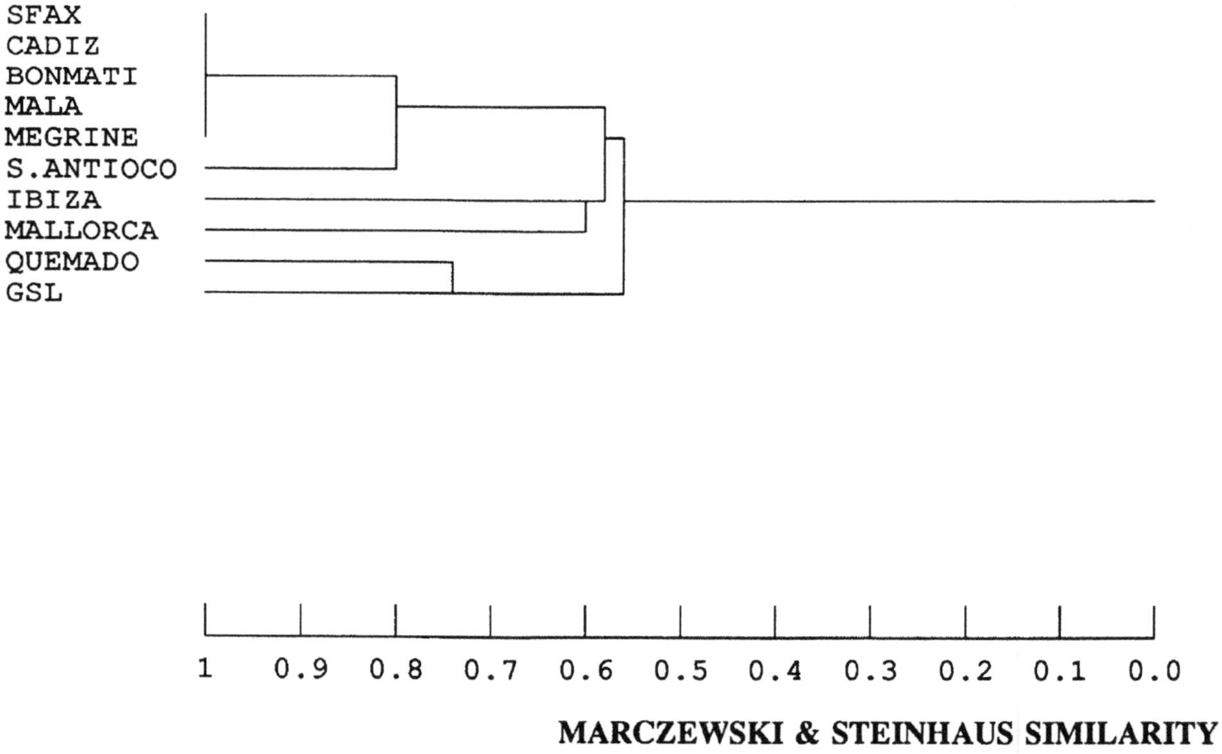


*Fig. 5.* Dendrogram drawn from the results of the Marczewski & Steinhaus statistic applied to the banding pattern obtained for PGI.

phism in the case of PGI isozymes, allowing further separation among populations of the same species (Figs. 4 & 5). Those isozymes may be related to fitness in response to different environmental factors (Requintina, 1988), given that multiple variants of an enzyme are able to catalyze the same reaction under different environmental circumstances (Hochachka & Clayton-Hochachka, 1973). In the case of *Artemia*, the special conditions in which this genus lives (i.e., drastic changes in salinity, food availability, oxygen...) might even speed up the interactions between the organism's genetic components and the environment.

Two factors, the higher resolution of IEF in respect to that of conventional electrophoresis (Ferguson, 1980), and the possibilities of testing many enzymatic systems in different pH gradients, make IEF a powerful tool for analyzing *Artemia* populations.

## Acknowledgements

This work has been carried out under support of grant 'MAR 88-0224' from the C.I.C.Y.T. J.C. Navarro has a predoctoral scholarship from the 'Generalitat Valenciana' and F. Hontoria from the 'Excma. Diputación de Castellón'.

## References

Abreu-Grobois, F. A. & J. A. Beardmore, 1980. International Study of *Artemia* II. Genetic characterization of *Artemia* populations – an electrophoretic approach. In G. Persoone, P. Sorgeloos, O. Roels & E. Jaspers (eds), The Brine Shrimp *Artemia*. Morphology, Genetics, Radiobiology, Toxicology. Universa Press. Wetteren: 133–146.

Angus, R. A., S. W. Hardwick & G. B. Cline, 1988. A computer program that calculates various similarity coefficients between electrophoregrams. J. Hered. 79: 74–75.

Barigozzi, C., 1980. Genus *Artemia*: problems of systematics. In G. Persoone, P. Sorgeloos, O. Roels & E. Jaspers (eds), The Brine Shrimp *Artemia*. Morphology, Genetics, Radiobiology, Toxicology. Universa Press. Wetteren: 147–153.

Bowen, S. T., M. L. Davis, S. R. Fenster & G. A. Lindwall, 1980. Sibling species of *Artemia*. In G. Persoone, P. Sorgeloos, O. Roels & E. Jaspers (eds), The Brine Shrimp *Artemia*. Morphology, Genetics, Radiobiology, Toxicology. Universa Press. Wetteren: 155–167.

Bowen, S. T. & G. Sterling, 1978. Esterase and malate dehydrogenase isozyme polymorphisms in 15 *Artemia* populations. Comp. Biochem. Physiol. 61B: 593–595.

Ferguson, A., 1980. Biochemical systematics and evolution. Wiley & Sons, New York. 194 pp.

Hochachka, P. W. & B. Clayton-Hochachka, 1973. Glucose-6-Phosphate dehydrogenase and thermal acclimation in the Mullet fish. Mar. Biol. 18: 251–259.

Lundström, R. C., 1977. Identification of fish species by thin layer polyacrylamide isoelectric focusing. Fish. Bull. 75: 571–576.

Navarro, J. C., P. Requintina, F. Amat & K. L. Simpson, 1987. Primeros datos sobre los patrones isozímicos de las cepas autóctonas de *Artemia*. Cuad. Marisq. Publ. Tec. 12: 549–551.

Requintina, J. P. D. & K. L. Simpson, 1987. International study on *Artemia* XLV. Geographical and developmental changes in isozymes of *Artemia* as separated by isoelectrofocusing. In P. Sorgeloos, D. A. Bengtson, W. Decleir & E. Jaspers (eds), *Artemia* research and its applications. Morphology, Genetics, Strain Characterization, Toxicology. Universa Press. Wetteren: 157–166.

Requintina, P. J., 1988. The isozymes of *Artemia* spp. as separated by isoelectrophocusing on polyacrylamide gels. Ph.D. Thesis. University of Rhode Island. 83 pp.

Seidel, C. R., 1983. The use of thin layer isoelectric focusing to differentiate *Artemia* spp. on the basis of geographical, temporal developmental variation, and homogeneity of sample. Master thesis. University of Rhode Island. 85 pp.

Seidel, C. R., J. Kryznowek & K. L. Simpson, 1980. International study on *Artemia* XI. Amino acid composition and electrophoretic protein patterns of *Artemia* from five geographical locations. In G. Persoone, P. Sorgeloos, O. Roels & E. Jaspers (eds), The Brine Shrimp *Artemia*. Ecology, Culturing, Use in Aquaculture. Universa Press. Wetteren: 375–382.

Seidel, C. R. & K. L. Simpson, 1984. Rapid differentiation of *Artemia* spp. populations by thin layer isoelectric focusing. Aquacult. Engineer. 3: 303–316.

Sneath, P. H. A., 1957. The application of computers to taxonomy. J. gen. Microbiol. 17: 201–226.

*Hydrobiologia* **212**: 187–193, 1991.
*D. Belk, H. J. Dumont & N. Munuswamy (eds), Studies on Large Branchiopod Biology and Aquaculture.*



# The filtration apparatus of Anostraca (Crustacea): species-specific setulation in the genus *Streptocephalus*

Johan Mertens[1], Nadesan Munuswamy,[2] Christine De Walsche[1]& Henri J. Dumont[1,*]
[1]*Institute of Animal Ecology, State University Ghent, Ledeganckstraat 35, B-9000-Ghent, Belgium (*author for correspondence)*; [2]*Department Zoology, Life Science Building, Quindy Campus, Madras 600 025, India*

*Key words:* Anostraca, *Streptocephalus*, filtration, limb structure

## Abstract

The structure of the setae, setules, spines and spinules on the limbs of males and females of *Streptocephalus torvicornis*, *S. rubricaudatus* and *S. proboscideus* are described, based on scanning electron micrographs. The data disclose significant variation among the three species in fine structure of the setules and inter-setular distance. The variation may be related to different trophic specialisations, facilitating the sympatric occurrence of congeneric species.

## Introduction

Like most phyllopods, Anostraca are mainly filter-feeders. The filtering apparatus and the filter mesh-sizes of different cladoceran species are well documented (Geller & Muller, 1981; Watts & Petri, 1981; Dumont & Pensaert, 1983; Beenderberger & Geller, 1985), but the fine surface structure of the filtering limbs, as described for *Branchinecta ferox* (Fryer, 1983), remains unknown for most other Anostraca. The diet of *Streptocephalus proboscideus* is composed of animal as well as of algal and detrital elements (Mertens *et al.*, 1990), and its mode of filtration is size- and shape selective. In the present paper we explore the possibility that species-specific ultrastructure of the filtration apparatus contributes to selectivity among prey types. The morphology of the limbs of fairy shrimps was described by Linder, 1941, and Daborn, 1979, but qualitative and quantitative details were overlooked.

## Materials and methods

The species investigated are *S. torvicornis* (Waga), *S. rubricaudatus* (Klunzinger), and *S. proboscideus* (Frauenfeld). All material was obtained from laboratory cultures. *S. torvicornis*, collected in La Zaida, Spain, was transported alive to the lab where it produced abundant offspring. For a description of its natural habitat, see Alonso (1985). *S. proboscideus* was obtained from continuous masscultures kept in the lab since 1985. The original material was cyst-containing dry mud from Al Gedid, a rainpool south of Khartoum, Sudan (see Rzoska, 1961). *S. rubricaudatus* was reared from dry mud from Jebel Uweinat (northeastern Sudan) (see De Ridder *et al.*, 1988).

Limbs from full-grown males and females of each species were dissected and either mounted for a study by light microscopy, or critical-point dried, and gold coated for SEM-work. Intersetule spacing was measured on calibrated S.E.M. photonegatives.

## Results

### *Gross Morphological features of the limbs*

Most fairy shrimps have phyllopodotype trunk limbs, with a flattened, leaf-like structure (Linder, 1941; Daborn, 1979). Each pyllopod ends in an exopodite and endopodite (Figs 1, 2), which consist of flattened lobes. The lobes are fringed by a row of setae, which shorten to the size of minute spines at both ends of the row. The inner edge of a limb is composed of five distinct endites (numbered 1 to 5). The appendages on these endites are organized according to a common pattern in the genus *Streptocephalus*. On endite 1 a frontal row, containing a spine and a spinule (Figs 3, 4) and one tactile sita occur (Fig. 5); posteriorly, a big comb of long, articulating, closely spaced setae is found (Fig. 3). A similar but shorter comb is found on endite 2, but not on 3, 4 and 5. In what follows, we will focus on the micro-structure of the setae on endite 1 and 2.

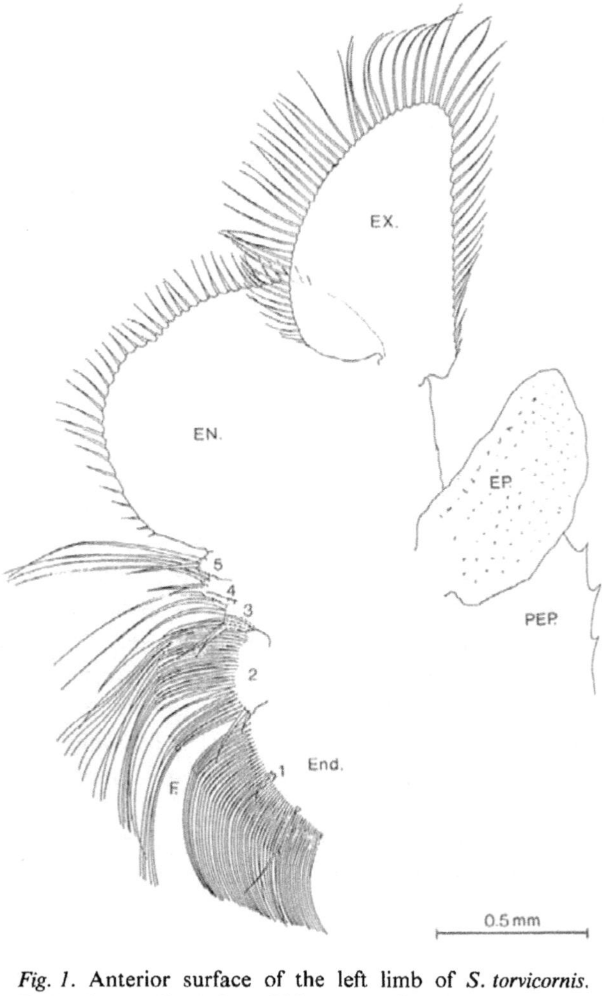


*Fig. 1.* Anterior surface of the left limb of *S. torvicornis*. End 1, 2...: endite 1, 2,...; EN.: endopodite; EX.: exopodite; EP.: epipodite; PEP.: pre-epipodite; F: food groove.

In contrast to the rigid appendages of the exo- and endopodite, all long setae on the endites have an articulation (Figs 1, 3). Especially on endites 1 and 2 they are arranged approximately in parallel to each other, and in a plane. This is enhanced by the interlocking of two rows of setules pertaining to adjacent setae (Fig. 6). Indeed, the proximal part of each seta (below the internal articulation) presents two rows of setules, at an angle of about 90° with each other, facing the food grove. The setulation of the distal part of a seta is plumose. (More details on micro-structure can be found in the next section).

Limbs 4–8 are structurally identical, with no apparent differences between the sexes. The limb armature in three different species of the genus *Streptocephalus* are summarized in Table 1. No interspecific differences were detected at this level of resolution. As seen in Fig. 2, the articulating setae of endite 1 and 2 arch over the median food groove. The setae of endite 1 partition the groove in such a way that their internal articulation reaches the medial margin of the opposing endite, and *vice versa*. A pair of limbs thus forms a filter comb across the food groove, while the mesh size of their setules defines a micro-filter. The distal parts of the articulating setae act as 'sweepers' for the proximal parts on the setae of the opposite limb.

### *The fine limb structure*

S.E.M. resolves the setulation of the endites to a level where interspecies variation can be

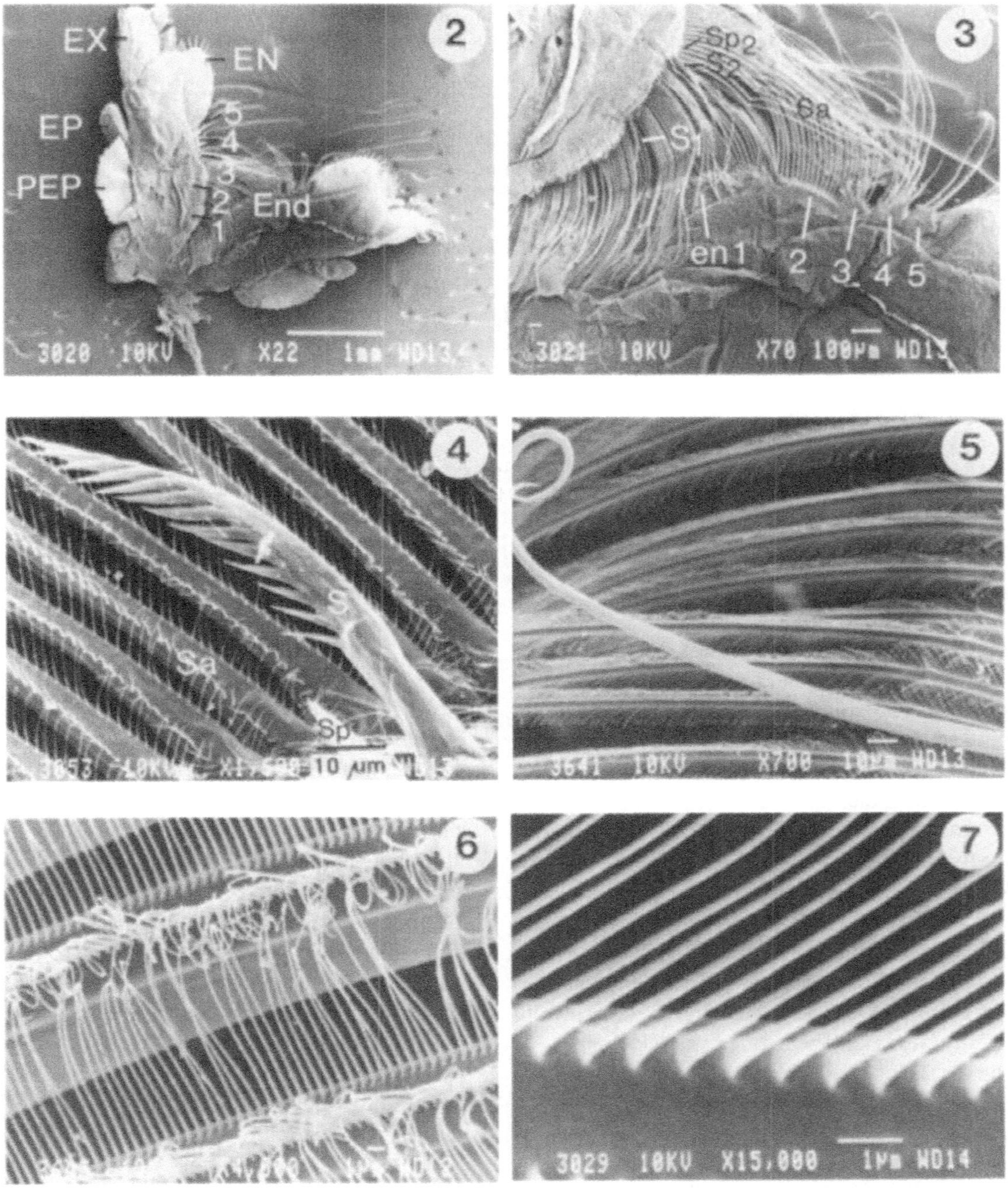


*Fig. 2–7.*

2. Anterior face of the 4th pair of limbs of *S. rubricaudatus* (abbreviations as in Fig. 1).
3. Detail of the 4th left limb of *S. torvicornis*. Overall view (anterior face) of endite 1 (distal half) and endite 2 (proximal part). S: spine; Sp: spinule; sa: articulating seta; st: tactile seta; index 1: of endite 1; index 2: of endite 2.
4. Detail of the endite 1 of *S. proboscideus* showing the spine (S), spinule (Sp) and setae (sa).
5. Tactile seta on endite 1 of *S. torvicornis*.
6. Proximal fragments of the articulating setae in *S. torvicornis*. Note the two rows of interlocking setules.
7. Setulation of *S. torvicornis*.

*Table 1.* Limb 4–8 armature in single specimens of *S. torvicornis* (S. p.) and *S. rubricaudatus* (S. r.). S: spine; Sp = spinule; sa: articulating seta; st = tactile setae; Ant = anterior row; Post = posterior row; Ex = exopodite; EN = endopodite; 1, 2..., 5 : endite 1, 2..., 5.

| Species | | EX | EN | . | 1 | 2 | 3 | 4 | 5 |
|---|---|---|---|---|---|---|---|---|---|
| S.t. | | | | Ant | 1S, 1Sp, st | 1S, 1Sp | 2S | 2S | 1S |
| | | 60–73 | 39–44 | Post | 116–130sa | 28–30sa | 3sa | 2sa | 2sa |
| | x̄ | 66 | 42 | | 123 | 29 | 3 | 2 | 2 |
| | s | 5 | 2 | | 5 | 1 | 0 | 0 | 0 |
| | n | 10 | 10 | | 10 | 10 | 10 | 10 | 10 |
| S.p. | | | | Ant | 1S, 1Sp, 1st | 1S, 1Sp | 2S | 2S | 1S |
| | | 62–75 | 36–43 | Post | 102–124sa | 20–25sa | 3sa | 2sa | 2sa |
| | x̄ | 67 | 39 | | 112 | 23 | 3 | 2 | 2 |
| | s | 5 | 2 | | 7 | 2 | 0 | 0 | 0 |
| | n | 10 | 10 | | 10 | 10 | 10 | 10 | 10 |
| S.r. | | | | Ant | 1S, 1Sp, 1st | 1S, 1Sp | 2S | 2S | 1S |
| | | 51–57 | 31–35 | Post | 101–115sa | 26–27sa | 3sa | 2sa | 2sa |
| | x̄ | 54 | 33 | | 110 | 26 | 3 | 2 | 2 |
| | s | 3 | 2 | | 7 | 0.4 | 0 | 0 | 0 |
| | n | 5 | 5 | | 5 | 5 | 5 | 5 | 5 |

measured. Figures 4, 8 and 9 show the two rows of setules on the proximal part of each long seta. Measuring the distance between the implantation of these setules (Table 2), a statistically significant difference ($p < 0.05$) between the three species, and between males and females of *S. torvicornis* was found. *S. proboscideus* stands out by the wide gap between individual setules (3.87 m), which makes it impossible for this species to capture most bacteria occurring in water, unless clumped or adhering to a larger particle. In *S. torvicornis* (Fig. 7) and *S. rubricaudatus* (Fig. 8), the setules are smooth. In *S. proboscideus* they bear a row of spiny structures on their inner side (Fig. 9).

The distal (post-articular) sector of the setae is armed differently from the proximal one. In *S. proboscideus*, short median spikes occur between the rows of setules. Although not equidistant, these spikes are arranged in a row. The setules themselves are armed with heavier spines than in the pre-articular zone (Figs 10, 11). This pattern is also found in *S. torvicornis*, although here the median spikes are not lined, but randomly distributed in a broad band (Fig. 12). Further its two rows of elastic setules are much longer here than proximal, conveying to the distal part of the setae their typical sweeper structure. In contrast to the proximal part, the setules on the distal part of the setae are not smooth but display spiky outgrowths in all species. These spikes are dense and spiny in *S. proboscideus* (c. 6.7 per setule), hairy and numerous in *S. rubricaudatus* (c. 5 per setule), and hairy and few (c. 3 per setule) in *S. torvicornis*. In contrast to the posterior row, containing only long plumose setae, the anterior row of endites is armed with spines and spinules (Fig. 3).

The endites 1 and 2 also have a long combed spine, flanked by a naked short spine (Fig. 4). These spines lack the internal articulation of the other setae on the endites. Endite 1, finally, bears

*Table 2.* Mean value ($X$) of intersetular distance (in m). Standard deviation ($s$), sample size ($n$) and absolute error ($a.e.$) in *S. torvicornis* (S.t.), *S. proboscideus* (S.p) and *S. rubricaudatus* (S.r.), males (♂) and females (♀) respectively.

| | S.t.♂ | S.p.♀ | S.r.♂ | S.t.♀ | S.p.♀ |
|---|---|---|---|---|---|
| $X$ | 0.874 | 3.871 | 0.655 | 0.758 | 3.851 |
| $s$ | 0.023 | 0.215 | 0.019 | 0.049 | 0.311 |
| $n$ | 30 | 32 | 28 | 40 | 42 |
| $a.e.$ | 0.045 | 0.108 | 0.042 | 0.046 | 0.172 |

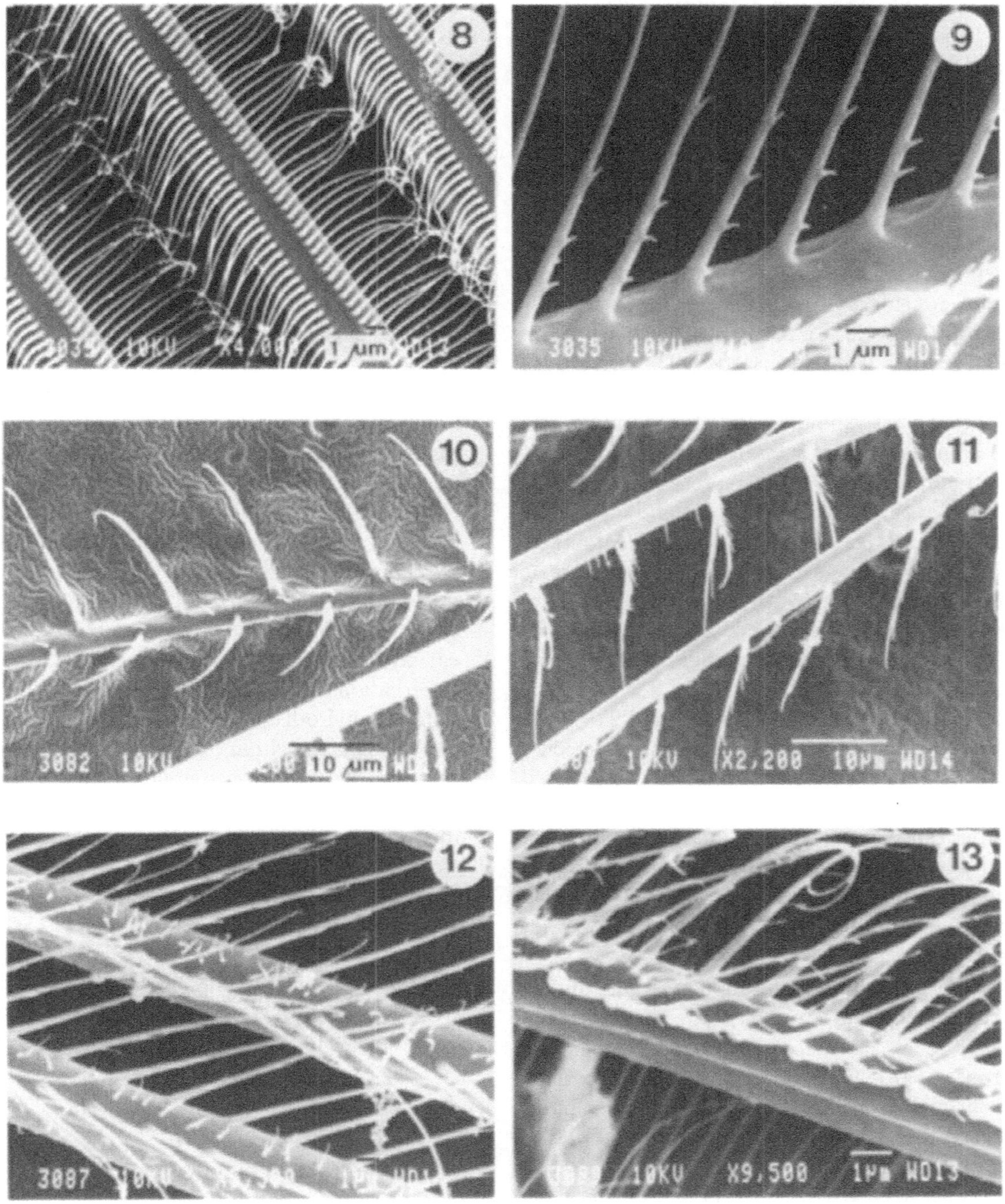


*Figs 8–13.*

8. Setulation of *S. rubricaudatus*.
9. Setulation of *S. proboscideus*.
10. Detail of distal part of a seta of *S. proboscideus*, median view.
11. The same setae as in Fig. 10, side view.
12. Detail of the distal part of a seta in *S. torvicornis*.
13. Detail of the distal part of a seta in *S. rubricaudatus*.

a single (tactile?) seta, found nowhere else on the limb (Figs 3, 5). Apart from a row of proximal, spaced, short spines, this seta has a smooth, flexible structure.

## Discussion

The structure of the medio-anterior row of appendages on all endites is identical for the three species considered. Recently, we confirmed this for a fourth species, *S. dichotomus*, also. The limb structure of *S. seali*, discussed by Daborn (1979) is insufficiently detailed to allow a distinction between spines, spinules and flexible setae. He figures the armature on endite 1 (Fig. 8) as three, and of endite 2 as two spines, respectively. If we really have one spine, one spinule and one tactile seta on endite 1, and one spine and one spinule on endite 2, we may assume that this armature is characteristic for the genus. According to Linder (1941) all the genera and species are similar for the structure of the medio-anterior row of the first two densities.

As to the comb size on endites 1 and 2 of P4-8, there is little hope of finding consistent specific characters here as well. Admittedly, we based setal counts (Table 1) on one specimen of each sex per species only, but an examination of another seven specimens of *S. proboscideus* produced a comb number range of 20–33 setae for endite 2, a complete overlap with *S. torvicornis* and *S. rubricaudatus*.

Like Daborn (1979), we found no sexual dimorphism in setation in *Streptocephalus*, in contrast to some other genera, particularly *Branchinecta* (Daborn, 1979; César, 1985). Yet, we identified one sexual dimorphism in *S. torvicornis*, i.e. intersetular distance in the proximal sector of the setae (Table 2).

As visualized by S.E.M., the structure of the setae on the endites is different, even in the closely related *S. torvicornis* and *S. rubricaudatus* (Hartland-Rowe, 1968). These criteria may be used as a taxonomic tool at the species level.

Since the proximal part of the setae of endites 1 and 2 builds the filter comb, a variation in setular distance corresponds to a variation in mesh size of the filter apparatus of the three species, and of males and females in *S. torvicornis*. In cladocerans, with their similar filtering apparatus, Geller & Muller (1981) grouped species according to mesh-size as 'fine, medium and coarse filter-feeders'. The S.E.M. structure of the filter comb of *Daphnia* resembles that of *Streptocephalus*. As in *Daphnia*, we can describe its filtering process in three steps. (1) Capturing particles by the setules on the proximal half of the setae. (2) Removing particles captured from the filter-combs by means of the sweeping setules on the distal half of the setae of the opposite limb. (3) Collecting particles in the mid-ventral food groove. It is not excluded that the combed spines act as scrapers, assisting the sweeping action of the plumose setae.

The difference in mesh-size within *Streptocephalus* may be related to different trophic specialisations. This may facilitate the sympatric occurrence of related species, such as *S. torvicornis* and *S. rubricaudatus* in the central and eastern Sahara (Mertens *et al.*, 1989). Field observations in southern Algeria even revealed different microdistributions of males and females of *S. torvicornis* within one waterbody. The males occupied the upper water layers, while females dwelled in the deeper strata, several meters below the males.

## Acknowledgements

We thank Mrs. S. Dinakis for the preparation of microscopic slides, and Mrs. S. Wellekens for assistance with the scanning electron microscopy.

## References

Alonso, M., 1985. A survey of the spanish Euphyllopoda. Misc. Zool. 9: 179–208.

Breendelberger, H. & W. Geller, 1985. Variability of filter structures in eight *Daphnia* species: mesh sizes and filtering areas. J. Plankton Res. 7: 473–486.

César, I. I., 1985. Estudio comparado de la estructura y dimorfismo sexual de los filopodos en algunos Anostraca (Crustacea). Rev. Ass. Cienc. nat. Litoral 16: 49–72.

Daborn, G. R., 1979. Limb structure and sexual dimorphism

in the Anostraca (Crustacea). Can. J. Zool. 57: 824–900.

De Ridder, M., J. Mertens & H. J. Dumont, 1988. Crustacea and Rotatoria from Jebel Uweinat (north-eastern Sahara). Biol. Jb. Dodonaea 56: 111–114.

Dumont, H. J.& J. Pensaert, 1983. A revision of the Scapholeberinae (Crustacea: Cladocera). Hydrobiologia 100: 3–45.

Fryer, G., 1983. Functional ontogenetic changes in *Branchinecta ferox* (Milne-Edwards) (Crustacea: Anostraca). Phil. Trans. r. Soc. Lond. B 303: 229–343.

Geller, W. & H. Muller, 1981. The filtration apparatus of Cladocera: filter mesh-sizes and their implications on food selectivity. Oecologia 49: 316–321.

Hartland-Rowe, R., 1968. On the identity of *Streptocephalus rubricaudatus* (Klunzinger, 1867) (Anostraca). Crustaceana 15: 319–321.

Linder, F., 1941. Contribution to the morphology and taxonomy of the Branchiopoda Anostraca. Zool. Bidr. Uppsala 20: 101–302.

Mertens, J. & H. Dumont, 1989. Confirmation of *Streptocephalus rubricaudatus* as a good species (Anostraca). Crustaceana 56: 211–212.

Mertens, J., N. Munuswamy, C. De Walsche & H. Dumont, 1990. On predatory tendencies in the feeding ecology of the fairy shrimp *Streptocephalus proboscideus* (Frauenfeld, 1873) (Crustacea: Anostraca). Hydrobiologia 198: 119–123.

Rzoska, J., 1961. Observations on tropical rainpools and general remarks on temporary waters. Hydrobiologia 17: 265–286.

Watts, E. & M. Petri, 1981. A scanning electron microscope study of the thoracic appendages of *Daphnia magna* Straus. J. nat. hist. 15: 463–473.

*Hydrobiologia* **212**: 195–202, 1991.
D. Belk, H. J. Dumont & N. Munuswamy (eds), *Studies on Large Branchiopod Biology and Aquaculture.*



# Structural differences between the cyst walls of *Streptocephalus dichotomus* (Baird), *S. torvicornis* (Waga), and *Thamnocephalus platyurus* (Packard) (Crustacea: Anostraca), and a comparison with other genera and species

Christine De Walsche, Nadesan Munuswamy* & Henri J. Dumont
*Institute of Animal Ecology, State University of Ghent, 9000-Ghent, Belgium; *present address: Department of Zoology, University of Madras, Guindy Campus, Madras 600 025 India*

*Key words:* Anostraca, *Streptocephalus*, Cyst wall structure, Alveolar layer

## Abstract

The surface topography and layering of the cyst wall of the fairy shrimps *Streptocephalus dichotomus*, *S. torvicornis* and *Thamnocephalus platyurus* are described, based on scanning electron micrographs. Cyst walls are not invariably bilayered, but may be composed of up to four layers. They are species-specific in all taxa so far studied. Cyst wall structure within single cysts may differ according to the topographical location on the cyst.

## Introduction

The majority of the Anostraca live in a physically and chemically fluctuating habitat, with recurrent phases of complete dessication. The eggs (cysts) of these species are protected by a thick wall. The layering of the wall (or shell) in *Artemia* has been studied by histological and TEM techniques (Morris & Afzelius, 1967; Anderson *et al.*, 1970), and by SEM techniques (Mazzini, 1978; Spotte & Anderson, 1988). In freshwater species, only few studies are available. Garreau de Loubresse (1974) made a detailed TEM study of the egg shell and its formation in *Tanymastix lacunae*, and Munuswamy & Subramoniam (1983, 1984) investigated the ultrastructure (by TEM) and histochemistry of the cyst wall in *Streptocephalus dichotomus*.

The most comprehensive study, however, is by Gilchrist (1978) who studied surface and three-dimensional structure of cyst walls in *Branchipus*, *Branchinecta*, and *Chirocephalus*. The present paper explores the cyst wall structure, using SEM, in *Streptocephalus dichotomus*, *S. torvicornis*, and in a related genus and species, *Thamnocephalus platyurus*.

## Material and methods

Brown-coloured, thick-walled cysts of the fairy shrimp *S. dichotomus*, collected near Madras, India, and of *S. torvicornis*, collected in Rabat Al-Sha'ry, Yemen, were isolated from mud samples. Cysts of *Thamnocephalus platyurus*, originally from Texas, were supplied by Dr D. Belk. To study their internal surface topography, cysts were cross-sectioned with a fine surgery knife, mounted on standard metal stubs, gold-coated (9 nm, Balzers Union SCD 040) and observed under a JEOL JSM 840 scanning electron microscope at an accelerating voltage of 10 kV.

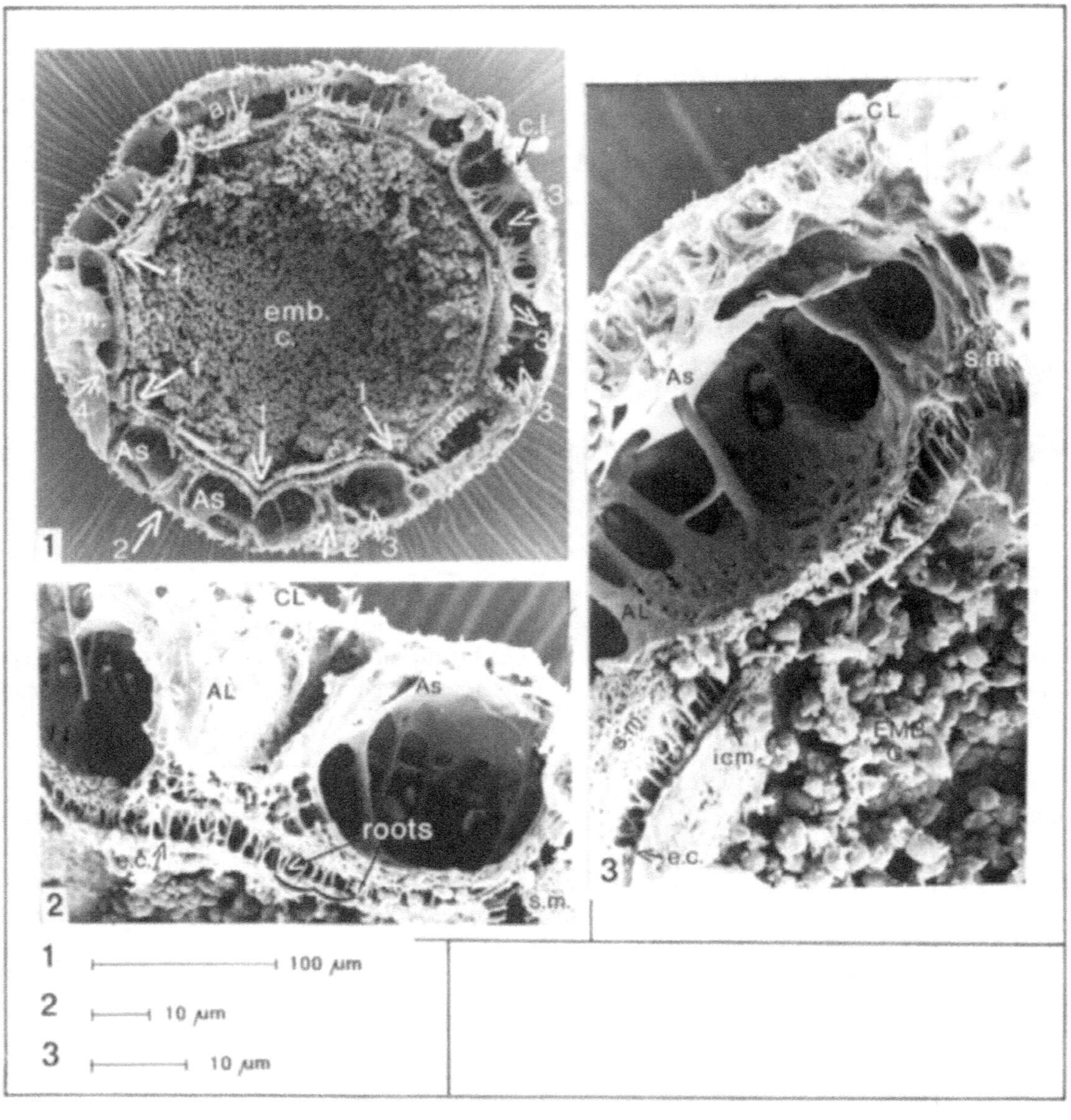


*Figs 1–3.* Scanning electron micrograph (SEM) of a dehydrated cyst of *S. dichotomus.* 1. Cross section of a cyst. 2–3. Enlarged sections of cyst wall. *Arrows*: '1': upturned section of internal layer (i.l.); '2': down falling curtains of cortical layer (C.L.); '3': 'thin' window sections of C.L.; '4': externally visible 'spokes' of a window section of C.L.. a.l.: alveolar layer; A.S.: covering fleece of alveolar layer; c.l.: cortical layer; e.c.: embryonic cuticle; emb.c.: embryonic cells; i.c.m.: internal cuticular membrane; i.l.: internal layer; p.m.: peripheral membrane; s.m.: spongy mass at foot of A.L..

---

*Figs 4–6.* 4. Cyst of *S. dichotomus* with, enlarged, a single pentagonal field, covered by an organic film, obliterating most of the pores (exceptions: arrow). 5. Cyst of *Chirocephalus diaphanus.* Enlarged, small section of the honeycombed cyst surface, showing the presence of numerous pores, restricted to the base of the scales. 6. *Thamnocephalus platyurus.* Enlarged, sector of meandering, 'rolling' crests. R.S.: crest and flank zones, covered with pores; D.Z.: depressed zone, free of pores. ⟶

4

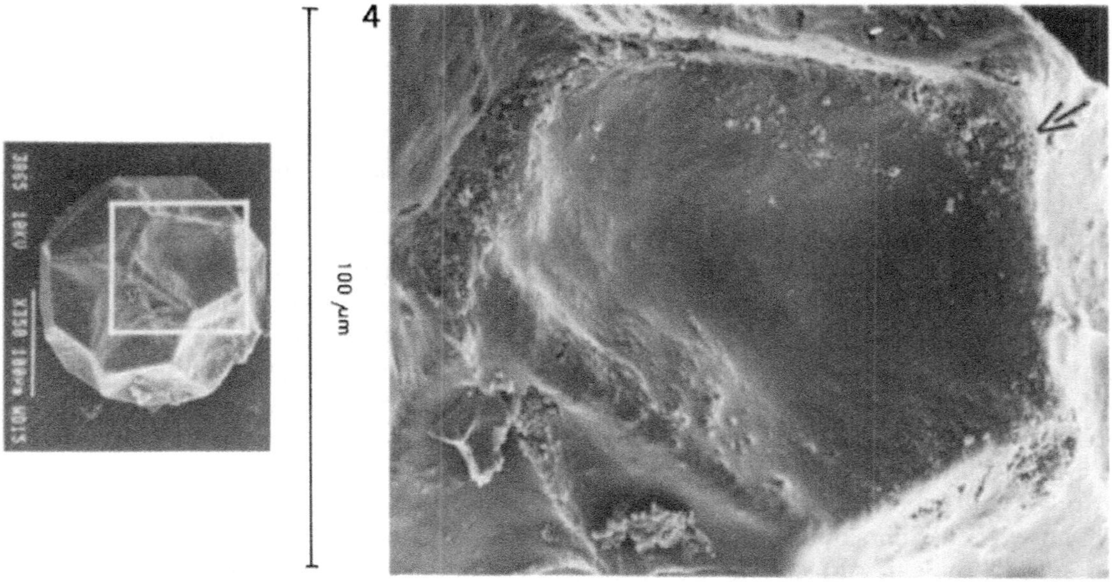


5

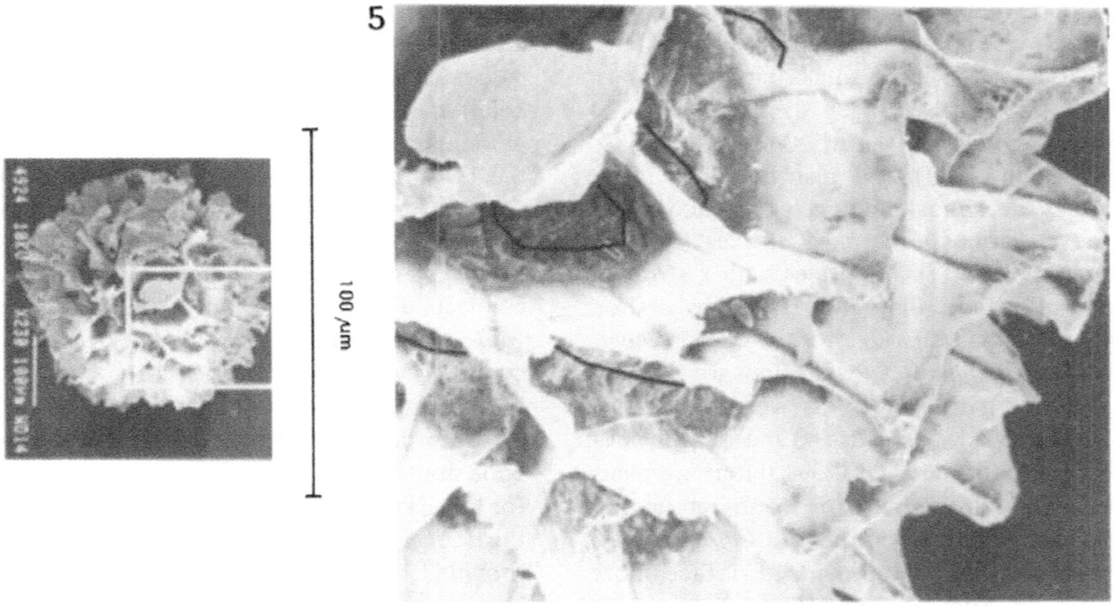


6

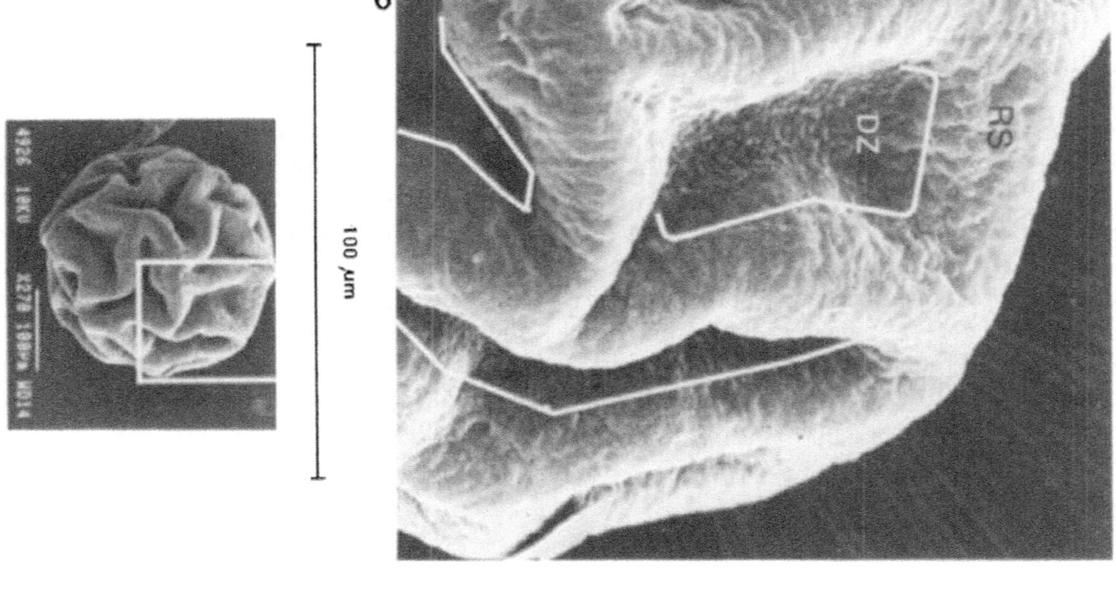

## Results

Dry cysts of Anostraca, with few exceptions (notably the genus *Artemia*), have an accidented surface, which is readily visible on fertilised cysts in the ovisac. In the species *Streptocephalus dichotomus* for example, a system of ridges defines a series of – mostly – pentagonal and diamond-shaped, depressed fields. The picture of Fig. 4 is an example of a cyst with only little relief; most cysts, even in this species, have more distinctly raised ridges and more deeply depressed fields (see Munuswamy & Subramoniam, 1984), and in related species, the cyst surface may be deeply wrinkled. In many Anostraca, numerous small pores are to be found on the cyst surface (e.g. Figs 5, 6, 8–9, 11), or a single, large pore may occur in the centre of each cyst field (e.g. in *Branchinella kugenamaensis*: Munuswamy & Subramoniam, 1983). The distribution of these pores may be almost homogeneous across the cyst, as in various *Branchipus* species (but see Gilchrist, 1978: 339). In species with a strongly wavy cyst surface, they cluster on top and/or along the flanks of the ridges. If the crest is wavy, like in *Thamnocephalus platyurus* Packard (Fig. 6), pores occur on the ridges, as well as on their flanks. If the ridge is narrow and sharply delimited, pores (of diameter 0.5–1.0 $\mu$m) are plentyfull on the flanks, but rare to absent on the ridges (Figs 8, 9: *S. torvicornis*). In *Chirocephalus*, where the cyst has a honeycombed surface, most published SEM-micrographs (Mura *et al.*, 1978; Mura, 1986; Alonso & Alcaraz, 1984; Thiery & Gasc, 1991) lack resolution, and do not reveal the pore pattern well. At variance with a statement by Gilchrist (1978), however, pores are certainly present, and again, they cluster around the scaliform base of the honeycombed outgrowths (Fig. 5).

Most published SEM-micrographs of *Streptocephalus*, e.g. (Alonso & Alcaraz, 1984) represent too low-magnifications to visualise pore patterns, but Brendonck's (1989) figures show that, in *S. proboscideus* (Frauenfeld) too, pores are concentrated along cyst ridges, with few of them appearing on the ridges themselves.

In *S. dichotomus*, pores are fewer than in *Thamnocephalus* or in *S. torvicornis* and *proboscideus*, and also cluster around the ridges. In the specimens shown in Figs 4 and 11–12, a film of organic matter mixed with anorganic particles obliterates most or part of them. However, upon drying, this film cracks or detaches from the cyst surface in many places, such that the pores still connect to the outside world.

In cross section, the cyst wall of *S. dichotomus* is composed of a number of distinct layers (Figs 1–3). The external zone is constructed from a series of vertical spokes, with broad basis, and narrowing outwardly, where they connect to the peripheral membrane (Fig. 3). Its floor consists of a thin, intact or perforated fleece (AS: Figs 1, 2, 3), which stretches in rags between supporting columns. This layer is thickest and most coherent at the level of the flanks of the external ridges. At the level of the poreless depressed fields, it is thinnest, and consists only of the external membrane, supported by a lattice of horizontally interwoven spokes (Fig. 1, arrows '1' and '3') and connects directly to the central, alveolar layer. It was also verified that the outwardly visible pores connect directly with this external zone, in this and the other species studied (Figs 11, 14).

The wide (20–30 $\mu$m across), central alveolar zone may best be described as a continuous gas (or liquid) mantle separating the outer zone from the deeper layers. Support to the outer layer is provided by a series of straight, branched, and sometimes anastomosed pillars (Figs 1–3).

The alveolar zone has a bottom which consists of spongy material (s.m.). This varies in width between c. 10–15 $\mu$m and virtual absence, in an apparently haphazard way (Fig. 1). In places where its looks extremely thin (Fig. 3), it forms a more coherent, although perforated (AL) layer, into which debouch the supporting pillars of the alveolar layer, and from which originates a dense forest of columns connecting it to a thin (1–2 $\mu$m), absolutely poreless, laminated (Fig. 2) barreer, the 'embryonic cuticle' (e.c.).

The partly anastomosed columns of this inner layer (i.l.) emanate from and belong to the outside cyst wall. In places, one may see their 'roots'

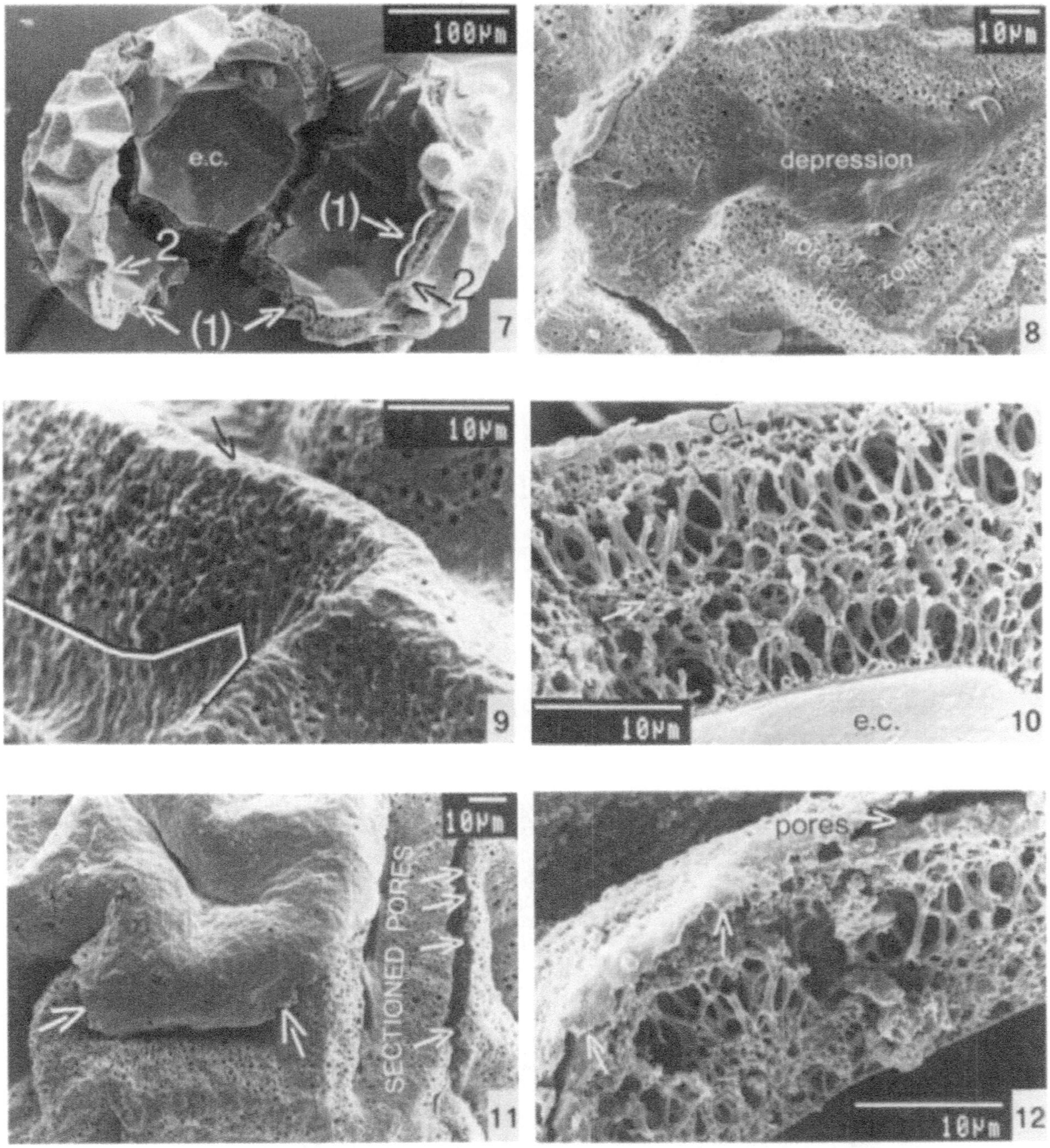


*Figs 7–12. Streptocephalus torvicornis.* 7. Low magnification view of a hatched cyst. *Arrows*: '1': ridge area, with central area of confluent alveoles; '2': depressed area, where cyst wall is distinctly narrowed. 8. Enlarged section, showing porefree depressions and ridges. 9. A single ridge (arrow), flanked by pore-rich section, and ending in pore-free depression. 10. Cross section of wall at the level of a ridge flank. C.L.: cortical zone; e.c.: embryonic cuticle; the intervening alveolar layer is more or less trilayered itself, and characterized by a central, dense zone (arrow). 11. Section of a cyst partly covered by a dirt film (arrows). A crack in the crest allows to see how pores (arrowed) communicate with the cortical space. 12. Section of a cyst wall at the level of a crest: a central zone of wide alveoles characterizes the alveolar zone. Arrows: dirt film overlying the crest but separated herefrom by a split (arrow). Two pores are also visible.

attach to the outside surface of the embryonic cuticle, without actually penetrating it (Figs 2, 14, 16). In places, one may also see a thinnest membrane peeling of from the e.m. (Fig. 3): the inner cuticular membrane (icm). As is well-known (e.g. Garreau de Loubresse, 1974), the embryonic cuticle is contributed directly by the blastomeres, while the sections of the wall outward of it are products of the egg-shell glands.

*S. torvicornis* shows a simpler wall structure (Figs 7–12), in which a zonation is less clear-cut, and in which zones are less strictly organised. The

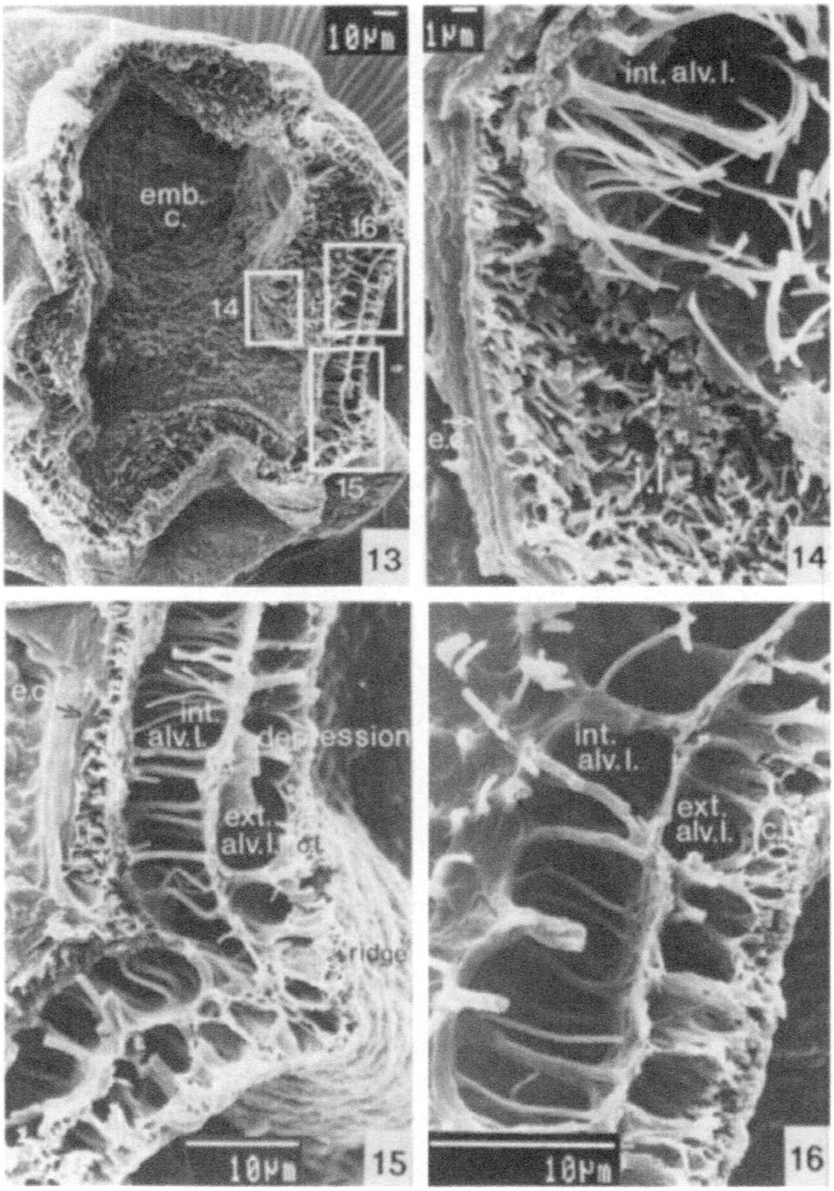


*Figs 13–16. Thamnocephalus platyurus.* 13. Cross-section of cyst, indicating sections enlarged in Figs 14–16. 14. Part of teared-off section of wall at level where i.l. opens, showing the frayed structure of the inner layer. 15. Section through the cyst wall at the level of a crest. The structure is invariant, whether in a depression or at a crest, and is composed of four layers. 16. Enlarged view of the wall with (upper left corner) part of the i.alv.l., ripped loose.

structure, as a whole, is spongy (Figs 10, 12). A denser cortical layer (c.l.) merges with a wide central zone (cz.), which, in areas of greatest thickness (ridge-flanking zones) may either appear trilayered, with a central zone of smaller alveoles flanked by two zones with wide alveoles (Fig. 10), or show a central area with large, confluent alveoles (Fig. 12). The latter situation is especially well expressed exactly under the ridges (Fig. 7, arrows '1'). No individualized basal spongy layer, sending a screen of pillars down to the embryonic cuticle, is present. At the level of the deep depressions between ridges, the spongy wall is simply much thinner (and with smaller alveoles) than elsewhere (Fig. 7, arrows '2'). This is a major difference with the otherwise closely related *S. dichotomus*, where the overall width of the cyst wall varies little. Here, at the level of the external ridges, the deep layer curves upward, to make direct contact with the external zone, and thus build a meandering, semicontinuous, though widely perforated wall (Fig. 1, arrows '1'). Where the distance between ridges is large, 'curtains' of fleecy material may descend from the external zone to the deep 'pillar' zone in reinforcement (Fig. 1, arrows '2'). A typical organisation between two walls of a pentagone may thus consist of two 'cells'. It is of interest to attract attention to the fact that the cavity of the cyst is not spherical, but presents a relief which is an ironed-out version of the external relief (cf. Fig. 7).

In *T. platyurus* (Figs 13–16), a structure more reminescent of *S. dichotomus* than of any other species studied, is found. A cortical layer is separated by a fleecy septum from a double 'alveolar' layer. The latter is a gas chamber, supported by numerous pillars that extend as far as the midway partition, but are clearly not continuous across the partition. The 'much thinner – internal layer (i.l.) is rather similar to that in *S. dichotomus*, and consists of a 'forest' of numerous short filaments attached to the e.c. at its lower end, to a granular septum separating it from the internal half of the 'alveolar' layer at its upper end (Fig. 16). The upper right half of Fig. 13 shows a section of the wall which was ripped off by the sectioning process, and is thus shown in frontal instead of transversal view. Fig. 14, showing an enlargement at the level of the origin of the crack, allows one to appreciate the elaborate nature of the structures involved. Fig. 16, left upper section, shows some transversally sectioned 'pillars' of the internal alveolar layer. They are circular and solid.

The lumen of the cyst, finally, is partly occupied by embryonic cells (EC). Like in Artemia, these cells are about 4 000 in number (Nakanishi *et al.*, 1962), confirming the long-held contention that all encysting Anostraca, upon reaching dormancy, arrest development at a late gastrula stage (Artom, 1909; Gross, 1935; Nourisson, 1959; Benesch, 1969).

## Discussion

The spongy nature and complexity of the cyst wall suggest roles other than, and additional to a mere physical protection, including thermal insulation (Gilchrist, 1978) of the embryo. As an egg is laid, and the extensive system of lacunae in the three layers of its wall are filled with liquid, it will sink to the bottom of its habitat. Once this dries out, the numerous aeropyles will ensure that liquid is rapidly replaced with air, and the cyst will float on all subsequent wettings, and become available for passive dispersal.

It is significant that differences in the structure of the cyst wall occur between all genera and species so far studied, i.e. the genera *Branchipus*, *Branchinecta*, *Chirocephalus* and *Artemia* (Gilchrist, 1978), and *Streptocephalus* and *Thamnocephalus* (present paper). Factors of variation include 1. the number of layers in the cyst wall; 2. the degree of continuity of the alveolar layer, varying from a continuous (?) mantle in *Branchinecta packardi* (Gilchrist, 1978), to an uninterrupted spongy mass in *Artemia sp.* (Mazzini, 1978); 3. the diameter of the wall, which may be quite uniform, or vary with the occurrence of ridges and depressions.

We therefore conclude that Gilchrist's (1978) contention that all cyst walls are bilayers, composed of a cortical and an alveolar layer, is invalid: cyst walls may be far more complex than that, and 'local' topographical differences across the cyst's three-dimensional surface may also be significant. Although one should beware of too hastily drawn generalisations, and explore cyst structure in its full three-dimensional complexity, it certainly appears justified to extend this type of studies to a score of other genera and species.

## Acknowledgements

N. Munuswamy thanks the Belgian Administration for Development Cooperation for financial assistance. We thank Dr L. Criel for useful comments.

## References

Alonso, M. & M. Alcaraz, 1984. Huevos resistentes de crustaceas eufilopodos no cladoceros de la peninsula Iberica: Observacion de la morfologia externa mediante tecnicas de microscopia electronica de barrido. Oecol. aquat. 7: 73–78.

Anderson, E., J. H. Lochhead, M. S. Lochhead & E. Huebner, 1970. The origin and structure of the tertiary envelope in thick-shelled eggs of the brine shrimp, *Artemia*. J. Ultrastruct. Res. 32: 497–505.

Artom, C., 1909. La maturazione, la fecondazione e i primi stadii di Sviluppo dell'uovo dell' '*Artemia salina*' Lin. di Cagliari. Biologica (Torino) 1: 495–515.

Benesch, R., 1969. Zur Ontogonie und Morphologie von *Artemia salina*. Zool. Jb. Anat. 86: 307–458.

Brendonck, L., 1989. Redescription of the fairy shrimp *Streptocephalus proboscideus* (Frauenfeld, 1873) (Crustacea: Anostraca). Bull. Kon. Belg. Inst. Natuurwet. Biol. 59: 49–57.

Garreau de Loubresse, N., 1974. Etude chronologique de la mise en place des enveloppes de l'œuf d'un crustacé phyllopode, *Tanymastix lacunae*. J. Microsc. 20: 21–38.

Gilchrist, B. M., 1978. Scanning electron microscope studies of the egg shell in some Anostraca (Crustacea: Branchiopoda). Cell Tiss. Res. 193: 337–351.

Gross, F., 1935. Die Reifungs- und Furchungsteilungen von *Artemia salina* im Zusammenhang mit dem Problem des Kernteilungsmechanismus. Z. Zellforsch. Mikrosk. Anat. 23: 522–565.

Mazzini, M., 1978. Scanning microscopes morphology and amino-acid analyses of the egg-shell of encysted brine shrimp *Artemia salina* Leach. (Crustacea, Anostraca). Monit. zool. ital. N.S. 12: 243–252.

Morris, J. E. & B. A. Afzelius, 1967. The structure of the shell and outer membranes in encysted *Artemia salina* embryos during cryptobiosis and development. J. Ultrastruct. Res. 20: 244–259.

Munuswamy, N. & T. Subramoniam, 1983. Scanning electron microscope studies on the egg shell of two fresh water fairy shrimps, *Streptocephalus dichotomus* and *Branchinella kugenumaensis* (Branchiopoda: Anostraca). Cytobios 37: 181–186.

Munuswamy, N. & T. Subramoniam, 1984. Egg envelopes of *Streptocephalus dichotomus* Baird: A structural and histochemical study. Hydrobiologia 114: 17–28.

Mura, G. F., 1986. S.E.M. morphological survey on the egg shell in the Italian anostracans (Crustacea, Branchiopoda). Hydrobiologia 134: 273–286.

Mura, G. F., F. Accordi & M. Rampini, 1978. Studies on the resting eggs of freshwater fairy shrimps of the genus *Chirocephalus:* biometry and scanning electron microscopic morphology (Branchiopoda: Anostraca). Crustaceana 35: 190–194.

Nakanishi, Y. H., T. Iwasaki, T. Okigaki & H. Kato, 1962. Cytological studies of *Artemia salina*. I. Embryonic development without cell multiplication after the blastula stage in encysted dry eggs. Annot. Zool. Jap. 35: 223–228.

Nourisson, M., 1959. Quelques données relatives au développement postembryonnaire du crustacé Phyllopode *Chirocephalus stagnalis* Shaw. Terre Vie 106: 174–182.

Spotte, S. & G. Anderson, 1988. Chemical decapsulation of resting cysts of the anostracans, *Artemia franciscana* and *Streptocephalus seali* as revealed by scanning electron microscopy. J. Crust. Biol. 8: 221–231.

Thiéry, A. & C. Gasc, 1991. Resting eggs of Anostraca, Notostraca and Spinicaudata (Crustacea, Branchiopoda) occurring in France: identification and taxonomical value. In Belk, D., H. J. Dumont & N. Munuswamy (eds), Studies on Large Branchipod Biology and Aquaculture. Developments in Hydrobiology 64. Kluwer Academic Publishers, Dordrecht: 245–259. Reprinted from Hydrobiologia 212.

*Hydrobiologia* **212**: 203–208, 1991.
D. Belk, H. J. Dumont & N. Munuswamy (eds), *Studies on Large Branchiopod Biology and Aquaculture.*



# Distribution and morphological variation of *Streptocephalus torvicornis* (Waga, 1842) in Northern Africa

Henri J. Dumont, Christine De Walsche & Johan Mertens
*Institute of Animal Ecology, State University of Ghent, K.L. Ledeganckstraat 35, 9000 Ghent, Belgium*

*Key words:* *Streptocephalus*, Anostraca, Sahara, Sahel, morphological variation

**Abstract**

The known range of *S. torvicornis* is extended to areas of the Western and Central Sahara and Sahel. Morphological variation between populations is important, and cannot be accomodated within the known Iberian and Maghrebian subspecific taxa, which appear untenable. A grouping of populations according to drainage basins appears, however, possible. Populations of the Hoggar, Air, and Tibesti waters, draining towards the Soudanian zone, are more closely related than populations from waters draining towards the Atlantic. Tassili-n-ajjer populations from Central Algeria have individual characteristics, and deserve further study.

## Introduction

The identification of species of the genus *Streptocephalus* relies on the structure of the second antenna (A II) of the male (Fig. 1). In *S. torvicornis* (Waga), repeated attempts have been made to define subspecific and even infrasubspecific categories, using characteristics of this structure.

Ignoring the Central European populations, and focusing on Northern Africa, we find that two subspecies are currently recognised (Daday, 1910): *S. t. torvicornis*, with a series of small, approximately equal-sized denticles on the thumb of the male A II, and *S. t. bucheti*, with denticles increasing in size towards the proximal end of the thumb. In addition, Thiéry (1987) finds that $a/b$, the length ratio of thumb to spur (Fig. 1), is $\leqslant 2.5$ in *S. t. bucheti*, and $\geqslant 3.5$ in *S. t. torvicornis*, based on a study of Iberian and North African populations. The known African range of *S. torvicornis* was limited to the Western and Central Maghrebian zone (Gauthier, 1928, 1931; Thiéry, 1987), and published information was not available about populations inland of the Atlas mountain chain. Since 1975, in the framework of several projects on relict aquatic faunas of the Sahara and Sahel, we collected Anostraca in a large number of localities, some of which contained sizeable

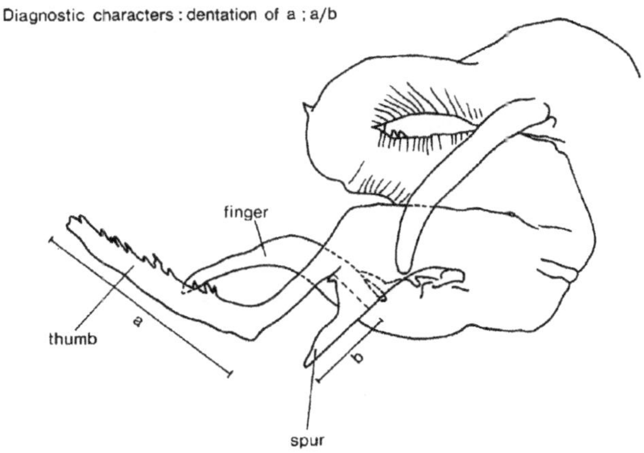


*Fig. 1.* Structures on antenna II of *S. torvicornis* male referred to in the text.

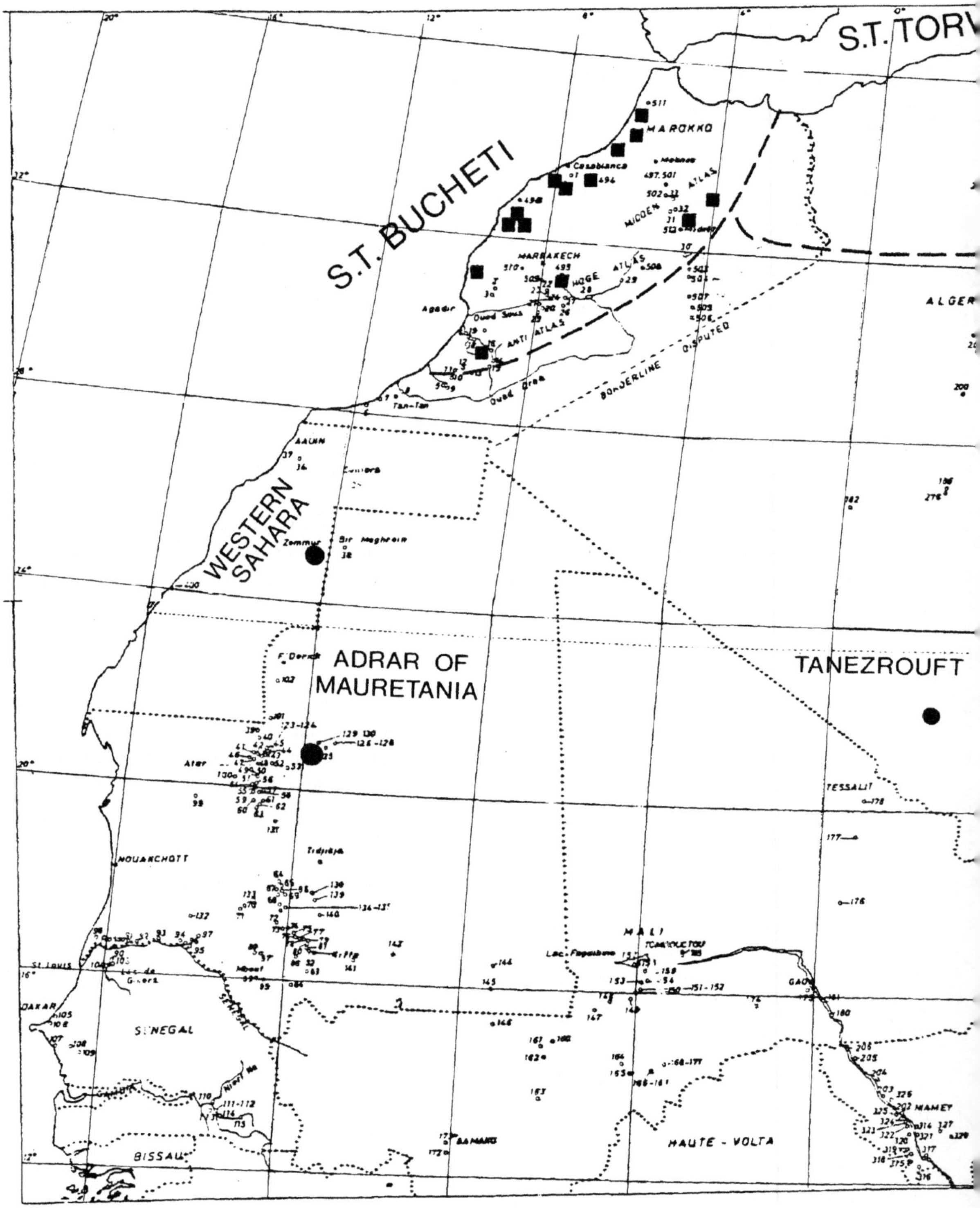


*Fig. 2.* Circles: New locality records for *S. torvicornis*; squares: published locality records. Range boundaries as previously defined for two regional subspecies indicated by dashed lines. The numbers of the map list all aquatic localities sampled within the framework of our exploration of the desert between 1975 and 1989.

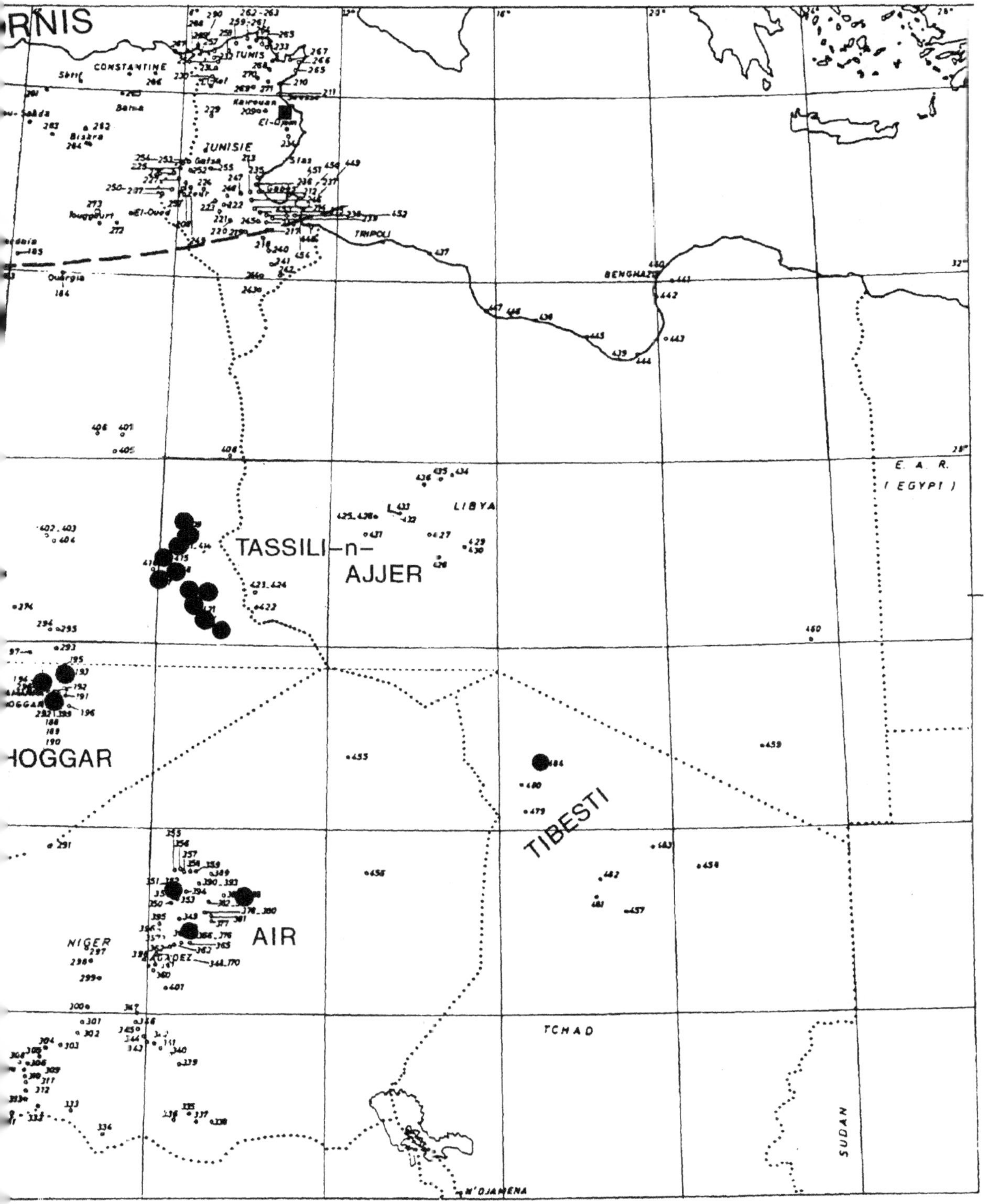

RNIS
CONSTANTINE
TUNIS
TUNISIE
TRIPOLI
BENGHAZI
Ouargla
Touggourt
El-Oued
Biskra
Batna
Kairouan
Sousse
Gafsa
Sfax
LIBYA
E. A. R.
( EGYPT )
TASSILI-n-AJJER
HOGGAR
TIBESTI
AIR
NIGER
AGADEZ
TCHAD
SUDAN
N'DJAMENA

populations of *S. torvicornis* (Table 1, Fig. 2). The aim of the present paper is to discuss the status of these populations.

*Table 1.* New locality records for *S. torvicornis* in Africa.

| | | |
|---|---|---|
| *Sahara Atlas (Algeria)* | | |
| Ain Ouarka | 32 50′N–0 10′W | 19/01/88 |
| *Tassilli n'Ajjer (Algeria)* | | |
| Guelta Adessei | 25 20′N–08 01′E | 06/06/78; 26/05/87 |
| Guelta Essendilene | 24 40′N–09 10′E | 20/05/87 |
| Guelta Touina | 24 40′N–09 10′E | 20/05/87 |
| Guelta In Tihaye | 25 10′N–08 50′E | 22/05/87 |
| Guelta Bei-Bei | 24 38′N–09 26′E | 09/06/78 |
| Oued Tabaraket | 24 39′N–09 28′E | 09/06/78 |
| Oued Tesselate | 26 10′N–08 18′E | 05/06/78 |
| Iherir | 25 23′N–08 25′E | 07/06/78 |
| Oued Djerat | 26 20′N–08 31′E | 01/06/78 |
| *Hoggar (Algeria)* | | |
| Ighahar | 22 52′N–05 23′E | 07/09/76 |
| Ehengas | 22 52′N–05 37′E | 05/09/76 |
| Guelta Imeleoulaouene | 22 53′N–05 40′E | 05/02/76; 03/05/79 |
| Edjif Mellene | 23 28′N–05 49′E | 05/09/76 |
| *Tanezruft (Algeria)* | | |
| Guelta Tamada | 24 35′N–03 13′E | 06/05/79 |
| *Tibesti (Chad)* | | |
| Bardai | 21 N– 18 E | 79 |
| *Air (Niger)* | | |
| Ighalabelabene | 17 41′N–08 49′E | 08/09/77 |
| Thagar | 17 41′N–08 49′E | 08/09/77 |
| Tamaza | 18 34′N–08 45′E | 20/09/78 |
| *Western Sahara* | | |
| Guelta Zemmur | 25 09′N–12 24′W | 28/05/75 |
| *Adrar of Mauretania* | | |
| Gueltet Zli | 19 31′N–12 47′W | 09/02/76 |

## Materials

Table 1 and Fig. 2 provide a list and map of all localities where *S. torvicornis* was observed between 1975 and 1989. However, only four localities, three of which are situated in the Hoggar mountains, and one in the Tassili-n-ajjer (both Algeria) yielded numbers of males sufficient for biometrical purposes (Table 2, Fig. 4). One of these, Guelta Adessei, is a semipermanent lakelet situated in a deep canyon protected from sunlight. *S. torvicornis* was observed here in 1978, 1979, and again in 1987, each time in considerable swarms. It appears likely that this population is present uninterruptedly for as many years as an aquatic cycle of the guelta lasts. Since first visited by us in 1978, to date, the guelta has never been seen dry. Two samples from this guelta are included in the present study. For all males, total length, the a/b ratio, and the correlation between both were calculated (Table 2).

## Results and discussion

The Southern boundary of the range of *S. torvicornis*, formerly believed to be the High Atlas and the Sahara Atlas (Thiéry, 1987; Gauthier, 1917, 1931) is now seen to be situated much deeper into Africa. We are in a position to add records of populations from the Northern Sahara, Mauretania, the Hoggar mountains, Tassili plateau, and Tanezrouft plain in Algeria, the Tibesti mountains in Northern Chad, and the Air mountains in

*Table 2.* Ratio of thumb: spur length ($a/b$) in relation to body length in *S. torvicornis* males. $x$ = mean; $s$ = standard deviation; $r^2$ = coefficient of determination; $n$ = sample size.

| Localities | $n$ | $a/b$ | | | | Body length | | | | $r^2$ |
|---|---|---|---|---|---|---|---|---|---|---|
| | | $x$ | $s$ | min | max | $x$ | $s$ | min | max | |
| G. Adessei 1978 | 15 | 3.0 | 0.5 | 1.9 | 4.2 | 18.8 | 1.1 | 16.5 | 20.4 | 0.0602 |
| G. Adessei 1987 | 15 | 3.4 | 0.5 | 2.8 | 4.0 | 17.5 | 1.0 | 15.8 | 18.9 | 0.0275 |
| G. Imeleoulaouene | 20 | 3.5 | 0.4 | 2.8 | 4.5 | 11.5 | 0.7 | 10.0 | 12.2 | 0.1661 |
| G. Ighahar | 12 | 3.2 | 0.6 | 2.3 | 4.0 | 14.3 | 1.2 | 13.6 | 15.3 | 0.0053 |
| G. Ehengas | 16 | 3.3 | 0.6 | 2.6 | 4.2 | 10.7 | 0.9 | 9.3 | 12.8 | 0.0743 |

Northern Niger. Its full range thus extends from Central Europe, across the Iberian Peninsula, to the West and Central Sahara and the Sahel. There is, as yet no confirmation of its presence in Libya, Egypt and Sudan, where the related *S. rubricaudatus* occurs (Mertens & Dumont, 1989).

As regards the *a/b* ratio of Thiéry (1987), a total absence of correlation with body length was established ($0.005 < r^2 < 0.166$, Table 2), confirming that *a/b* is a valid variable for discriminating between populations of unequal body size. This discriminator was tested in time for the population of Guelta Adessei and was 3.0 in 1978, 3.4 in 1987. These mean values are not significantly different ($p < 0.05$). Extending this exercise to all four localities, no significant differences could be detected between any of their respective populations ($P < 0.05$, Table 2), and in all cases $2.5 < a/b < 3.5$. It follows that none of these populations can with certainty be assigned to one of the presumed subspecies. Using Thiéry's criterion, a fair number of individuals (30) fall into the *S. t. torvicornis* range, and only few (3) in the *S. t. bucheti* range. The majority (41) fall between what now seem to be meaningless ratios as far as species characterization is concerned.

As regards the denticulation of the thumb (Fig. 3), interpopulation differences were found to exist, and a grouping of populations into several clusters with particular shapes appears possible. Atlantic populations, such as found between Spain (Fig. 3 : 1), Morocco (e.g. Thiéry, 1979, p. 150), and Zemmur (Fig. 3 : 2) are characterised by strong denticles at the base of the thumb. A clinal variation, with a maximum size difference between basal and apical denticles in Spain, and a minimum in the Western Sahara, seems to be characteristic of all these Western populations.

The population from the Central and South Sahara mountains of Hoggar, Air, and Tibesti, are also quite similar *inter se*. Their thumb is characterized (Fig. 3 : 4–6) by numerous small denticles, all of similar size. These mountains drain intermittently towards the Niger River and Lake Chad. Since until the middle Holocene both basins have repeatedly overflown into each other

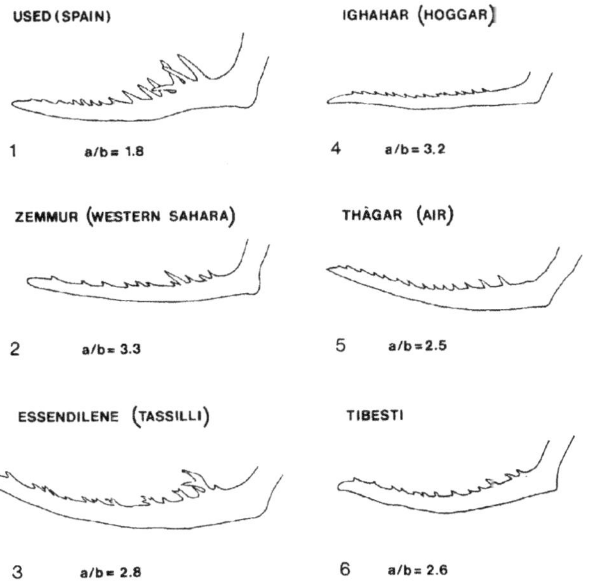


*Fig. 3.* Variation in denticulation of thumb on antenna II of *S. torvicornis* males from different localities. The *a/b* ratio is also given.

(Beadle, 1981; Dumont, 1979), this uniformity in structure might still reflect a situation of the past. It also indicates that passive dispersal of cysts between adjacent river basins may not be as effective on a short time scale as is often thought. The Tassili-n-ajjer populations, finally, stand out by their individual characters: rather large denticles at the base, and small denticles at the tip of the thumb, with frequent bifid and trifid structure (Fig. 3 : 3). Since the Tassili has a complex drainage, partly towards the South via the Oued Tafassasset, partly to the mediterranean via several oueds, and partly endorheic, a particular sampling effort on as many populations from this area as possible seems indicated in order to better define their status.

## Acknowledgements

The field work for collecting the various *Streptocephalus* populations discussed was made possible by grants 2.0009/75 and 32.0010.88 of the Foundation for Collective Fundamental Research (FKFO), Belgium.

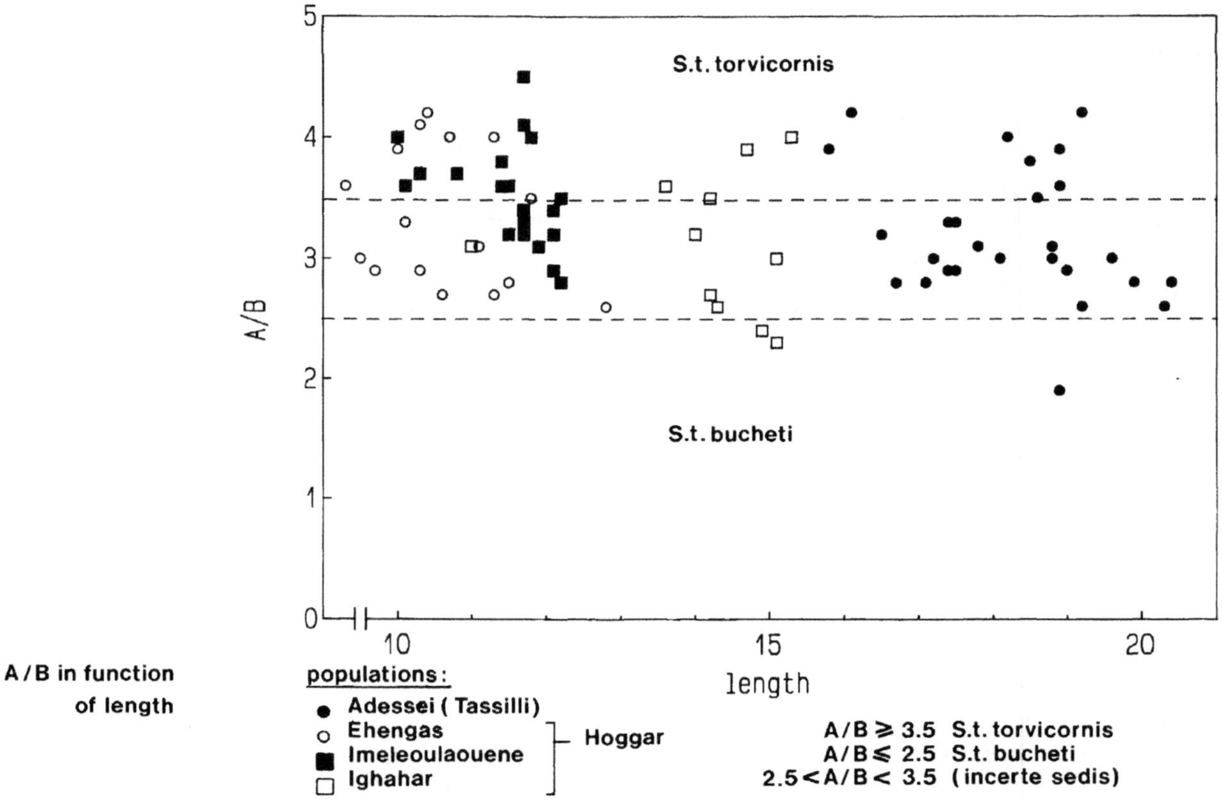


*Fig. 4.* plot of *a/b* ratio as a function of body length. Populations: ● Adessei (Tassilli) ○ Ehengas (Hoggar) ■ Imeleoulaouene (Hoggar) □ Ighahar (Hoggar)
$a/b \geqslant 3.5$ *S. t. torvicornis*
$a/b \leqslant 2.5$ *S. t. bucheti*
$2.5 < a/b < 3.5$

## References


Beadle, L. C., 1981. The inland waters of tropical Africa. Longman, London and N.Y., 475 pp.

Daday, E., 1910. Monographie systematique des Phyllopodes Anostraces. Ann. Sci. Nat. 11: 91–489.

Dumont, H. J., 1979. Limnologie van Sahara en Sahel. Unpublished D. Sc. Thesis, University of Ghent, 557 pp.

Gauthier, H., 1928. Recherches sur la faune des eaux continentales de l'Algerie et de la Tunisie. These, Paris. 419 pp.

Gauthier, H., 1931. Faune aquatique du Sahara central. Recoltes de M. L. G. Seurat au Hoggar en 1928. Bull. Soc. Hist. nat. Afr. N. 22: 350–400.

Mertens, J. & H. J. Dumont, 1988. Confirmation of *Streptocephalus rubricaudatus* as a good species (Anostraca). Crustaceana 56: 211–212.

Thiéry, A., 1987. Les crustaces branchiopodes Anostraca, Notostraca & Conchostraca des milieux limniques temporaires (dayas) au Maroc. These Dr. Sc. Univ. d'Aix-Marseille, 403 pp.

*Hydrobiologia* **212**: 209–219, 1991.
*D. Belk, H. J. Dumont & N. Munuswamy (eds), Studies on Large Branchiopod Biology and Aquaculture.*



# Distribution of species of Anostraca, Notostraca, Spinicaudata, and Laevicaudata in Mexico

Alejandro M. Maeda-Martinez
*Escuela Superior de Biologia, Universidad Juarez del Estado de Durango, apartado postal 329-B, Gomez Palacio, Dgo. Mexico; present address: University of Ghent, Institute of Ecology, K. Ledeganckstraat 35, B-9000 Gent, Belgium*

*Key words:* Anostraca, Notostraca, Spinicaudata, Laevicaudata, Branchiopoda, Mexico, distribution, conchostracan

## Abstract

We inventorize the known distribution of the species of Anostraca, Notostraca, Spinicaudata, and Laevicaudata present in Mexico. Determinations of the species and their distribution involved: 1) Examination of Mexican collections of phyllopods, 2) Extensive sampling in 17 states, and 3) Study of laboratory cultures. The number of phyllopod species known to occur in Mexico is 27: 14 of Anostraca, including three undescribed species, 1 of Notostraca, 10 of Spinicaudata, and 2 of Laevicaudata. These results also include 44 new state records. The records of *Branchinecta packardi* are the first for Mexico. Some species have extensive distributions, not limited to Mexico: *Streptocephalus mackini*, *S. texanus*, *Thamnocephalus platyurus*, *Triops longicaudatus* and *Leptestheria compleximanus*. The species known only from Mexico are *Streptocephalus kargesi*, *S. moorei*, *Paralimnetis mapimi*, two undescribed species of *Branchinecta*, and an undescribed species of *Streptocephalus*. The largest number of species found to occur in the same pond at one time was eight.

## Introduction

Like the brine shrimp *Artemia*, freshwater phyllopods have possibilities for use as food for fishes and crustaceans of economic importance. Nevertheless, the anostracans, notostracans and conchostracans of Mexico are poorly known, even though some of the species were collected in the last century. For these reasons I planned a project that has the following main objectives: 1) Understanding of the geographical distribution and specific localities of the species of Mexico, 2) Form a stock of eggs of Mexican species, and 3) Design the methods of culture for freshwater phyllopods based on ecological studies. The present work relates to the first objective.

## Antecedents

According to Packard (1883), the first phyllopod reported from Mexico was *Cyzicus mexicanus* (Claus), a species described in 1860 and native to Zimapan, Hidalgo. Packard (loc. cit.) indicated that *Triops longicaudatus* (LeConte), described in 1871 as *Apus aequalis*, occurs in Matamoros, Coahuila, and in the state of Guanajuato, and as *A. lucasanus* in Cabo San Lucas, Baja California Sur. Richard (1895) reported from Baja California Sur *Artemia* sp. (cited as *Artemia gracilis* but probably *Artemia franciscana* (Kellogg)), collected on San Jose Island, *Triops longicaudatus*, named *A. aequalis*, from Espiritu Santo Island; from La Purisima creek, *Leptestheria compleximanus*

(Packard) cited as *Estheria compleximanus*, and *Eocyzicus digueti* (Richard) sub *Estheria digueti*. *Eocyzicus vanhoffeni* Daday, was described in 1915, although the exact locality is uncertain (Mattox, 1954a). In 1926, Daday reported *Eulimnadia antillarum* (Baird) from the Caribbean and Mexico and described *Eulimnadia geayi* from Colombia, Venezuela, and Mexico (Martin, 1989) in the state of Coahuila (Tash & Shaffer, 1964). Creaser published in 1931 that *Streptocephalus seali* Ryder was collected in the state of Veracruz (Moore, 1966). In 1935, Pratt reported *Thamnocephalus platyurus* Packard from the state of San Luis Potosi (Moore, 1964). Linder (1941) described *T. platyurus mexicanus*, based on material collected in 1910 from Mohovano, Coahuila. Later, Linder (1952) indicated that *Triops longicaudatus* (sub *Apus longicaudatus*), was found in Azcapotzalco, Distrito Federal, plus the localities cited by Packard (1883).

Dexter (1953) mentioned that *Streptocephalus texanus* Packard was collected in 1878 from San Luis Potosi. Moore (1958) recorded *Streptocephalus similis* Baird from Nuevo Leon, San Luis Potosi and Tamaulipas, and *Thamnocephalus platyurus* from Nuevo Leon. Mattox (1954b, 1959) pointed out the presence in Mexico (without mentioning specific localities) of the conchostracans *Lynceus brevifrons* (Packard), *Eulimnadia antillarum*, *Leptestheria compleximanus*, and *Cyzicus mexicanus*. Tash & Shaffer (1964) indicated that *E. antillarum* and *L. compleximanus* occur in Chihuahua. Moore & Young (1964) elevated *T. p. mexicanus* to species status *T. mexicanus*, and at the same time indicated that its distribution in Mexico was Coahuila and Nayarit. For *T. platyurus*, they listed the state of Coahuila as a new state record. Moore (1965) reported *Eulimnadia texana* (Packard) from Mexico and later (Moore, 1966), described *Streptocephalus mackini* from Coahuila, Durango, Nuevo Leon, Distrito Federal, Puebla, Estado de Mexico, Queretaro, Guanajuato, San Luis Potosi, Jalisco, Zacatecas, and Tizayuca, Mexico (probably in the state of Hidalgo). In the same paper, he described *Streptocephalus linderi* giving its distribution as Tamaulipas, Nuevo Leon, and Coahuila. Belk (1973) described *Streptocephalus moorei* based on specimens collected in a pond in Chihuahua, cohabitating with *S. mackini*, *T. platyurus*, *Eocyzicus digueti*, *L. compleximanus*, and *T. longicaudatus*. In a subsequent paper, Belk (1975) points out Coahuila and Oaxaca as new states for *Artemia* sp. (cited as *A. salina*).

The first record of *Caenestheriella setosa* (Pearse) in Mexico was made by Belk & Belk (1975), on specimens from Chihuahua. Clark & Bowen (1976) indicated the occurrence of populations of *Artemia franciscana* on the Pichilingue Island in the Gulf of California near La Paz, Baja California Sur, and in a natural lake in San Quintin on the west cost of Baja California Norte. *Branchinecta lindahli* Packard has been collected on the Guadalupe Island, Baja California Norte (Belk & Lindberg, 1979). Abreu-Grobois & Beardmore (1980) showed that *A. franciscana* is known from Yavaros, Sonora. Castro (1980) points out that populations of *Artemia* sp. (cited as *A. salina*) occur in Sinaloa and Chiapas and later (Castro, 1982) added the states of Hidalgo, Estado de Mexico, San Luis Potosi and Yucatan. Spicer (1982) mentioned Coahuila as a new state record for *Streptocephalus texanus* and Belk (1983) indicated that *B. lindahli* is known from Baja California Sur, cohabiting with *S. texanus*. Spicer (1985) described *Streptocephalus kargesi* using material from Veracruz, and Maeda-Martinez (1987) described the conchostracan *Paralimnetis mapimi* based on specimens from Chihuahua and Durango. This was the first record of the genus *Paralimnetis* in North America. Martin (1989) described *Eulimnadia belki* from material collected on the Cozumel Island, State of Quintana Roo. Belk (1989) described *Eulimnadia cilindrova* giving localities in Mexico of Chihuahua, Jalisco, and San Luis Potosi.

## Materials and methods

The determination of the species and their distribution involved: 1) Examination of collections of phyllopods at the Facultad de Ciencias Biologicas, Universidad Autonoma de Nuevo Leon (FCB); Escuela Superior de Biologia, Universidad Juarez del Estado de Durango (UJED) and Instituto de Biologia, Universidad Nacional Autonoma de Mexico (UNAM); 2) Extensive collection in 17 states and 3) Culture in the laboratory for the corroboration of an undescribed species of *Branchinecta* from Coahuila and the demonstration of the presence of *Triops longicaudatus* in Baja California Sur.

The determinations of species were based on the keys and publications of Belk (1973, 1975, 1989), Linder (1941, 1952), Martin (1989), Mattox (1965), Moore (1958, 1966), Packard (1883) and Spicer (1985). The list of the inventory of the Mexican species and the data of specific localities are deposited at Universidad Juarez del Estado de Durango, Escuela Superior de Biologia, Gomez Palacio, Durango, Mexico, and University of Gent, Institute of Ecology, Gent, Belgium.

*Argillotrophic culture*

I developed a laboratory argillotrophic culture method employing bacterial production through the decomposition of cattle manure. The principle of this culture method is based on my observation that habitats with argillotrophic conditions are the most usual ones inhabited by mexican freshwater species. The term 'argillotrophic' was coined by Hutchinson (1937, Daborn, 1975) to describe certain lakes and ponds of semiarid regions that are characterized by persistent and extreme turbidity, and absence of phytoplankton, and a zooplankton association of low diversity, the fauna being dependent upon suspended particulate matter for food.

After several tests, I arrived at this procedure: The eggs were removed from the dry soil and placed in a container of plankton gauze of 0.1 mm to rehydrate in demineralized water, previously areated for one hour. The hydration lasted two

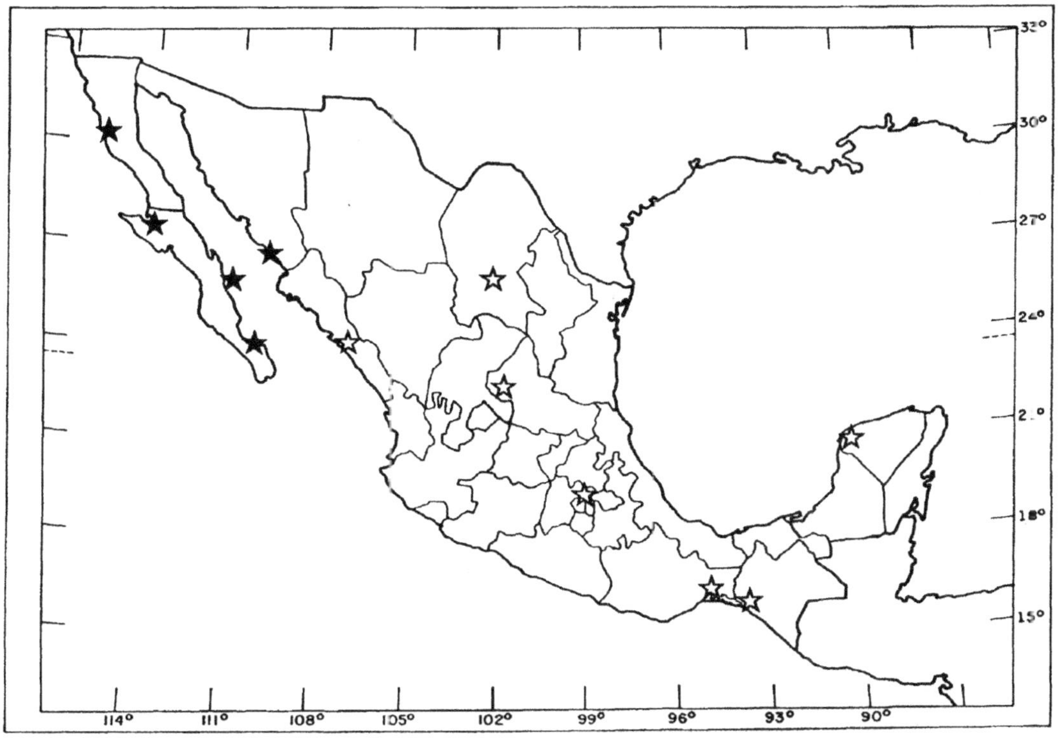


*Fig. 1.* Distribution of *Artemia franciscana* (closed stars) and *Artemia* sp. (open stars).

hours under continnous aeration and then the eggs were placed in eclosion flasks with areation and light from two fluorescent lamps of 40 watts each. The nauplii were placed in the argillotrophic culture medium. This culture medium was prepared with boiled and aged tap or demineralized water. One day before adding nauplii, a suspension of food-substratum, filtered with planktonic mesh, was added in a proportion of 10 ml per 150 ml of areated culture medium. In my test, the medium had between 500 and 1000 $\mu$S of conductivity (18–26 °C). The food-substratum suspension was prepared with 20 g of macerated cattle manure and 10 g of sterile clay soil in four liters of tap water. This preparation was boiled for 10 minutes, filtered with a mesh of 1 mm, and left to work for two days. After three days, uncountables numbers of bacteria, principally *Escherichia coli* and *Enterobacter aerogenes*, were present. This food-substratum suspension was filtered over plankton gauze and was added in a proportion of 5 ml to each 150 ml every 12 hours. The animal density in my cultures was two shrimps per 150 ml of culture medium.

## Results and discussion

In Mexico, *Artemia franciscana* is known from five localities in three states: Baja California Norte, Baja California Sur and Sonora (Fig. 1). Other populations of brine shrimp (*Artemia* sp.) occur in seven states: Coahuila, Chiapas, Estado de Mexico, Oaxaca, San Luis Potosi, Sinaloa, and Yucatan (Fig. 1). However, the specific status of these populations has not been determined. *Branchinecta lindahli* described from the USA (Packard, 1883) has a known distribution in Mexico of six localities in three states: Baja California Norte, Baja California Sur and the new state record Coahuila (Fig. 2). El Arenoso, Baja California Sur was considered as the most

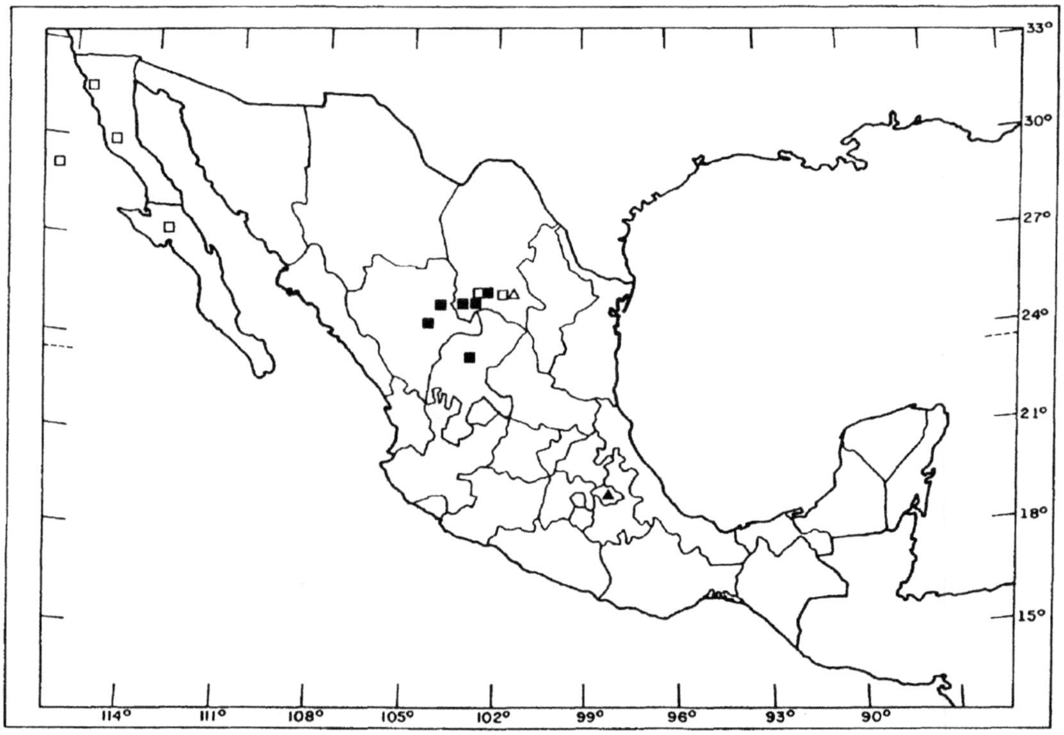


*Fig. 2.* Distribution of *Branchinecta lindahli* (open squares), *Branchinecta packardi* (closed squares), *Branchinecta* sp. A (open triangle) and *Branchinecta* sp. B (closed triangle).

southern occurrence of *B. lindahli* (Belk. 1983). The populations of Coahuila represent a range extention to the south. *Branchinecta packardi*, known before only from the United States (Belk, 1975), occurs in the Mexican Republic in ten localities in three states: Coahuila, Durango and Zacatecas (Fig. 2). These samples represent the first records for Mexico. The population of Zacatecas (Tropic of Cancer) is now the southern range limit for *B. packardi*. *Branchinecta* sp. A is an undescribed species, endemic for the El Dorado, Coahuila area (Fig. 2) and *Branchinecta* sp. B, also a undescribed species, is known only from Tlaxco, Tlaxcala and this record is probably the most southern limit of distribution of the genus in North America (Fig. 2).

*Streptocephalus kargesi* is endemic to Veracruz and known only from the type locality (Fig. 3). Besides Texas, USA, *Streptocephalus linderi* occurs in six localities in Coahuila, Nuevo Leon, and Tamaulipas (Moore, 1966; Fig. 3). The most common phyllopod in Mexico is *Streptocephalus mackini* with 97 localities in 19 states. Moore (1966), indicated for this species the states of Coahuila (type locality in this state), Distrito Federal, Durango, Estado de Mexico, Guanajuato, Hidalgo, Jalisco, Nuevo Leon, Puebla, Queretaro, San Luis Potosi, and Zacatecas. Belk (1973, 1986) added Chihuahua and Aguascalientes. Also, *S. mackini* has been reported from Sonora (Bowman, 1985). In this paper I add three new states: Morelos, Sinaloa and Tlaxcala (Fig. 4). *S. mackini* has been collected from USA (Moore, 1966, Belk, 1975). *Streptocephalus moorei* is endemic to Chihuahua in the Mapimi Bolson, Chihuahuan – Desert, with one locality (Fig. 3). Probably the first streptocephalid reported from Mexico was *Streptocephalus seali*, a species widely distributed in the USA. It was recorded from Veracruz by Creaser in 1931 (Moore, 1966).

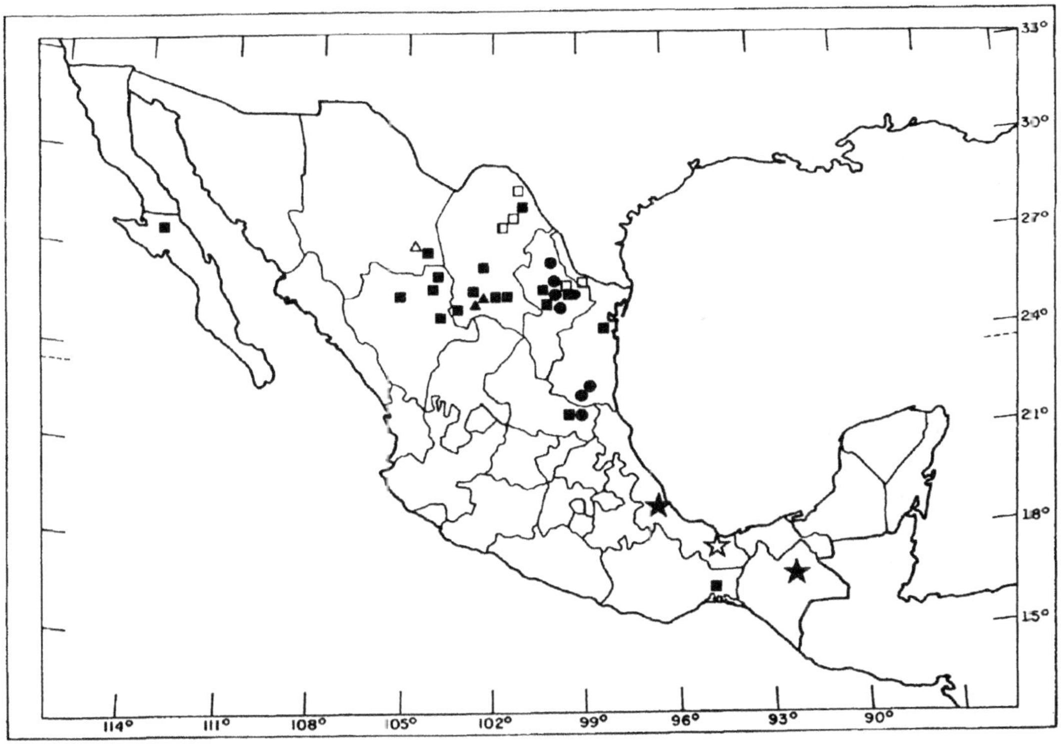


*Fig. 3.* Distribution of *Streptocephalus kargesi* (open star), – *Streptocephalus linderi* (open squares), *Streptocephalus moorei* (open triangle), *Streptocephalus seali* (closed stars), – *Streptocephalus similis* (closed circles), *Streptocephalus texanus* (closed squares) and *Streptocephalus* sp. (closed triangle).

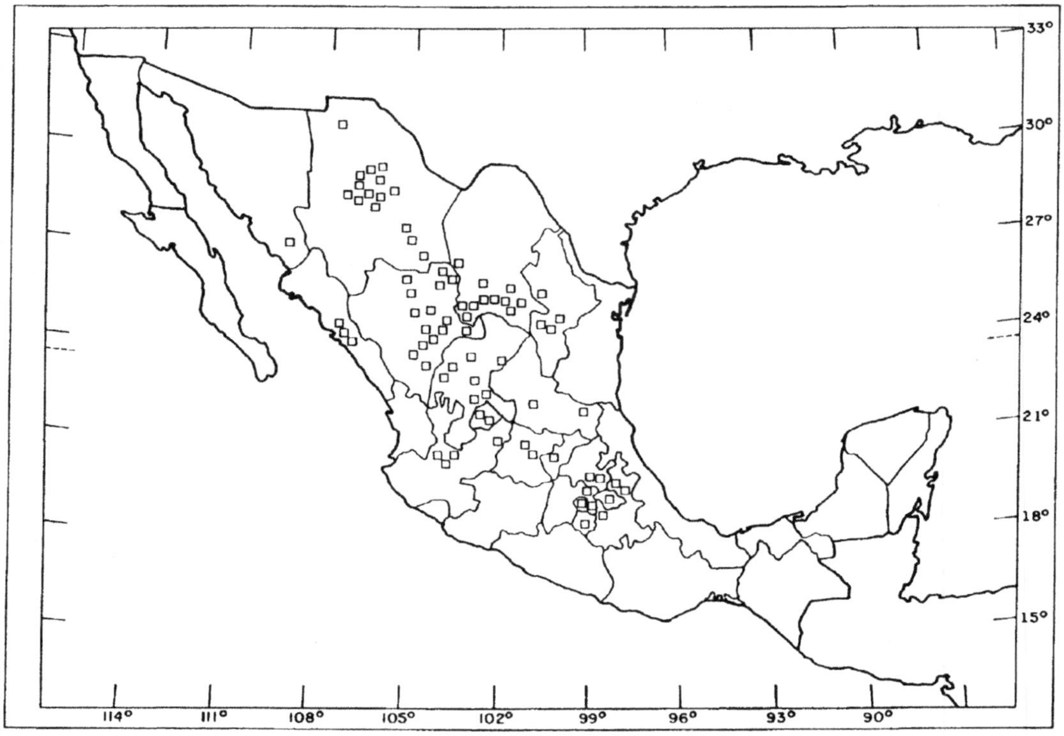


*Fig. 4.* Distribution of *Streptocephalus mackini* (open squares).

However, there is only one other record of it in Mexico. This is Belk's (1986) report of *S. seali* from Chiapas (Fig. 3).

*Streptocephalus similis*, a species known from Texas, Jamaica, Puerto Rico, and Hispaniola (Belk, 1975), has been collected in the Mexican Republic from ten localities in three states: Nuevo Leon, San Luis Potosi and Tamaulipas (Moore, 1958; Fig. 3). *Streptocephalus texanus* is the second most widely distributed streptocephalid in Mexico with 19 separate localities. It was collected from six states: Baja California Sur, Coahuila, Nuevo Leon, Oaxaca, San Luis Potosi, and Tamaulipas (Belk, 1975, 1983, 1986; Bowman, 1985; Dexter, 1953; Spicer, 1982). This paper presents Chihuahua and Durango as new state records (Fig. 3). An undescribed species of *Streptocephalus* is endemic to Coahuila with two localities in the Mapimi Bolson (Fig. 3).

The distribution of *Thamnocephalus mexicanus* includes ten localities in five states: Coahuila (type locality in this state), Nayarit, Sonora (Bowman, 1985; Linder, 1941; Moore, 1964, 1966) and my new state records of Durango and Nuevo Leon (Fig. 5). *Thamnocephalus platyurus* occurs in 28 localities in six states: Coahuila, Chihuahua, Durango, Nuevo Leon, San Luis Potosi (Belk, 1973, 1986; Moore, 1958, 1964, 1966) and Zacatecas (new state record) (Fig. 5).

The notostracan *Triops longicaudatus* is a phyllopod with a wide distribution in Mexico. Populations with males and females, and others without males are present. Eleven localities in the states of Coahuila, Chihuahua, Durango, Estado de Mexico, San Luis Potosi, and Zacatecas, have bisexual populations. Twenty eight localities in Baja California Sur, Chihuahua, Coahuila, Distrito Federal, Durango, Estado de Mexico, Guanajuato, Hidalgo, and Nuevo Leon have populations without males (Fig. 6). The new state records for this species are Durango, Estado de Mexico, Guanajuato, Hidalgo, Nuevo Leon, San Luis Potosi and Zacatecas. There is a reference in Rioja, 1971 about the occurrence of *Lepidurus*

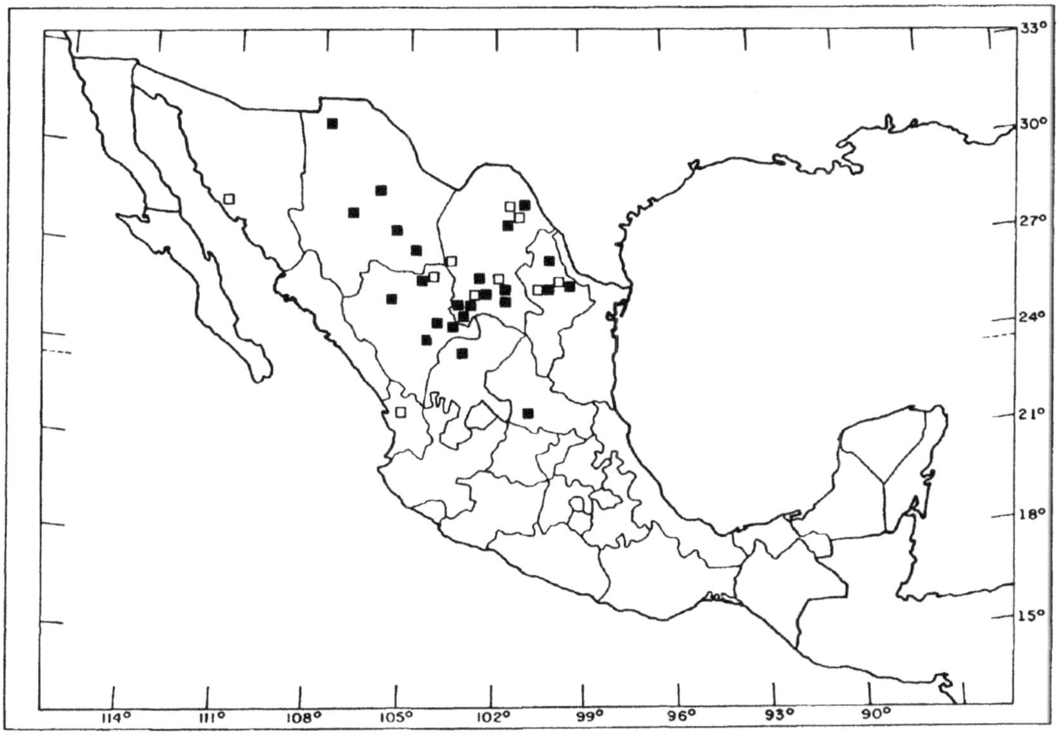


*Fig. 5.* Distribution of *Thamnocephalus mexicanus* (open squares) and *Thamnocephalus platyurus* (closed squares).

(*L. bilobatus* Packard and *L. lynchi* Linder) in Mexico. It seems that this record is not based in any deposited material, consequently *Lepidurus* is not included in the present inventory.

The known distribution in Mexico of *Caenestheriella setosa* includes ten localities in the states of Chihuahua, Durango, San Luis Potosi, and Zacatecas (Fig. 7). The last three of these are new locality records. *Caenestheriella setosa* has been reported from USA (Mattox, 1965; Pennak, 1978; Belk & Belk, 1975, and Saunders & Wu, 1984). One little species in the Mexican Republic is *Cyzicus mexicanus*, in spite of the fact that it was the first mexican species described. It has been recorded from 16 states of USA and two Canadian provinces (Mattox, 1959, Pennak, 1978; Saunders & Wu, 1984). Apparently, since Claus's description in 1860 this species had not been reported from any new localities in Mexico before my study. Now we know that besides Zimapan, Hidalgo (type locality), *Cyzicus mexicanus* is present in four more localities in four states: Distrito Federal, Estado de Mexico, Hidalgo and Puebla (Fig. 7).

Since its description, *Eocyzicus vanhoffeni* has not been recorded again. Mattox (1954) only mentioned and compared this species with specimens of *E. digueti* and *E. concavus* from USA. *Eocyzicus digueti* is known to occur in eight localities in six states: Baja California Sur (Richard, 1895), Chihuahua (Belk, 1973), and the new state records Coahuila, Durango, Estado de Mexico, and Sonora (Fig. 7). *Leptestheria compleximanus* is the conchostracan with the largest known distribution in Mexico. Populations of this species occur in 21 localities including some from Baja California Sur (Richard, 1895), Chihuahua (Belk, 1973, 1986; Tash & Shaffer, 1964) and Distrito Federal (Bowman, 1985), and others from the new state records of Coahuila, Durango, Estado de Mexico, San Luis Potosi, Sonora, and Zacatecas (Fig. 7).

*Eulimnadia antillarum* and *E. geayi* were recorded from Mexico (Daday, 1926 in Martin,

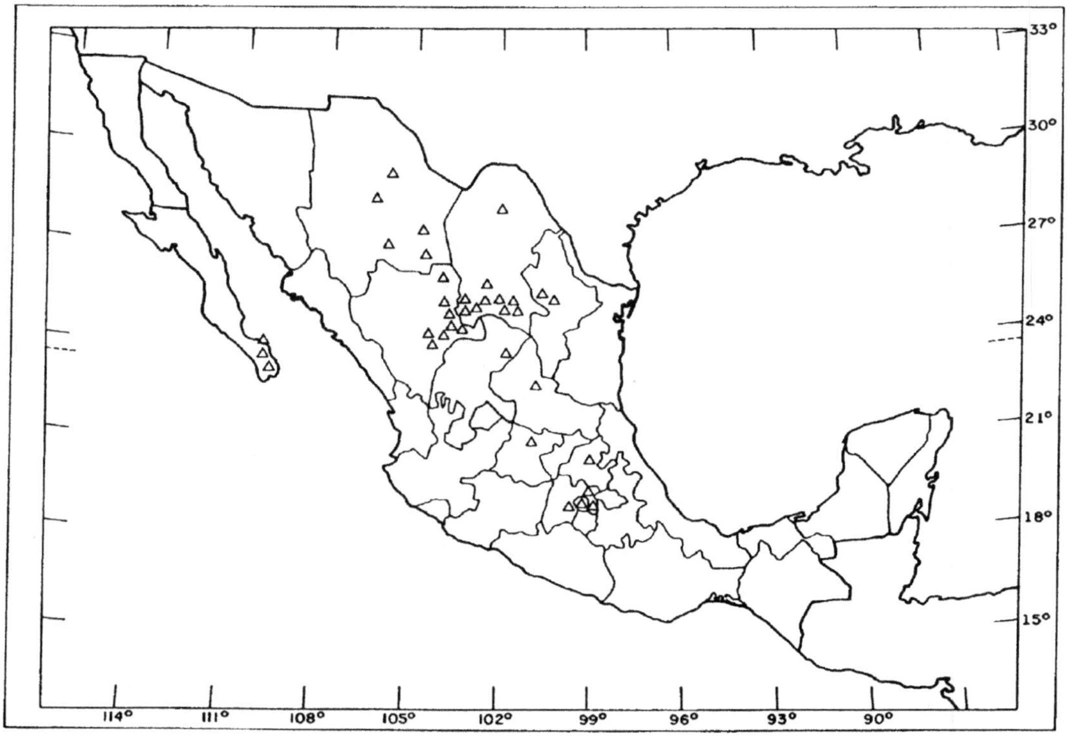


*Fig. 6.* Distribution of *Triops longicaudatus* (open triangles).

1989); but I have not been able to identify specific localities. Bowman (1985) listed *E. antillarum* from Durango, and Tash & Shaffer (1964) from Chihuahua. Nevertheless, the validity of this species is subject to question because of the absence of studies dealing with its egg morphology (Belk, 1989; Martin, 1989). I analized the available material of *Eulimnadia* using egg morphology and I did not recognize eggs as belonging to *E. geayi*. In Mexico, *Eulimnadia belki* has been collected only from the type locality in Quintana Roo. However, it is also known from Venezuela (Martin, 1989; Fig. 8). *Eulimnadia cilindrova*, known in two states of the USA, occurs in four localities in Chihuahua (type locality in this state). Jalisco and San Luis Potosi (Belk, 1989), and Coahuila, a new state record (Fig. 8). The known distribution of *Eulimnadia texana* includes 15 localities in six states: Coahuila, Chihuahua (Belk, 1989), and the new state records Durango, Morelos, Nuevo Leon, and Tamaulipas (Fig. 8).

*Lynceus brevifrons* was reported from Mexico without reference to localities (Mattox, 1959). It is now known from eight localities in three states: Durango (Bowman, 1985), and the new state records Chihuahua and San Luis Potosi (Fig. 8). Populations of *Paralimnetis mapimi* seem to occur only in the pools of Chihuahua (type locality) and Durango reported by Maeda-Martinez (1987; Fig. 8).

## Conclusion

The number of phyllopod species known to occur in Mexico amounts to 27: 14 Anostraca (three undescribed species are being studied by the author), 1 Notostraca, 10 Spinicaudata and 2 Laevicaudata. Conchostracans present more taxonomic problems than the other groups, due the lack of material and the absence of a systematic revision of the family Cyzicidae.

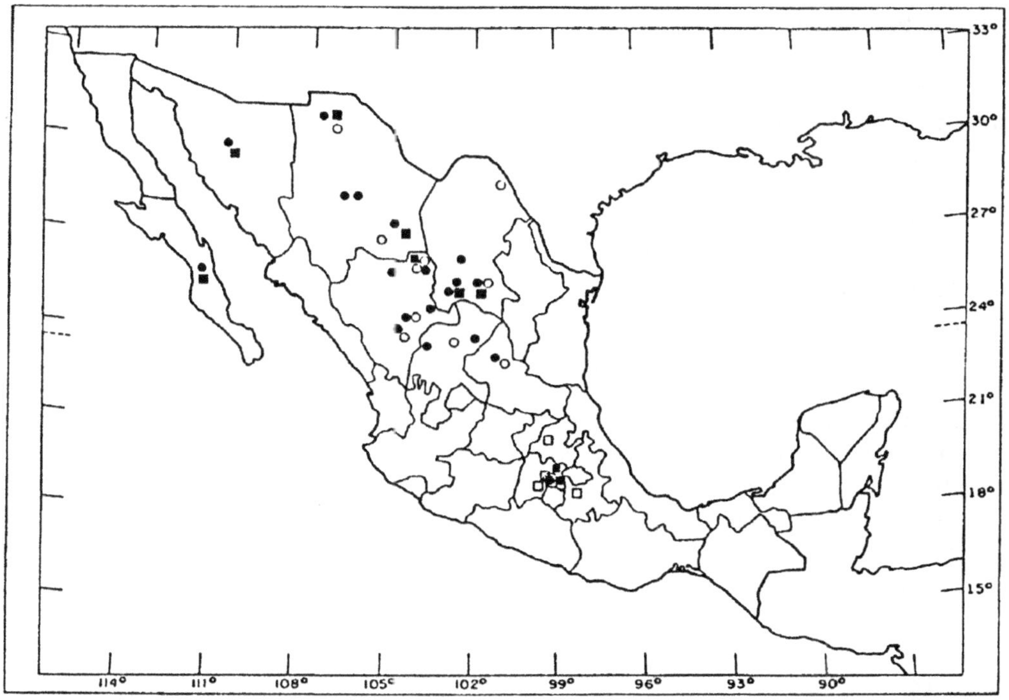


*Fig. 7.* Distribution of *Caenestheriella setosa* (open circles), *Cyzicus mexicanus* (open squares), *Eocyzicus digueti* (closed squares) and *Leptestheria compleximanus* (closed circles).

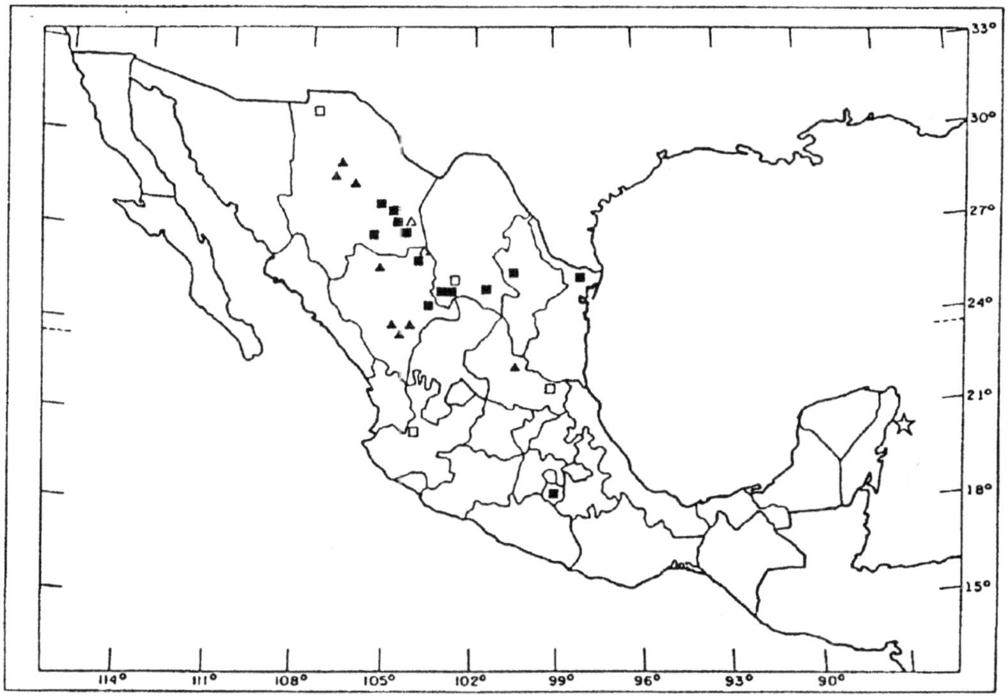


*Fig. 8.* Distribution of *Eulimnadia belki* (open star), *Eulimnadia cilindrova* (open squares), *Eulimnadia texana* (closed squares), *Lynceus brevifrons* (closed triangles) and *Paralimnetis mapimi* (open triangles).

The results of this inventory include 44 new state records. The records for *Branchinecta packardi* are the first for Mexico. The species with largest distribution are *Streptocephalus mackini*, *S. texanus*, *Thamnocephalus platyurus*, *Triops longicaudatus*, and *Leptestheria compleximanus*. The species known only from this country are *Branchinecta* sp. A. (Coahuila), *B.* sp. B. (Tlaxcala), *Streptocephalus kargesi* (Veracruz), *S. moorei* (Chihuahua), *S.* sp. (Coahuila) and *Paralimnetis mapimi* (Chihuahua and Durango).

From 183 localities with phyllopods, 122 had only one species present. The rest had two or more co-occurring species; nine or ten species are known to inhabit the same pool in two localities of Coahuila, and eight was the largest number of species living together in the same pool at one time, in only one locality of Durango.

## Note added in proof

We note the recent records of *Branchinecta packardi* from San Luis Potosi by Strenth N. E. & T. G. Littleton (1990, 'First record of *Branchinecta packardi* Pearse (Crustacea, Anostraca) from Mexico', Tex. J. Sci. 42: 411–412), and of *Streptocephalus mackini* and *Cyzicus mexicanus* from Estado de Mexico by Villaclara, G., V. Sladecek & J. Brtek (1990 'An artificial reservoir with the dominance of 'Eubranchiopoda'. Arch. Hydrobiol. 119: 115–116).

## Acknowledgements

I thank Dr. Denton Belk for his help in the preparation of the manuscript, Biol. Humberto Garcia V., Biol. Oscar Montanez R., Sr. Gerardo Garcia, Biol. Miguel Castellanos, Biol. Esperanza Borjas, Biol. Livier Lopez, Mr. Eduardo Guzman L., Q. F. B. Enrique Salas P. and Q. F. B. Luis Angel Cervantes for their assistance in the collection and cleaning of samples. I also thank M. C. Hortencia Obregon B., M. C. Raul Diaz M. and Dr. T. Bowman for their aid with this paper. This research was supported by Universidad Juarez del Estado de Durango, Consejo Nacional de Ciencia y Tecnologia (PCECBNA-021463) and Instituto de Investigaciones Cientificas del Estado de Coahuila, Unidad Laguna.

## References

Abreu-Grobois, F. A. & J. A. Beardmore, 1980. Genetic characterization of *Artemia* populations – an electrophoretic approach. In G. Persoone, P. Sorgeloos, O. Roels, and E. Jaspers (eds), The brine shrimp *Artemia*. Vol. 1, Morphology, genetics, radiobiology, toxicology: 133–146. Universa Press, Wetteren Belgium.

Belk, D., 1973. *Streptocephalus moorei* n.sp., a new fairy shrimp (Anostraca) from Mexico. Trans. am. microsc. Soc. 92: 507–512.

Belk, D., 1975. Key to the Anostraca (fairy shrimps) of North America. S. West. Nat. 20: 91–103.

Belk, D., 1983. New fairy shrimp distribution records among collections at the California Academy of Sciences. S. West. Nat. 28: 380–381.

Belk, D., 1986. A list of Mexican anostracans and conchostracans deposited in the collection of Denton Belk (DB), San Antonio, Texas, USA. Unpublished.

Belk, D., 1989. Identification of species in the conchostracan genus *Eulimnadia* by egg shell morphology. J. Crust. Biol. 9: 115–125.

Belk, D. & M. S. Belk, 1975. Hatching temperatures and new distributional records for *Caenestheriella setosa* (Crustacea, Conchostraca). S. West. Nat. 20: 409–411.

Belk, D. & D. R. Lindberg, 1979. First freshwater animal reported for Isla de Guadalupe represents a southern range extension for *Branchinecta lindahli* (Crustacea: Anostraca). S. West. Nat. 24: 390–391.

Bowman, T. E., 1985. A list of Mexican anostracans, notostracans and conchostracans deposited at the National Museum of Natural History, Smithsonian Institution (USNM). Unpublished.

Castro, T., 1980. Distribucion geografica e importancia de *Artemia* en Mexico y evaluacion de la poblacion en el sur de la Bahia de Ceuta, Sinaloa, Mexico. Rep. de Inv. 6. Univ. Aut. Met.-Xochimilco, Mexico, D.F.

Castro, T., 1982. Distribucion de *Artemia* en la Republica Mexicana. Memorias del curso *Artemia* sp. en la acuicultura, la investigacion y la docencia. Univ. Aut. Met.-Xochimilco, Mexico, D.F.

Clark, L. S. & S. T. Bowen, 1976. The genetics of *Artemia salina*. VII. Reproductive isolation. J. Heredity 67: 385–388.

Daborn, G. R., 1975. The argillotrophic lake system. Verh. int. Ver. Limnol. 19: 580–588.

Dexter, R. W., 1953. Studies on North American fairy shrimps with the description of two new species. Am. Midl. Nat. 49: 751–771.

Linder, F., 1941. Contributions to the morphology and the taxonomy of the Branchiopoda Anostraca. Zool. Bidrag Uppsala 20: 101–302.

Linder, F., 1952. Contributions to the morphology and taxonomy of the Branchiopoda Notostraca, with special reference to the North American species. Proc. U.S. Nat. Mus. 102: 1–69.

Maeda-Martinez, A. M., 1987. *Paralimnetis mapimi* nueva especie (Crustacea: Conchostraca), un nuevo camaron almeja del Desierto Chihuahuense en el Norte de Mexico. S. West. Nat. 32: 67–74.

Martin, J. W., 1989. *Eulimnadia belki*, a new clam shrimp from Cozumel, Mexico (Conchostraca: Limnadiidae), with a review of Central and South American species of the genus *Eulimnadia*. J. Crust. Biol. 9: 104–114.

Mattox, N. T., 1954a. Description of *Eocyzicus concavus* (Mackin) with a review of other North American species of the genus (Crustacea: Conchostraca). J. Wash. Acad. Sci. 44: 46–49.

Mattox, N. T., 1954b. A new *Eulimnadia* from the rice fields of Arkansas with a key to the American species of the genus (Conchostraca, Limnadiidae). Tulane Stud. Zool. 2: 3–10.

Mattox, N. T., 1959. Conchostraca. Chapter 26, pp. 577–586. In: W. T. Edmonson (ed.), Freshwater Biology. 2nd edn. Wiley, N.Y. 1248 p.

Moore, W. G., 1958. On the occurrence of *Streptocephalus similis* Baird in Mexico and the United States. J. Wash. Acad. Sci. 48: 169–175.

Moore, W. G., 1965. New distribution records for Conchostraca (Crustacea, Branchiopoda), with an account of their occurrence in Louisiana. Proc. La. Acad. Sci. 28: 41–44.

Moore, W. G., 1966. New world fairy shrimps of the genus *Streptocephalus* (Branchiopoda, Anostraca). S. West. Nat. 11: 24–48.

Moore, W. G. & J. B. Young, 1964. Fairy shrimps of the genus *Thamnocephalus* (Branchiopoda, Anostraca) in the United States and Mexico. S. West. Nat. 9: 68–77.

Packard, A. S., 1883. A monograph of the phyllopod Crustacea of North America, with remarks on the order Phyllocarida. 12th Annual Rept. U.S. Geol. Geogr. Survey of the Territories, Part I: 295–592.

Pennak, R. W., 1978. Freshwater Invertebrates of the United States. 2nd edn. Wiley, N.Y. XV + 803 pp.

Rioja, E., 1971. Crustaceos. In L. Cendrero (ed.), Zoologia Hispanoamericana. Invertebrados. Porrua, edit. Mexico, D.F.

Richard, M. J., 1895. Sur les crustacés phyllopodes recueillis par M. Diguet dans la Basse-Californie. Bull. Mus. Hist. nat. Paris 1: 107–108.

Saunders, J. F. & S. K. Wu, 1984. Eubranchiopoda of Colorado, Part 3. Conchostraca. Nat. Hist. Invent. Color. 8: 1–19.

Spicer, G. S., 1982. A new state record of *Streptocephalus texanus* Packard (Anostraca) from Mexico. S. West. Nat. 27: 244–245.

Spicer, G. S., 1985. A new fairy shrimp of the genus *Streptocephalus* from Mexico with a phylogenetic analysis of the North American species (Anostraca). J. Crust. Biol. 5: 168–174.

Tasch, P. & B. L. Shaffer, 1964. Conchostracans: living and fossil from Chihuahua and Sonora, Mexico. Science 143: 806–807.

*Hydrobiologia* **212**: 221–230, 1991.
D. Belk, H. J. Dumont & N. Munuswamy (eds), *Studies on Large Branchiopod Biology and Aquaculture.*



# *Branchipus cortesi* n. sp.: a new anostracan from western Spain (Crustacea, Branchiopoda)

M. Alonso & D. Jaume
*Department d'Ecologia, Facultat de Biologia, Universitat de Barcelona, Avda. Diagonal 645, 08028 Barcelona, Spain*

*Key words:* Anostraca, taxonomy, biogeography, ecology, *Branchipus*, Spain

## Abstract

*Branchipus cortesi*, n. sp. (Anostraca, Branchipodidae) is characterized by the broadly enlarged distal segments of male antenna 2, and a combination of morphological features concerning thoracic limbs, abdominal segments and egg morphology. The species occurs in temporary fresh water bodies in flatlands with temperate Mediterranean climate; it is distributed in the south-western part of Spain. Our study includes a morphological analysis using optical and scanning electronic microscopes. Questions about ecology and distribution of the new species are also discussed.

## Introduction

Mediterranean countries are rich in anostracean species (Cottarelli & Mura, 1983; Alonso, 1985; Thiéry, 1986; Nourisson & Thiéry, 1988). Although major diversification occurs apparently at generic level, accurate taxonomical studies have enabled to distinguish between closely related taxa (Thiéry, 1986; Thiéry & Champeau, 1988). Up to now, four species of *Branchipus* had been reported from Mediterranean countries: *B. blanchardi* Daday 1908 (Emend. Alonso, 1989) from the Alps, *B. visnyai* Kertész 1956 from Hungary and Italy, *B. pasai* Cottarelli 1968 from North Africa and Italy, and *B. schaefferi* Fischer 1834, the most widespread one.

## Materials and methods

The differential diagnosis compares the new species with several populations of *Branchipus schaefferi* and the type populations of *B. blanchardi*. Material examined: 1) *Branchipus cortesi* n. sp.: 50 males and 25 females from Charca Gris (type locality), Almaraz, Cáceres, Spain, February 1987; 2 males and 2 females from Gamonal (Toledo); 3 males and 5 females from Grande de la Albuera lagoon, Badajoz, April 1979; 50 males and 25 females from Coto Doñana National Park, Huelva, Spain, March 1987 (C. Montes & M. Bernues *leg.*); 2) *Branchipus schaefferi* Fischer 1834: 20 males and 10 females from La Zaida lagoon, Zaragoza, Spain, August 1984; 3 males and 5 females from Casbuso lagoon, Majorca, Spain, January 1987; 3 males and 11 female from Dubovec, South Slovakia, April 1970 (Dr. J. Brtek's collection, n° VII/a-2295); 3) *Branchipus blanchardi* Daday 1908 (Emend. Alonso, 1989): 120 males and 20 females from Cristol lake (type locality), Nevache, Hauts Alpes, France, August 1987.

Samples were taken with a 100 $\mu m$ mesh sampling net and fixed with 4% formaline. Camera lucida drawings were made with a Olympus BH-2 compound microscope provided with phase-contrast optics. A SEM was used for photographs; specimens were critical-point dried and sputter-coated with gold-palladium. Terminology follows Linder (1941) and Fryer (1983).

*Branchipus cortesi* n. sp.

*Etymology:* The 'terra typica' of this species is Extremadura, the native land of many Spanish conquerors. This species is named after Hernán Cortés.

*Type locality.* Charca Gris, Almaraz, Cáceres, Spain. It is a shallow, temporary water body with a surface area of about 100 $m^2$ surrounded by a scrub (*Cistus ladaniferus*, *Lavandula*). The bedrock is a quaternary terrace-gravel of the Tajo river. Water is poorly mineralized (60 $\mu S\ cm^{-1}$ of conductivity, 0.17 meq $l^{-1}$ of total alkalinity, 3.5 mg $l^{-1}$ of Calcium). Water colour is grey and turbidity is high, due to suspended clay particles. Macrophytic vegetation is scarce, only *Ranunculus* and *Isoetes*.

*Type series.* All the specimens collected in the Charca Gris in February 1987.

*Holotype.* A mature male 17 mm in length (not including cercopods) preserved in 4% formalin with glycerol has been deposited in the National Museum of Natural Sciences in Madrid; collection of non-insect invertebrates, catalog number 20.4.292.

*Allotype.* A mature female 15 mm in length (not including cercopods) preserved in 4% formalin with glycerol has been deposited in the National Museum of Natural Sciences in Madrid; collection of non insect invertebrates, catalog number 20.4.292.

*Paratypes.* Five mature males and five mature females preserved in 4% formalin with glycerol have been deposited in the Museum of National Sciences in Madrid; collection of non-insect invertebrates, catalog number 20.4.292. Five mature males and five mature females preserved in 4% formalin with glycerol have been deposited in the Department of Ecology, Barcelona University, catalog number E.M.A. P 009.

**Description**

*Female*

*Length to end of cercopods 8.9–18 mm.* Population of Coto Doñana National Park has the smallest specimens. Ratio of head and thorax length to abdomen and cercopods length about 0.9. Body not pigmented.

*Head* (Figs. 1a, 4a). Nuchal organ elliptic. Antenna 1 biarticulate with three long sub-terminal setae and eight aesthetascs (Fig. 1b). Antenna 2 conical with sharp end; two tufts of sensory setae, one proximal and the other subdistal (Fig. 1a). Labrum (Figs. 1c, 1d) without apical protuberances (Fig. 1d), ending in a small right angle lamella; spinous pad (SP) sharp. Maxilla 1 provided with 28–31 setae and a small posterior ventral spine (Figs. 1e, 1e′, 1e″). Maxilla 2 (Fig. 1f) with swollen proximal part provided with two soft setae and one small spinule; finger-like distal part with long plumose seta at tip.

*Thoracic limbs.* First then thoracic limbs (Fig. 2a) with single praeepipodite without deeply serrated margin and an incision at half the height. Epipodite with smooth margin. Exopodite elongated with 7–10 proximal spine-like setae on outer margin. Endopodite enlarged in distal part with several spine-like setae on proximal inner margin (Figs. 2c–2l). These setae unequal in length, the first proximal one, almost always the shorter. Endites 3,4 and 5 with 2,2,1 (endite 5 may have also 2–3) anterior short setae and 3,2,2, long posterior setae (Figs. 2c–2l). Anterior spine of proximal part of endite 2 almost twice anterior spine of endite 1, both with small, rough chitinized denticle at the base (Fig. 2a″). About 17–18 setae, each 100 $\mu m$ long, in filtering comb of endite 2.

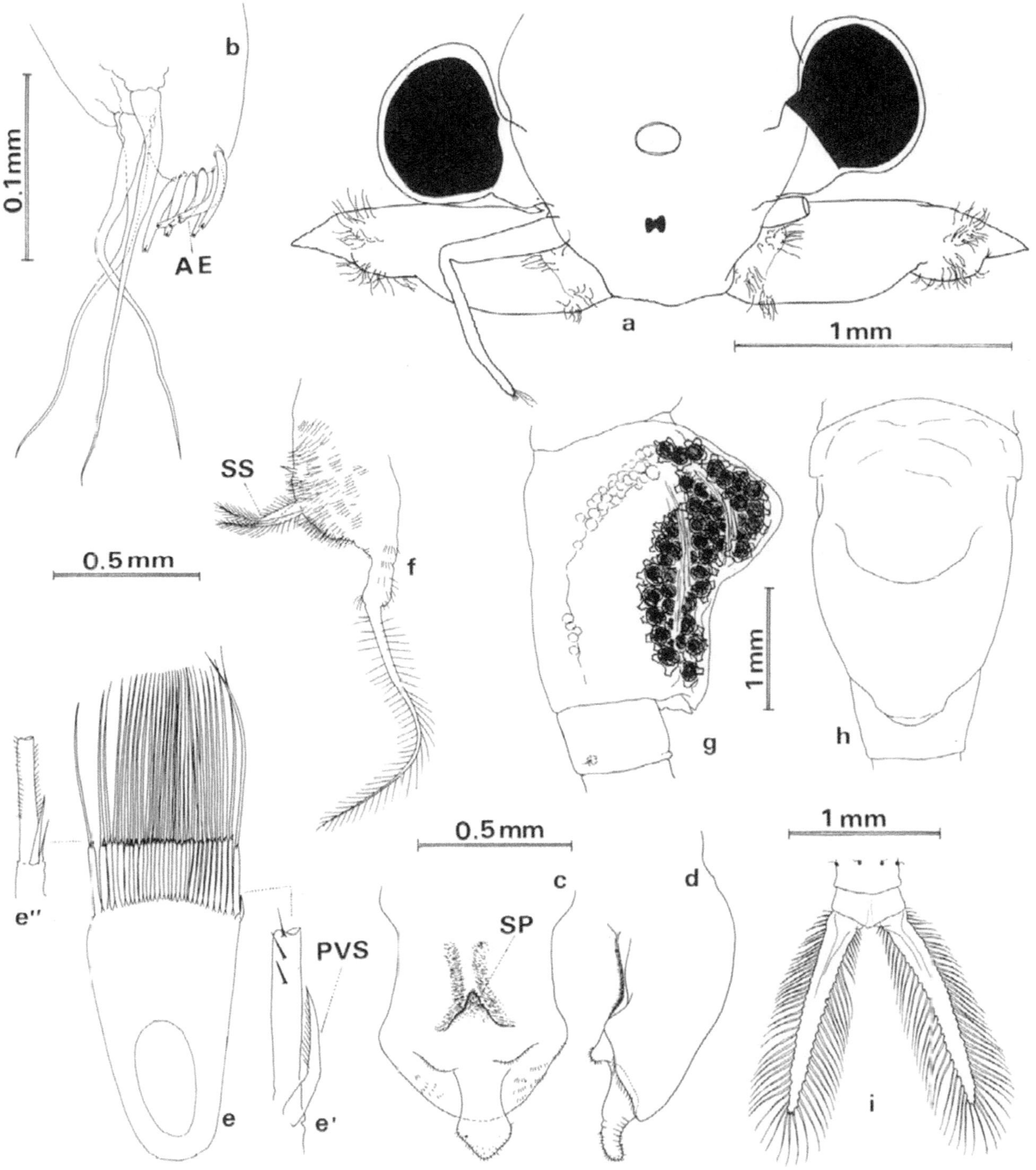


*Fig. 1.* *Branchipus cortesi* n. sp. from Charca Gris (Almaraz, Spain). Female: a, dorsal view of head; b, distal end of antenna 1 (AE = aesthetascs); c, ventral view of labrum (SP = spinous pad); d, lateral view of labrum; e, maxilla 1; e', detail of posterior ventral spine = PVS, e'', detail of seta of maxilla 1; f, maxilla 2 (SS = soft setae); g, lateral view of ovisac; h, ventral view of ovisac; i, ventral view of cercopods and telson.

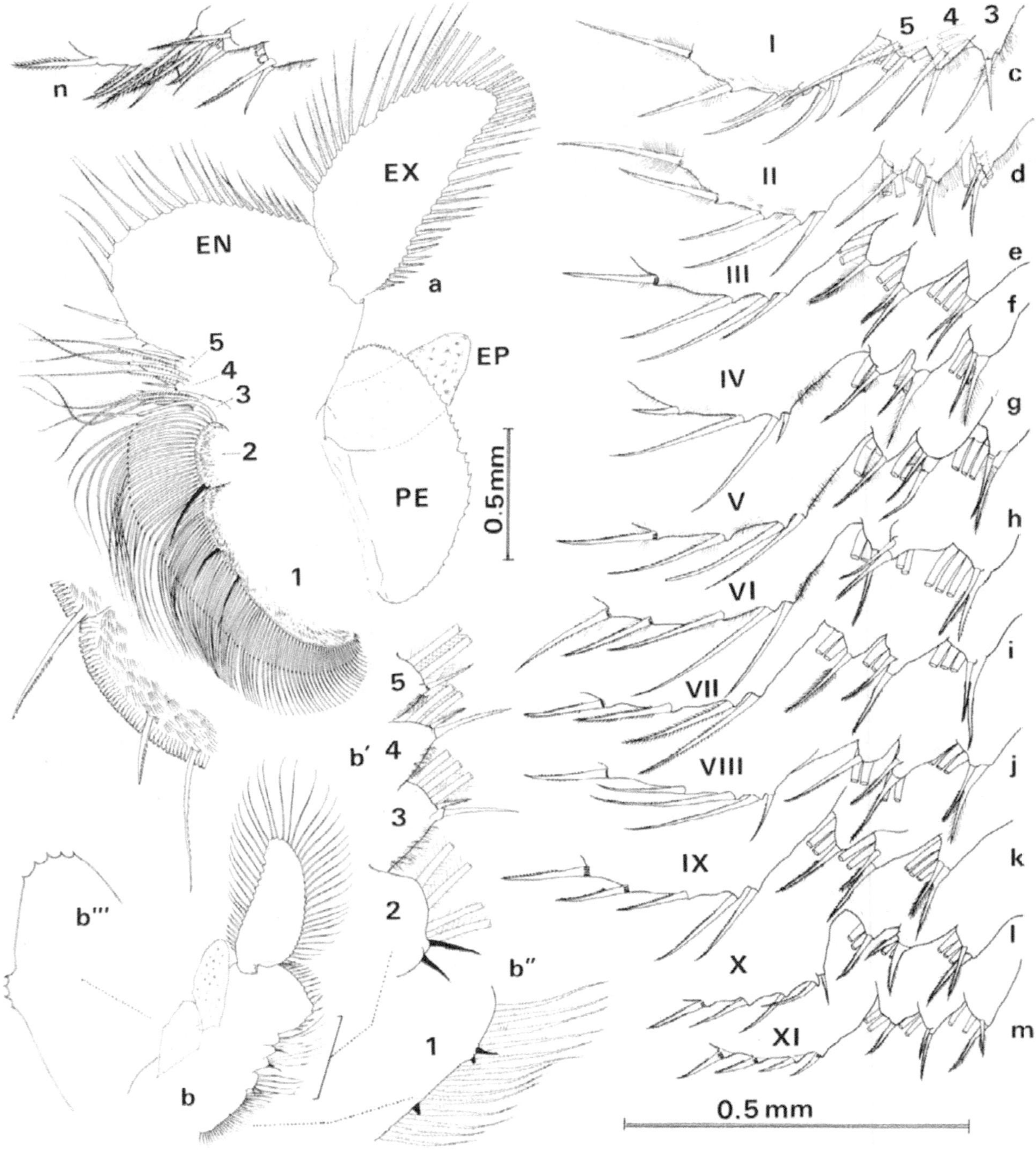


*Fig. 2. Branchipus cortesi* n. sp. from Charca Gris (Almaraz, Spain). Female: a, fifth thoracic limb (numbers correspond to different endites; EN = endopodite; EX = exopodite; EP = epipodite; PE = praeepipodite); b, eleventh thoracic limb; b′, detail of endites 2–5 of eleventh thoracic limb; b″, detail of anterior spines of endite 1 of eleventh thoracic limb; b‴, detail of praeepipodite of eleventh thoracic limb; c–m, endites 3,4 and 5, and setae of the proximal part of endopodite of thoracic limbs I to XI. Male: n, endites 3,4 and 5, and setae of proximal part of endopodite of first thoracic limb.

*Last (eleventh) thoracic limb* (Figs. 2b, 2b′, 2b″, 2b‴. Margin of praeepipodite only serrated on distal part and sometimes near the medial incision. Epipodite with margin serrated at tip. Endites 2,3,4 and 5 of about same size and globular. Anterior spine of endite 2 small with a smaller seta close to its base. Anterior spines of endite 1 substituted by small denticles.

*Ovisac* (Figs. 1g, 1h) Short and not expanded laterally; ventral-posterior elongation short, extending only to middle of thirth abdominal segment. Ventral anterior lobe as high as abdominal segment, scarcely extending beyond first abdominal segment. Number of resting eggs greater in larger specimens, as usual with anostraceans. Resting eggs (Figs. 4b, 4c) polyhedral with several facets (15) delimited by crests; approximate diameter 0.25 mm.

*Abdominal segments* smooth. Ventral sensory setae on segments 4, 6 and 8; lateral sensory setae on segments 3, 5, 7 and 8; dorsal sensory setae only on segment 8.

*Cercopods* (Fig. 1i) as long as last three abdominal segments plus telson, provided with long setae on inner and outer side.

*Male*

Length to end of cercopods 8.7–18 mm. Body not pigmented.

*Antenna* 1 as in female. *Antenna 2* (Figs. 3a–3e, 4e, 4f). Basal segments of antenna 2 completely joined medianly forming the clypeus as typical for family *Branchipodidae*. Anterior chitinous outgrowths of clypeus with one pyramidal lobe, enlarged on middle; tip of lobes scarcely exceeding distal end of clypeus. Two dorsal, soft, setulated outgrowths on each side of clypeus; proximal ones near the base of frontal appendages barely protruding; distal ones more elongated, finger-shaped. Frontal appendages consisting of two whip-shaped formations arising from dorsal, proximal, portion of clypeus as characteristic for the genus; proximal portion broad and compressed; distal portion with elliptical section and small longitudinal keel; tip roughly exceeds antenna 2 in length and appears curved ventralward. Distal segments almost as long as clypeus; they appear curved inward in dorsal and ventral view, and curved ventralward in lateral view; medial part strongly enlarged as in Fig. 3e and in the cross-sections (Figs. 3e′, 3e″); distal end with two tips, ventral one longer and sharper than dorsal one; three small areas provided with small papillae, one on ventral tip, and two other on anterior ventral lobes of the expansion.

*Thoracic limbs*. General structure as in female. Endite 5 of first thoracic limb (Fig. 2n) provided with 1–4 anterior setae.

*Genital organs*. Genital segments higher than other segments (Fig. 3g, 3h) but not significantly broader. *Vesicula seminalis* not differentiated from vas deferens. Basal part of penis joined with a medio-ventral pore-like structure (P), with unknown function. In inner side of basal part of penis a curved spine-like outgrowth is developed. Retractile part of penis clearly differentiated from basal part and somewhat longer, provided with two ventral longitudinal rows of small spines, outer row with 16–18 evenly spaced spines, inner row with 8–10 irregularly spaced spines (Fig. 3f).

*Abdominal segments* (Figs. 3g–3h). Smooth. Sensory setae distributed as in female.

*Cercopods* as long as three last abdominal segments plus telson, strongly curved inward; inner and distal lateral setae reduced in length (Fig. 3i).

*Geographic distribution and ecology*

There are four localities of *Branchipus cortesi* in Spain, situated in the Tajo (Charca Gris and Gamonal), Guadiana (Grande de Albuera lagoon) and Guadalquivir (Coto Doñana) basins. All are situated between 10 and 350 m above sea

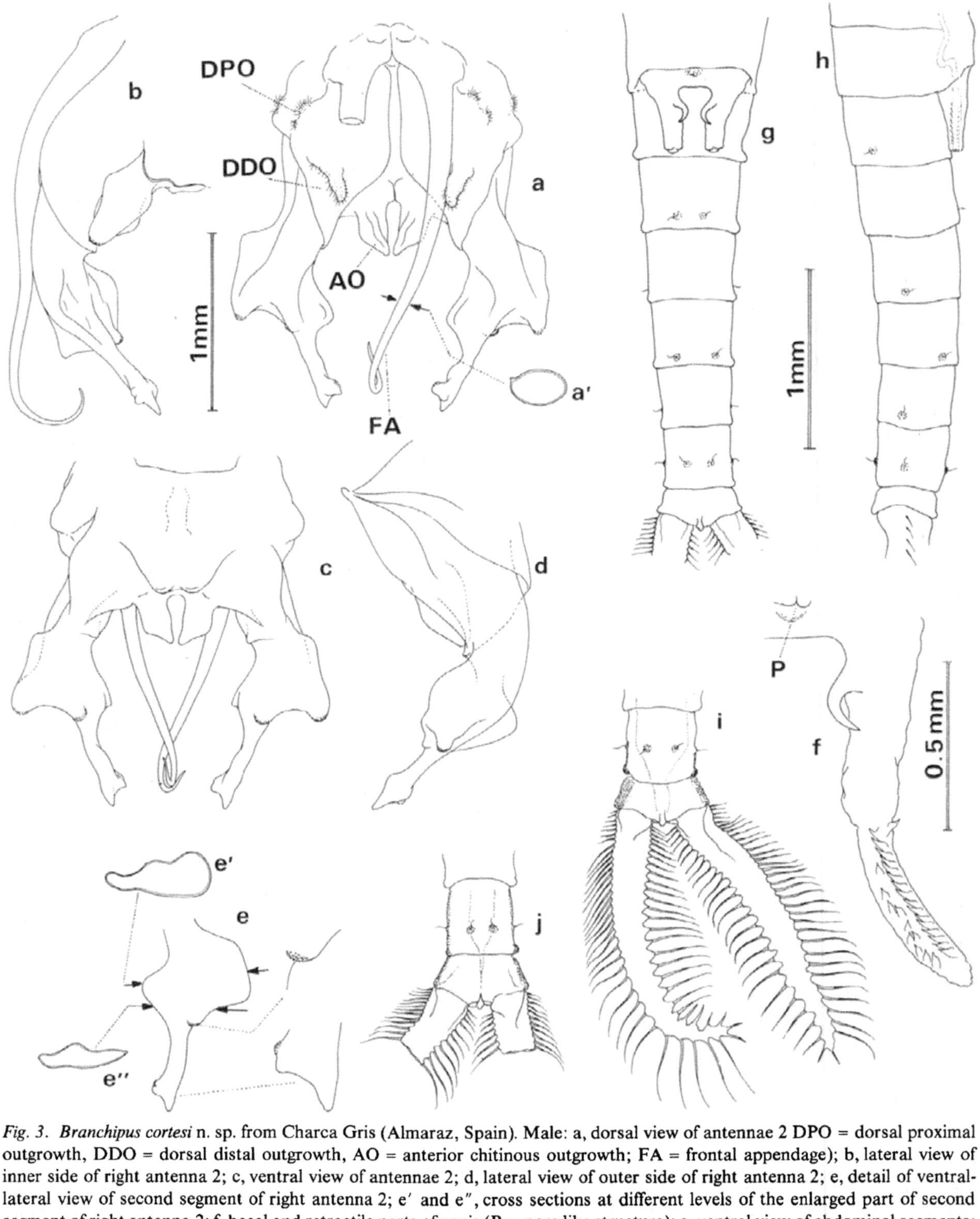


*Fig. 3. Branchipus cortesi* n. sp. from Charca Gris (Almaraz, Spain). Male: a, dorsal view of antennae 2 DPO = dorsal proximal outgrowth, DDO = dorsal distal outgrowth, AO = anterior chitinous outgrowth; FA = frontal appendage); b, lateral view of inner side of right antenna 2; c, ventral view of antennae 2; d, lateral view of outer side of right antenna 2; e, detail of ventral-lateral view of second segment of right antenna 2; e′ and e″, cross sections at different levels of the enlarged part of second segment of right antenna 2; f, basal and retractile parts of penis (P = pore like structure); g, ventral view of abdominal segments; h, lateral view of abdominal segments; i, ventral view of cercopods; j, dorsal view of telson.

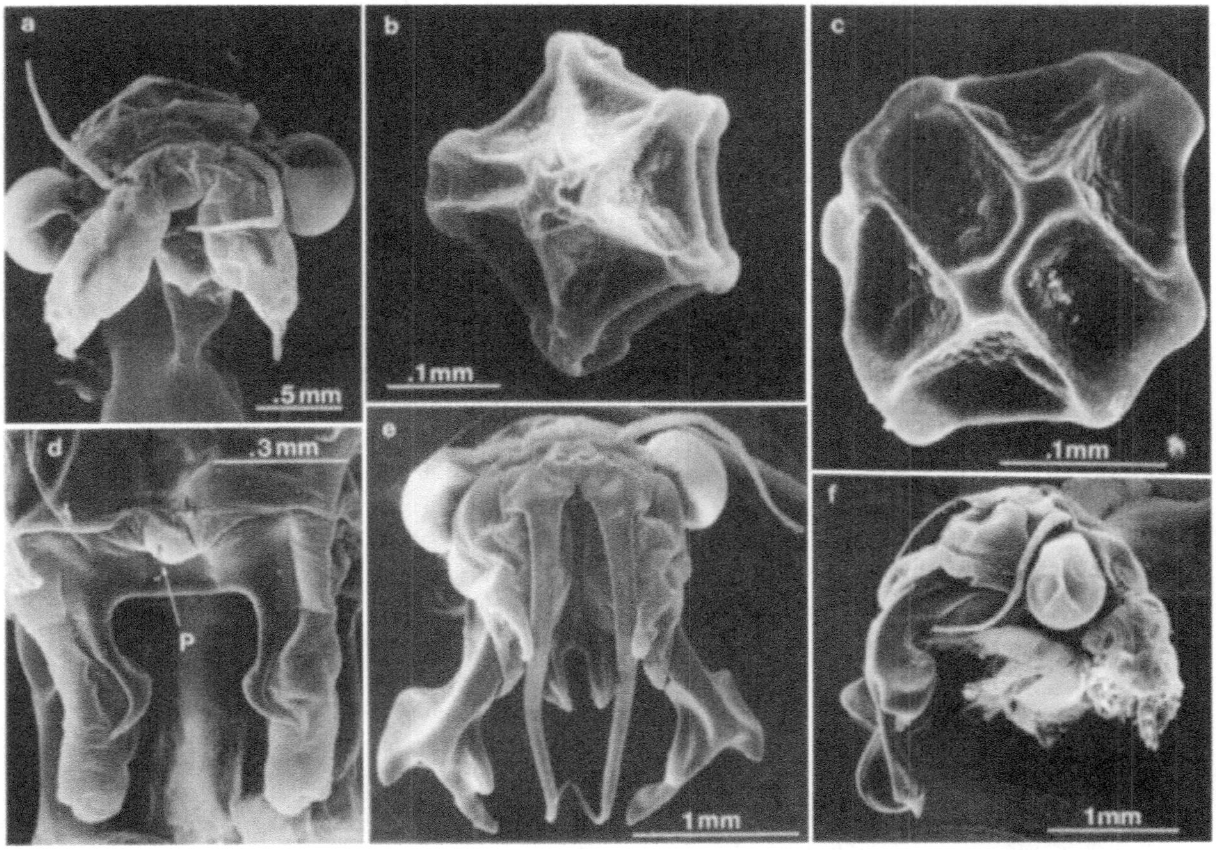


*Fig. 4. Branchipus cortesi* n. sp. from Charca Gris (Almaraz, Spain). Female: a, dorsal view of head; b and c, two different views of resting eggs. Male: d, basal parts of penis (P = pore-like structure); e, dorsal view of head; f, lateral view of head.

level in areas with rainfall below 700 mm. The water bodies are shallow (< 1 m deep) and flooded from December to May. Substrate is clayish. Water mineralization is low (Conductivity, 60–500 $\mu$S; total alkalinity 0.17–0.40 meq $l^{-1}$; Calcium, 0.04–3.5 mg $l^{-1}$). Aquatic vegetation is scarce, with as most frequent taxa: *Isoetes* (*I. setaceum*), *Marsilea*, *Callitriche* and *Ranunculus*. Populations develop at the beginning of inundation when water is turbid by suspended clay.

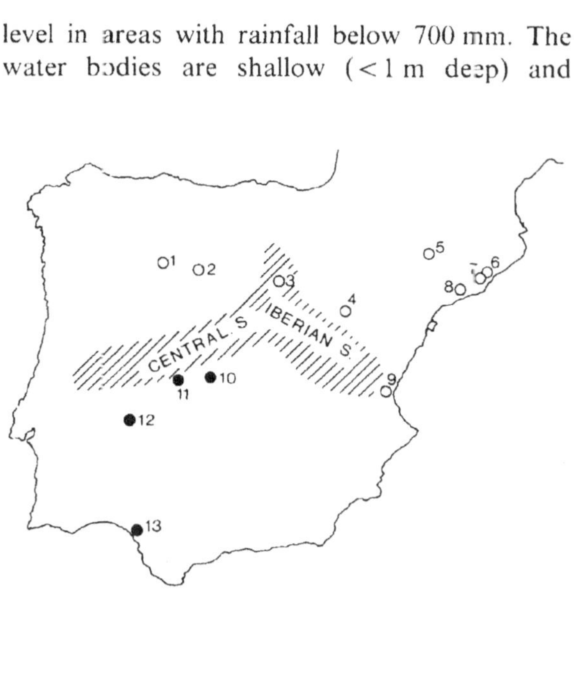


*Fig. 5.* Distribution of *Branchipus schaefferi* Fisher, 1834 and *B. cortesi* n. sp. in the Iberian Peninsula. Shaded area symbolize the Central and Iberian Systems. o *B. schaefferi*: 1, Bamba (Zamora); 2, Villanubla (Valladolid); 3, Puerto del Madero (Soria); 4, La Zaida (Zaragoza); 5, Bellver (Lleida); 6, Maçanes (Barcelona); 7, Molins de Rei (Barcelona); 8, Tarragona; 9, Valencia. • *B. cortesi*: 10, Gamonal (Toledo); 11, Almaraz (Cáceres); 12, La Albuera (Badajoz); 13, Doñana (Huelva).

## Discussion

*Branchipus cortesi* is a well defined species. At first glance it seems related with *B. blanchardi* because of the enlargement of the male second antennae; however, a detailed morphological analysis shows that *B. cortesi* is closer to *B. schaefferi* and related species (*B. pasai* and *B. visnyai*). The characteristics useful in separating these three species are summarized in Table 1.

The differences between *B. cortesi* and *B. blanchardi* (see Alonso, 1989) are the following: 1) Distal segments of the male second antennae more enlarged in *B. cortesi*, mainly on the outer margin; the cross-section along the largest part of these segments shows that the thickness is greater in the outer margin in *B. cortesi* and in the middle in *B. blanchardi*, 2) Anterior chitinous outgrowths of the clypeus with only one lobe in *B. cortesi* and two lobes in *B. blanchardi*; such outgrowths are shorter in *B. cortesi*. In the latter species they roughly exceed the distal end of the clypeus while in *B. blanchardi* they exceed the distal end of the clypeus. 3) Proximal 2–3 setae on inner margin of the endopodite of the first ten thoracic limbs long, slender and longer than the following setae in *B. cortesi*; in *B. blanchardi* such setae are short, spine-like and shorter than the following setae. 4) Basal spine (or denticle) on endite 2 of eleventh thoracic limb longer in *B. cortesi*. 5) Male genital segments lower in *B. cortesi*. 6) Male abdominal segments smooth in *B. cortesi*, while these segments have two ventral-lateral cuticular outgrowths in *B. blanchardi*.

*B. cortesi* can be differentiated from *B. schaefferi* and related species (see Nourisson & Thiéry, 1988; Cottarelli, 1968) by several characters namely: 1) Distal segments of the male sec-

*Table 1.* Major morphological differences between *B. schaefferi*, *B. blanchardi* and *B. cortesi*.

| Character | *B. schaefferi* | *B. blanchardi* | *B. cortesi* |
|---|---|---|---|
| Male second antenna | | | |
| Anterior outgrowths | With one lobe. Enlarged near base | With two lobes | With one lobe. Enlarged on middle |
| Distal segments | Not enlarged in medial part | Enlarged in medial part | More strongly enlarged on medial part |
| Frontal appendages | Tip largely exceeds antenna 2 in length | Tip roughly exceeds antenna 2 in length | Tip roughly exceeds antenna 2 in length |
| Endopodite of ten first thoracic limbs | 2–3 proximal setae on inner margin long, roughly equal in length and longer than following distal ones | Proximal setae very short, spine -like and shorter than following distal ones | 2–3 proximal setae on inner margin long. First one often shorter. Approximately of same length of following distal ones |
| Eleventh thoracic limb | Margin of epipodite and praeepipodite almost entirely serrated | Margin of epipodite and praeepipodite only serrated on distal part | Margin of epipodite only serrated on distal part. Margin of praeepipodite serrated on distal part and on medial incission |
| Ovisac | Ventral anterior lobe extends ventrally until middle of second abdominal segment | Ventral anterior lobe scarcely extends beyond first abdominal segment | Ventral anterior lobe scarcely extends beyond first abdominal segment |
| Male abdominal segments | Smooth | With two ventral-lateral cuticular outgrowths | Smooth |

ond antennae strongly wider in *B. cortesi.* 2) Anterior chitinous outgrowths of the clypeus enlarged on the middle and have a trapezoid shape; this character is shared with *B. pasai.* In *B. schaefferi* and *B. visnyai* such outgrowths are longer (their tip exceed largely the distal end of the clypeus) and cuneiform. 3) Frontal appendages shorter in *B. cortesi.* 4) First proximal seta on the inner margin of the endopodite often shorter than the following one in *B. cortesi*; in *B. schaefferi* both two setae are roughly equal in length. 5) Margin of the epipodite and praeepipodite of the eleventh thoracic limb only serrated on distal part in *B. cortesi*; such margins appear almost entirely serrated in *B. schaefferi.* 6) Ventral anterior lobe of the ovisac short in *B. cortesi*, extending only a little beyond the first abdominal segment; in *B. schaefferi* this lobe extends ventrally until the middle of the second abdominal segment.

Egg morphology is uniform within the genus *Branchipus* and comparisons between species are not conclusive. Eggs of *B. cortesi* seem to be more polyhedral than those of the other European *Branchipus*; in this species the edges are well marked and straight, defining clear facets; in the other species the eggs are more or less spherical with surface unevenly wrinkled (see Mura, 1986; Mura & Thiéry, 1986).

Only two species of *Branchipus* have been found in Spain: *B. schaefferi* and *B. cortesi.* Both have a similar autoecology but are allopatric (Fig. 5). The boundary of their distribution are the Central and Iberian Systems, a east-west ridge that separate the Duero and Ebro basins, where *B. schaefferi* occurs, from the Tajo, Guadiana and Guadalquivir basins, where *B. cortesi* has been found. As *B. schaefferi*, *B. cortesi* inhabits temporary, muddy freshwaters where it coexists with a community characteristic of these kinds of environments in the Mediterranean steppes (Gauthier, 1928; Alonso & Comelles, 1984). Its most representative species are *Triops cancriformis mauritanicus* (Ghigi, 1921), *Mixodiaptomus incrassatus* (G.O. Sars, 1903), *Metacyclops minutus*, *Hemidiaptomus roubaui* (Richard, 1888), *Moina brachiata* Jurine, 1820, *Ephemeroporus phintonicus* Margaritora, 1969 and *Chydorus pizzari* Alonso, 1988.

Biogeography of Iberian peninsula has been affected by a weak influence of Pleistocene glaciations (glaciers only developped in high mountains) and the existence of barriers (mountains and seas) which impeded north-south Pleistocene migrations (Miracle, 1982). Both factors allowed the existence of areas in which populations were successively isolated. Isolation during the Pleistocene produced differentiation in plastic taxa such as the diaptomids, which have a characteristic west-east and north-south postglacial pattern of distribution. Branchiopods are more conservative and perhaps their differentiation is older. Thus, the Iberian endemics should be interpreted as Tertiary relics rather than of Pleistocene age. The non-glaciated areas could have been refuges for species widespread in the Tertiary, when aridity was more generalized. These species could gone extinct across most of their range by the climatic fluctuations of the late Tertiary and Pleistocene (Margalef, 1947). This hypothesis is supported by the existence of wide disjunctions and endemic species. Among the former, *Branchinecta orientalis* is the most illustrative; this species occurs in one locality in Spain, is missing from most of Europe and appears again in Hungary. *Branchipus cortesi* could be an example of species with the southwestern part of Iberian Peninsula as their last refuge.

## Acknowledgements

This work has been supported by the DGICYT grant PB 87–0397 (Proyecto Fauna Ibérica).

## References

Alonso, M., 1985. A survey of the Spanish Euphyllopoda. Misc. Zool. 9: 179–208.

Alonso, M., 1989. *Branchipus blanchardi* Daday, 1908 in the Alps: redescription from type locality and synonymy with *B. alpinus* Colosi 1922 (Crustacea, Anostraca). Ann. Limnol. 25: 47–53.

Alonso, M. & M. Comelles, 1984. A preliminary grouping of the small epicontinental water bodies in Spain and distribution of Crustaceans and Charophyta. Verh. int. Ver. Limnol. 22: 1699–1703.

Cottarelli, V., 1968. Osservazioni tassonomiche su *Branchipus visnyai* Kertész e diagnosi preliminare di *Branchipus pasai* n. sp. (Crustacea, Euphyllopoda, Anostraca). Fragmenta ent. 6: 63–68.

Cottarelli, V. & G. Mura, 1983. Anostraci, Notostraci, Concostraci (Crustacea: Anostraca, Notostraca, Conchostraca). Guide per il riconoscimento delle specie animali delle acque interne italiane. 18. Consiglio Nazionale delle Ricerche. 71 pp.

Fryer, G., 1983. Functional ontogenic changes in *Branchinecta ferox* (Milne-Edwards) (Crustacea: Anostraca). Phil. Trans. r. Soc. Lond. B 303: 229–343.

Gauthier, H., 1928. Recherches sur la faune des eaux continentales de l'Algérie et de la Tunisie. Imp. Minerva, Alger, 419 pp.

Linder, F., 1941. Contributions to the morphology and the taxonomy of the Branchiopoda Anostraca. Zool. Bidr. Uppsala 20: 101–302.

Margalef, R., 1947. Estudios sobre la vida en las aguas continentales de la región endorréica manchega. P. Inst. Biol. Apl. 4: 5–51.

Miracle, M. R., 1982. Biogeography of the freshwater zooplanktonic communities of Spain. J. Biogeogr. 9: 455–467.

Mura, G., 1986. SEM morphological survey on the egg shell in the Italian Anostracans (Crustacea, Branchiopoda). Hydrobiologia 134: 273–286.

Mura, G. & A. Thiéry, 1986. Taxonomical significance of scanning electron microscopic morphology of the Euphyllopods' resting eggs from Morocco. Vie Milieu 36: 125–131.

Nourisson, M. & A. Thiéry, 1988. Crustacés Branchiopodes (Anostracés, Notostracés, Conchostracés). in: Introduction pratique à la systematique des organismes des eaux continentales francaises 9. Bull. Soc. limn. Lyon 57: 75–135.

Thiéry, A., 1986. Les Crustacés Branchiopodes (Anostraca, Notostraca et Conchostraca) du Maroc occidental. I. Inventaire et répartition. Bull. Soc. Hist. Nat. Toulouse 122: 145–155.

Thiéry, A., 1986. *Linderiella africana* n. sp., premier représentant des Linderiellidae (Branchiopoda, Anostraca) en Afrique du Nord (Maroc). Crustaceana 51: 155–162.

Thiéry, A. & A. Champeau. *Linderiella massaliensis*, new species (Anostraca: Linderiellidae), a fairy shrimp from Southeastern France, its ecology and distribution. J. Crust. Biol. 8: 70–78.

*Hydrobiologia* **212**: 231–233, 1991.
*D. Belk, H. J. Dumont & N. Munuswany (eds), Studies on Large Branchiopod Biology and Aquaculture.*



# A note on the genus *Artemia* in Algeria

A. Zemmouri
*139 Ter Bd.S. Bonakouir, 1600 Alger, Algeria*

*Key words:* Algeria, Artemia, distribution, predation, salinity

## Abstract

Five new locality records for *Artemia* in Algeria are given. Some physical and chemical characteristics of their environment are recorded. Predation on *Artemia* by the fish *Aphanius fasciatus* occurs where the river Khrouf debouches in Chott Merouan but not elsewhere.

## Introduction

*Artemia*, known as doud ('worm') in certain regions of Southern Algeria and Libya (Monod, 1969), was repeatedly reported from Algeria (Blanchard & Richard, 1890; Blanchard, 1891; Gurney, 1909; Seurat, 1922; Gauthier, 1928; Beadle, 1943; Dumont, 1979). Among the sites where it was recorded are several smallest waterbodies (Gauthier, 1928; Gauthier-Lièvre, 1931) which are no longer in existence.

Nauplii of *Artemia* are an important life feed in the aquaculture of fish and crustaceans; adults are considered a luxury food on account of their cost (Sorgeloos, 1980). Faced with a growing demand, and until aquaculture, methods are sufficiently advanced to permit unlimited production, a search for unrecorded natural populations is needed. The object of the present paper is to bring on record several new *Artemia* biotopes in Algeria.

## The sites (Fig. 1A)

Inland saline basins in Algeria differ according to their topography, and have received the names Mellaha, Sebkha and Chott. A Mellaha (from the arab melh = salt) is a salt works, i.e. a saline soil into which artificial ponds are dug for salt extraction purposes. A sebkha is a closed basin with an impermeable clay bottom. Few are in excess of 25 km$^2$ of surface area; they are rainfed, and dry out during the hot season. A chott is a large, marshy depression, fed from various sources. *Artemia* has been recorded in each of these three types of waterbodies (Table 1).

In the particular case of Mellaha Guergour el-Amri, composed of a series of crystallizing ponds on a Sodium chloride-rich soil, into which water is pumped from a well as of April, we could ascertain that *Artemia* first appeared here in 1983. This corresponds to a time when large quantities of salt from Chott Merouan were temporarily brought here for packing. Direct inoculation of *Artemia* cysts seems to have occurred during this process, since Mellaha Guergour is not frequented by migrating birds.

## Population densities

Average population densities across time were estimated at one or two (Chott Merouan – Fig. 1b) stations. Three depth levels (surface, center, bottom) were averaged. It is seen that den-

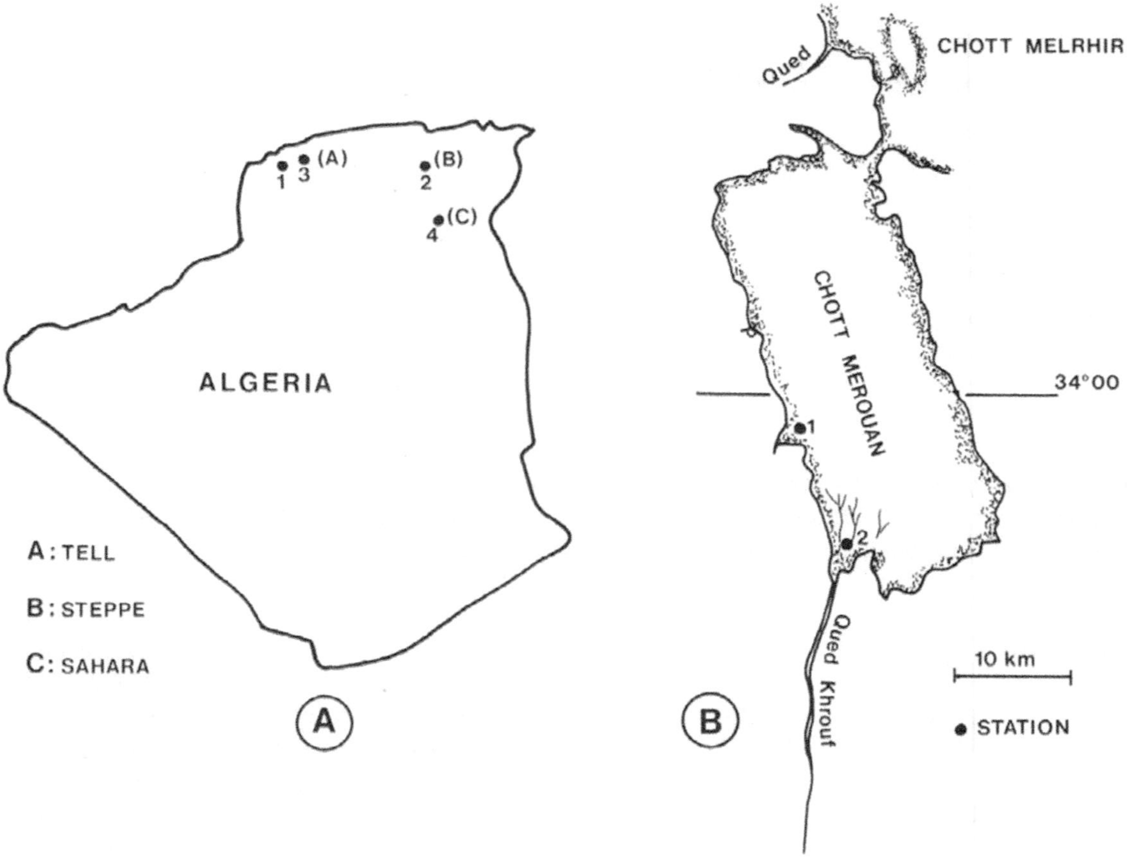


sities differ greatly between sites, and are particularly low in the presence of aquatic predators (see further).

**Temperature variation**

Because of their shallowness, we did not look for differences between the surface and bottom of our biotopes. Temperatures, while *Artemia* is present, varied widely in time and between sites. In Sebkha, 13 °C was recorded in February, 27 °C in May, almost identical as in Sebkha Sidi Bouzian (14° and 27 °C; respectively). In Sebkha N'zouri, only 6 °C was found in February, but 30 °C in June. In Chott Merouan differences between stations were also observed, and the general amplitude was from 9 °C to 35 °C, with diurnal fluctuations of 5 °C in summer.

*Table 1.* New sites of *Artemia* in Algeria.

| N° | Type | Name | Coordinates | Surface area | Altitude | Aquatic predation |
|---|---|---|---|---|---|---|
| 1 | Sebkha | Azrew | 35° 43′ N; 00° 08′ W | 25 km$^2$ | <30 m | – |
| 2 | Sebkha | N'zouri | 35° 50′ N; 06° 35′ E | 5 km$^2$ | 900 m | – |
| 3 | Sebkha | Sidi Bouzian | 35° 52′ N; 00° 34′ E | 12 km$^2$ | <100 m | – |
| 4 | Chott | Merouan | 34° 00′ N; 06° 10′ E | >250 km$^2$ | –35 m | + |
| 5 | Mellaha | Guergour El-Amri | 35° 59′ N; 05° 15′ E | <1 km$^2$ | >850 m | – |

*Table 2. Artemia* phenology and estimation of density.

| N° site | Phenology of Artemia | Depth cm | Average density | Remarks |
|---|---|---|---|---|
| 1 | November–May | 80 | 76 ind $L^{-1}$ | |
| 2 | February–May | 63 | 0.1 ind $L^{-1}$ | |
| 3 | November–May | 90 | 28 ind $L^{-1}$ | |
| 4, st 1. | October–March | 55 | 15 ind $L^{-1}$ | |
| st 2. | October–March | 135 | 0 ind $L^{-1}$ (rare) | Predators present |
| 5 | April–August | 70 | Not estimated | |

## Salinity

The Sebkhas Azrew and Sidi Bouzian varied in parallel, and reached a maximum salinity in May of 160‰ and 170‰, respectively. For Chott Merouan, we obtained a maximum salinity values 250‰. Its salinity is mainly due to Sodium chloride, with some admixture of sulfate. The three sebkhas have an almost identical chemical composition, which is as follows:

| | |
|---|---|
| NaCl | 99.19% |
| $MgCl_2$ | 0.349% |
| $CaCl_2$ | 0.039% |
| $CaSO_4$ | 0.375% |

particulate matter: 0.035%

## Biological aspects

*Artemia* usually monopolizes the biotopes where it lives. A notable exception was station 2 at the mouth of Chott Merouan, where numerous rotifers and copepods co-occurred with it. Here the river Khrouf's water enters the Chott, and thus salinity is much lower than inside the Chott.

At this same station, some aquatic vertebrates were present as well, and particularly the salt-tolerant Cyprinodont fish *Aphanius fasciatus*. The feeble density of *Artemia* at this station, as compared to station 1 and other sites as well, is ascribed to predation by this fish. That *Artemia* survived in the Chott is probably due to the higher salinity in other parts of it, to which the *Artemia* still has access, but the fish has not. It may, finally, be added that all *Artemia* discussed here were bisexual, and that maximum size recorded was 8 mm.

## Acknowledgement

I thank Prof. H. Dumont for guidance and help with this manuscript.

## References

Beadle, L. C., 1943. An ecological survey of some inland saline waters of Algeria. J. linn. Soc. 41: 218–242.

Blanchard, R., 1891. Resultats d'une excursion zoologique en Algérie. Mem. Soc. zool. Fr. 4: 208–245.

Blanchard, R. & J. Richard, 1890. Sur les crustacés de Sebhkas et de Chotts d'Algérie. Bull. Soc. Zool. Fr. 15: 136–138.

Dumont, H. J., 1979. Limnologie van Sahara en Sahel. Unpublished Thesis, RU Gent, 557 pp.

Gauthier, H., 1928a. Recherches sur la Faune des eaux continentales de l'Algérie et de la Tunisie. Minerva, Alger, 419 pp.

Gauthier, H., 1928b. Description du *Tanymastix mozabica*, nouveau phyllopode anostracé du Sahara et liste des anostracés récoltés par l'auteur en Algérie et en Tunisie. Bull. Soc. Hist. nat. Afr. N. 19: 87–91.

Gauthier-Lièvre, L., 1931. Recherches sur la flore des eaux continentales de l'Algérie et de la Tunisie. Minerva, Algér, 299 pp.

Gurney, R., 1909. On the Fresh water crustacea of Algeria and Tunisia. J. r. microsc. Soc. 1909: 273–305.

Monod, T., 1969. A propos du Lac des Vers ou Bahr-el-Düd (Libye). Bull IFAN, A 31: 25–41.

Seurat, L. G., 1922. Faune des eaux continentales de la Berbérie. Alger, publication de l'Université, Fac Sci., Lab. Zool. appl., 66 pp.

Sorgeloos, P., 1980. The use of brine shrimp *Artemia* in aquaculture. In: The brine shrimp *Artemia*. vol. 3, Ecology, culturing, use in aquaculture. Universa, Wetteren, Belgium, 3: 25–46.

Zemmouri, A., 1986. L'élevage larvaire d'*Artemia*. Mém. Fin d'étude. Inst. Sci. mer Amménag. Littoral, 1986.

*Hydrobiologia* **212**: 235–240, 1991.
*D. Belk, H. J. Dumont & N. Munuswamy (eds), Studies on Large Branchiopod Biology and Aquaculture.*



# The geographical distribution of *Triops australiensis* (Crustacea: Notostraca) in Australia: a biogeoclimatic analysis

W. D. Williams[1] & J. R. Busby[2]
[1] *Department of Zoology, University of Adelaide, G.P.O. Box 498, Adelaide, S.A. 5001, Australia;*
[2] *Australian Biological Resources Study, G.P.O. Box 1383, Canberra, A.C.T. 2601, Australia; present address: ERIN Unit, Australian National Parks and Wildlife Service, G.P.O. Box 1383, Canberra, ACT 2601, Australia*

## Abstract

Until relatively recently many species of entomostracan crustaceans were thought to have widespread, even cosmopolitan distributions. Evidence now suggests that this is far less the case than thought. However, an exception appears to be provided by the crustaceans of episodically filled waterbodies. Typically these include Anostraca, Notostraca and Conchostraca. The paper considers the distribution of *Triops australiensis* (Branchiopoda: Notostraca) in Australia. There, it is a frequent inhabitant of episodically filled temporary waters. It is absent from such waters in northern Australia. A biogeoclimatic analysis using the program BIOCLIM indicates that this is because of the absence of suitable localities. BIOCLIM enables the preparation of maps which provide statistical predictions of climates suitable for the survival of a particular taxon. These predictions are based upon the known climate of localities where the taxon in question actually occurs. It is suggested that for the biota of shallow ephemeral waters (such as those typically inhabited by Anostraca, Conchostraca and Notostraca) biogeoclimatic analysis has a powerful predictive value.

## Introduction

Until relatively recently, many species of entomostracan crustaceans were believed to be widely distributed, and even cosmopolitan. Refutation of this belief began slowly, but has gathered momentum in recent years. A seminal catalyst in this process has been the paper by Frey (1982). Even in localities where the occurrence of widely spread species could be expected *a priori*, viz. temporary waters, there is increasing evidence that selection has been for local adaptation leading to restricted geographical distributions (Williams, 1988). Indeed, many features previously thought to be adaptations *enabling* dispersal may well be adaptations *hindering* it, e.g. hooks on cladoceran ephippia.

These ideas are generally supported by recent studies on Australian temporary fresh and saline waters (Williams, 1985, 1988). There is, however, one sort of environment which, it has been suggested (Williams, 1984; Williams & Kokkinn, 1988), provides an exception, and which is populated by widely distributed and easily dispersed species. The exception is episodically filled temporary bodies of water. Here, it is suggested, the irregularity of filling precludes the evolution of local adaptations and thus speciation; consequently their biota does indeed comprise widely distributed and easily dispersed forms. Support for this view is provided by a recent study of the fauna of Lake Eyre, a large salt lake in central Australia which has filled only three times this century (1949–52, 1974–76, 1984–85). This study

(Williams & Kokkinn, 1988) strongly suggested that only widely distributed forms were present.

On the same basis, the fauna of episodically filled fresh waters in the interior of the Australian continent should be populated by widely distributed and easily dispersed species. Unfortunately, we know very little about the fauna of such localities. There is, however, one species which is frequently collected: *Triops australiensis*. The relatively large size and distinctive appearance of this animal has long attracted attention and numerous collections of it exist in museums. Williams (1968) analysed these and other collections and was able – with some confidence – to map the geographical distribution of the species. He was unsure, however, if the presumed absence of *T. australiensis* from northern Australia reflected unsuitable conditions there or simply the absence of collections. In the intervening two decades, many collections have been made from temporary bodies of fresh water in northern Australia (especially by M.J. Tyler and M. Davies in connection with their study of northern Australian Amphibia: > 150). None of these collections has ever contained *T. australiensis*. It may be reasonably concluded, therefore, that the species is indeed absent from northern Australia.

It is now of interest to consider if this absence is due to the absence of suitable localities. Clearly, if it is not, the absence is inconsistent with the hypothesis that the fauna of episodically filled waters comprises widely distributed and easily dispersed species. But how can the occurrence of suitable conditions be determined in remote areas? It is suggested that, at least so far as ephemeral waters are concerned (and for these more than for any other sort of aquatic habitat), it can be predicted by the analysis of certain climatic parameters. Clearly, the absence or presence of such waters reflect rainfall patterns, and longevity and many important physical features reflect other climatic features. To that end, a biogeoclimatic analysis of the known distribution of *T. australiensis* was undertaken utilizing 16 climatic parameters. The analysis was used to predict where suitable conditions occurred on the basis of conditions where the species is known to occur. This paper reports the results of the analysis and discusses them.

## Methods

Use was made of the computer programme BIOCLIM (Bioclimatic Prediction System) developed by one of us (JRB) in collaboration with H.A. Nix (Busby, 1986a). This programme, using distributional data for *T. australiensis*, produced climate profiles for this species. These climate profiles were then used to indicate areas in Australia that were apparently climatically suitable for *T. australiensis*.

All distributional data in Williams (1968) together with some additional data were assigned geographical coordinates (latitude and longitude) and an elevation (total 101 localities). Climate data were derived from continent-wide surfaces of monthly mean minimum and maximum temperatures and 19 regional surfaces for monthly precipitation, these surfaces themselves being derived from Laplacian smoothing spline functions fitted to data measured at meteorological stations. The surfaces were used to obtain estimates of climate values as functions of latitude, longitude and elevation for any part of the continent. From derived climate values, 16 selected parameters for each data point were obtained, namely annual mean temperature, minimum temperature of the coolest month, maximum temperature of the warmest month, annual temperature range, mean temperature of the coolest quarter, mean temperature of the warmest quarter, mean temperature of the wettest quarter, mean temperature of the driest quarter, annual mean precipitation, precipitation of the wettest month, precipitation of the driest month, coefficient of variation of monthly precipitation, precipitation of the wettest quarter, precipitation of the driest quarter, precipitation of the coolest quarter and precipitation of the warmest quarter.

The climate profile of *T. australiensis* was obtained from these parameters, and comprised minimum, 5, 25, 75 and 95 percentile and maximum values for each parameter. Further details of the method are given by Busby (1986a, b) and Nix (1986).

## Results

The climate profile of *T. australiensis* is indicated in Table 1. The predicted distribution of the species based on the similarity of climates at points on a 0.5-degree latitude–longitude grid to the climate profile is plotted in Fig. 1. Here, points failing to match all 16 climate parameters within the total range of the profile of *T. australiensis* are not plotted, points matching all 16 parameters within the total range are plotted as ' – ', points matching all 16 parameters within the 90-percentile (5 to 95 percent) range as ' + ', points matching 13–15 of the parameters within the interquartile range (25 to 75 percent) range as '0', and points matching all 16 parameters within the interquartile range as open squares. Where the climate matches acceptance criteria, the localities have apparently suitable climates for *T. australiensis*, and the closer the match (i.e. from ' – ' through to open squares), the more suitable the climate. Localities where *T. australiensis* has been collected are shown as solid circles. Thus Fig. 1 provides a statistical prediction of where *T. australiensis* should (and does) occur. Climatic parameters which appear to be the most significant in the sense that more points on the 0.5-degree latitude–longitude grid fail to match the species' profile with these parameters than with others are: precipitation of the coolest quarter, followed by mean temperature of the coolest quarter and the mean minimum temperature of the coolest month.

## Discussion

The results of the biogeoclimatic analysis suggest that *T. australiensis* is not absent from northern Australia because it is unable to reach suitable environments there; it is absent because no suitable environments for this species apparently exist in northern Australia. The geographical distribution of *T. australiensis* in Australia, therefore, is consistent with the hypothesis that the fauna of episodically filled aquatic localities is widely distributed and its components easily dispersed.

The absence of *T. australiensis* from northern

*Table 1.* Climate profile for *Triops australiensis.*

| Climatic parameter | Min. Value | Percentile | | | | Max. Value |
|---|---|---|---|---|---|---|
| | | 5 | 25 | 75 | 95 | |
| Annual mean temp. (°C) | 13.3 | 15.5 | 18.0 | 21.9 | 26.0 | 27.2 |
| Min. temp. coolest month (°C) | – 1.1 | 2.4 | 3.8 | 6.0 | 12.1 | 13.0 |
| Max. temp. warmest month (°C) | 25.6 | 28.5 | 33.2 | 37.5 | 39.0 | 40.4 |
| Annual temp. range (°C) | 20.1 | 23.8 | 28.3 | 31.8 | 33.0 | 33.8 |
| Mean temp. coolest quarter (°C) | 6.7 | 9.0 | 11.2 | 13.7 | 19.6 | 22.1 |
| Mean temp. warmest quarter (°C) | 18.9 | 21.6 | 24.9 | 28.9 | 31.3 | 32.3 |
| Mean temp. wettest quarter (°C) | 9.0 | 11.0 | 18.8 | 28.3 | 31.1 | 31.9 |
| Mean temp. driest quarter (°C) | 8.1 | 10.7 | 13.6 | 21.1 | 23.4 | 26.8 |
| Annual mean precipitation (mm) | 139 | 152 | 224 | 385 | 863 | 3203 |
| Precipitation wettest month (mm) | 18 | 20 | 28 | 61 | 124 | 674 |
| Precipitation driest month (mm) | 1 | 2 | 7 | 18 | 37 | 72 |
| Coeff. variation monthly precip. (%) | 8.0 | 10.8 | 23.3 | 53.8 | 98.0 | 119.5 |
| Precipitation wettest quarter (mm) | 49 | 52 | 76 | 165 | 339 | 1761 |
| Precipitation driest quarter (mm) | 7 | 8 | 25 | 61 | 122 | 223 |
| Precipitation coolest quarter (mm) | 16 | 27 | 41 | 79 | 175 | 372 |
| Precipitation warmest quarter (mm) | 34 | 48 | 60 | 146 | 287 | 893 |

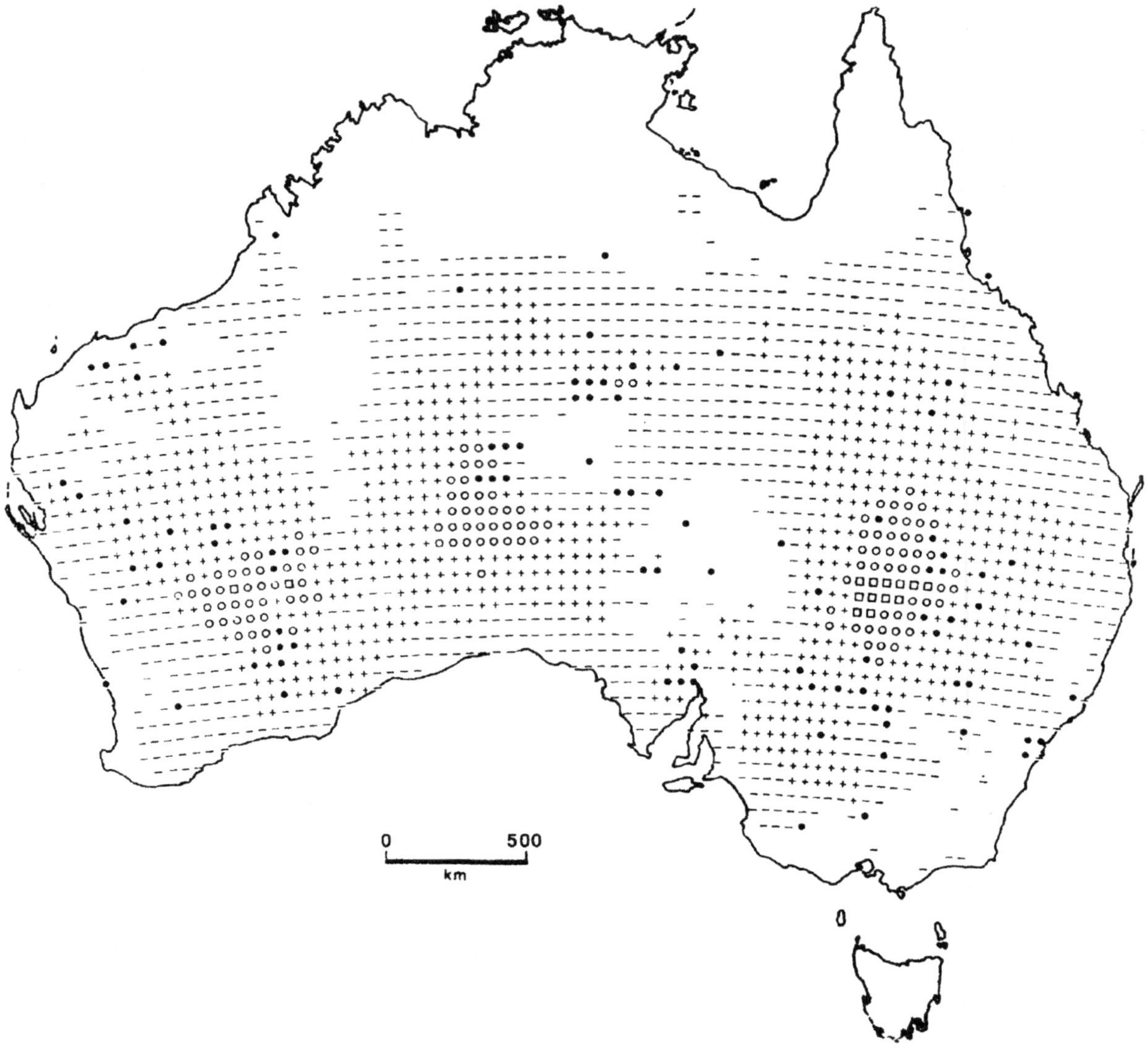


*Fig. 1.* Predicted and actual distribution of *T. australiensis* in Australia. For explanation of symbols see text.

Australia does not, of course, provide any *direct* evidence concerning its ability to disperse, and it is conceivable that, within the area of suitable environments, it does not disperse far. We consider this unlikely, but in this connection note that in a recent and careful study of the functional morphology of the Notostraca, Fryer (1988) found morphological differences of a sort which led him to propose that 'such differences suggest that widely distributed taxa of the Notostraca are perhaps broken up into at least more races than has been supposed, or may even consist of several cryptic species'.

It may be that such races or cryptic species of *Triops* do occur in ephemeral waters which are predictably filled, but so far as populations of *Triops* inhabiting *episodically* filled waters of inland Australia are concerned, it remains difficult to see how speciation could occur. It seems much more likely to us that a single widely distributed and easily dispersed species of *Triops* occurs in such waters. In any event, over most if

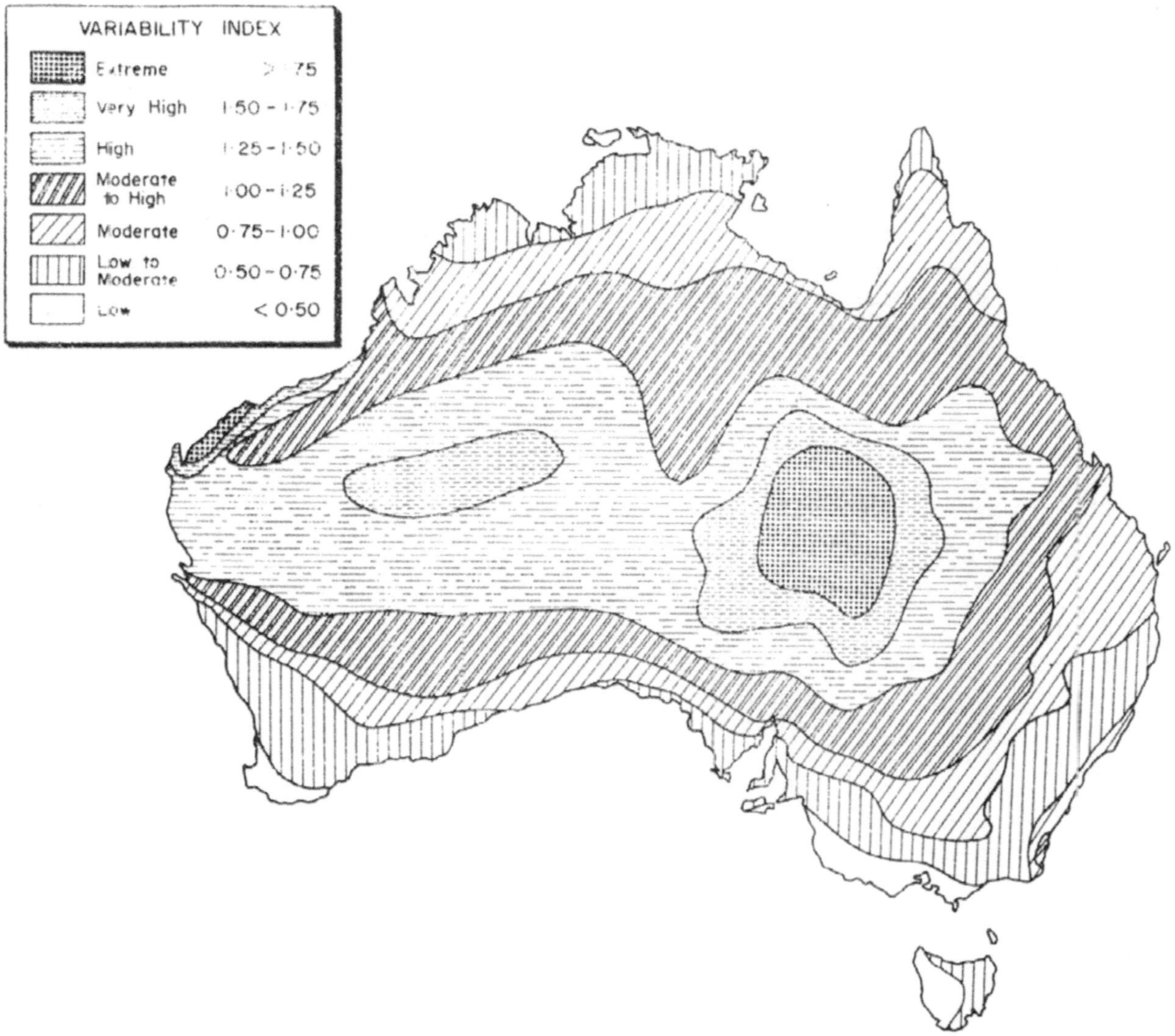


*Fig. 2.* Annual rainfall variability in Australia, as indicated by the variability index (see text). Redrawn and modified from Gaffney (1975).

not all of the range of distribution of *T. australiensis*, the climate displays considerable variability from year to year. Such climatic variability can be indicated in various ways, but one way of doing so is by using the so-called *variability index* (V.I.) based on 90, 50 and 10 percentiles, viz. V.I. = (90 – 10)/50. Using this index, Fig. 2 plots the variability of annual rainfall in Australia, a climate parameter of obvious importance in the present connection. Comparison of Figs. 1 and 2 clearly shows that over most of the range of *T. australiensis* the variability of annual rainfall is from moderate to extreme.

Only close taxonomic examination of *T. australiensis*, involving scanning electron microscopy and electrophoresis as suggested by Fryer (1988), can unequivocally determine whether more than one species or several races of *Triops* occur in Australia. However, we believe that for material from episodically filled waters such an examination will prove negative.

## Acknowledgements

We are grateful to A/Prof. M. J. Tyler and Dr M. Davies, University of Adelaide, for placing at our disposal numerous collections of material from temporary bodies of water in northern Australia. Dr Geoffrey Fryer FRS, Windermere, was kind enough to comment on a draft manuscript. This paper was prepared when WDW was supported by the Australian Water Research Advisory Council. That support is acknowledged.


## References


Busby, J. R., 1986a. Bioclimate Prediction System (BIOCLIM). User's Manual. Version 2. Bureau of Flora and Fauna, Canberra.

Busby, J. R., 1986b. A biogeoclimatic analysis of *Nothofagus cunninghamii* (Hook.) Oerst. in southeastern Australia. Aust. J. Ecol. 11: 1–7.

Frey, D. G., 1982. Questions concerning cosmopolitanism in Cladocera. Arch. Hydrobiol. 93: 484–502.

Fryer, G., 1988. Studies on the functional morphology and biology of the Notostraca (Crustacea: Branchiopoda). Phil. Trans; Roy. Soc. B, 321: 27–124.

Gaffney, D. O., 1975. Rainfall deficiency and evaporation in relation to drought in Australia. Paper to ANZAAS Conference, Canberra, 1975: 17 pp.

Nix, H. A., 1986. A biogeographic analysis of Australian elapid snakes. In R. Longmore (ed.) Atlas of Elapid Snakes of Australia. AGPS, Canberra: 4–15.

Williams, W. D., 1968. The distribution of *Triops* and *Lepidurus* (Branchiopoda) in Australia. Crustaceana 14: 119–126.

Williams, W. D., 1984. Chemical and biological features of salt lakes on the Eyre Peninsula, South Australia, and an explanation of regional differences in the fauna of Australian salt lakes. Verh. int. Ver. Limnol. 22: 1208–1215.

Williams, W. D., 1985. Biotic adaptations in temporary lentic waters, with special reference to those in semi-arid and arid regions. Hydrobiologia 125: 85–110.

Williams, W. D., 1988. Limnological imbalances: An antipodean viewpoint. Freshwat. Biol. 20: 407–420.

Williams, W. D. & M. J. Kokkinn, 1988. The biogeographical affinities of the fauna in episodically filled salt lakes: A study of Lake Eyre South, Australia. Hydrobiologia 158: 227–236.

*Hydrobiologia* **212**: 241–243, 1991.
D. Belk, H. J. Dumont & N. Munuswamy (eds), *Studies on Large Branchiopod Biology and Aquaculture.*



# Branchiopods (non-cladocerans) of the Maltese Islands (central Mediterranean)

Sandro Lanfranco[1], Christine De Walsche[2], Patrick Schembri[1] & Johan Mertens[2]
[1]*Department of Biology, University of Malta, Msida, Malta*; [2]*Institute of Animal Ecology, State University of Ghent, K.L. Ledeganckstraat 35, B-9000 Ghent, Belgium*

*Key words:* Maltese Islands, Anostraca, Notostraca, Conchostraca, temporary pools

## Abstract

Four branchiopod species occur on the Maltese Islands, viz. *Branchipus schaefferi, B. visnyai, Cyzicus tetracerus* and *Triops cancriformis.* All four live in freshwater pools, but it is noteworthy that beside *B. schaefferi, T. cancriformis* was recorded from a mesohaline, coast-fringing habitat.

## Introduction

The Maltese Archipelago is comprised of the central mediterranean islands Malta and Gozo, together covering an area of 315 km$^2$. The islands are composed entirely of marine deposits, mainly limestone and marls. The seasonal distribution of rainfall defines two ecological seasons: a wet season (October to March) during which c. 70% of total annual rainfall falls (yearly average 530 mm) and a dry season (April to September).

The only data on Euphyllopods from the Maltese Islands are those of Gulia (1873), who recorded *Limnadia melotensis*, without describing it, and of Baldacchino (1983), who recorded *B. stagnalis* (= *B. schaefferi*) and *Eocyzicus orientalis* (= *C. tetracerus*). The information included here comes from collections made from the seventies to date.

Euyphyllopods live in two types of habitats on Malta: freshwater rainpools in inland depressions, and pools close to the sea. Both habitats dry out rapidly with the onset of the dry season. It also often happens that during a single wet season pools undergo several cycles of wetting and drying out, depending on the frequency of precipitation and the size of the pools.

Besides an occasional survey since the 1970ies, Branchiopods were studied in more detail in three groups of pools. The San Pawl tat-Targa pool is an inland freshwater rainpool. Oligo- and mesohaline pools are found at Ghadira-s-Safra, situated less than a metre above sea-level approximately 15 m inward of the narrow mediolittoral zone, and at Il-Qaliet, 6 km S.E. of Ghadira-s-Safra and also at 15 m from the shoreline. The flora at both sites includes numerous obligate halophytes. *Crypsis aculeata* (Poaceae), an endemic species, is restricted to this locality and the Amphipod *Orchestia mediterranea*, another indicator of the saline conditions in and around the pools, is abundant. For an overview of all localities where Branchiopods have been found, see Map 1.

## Materials and methods

Animals were collected using a standard handnet, preserved in 4% formaline and identified using

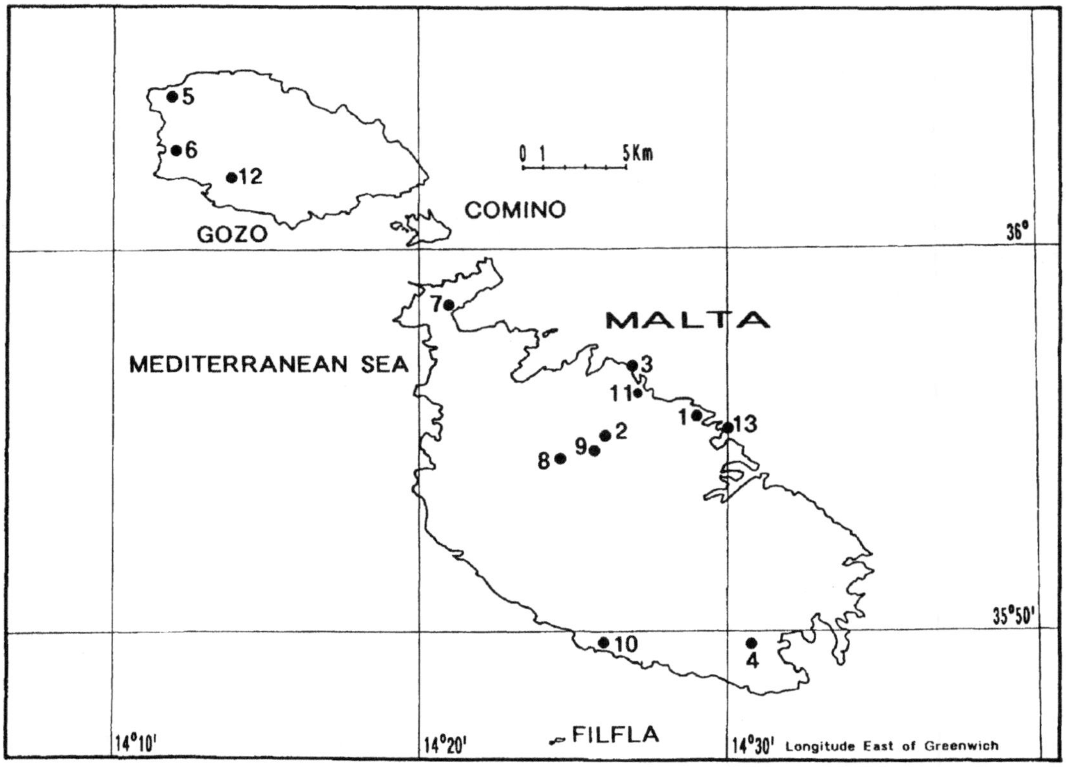


*Map 1.* numbers: localities from which Branchiopods have been recorded. 1 St. Andrew's – 2 San pawl tat-Targa – 3 Ghadira-s-Safra – 4 Birzebbuga – 5 Wied ir-Raheb – 6 Dwejra – 7 Ghadira (Mellieha Bay) – 8 Wied il-Ghasel (Mosta) – 8 Tal-Wej (Mosta) – 10 Ghar Lapsi – 11 Maghtab – 12 Xlendi Valley (Munxar) – 13 Il Qaliet.
localities for *Branchipus schaefferi*: 1, 2, 3, 4, 5, 6, 10, 11, 12, 13.
localities for *B. visnyai*: 8.
localities for *Cyzicus tetracerus*: 2, 7, 8.
localities for *Triops cancriformis*: 2, 3, 5, 9, 13.

the available literature. Electrical conductivity was measured at Ghadira-s-Safra and Il Qaliet, using a Hanna Instruments HI 8633 Portable Conductivity meter.

## Results

The Anostraca on the island belong to two species. The most widespread is *B. schaefferi*, which was found on Malta and Gozo in rainwater as well as in brackish waterpools. The conductivity data of the brackish pools are summarized in Table 1. *B. visnyai* was restricted to temporary rainwater pools on the island of Gozo. One species of Conchostraca (*C. tetracerus*) and one species of Notostraca (*T. cancriformis*) were found. *C. tetracerus* occurs on Malta and Gozo in rainwater pools, often syntopic with *B. schaefferi*. *T. cancriformis* has, so far, only been found on the island of Malta where it occurs in rainwater pools and in brackish-water pools, frequently syntopic with *B. schaefferi*.

*Table 1.* Electrical conductivity ($\mu S . cm^{-1}$) of pools at Ghadira-s-Safra (G.S.) and at Il-Qaliet (I.Q.).

| Date | G.S. | I.Q. |
|---|---|---|
| 08.12.89 | 1.350 | 24.370 |
| 12.01.90 | 700 | 17.270 |
| 16.01.90 | 620 | 18.420 |

## Discussion

*B. visnyai*, new to the Maltese Archipelago, is so far known only from a single locality in Gozo, a rainwater pool in Wied-ir-Raheb where it is sympatric with *B. schaefferi*. *B. schaefferi* is a euryhaline species, living in two types of habitat (fresh and saline pools), frequently together with *T. cancriformis* and *C. tetracerus*. *C. tetracerus* is common in rainwater pools. According to the literature, *T. cancriformis* is usually found in little mineralized waters, like in Spain (Alonso, 1985), Italy (Cottarelli *et al.*, 1983), Algeria and Morocco (Gauthier, 1934). Only Flössner (1972) gives records from brackish pools in Scotland. On Malta, *T. cancriformis* is found in saline marshland as well as in freshwater pools.

All Maltese species are known from Italy and Algeria. On the other hand, the absence of some species is remarkable. Thus, *Streptocephalus torvicornis*, widespread in Spain, in Northern Africa (Dumont *et al.*, 1990) and in Yugoslavia (Petrov *et al.*, 1990) is missing on the Maltese Islands as well as in Italy (Cottarelli *et al.*, 1983). The genus *Chirocephalus*, conversely, which desplays a notable diversification on the Italian Peninsula (Cottarelli *et al.*, 1983) but less so in North Africa, has not been recorded on the Maltese Islands either.

## References


Alonso, M., 1985. A survey of the spanish Euphyllopoda. Misc. Zool. 9: 179–208.

Baldacchino, A. E., 1983. A preliminary list of freshwater crusteans from the Maltese Islands. Central medit. Nat. 1: 49–50.

Cottarelli, V. & G. Mura, 1983. Guida per il riconoscimento delle specie animali delle acque interne ilaliane: Anostraci, Notostraci, Concostraci. Consiglio Nazionale delle Richerche (AQ.), 1: 194 pp.

Dumont, H. J., C. De Walsche & J. Mertens, 1991. Distribution and morphological variation in *Streptocephalus torvicornis* (Waga, 1842) in Northern Africa. In Belk, D., H. J. Dumont & N. Munuswany (eds), Studies on Large Branchiopod Biology and Aquaculture. Developments in Hydrobiology 64. Kluwer Academic Publishers, Dordrecht: 203–208. Reprinted from Hydrobiologia 212.

Flössner, D., 1972. Krebstiere, Crustacea. Kiemen- und Blattfüsser, Branchiopoda, Fischläuse, Branchiura. Die Tierwelt Deutschlands 60: 502 pp.

Gauthier, H., 1934. Contribution à l'étude de l'*Apus cancriformis* et de ses variations dans l'Afrique du Nord (Euphyllopode Notostracé). Bull. Soc. Sci. nat. Maroc. 14: 1–15.

Gulia, G., 1873. Fauna maltese: indice dei crostacei. il Barth, 15/16: 314–315.

Petrov, B. & M. Marinček, 1991. On the Anostraca (Crustacea) of Yugoslavia. In Belk, D., H. J. Dumont & N. Munuswamy (eds), Studies on Large Branchiopod Biology and Aquaculture. Developments in Hydrobiology 64. Kluwer Academic Publishers, Dordrecht: 267–272. Reprinted from Hydrobiologia 212.

*Hydrobiologia* **212**: 245–259, 1991.
*D. Belk, H. J. Dumont & N. Munuswamy (eds), Studies on Large Branchiopod Biology and Aquaculture.*



# Resting eggs of Anostraca, Notostraca and Spinicaudata (Crustacea, Branchiopoda) occurring in France: identification and taxonomical value

Alain Thiéry & Charles Gasc
*Laboratoire de Biologie animale-Hydrobiologie, Faculté des Sciences, 33, rue Louis Pasteur, Université d'Avignon, F-84000 Avignon, France*

*Key words:* Anostraca, Notostraca, Spinicaudata, resting eggs, external morphology, oötaxonomy

## Abstract

The external egg morphology of the phyllopod species recorded in France is described by means of S.E.M. Differences in diameter and ornamentation of envelope are stressed. We propose a dichotomous key to allow the identification of species even when their biotope is dry. Results are compared with previous studies on Anostraca, Notostraca and Spinicaudata and their taxonomical value is discussed.

## Introduction

In a guide for the Anostraca, Notostraca and Spinicaudata of France, Nourisson & Thiéry (1988) listed 15 species, identified using characteristics of the antennae, legs, penes, and broodpouches. In this paper we evaluate the external envelope of resting eggs (or cysts), as these are often sculptured in species-specific ways. For a review of the formation and structure of the tertiary shell, see Linder (1960), Garreau de Loubresse (1974), Gilchrist (1978), Trentini & Sabelli Scanabissi (1982), Thiéry (1985), Tommasini *et al.* (1989), and Tommasini & Sabelli (1989).

During the last decade, a number of authors discussed cyst morphology. Most of them studied surface patterns of described species, or used morphologic characters of the eggs in the diagnosis of new species (Mura *et al.*, 1978; Munuswamy & Subramoniam, 1983; Alonso & Alcaraz, 1984; Cottarelli & Mura, 1984; Munuswamy *et al.*, 1985; Samyiah *et al.*, 1985; Mura, 1986; Mura & Thiéry, 1986; Thiéry, 1987; Thiéry & Champeau, 1988; Belk, 1989; Martin, 1989; Brendonck *et al.*, 1990; De Walsche *et al.*, 1991, this volume). In all cases, this new way of studying phyllopods has proved to be of great interest.

Besides examining the specificity of resting egg morphology, the present study aims at developing an easy way of cyst identification.

This oötaxonomical study of the French species complements previously published diagnose work of Nourisson & Thiéry (1988). It also adds three new records of Anostraca, thus increasing the known phyllopod species in France to 18.

## Material and methods

a) Source of animals in Table 1. For S.E.M. studies, adult females were captured from their natural environment and cysts were removed from their broodpouches. Such a method provides clean eggs. However, when populations were unavailable during dry periods of their habitat or were still immature, eggs were sampled from surface sediments (noted in Table 1). The collec-

*Table 1.*

| | |
|---|---|
| 1 *Tanymastix stagnalis* (L., 1758) | rock-pool, Fontainebleau Forest + N.M.N.H. Paris, no reference: dehydrated eggs. |
| 2 *Tanymastix sp.* (aff.) *stellae* Cottarelli, | temporary pond, North of Bonifacio, Southern Corsica: dehydrated eggs. |
| 3 *Siphonophanes grubei* (Dybowski, 1860) | N.M.N.H. Paris, n° Bp 161 et 260. |
| 4 *Chirocephalus spinicaudatus* Simon, 1886 | N.M.N.H. Paris, n° 157. |
| 5 *Chirocephalus diaphanus* Prévost, 1803 | – temporary ditch, Chevilly, Orléans Forest.<br>– temporary pond at Saint Maximin, Provence. |
| 6 *Chirocephalus salinus* Daday, 1913 | "padoules" North of Bonifacio, Southern Corsica. |
| 7 *Artemia sp* (aff.) *tunisiana* | bisexual, salt-marshes of Sète-Villeroy. |
| 8 *Artemia sp.* (parthenogenetic strains) | salt-marshes of Aigues-Mortes and Fangassier pond, Camargue. |
| 9 *Branchinella spinosa* (M. Edw., 1840) | Fangassier pond, Camargue: eggs laid but adults absent. |
| 10 *Linderiella massalliensis* Thiéry & Champeau, 1988 | Locus typicus, temporary pond, Saint-Maximin, Provence. |
| 11 *Branchipus schäefferi* Fischer, 1834 | temporary pond, North of Opoul, Corbières. |
| 12 *Branchipus blanchardi* Daday, 1910 | Locus typicus, Briançon, High Alps: dehydrated eggs. |
| 13 *Triops cancriformis canciformis* (Bosc, 1801) | Opoul temporary pond, Corbières. |
| 14 *Lepidurus apus* (L., 1758) | Saint Pierre Le Moutier temporary pond, Allier. |
| 15 ——— | temporary ditch at Rochefort du Gard, Gard. |
| 16 ——— | karstic temporary lake near Les Rives, Larzac Plateau: dehydrated eggs. |
| 17 *Eoleptestheria ticinensis* (Balsamo-Crivelli, 1859) | Le Sambuc ricefield, Camargue. |
| 18 *Cyzicus tetracerus* (Krynicki, 1830) | Gavoti flooded vineyard, Provence. |
| 19 *Imnadia yeyetta* Hertzog, 1935 | "La Galère" temporary pond, Crau, Eastern Camargue: dehydrated eggs. |
| 20 *Limnadia lenticularis* (L., 1761) | N.M.N.H. Paris, no reference. |

tion locations of the various species are presented in Fig. 1 and Table 1.

A few species could not be collected from natural biotopes. For the anostracans *Siphonophanes grubei* and *Chirocephalus spinicaudatus*, and the Spinicaudata *Limnadia lenticularis*, cysts were taken from females in the collections of the National Museum of Natural History at Paris. The museum reference numbers are given in Table 1.

b) Fixation for S.E.M. observation. Cysts were kept in formaldehyde 10%, and then fixed in a bath of K.A.A.D. (30% Kerosene, 60% absolute ethanol, 5% acetic acid, 5% dioxane) for 24 hours. This procedure favorably preserves the natural shape of the tissues (Gasc *et al.*, 1984). The use of K.A.A.D. is particularly necessary for Spinicaudata (Cyzicidae and Leptestheriidae) eggs which have a thin tertiary envelope that some authors as Tommasini & Scanabissi Sabelli (1989) consider as a primary envelope, called also vitelline envelope, surrounded by a mucus envelope.

For Limnadiidae eggs, the situation is different. Belk (1989) does not consider the shell of

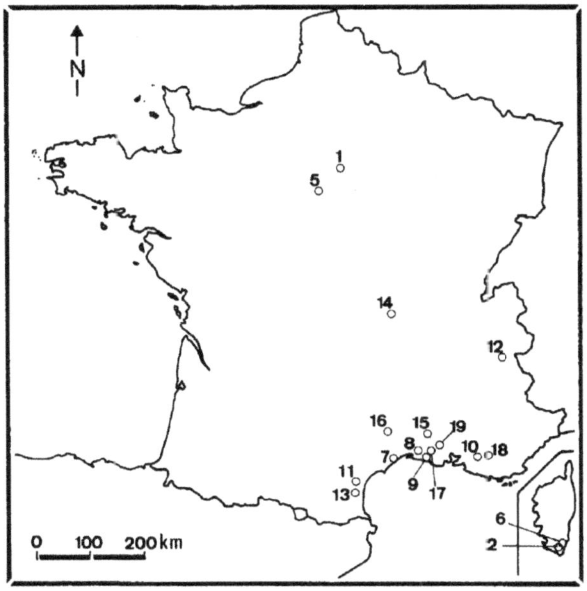


*Fig. 1.* Map showing the collecting sites of the samples analyzed in this study (circles). Numbers correspond to the list of locations given in Table 1 (numbers 3, 4 and 20 are lacking because they correspond to specimens from the National Museum of Natural History in Paris).

*Eulimnadia antlei* to be subject to deformation, so none of these eggs were critical-point dried. On his side, Bishop (1968) considers the eggs of all Limnadiidae and particularly *Limnadia stanleyana* to be resistant to deformation as those of the Anostraca. Despite preliminary critical-point drying, vacuum treatment induces deformations that alter the morphology of the eggs surface and their size. Figures 25, 26 and 27 allow a comparison of egg morphology in *Eoleptestheria ticinensis* without treatment by K.A.A.D. and critical-point-drying, without K.A.A.D. but with critical-point drying, and finally with K.A.A.D. and critical-point drying.

In general, eggs were dehydrated by the $CO_2$ critical point drying technique, coated with gold, and then observed using a 360 Cambridge. Stereoscan.

c) Representativeness of measurements. As noted by Gilchrist (1978) in *Artemia* and by Mura & Thiéry (1986) in *Tanymastigites jbiletica*, *T. brteki*, *Branchipus schaefferi*, and *Branchinecta ferox*, dehydratation of anostracan eggs during dry periods in the habitat, does not have a significant impact on their morphology. In our study, however, a variation in diameter of notostracan eggs could be observed. The variation between eggs which were protected from dehydratation, and eggs collected from dry sediments, reached 8–9% of the maximum diameter for *Lepidurus apus* (Rochefort du Gard): on May 12th, 1988 (flooded biotope, eggs protected from dehydration) they has a diameter of $518 \pm 2.47$ μm ($n = 42$); on August 27th, 1988 (biotope dry, eggs collected from dry sediments) the diameter was $474 \pm 4.07$ μm ($n = 46$). We suspect that this difference results from a compression of the alveolar layer of the tertiary envelope. No comparable data are available for conchostracan eggs.

We noted that the diameter of anostracan, notostracan, and conchostracan eggs are not related to the size of the mature female. This is in agreement with observations of Belk (1977) and Mura & Thiéry (1986). We did observe that larger adults tend to lay more eggs in each clutch than smaller adults (Thiéry, 1987).

All figures show whole eggs of anostracans and conchostracans presented to the same scale, to visualize their difference in diameter. The eggs of notostracans are more reduced, because of their bigger size.

## Results

### *a) Diameter and morphology of eggs*

Important differences in the diameter of eggs between orders can be observed, prompting us to establish 3 categories: (1) eggs with a diameter of more than 400 μm for Notostraca, (2) diameter between 220 and 380 μm for Anostraca, (3) eggs below 200 μm for Spinicaudata (Table 2).

Anostracan eggs fall in two main groups: lenticular cysts, and spherical or subspherical cysts. Their further differentiation is based on the ornamentation of the tertiary envelope (spines, ridges, ribs, smooth surface, etc.), and, when ornamentation is similar, on diameter. The

*Table 2.*

| Key | Species | Diameter (± standard error) |
|---|---|---|
| **Egg > 380 μm NOTOSTRACA** | | |
| Smooth egg with no distinct ornamentation — diameter — 430-520 μm | *Lepidurus apus* | 518.1 ± 2.5 μm (*n* = 82) |
| Smooth egg with no distinct ornamentation — diameter — 360-400 μm | *Triops cancriformis cancriformis* | 368-372 μm |
| **200 μm < egg < 380 μm ANOSTRACA** | | |
| lenticular — thin equatorial callus | *Tanymastix stagnalis* | 335-348 ± 3 μm (*n* = 50) |
| lenticular — thick equatorial callus | *Tanymastix aff. stellae* | 375-385 μm |
| spherical or subspherical — spherical — surface — verrucous | *Chirocephalus spinicaudatus* | 280-300 μm |
| spherical or subspherical — spherical — surface — smooth | *Artemia* — bisexual | 210-230 μm |
| | *Artemia* — partenogenetical | 220-270 μm |
| spherical or subspherical — spherical — ornamented surface — spiny | *Linderiella massaliensis* | 240-260 μm |
| spherical or subspherical — spherical — ornamented surface — without spines — with ripples | *Branchinella spinosa* | 260.1 ± 6.1 μm (*n* = 7) |
| spherical or subspherical — spherical — ornamented surface — without spines — with ridges — low | *Siphonophanes grubei* | 300 μm |
| spherical or subspherical — spherical — ornamented surface — without spines — with ridges — protruding | *Chirocephalus diaphanus* | 360-420 ± 5.1 μm (*n* = 9) |
| spherical or subspherical — spherical — ornamented surface — without spines — with ridges — protruding | *Chirocephalus salinus* | 341-360 ± 5.5 μm (*n* = 1) |
| spherical or subspherical — angular — irregular | *Branchipus schaefferi* | 246 ± 6.2 μm (*n* = 34) |
| spherical or subspherical — angular — regular with convergent ridges | *Branchipus blanchardi* | 300 ± 8.0 μm (*n* = 58) |
| **Egg < 200 μm SPINICAUDATA** | | |
| smooth and spherical — diameter range: 125-130 μm | *Eoleptestheria ticinensis* | 127 ± 1.2 μm (*n* = 42) |
| smooth and spherical — diameter range: 145-150 μm | *Cyzicus tetracerus* | 144 ± 2.9 μm (*n* = 42) |
| non-spherical — spiral | *Imnadia yeyetta* | 185/182 μm |
| non-spherical — biconvex | *Limnadia lenticularis* | 200/208 μm |

dichotomous key, thus derived (Table 2) allows for an easy and rapid identification of different species.

*Notostraca*

*Triops* Schrank, 1803 and *Lepidurus* Leach, 1816 (Fig. 2, 3, 4, 5, 6)

Eggs spherical and smooth. The surface is finely riddled with pores. Only the diameter can be used to distinguish between eggs of *Triops cancriformis* and *Lepidurus apus* (Table 2). The diameter of *Lepidurus apus* eggs does not vary in a given population, regardless of the age of the female, but shows slight differences between populations. Eggs of northern populations (Saint Pierre le Moutier) are 10–15% bigger than eggs of meridional populations (Gavoti, Rochefort du Gard). Such a difference had already been observed by Thiéry & Mura (in preparation), in North Africa between cysts of *Triops cancriformis mauritanicus* from a High Atlas population, and cysts of a lowland population.

*Anostraca*

Lenticular eggs

*Tanymastix* Simon 1886
(Fig. 17, 18, 19, 20, 29, 30, 33, 34, 35, 36, 37)

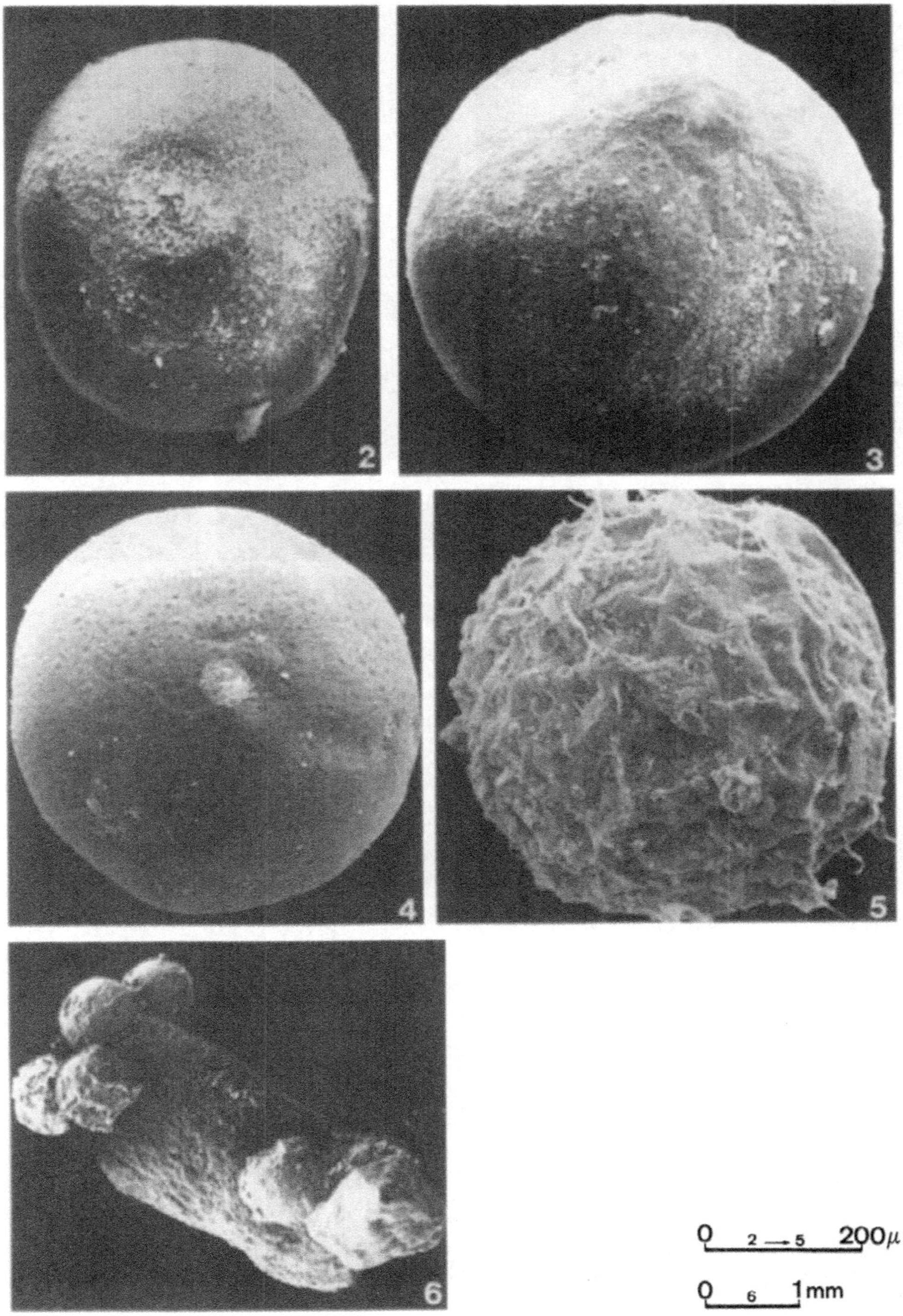


*Fig. 2–6.* Eggs of Notostraca. 2. – ***Triops cancriformis cancriformis*** from Opoul; 3. – ***Lepidurus apus*** from St Pierre le Moutier; 4. – ***Lepidurus apus*** from Rochefort du Gard; 5. – ***Lepidurus apus*** dry egg from the Larzac plateau; 6. – ***Triops cancriformis*** stuck on a gravel.

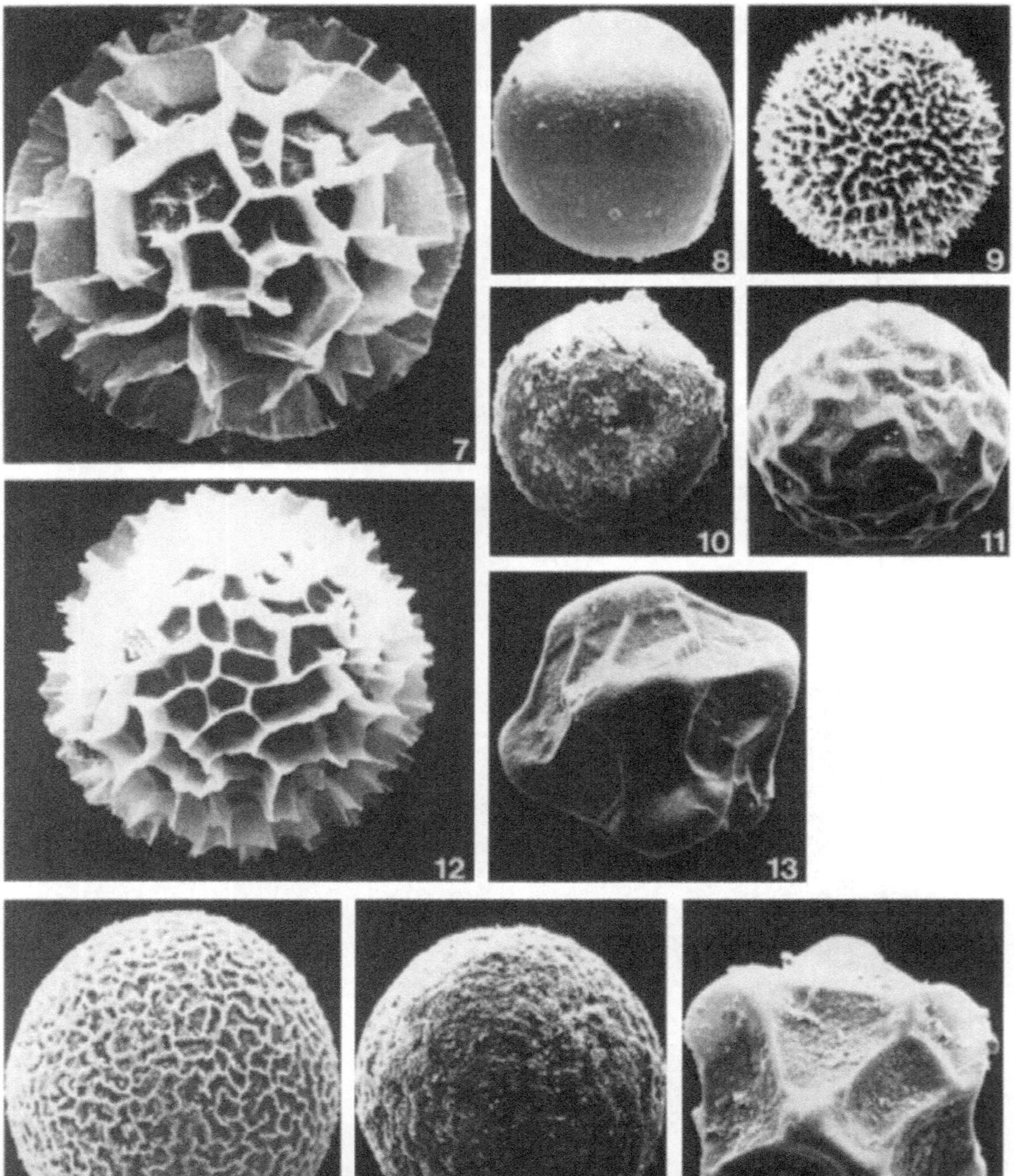


*Fig. 7–16.* Eggs of Anostraca. 7. – *Chirocephalus diaphanus*; 8. – a parthenogenetic *Artemia* (Camargue); 9. – *Linderiella massaliensis*; 10. – *Artemia* (aff.) *tunisiana* (Sète-Villeroy); 11. – *Branchinella spinosa*; 12. – *Chirocephalus salinus*; 13. – *Branchipus schaefferi*; 14. – *Siphonophanes grubei*; 15. – *Chirocephalus spinicaudatus*; 16. – *Branchipus blanchardi*.

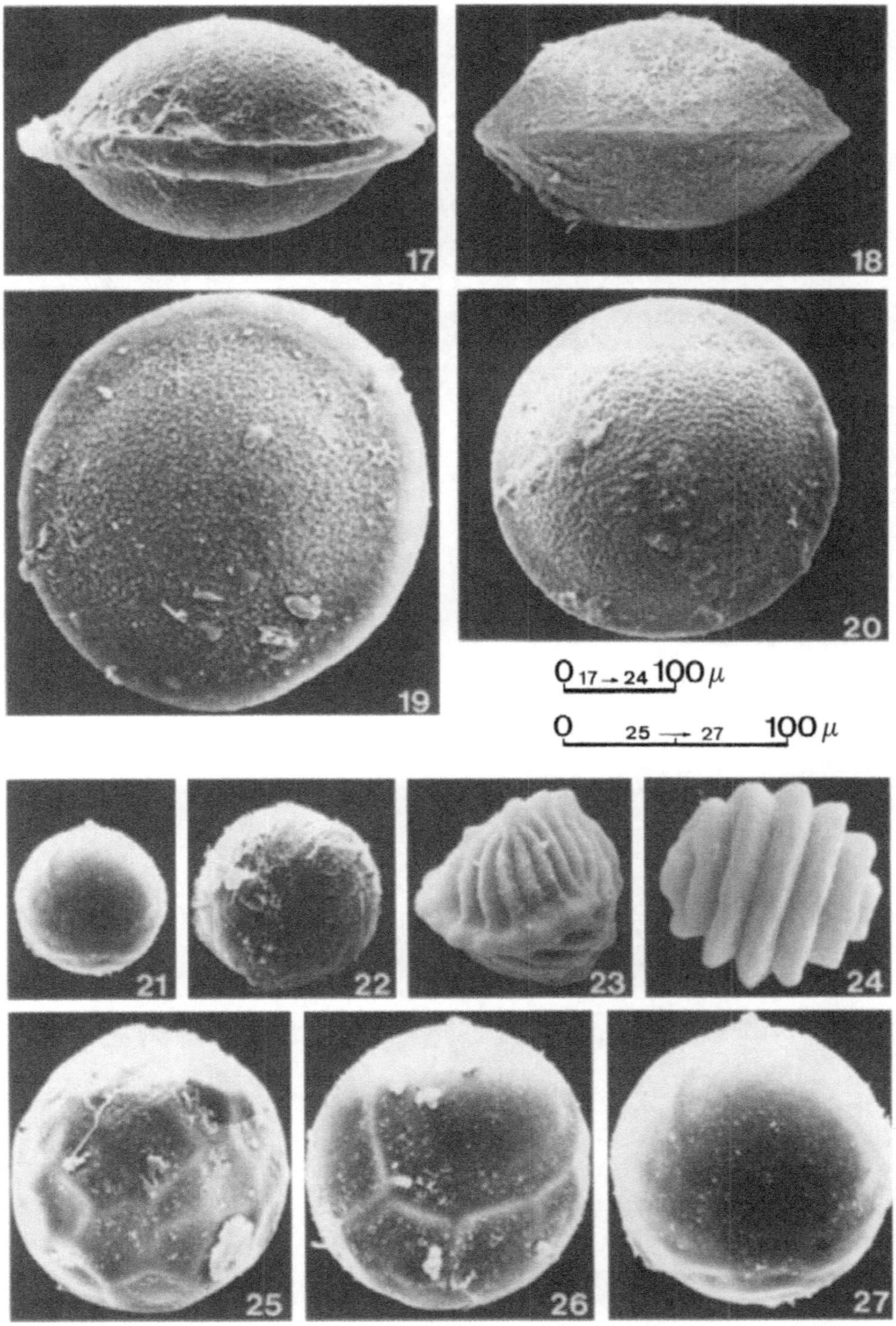


*Fig. 17–27.* Eggs of Anostraca and Spinicaudata (Fig. 17 to 24 are at the same scale as Fig. 7 to 16). 17. – *Tanymastix stagnalis*, lateral view 18. – *Tanymastix* (aff.) *stellae*, lateral view; 19. – *T. stagnalis* dorsal view; 20. – *T.* (aff.) *stellae*, dorsal view; 21. – *Eoleptestheria ticinensis*; 22. – *Cyzicus tetracerus*; 23. – *Limnadia lenticularis*; 24. – *Imnadia yeyetta*; 25. – *Eoleptestheria ticinensis* without K.A.A.D. treatment or critical point drying; 26. – idem, with critical point drying; 27. – idem, with K.A.A.D. fixation and critical point drying.

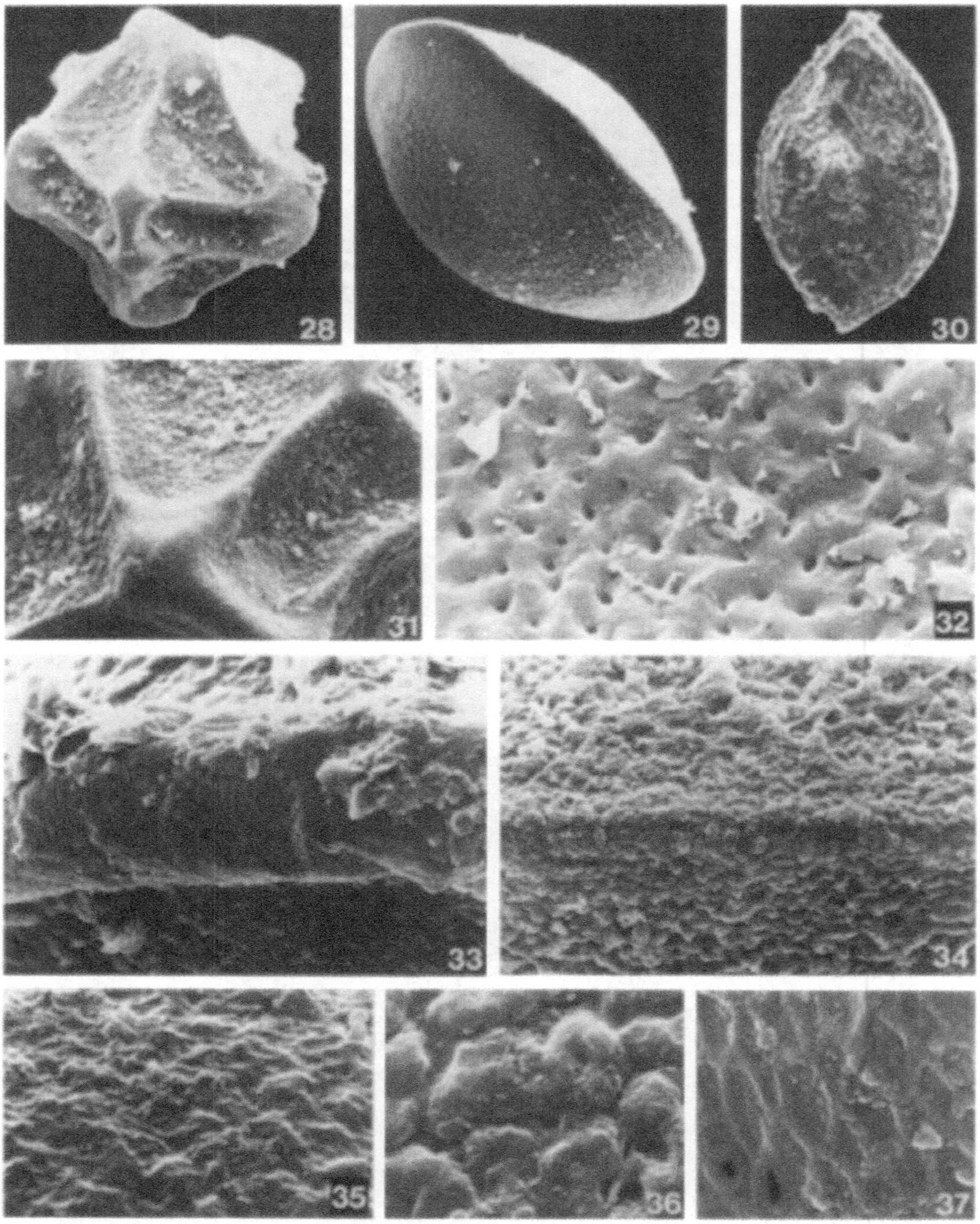


*Fig. 28–37.* 28. – whole egg of ***Branchipus blanchardi***; 29. – egg of *Tanymastix stagnalis*, diagonal view; 30. – longitudinal section through a cyst of *Tanymastix stagnalis* showing the biconvex form of the resting egg (characteristic of the genus *Tanymastix*) and the external envelope; 31. – cyst of ***Branchipus blanchardi***, detail enlarged; 32. – idem, with detail of outer surface showing pores; 33. – cyst of *Tanymastix* (aff.) *stellae*, detail of the equatorial callus encircling the biconvex egg; 34. – egg of *Tanymastix stagnalis*, detail of the equatorial callus, same scale as in Fig. 33; 35. – detail of the surface ornamentation of *Tanymastix* (aff.) *stellae* egg; 36. – detail of the surface ornamentation of the egg of *Tanymastix stagnalis*, same scale as in Fig. 35; 37. – detail of the surface of the equatorial callus of *Tanymastix* (aff.) *stellae* egg.

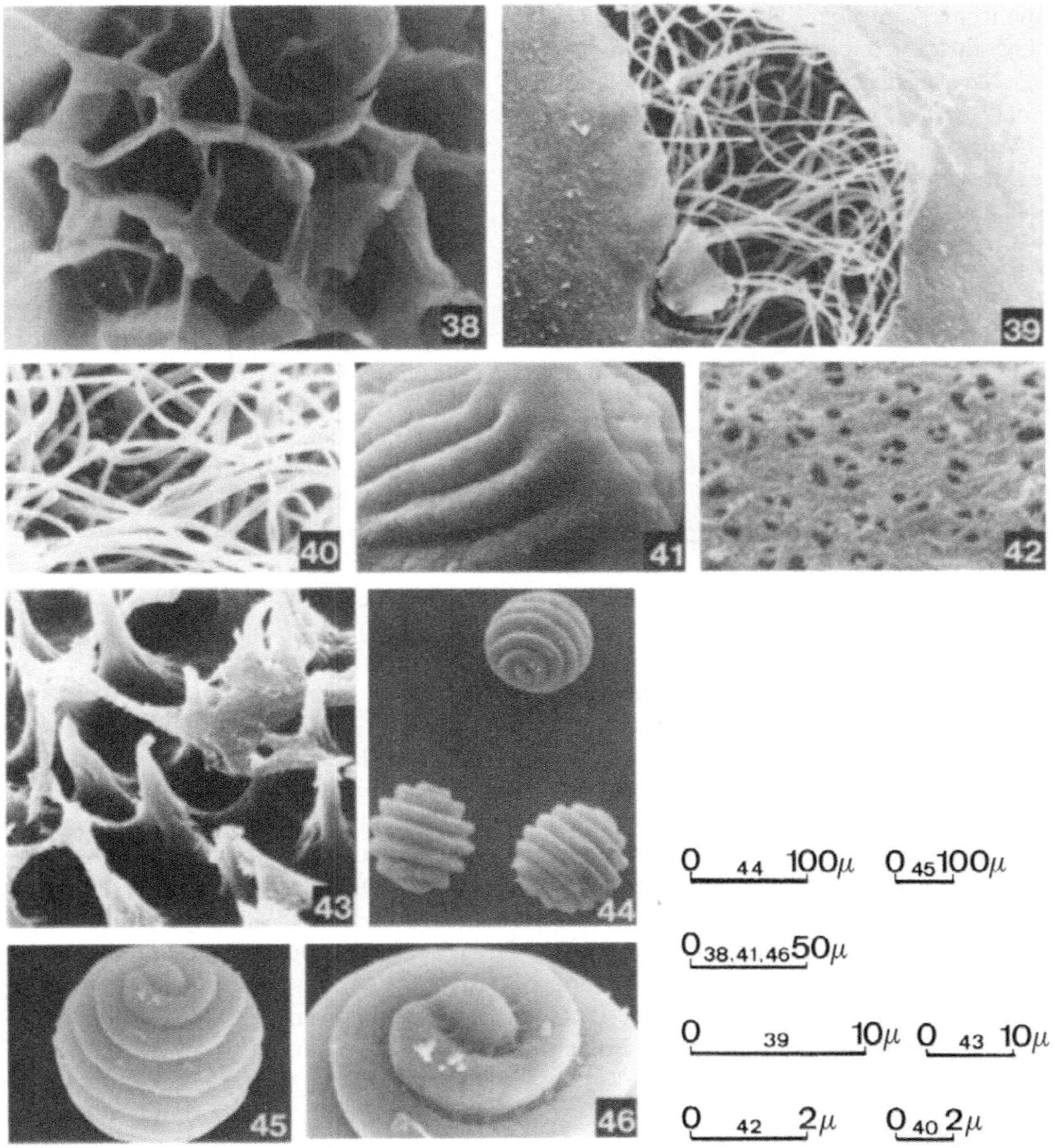


*Fig. 38–46.* 38. – Egg of *Chirocephalus salinus*, detailed enlarged; 39. – egg of *Cyzicus tetracerus* showing throught a craked menbrane the fine hairy structure; 40. – idem, detail enlarged of 'hairy' layer; 41. – egg of *Limnadia lenticularis*, detail enlarged; 42. – idem, detail of the surface showing the pores; 43. – cyst of *Linderiella massaliensis*, detail of spines; 44. – view of 3 cysts of *Imnadia yeyetta*; 45. – cyst of *Imnadia yeyetta*, apical view; 46. – idem, detail enlarged.

It is generally accepted that lenticular eggs are unique to the genus *Tanymastix*. Our data confirm the results of Orghidan (1945) and Botnariuc & Orghidan (1953) for *T. motasi*, of Alonso & Alcaraz (1984) and of Freiner & Grüttner (1984) for *T. stagnalis*, of Mura (1986) for *T. stagnalis* and *T. stellae* and of Mura & Thiéry (1986) for *T. bucheti*.

Previously only one species, *Tanymastix stagnalis*, was known to occur in France. However, we found that eggs of the *Tanymastix* population from Corsica differ in diameter as well as in shape from *T. stagnalis*. The equatorial callus encircling their biconcave eggs is thicker. The Corsian *Tanymastix* eggs are most similar to those of *T. stellae*, endemic to Sardinia (Mura, 1986). Although no adults were collected in Corsica, the geographical proximity between the Corsican and the Sardinian populations, suggests that the Corsican cysts belong to *T. stellae* which could then become a new member of the French fauna.

Spherical or subspherical eggs

*Branchipus* Schaeffer, 1766
(Fig. 13, 16, 28, 31, 32)

Figure 13 shows the ornamentation and egg diameter of *Branchipus schaefferi*. These data agree with the published descriptions of various authors on *Branchipus schaefferi* cysts: Gilchrist (1978) and Mura & Thiéry (1986).

*Branchipus blanchardi* eggs are described here for the first time (Fig. 16, 28). By comparison of eggs of *B. blanchardi* Daday 1910 and of *B. alpinus* Colosi 1922, illustrated in Mura (1986, p. 282), we synonymize both. Cottarelli (1968) studied the morphology of adults and observed a similarity between the two species. Recently, Nourisson & Thiéry (1988) reached the same conclusion on the basis of observations on specimens deposited at the National Museum of Natural History at Paris. This *B. blanchardi* Daday 1910, becomes an endemic species of both the French and Italian High Alps.

*Chirocephalus* Prévost, 1803
(Fig. 7, 12, 15, 38)

Observations on cysts of *Chirocephalus salinus* (Fig. 12) confirm results by Mura *et al.* (1978). The mean diameter is smaller than *Chirocephalus diaphanus* eggs. Nevertheless, the mean diameter in the Corsican population (342 $\mu$m) is larger than that measured by Mura *et al.* (1978) on a Sardinian population (256 $\mu$m).

Cysts of *Chirocephalus diaphanus* (Fig. 7) are larger but have the same shape as in *C. salinus*. Despite slight geographical variations in diameter, these eggs have great morphological stability (Gilchrist, 1978; Mura & Thiéry, 1986; Mura, 1986).

The egg of *Chirocephalus spinicaudatus* is illustrated here for the first time (Fig. 15). Its surface is almost smooth, with small warts. Its diameter varies from 280 to 300 $\mu$m, which is slightly less than the values of 330 $\mu$m reported by Botnariuc & Orghidan (1953). The egg, lacking ridges forming a polygonal pattern is very different from the eggs of *Chirocephalus diaphanus* and *Chirocephalus salinus*. It is to be placed in the same group as the *Chirocephalus marchesonii* egg in the Italian fauna (Mura *et al.*, 1978; Mura, 1986). Further study is needed to check the phylogenetic connection between these two taxa.

*Artemia* Leach, 1819
(Fig. 8, 10)

*Artemia* cysts are smooth. Two types can be distinguished on the basis of size: those with a diameter less than 240 $\mu$m and those with a diameter greater than 240 $\mu$m. Although no conclusion could be reached on egg morphology to differentiate bisexual from parthenogenetic populations, our data agree with results of Vilela & Castelo Branco (1987) on Portugal populations: eggs of parthenogenetic strains are slightly smaller than eggs of sexual strains (parthenogenetic *Artemia*, in France 238.8 $\pm$ 4.8 $\mu$m, in Portugal 241.2 $\pm$ 10.2 $\mu$m; bisexual *Artemia*: In France-Sète 245.3 $\pm$ 18.6 $\mu$m, in Portugal 268.0 $\pm$ 15.0 $\mu$m). A 50% higher standard error occurs in cysts of bisexual populations than in parthenogenetic populations. Among parthenogenetic populations, however, some variability in diameter is also observed. Vanhaecke & Sorgeloos (1980), collected eggs at Lavalduc with a diameter of 259.6 $\mu$m. The diameter of *Artemia* cysts collected at Sète-Villeroy is similar to that of bisexual populations from Tunisia (Van Ballaer *et al.* 1987).

*Branchinella* Sayce 1903
(Fig. 11)

Mixed with eggs of *Artemia*, some other cysts were collected in the Fangassier pond in Camargue. By their diameter and external ornamentation they seem close to *Branchinella spinosa* eggs (See Mura & Thiéry, 1986: 128 and Mura, 1986: 278), an euryhaline species living in the salt-marshes of Sardinia, Spain and northern Africa, mostly together with *Artemia* (Mura, 1988; Saadi & Champeau, 1987). Its discovery in Camargue together with *Artemia* is not surprising if passive dispersal of anostracan cysts *via* the intestinal tract of birds and particularly Flamingoes (*Phoenicopterus ruber roseus* Pallas) is accepted (Proctor & Malone, 1965; Persoone & Sorgeloos, 1980; Thiéry *et al.*, 1990). Adults have, however, not yet been discovered.

*Linderiella* Brtek, 1964
(Fig. 9, 43)

*Linderiella massaliensis* eggs have a spiny appearance. Except for one Chirocephalidae species, *Chirocephalus ruffoi*, which has also spiny eggs (Mura, 1986), this pattern seems to be characteristic of the genus *Linderiella*. (Thiéry, 1986; Thiéry & Champeau, 1988).

*Siphonophanes* Simon, 1886
(Fig. 14)

Eggs of *Siphonophanes grubei* are covered with low, thin ribs arranged in a dense complex network. Its diameter is 300–310 μm, slightly smaller than given by Botnariuc & Orghidan (1953).

*Spinicaudata*
*Imnadia* Hertzog, 1935
(Fig. 24, 44, 45, 46)

The eggs of *Imnadia yeyetta* have a spiral aspect (Fig. 24). Their size is less than 200 μm. The illustrations of *I. voitestii* Botnariuc & Orghidan ( = *I. yeyetta*) by Botnariuc (1947: pl I, Fig. 15) and by Brtek (1957) reveal the same characteristics.

*Limnadia* Brongniart, 1820
(Fig. 23, 41, 42)

The size of eggs of *Limnadia lenticularis* (Fig. 23) is close to those of *I. yeyetta*, but they are pyramid-shaped and biconcave, with ridges converging to the same point (called 'twisted eggs' by Martin, 1989). Sars (1896) and Daday (1925) illustrated these eggs from european samples, while Martin (1989) used material from Florida (USA). Their illustrations are similar to ours, but no dimensions are given. The data presented by Zaffagnini (1967) on *L. lenticularis* egg sizes (maximum diameter range 290–370 μm) are slightly different from ours.

Morphological differences between *Limnadia* Brongniart and *Eulimnadia* Packard eggs (Martin, 1989; Belk, 1989) may provide a distinction between these two genera. Webb & Bell (1979) consider them as synonyms.

*Eoleptestheria* Daday, 1913
(Fig. 21, 25, 26, 27)

*Eoleptestheria ticinensis* eggs are spherical (range 125–130 μm) with a smooth envelope as in the Leptestheriidae (see Tommasini & Sabelli, 1989).

*Cyzicus* Audoin, 1837
(Fig. 22, 39, 40)

The eggs of *Cyzicus tetracerus* are spherical and smooth, with diameter 145–150 μm. Their tertiary covering looks like a network of thin radiating and entangled hairs (0.3–0.5 μm thick and 10–15 μm long), covered by a particularly thin (< 1 μm) and fragile membrane. The frequent ripping of this membrane during fixation (Fig. 39) causes their hairy look. Daday (1914), observing this covering under light microscope, described it as 'Ova membrana bacillis radialiter dispositis armata tecta'. The illustrations by Daday (1914), Alonso & Alcaraz (1982: 77) (*Cyzicus grubei*), Thiéry & Mura (in preparation) (*Cyzicus bucheti*), and our recent observations (*Cyzicus tetracerus*) suggest that the hairy structure of the vitelline envelope (*sensu* Tommasini & Sabelli, 1989) is typical for eggs of *Cyzicus*.

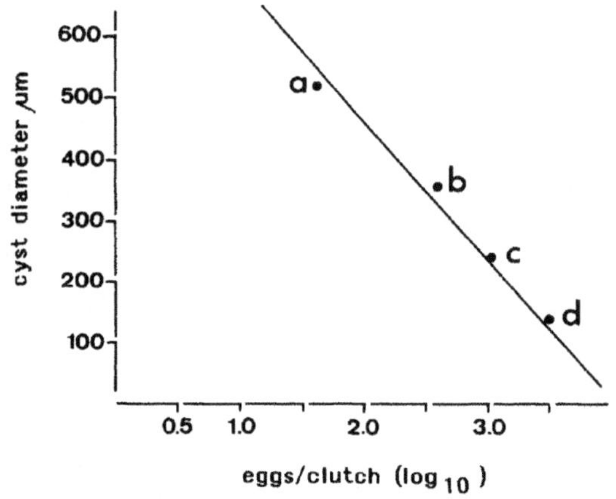


*Fig. 47.* Relationship between mean diameter of eggs (µm) and clutch size of larger mature females ($Log_{10}$) in branchiopods. a) *Lepidurus apus*, b) *Chirocephalus diaphanus*, c) *Branchipus schaefferi*, d) *Cyzicus tetracerus*. Diameter (µm) = − 223.24 $Log_{10}$ Clutch size + 931.59 ($n = 4$, $r^2 = 0.995$ $P < 0.001$).

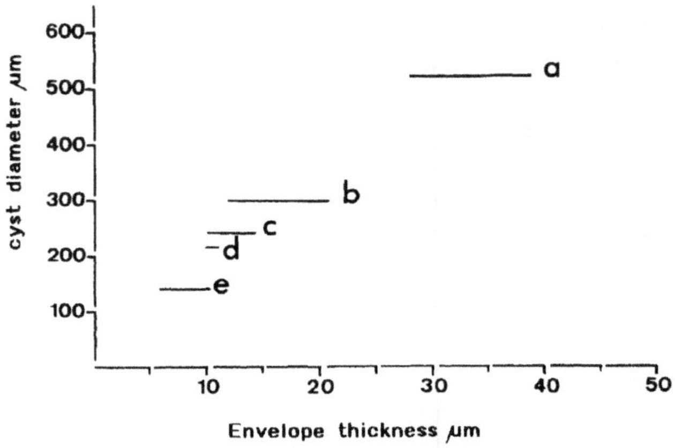


*Fig. 48.* Relationship between mean diameter of eggs (µm) and thickness of alveolar envelope (µm). a) *Lepidurus apus*, b) *Tanymastix stagnalis*, c) *Branchipus schaefferi*, d) *Artemia* sp., e) *Cyzicus tetracerus*. Diameter (µm) = 19.08 × thickness of envelop # + 9.64 ($n = 5$, $r^2 = 0.964$, $p < 0.001$).

### *b) Size of clutch*

The number of eggs per brood decreases when their diameter increases (Fig. 47). The equation fitted in a linear form after a $Log_{10}$ transformation is: Diameter (µm) = − 223.24 $Log_{10}$ clutch size + 931.59 ($n = 4$, $r^2 = 0.995$, $P < 0.001$).

### *c) Thickness of envelope*

A relation is also observed between egg diameter and thickness of the tertiary covering (Fig. 48). The equation for it is: Diameter (µm) = 19.08 thickness of alveolar envelop + 9.64 ($n = 5$, $r^2 = 0.964$, $P < 0.001$). A gradient from Notostraca to Spinicaudata through Anostraca can be observed. Notostracan eggs are covered by an alveolar tertiary envelope which is always thicker than 25–30 µm, sometimes 40 µm (Gilchrist, 1978; Trentini & Sabelli Scanabissi, 1982; Thiéry, 1985, this study). Anostracan eggs are only covered by a thin alveolar layer of 1.5–8 µm (Linder, 1960; Garreau de Loubresse, 1974; Gilchrist, 1978, and this study). The ratio between tertiary envelope thickness and egg diameter is, however, constant for the three groups in this study and is close to 4.5–5%.

## Discussion

The SEM morphology of eggs shows that many species present a stability of their egg characteristics, pattern and diameter, whatever their geographical location. For example, eggs of *Chirocephalus diaphanus*, *Branchipus scaefferi*, of *Branchinella spinosa*, of *Imnadia yeyetta* and *Limnadia lenticularis* seem to be constant in their zone of occurrence (Europe and (or) Mediterranean basin).

Our observations complete some other data on eggs and show that in some cases characteristics could be considered as generic, as for the hairy look of the viteline envelop of *Cyzicus*, the lenticular aspect of *Tanymastix* and the twisted shape of *Limnadia*. In few cases, the situation is less clear, as for *Linderiella*. All species of this genus have a spiny look, but one species of Chirocephalidae, *Chirocephalus ruffoi*, also has an egg with spines (see Mura, 1986: 281). This feature raises the problem of the link between Linderielidae and Chirocephalidae, two families that Brtek (in litt.) considers as closely related. Is the spiny egg a morphologic convergence or a phylogenetic link?

At last this study shows that there is a relation between the diameter of eggs and their number in clutches. Notostraca have bigger eggs and smaller clutches, while Spinicaudata have smaller eggs and larger clutches. Anostraca take an intermediate position. The relation between envelope thickness and egg diameter shows that there is also a gradient from Notostraca to Spinicaudata, across the Anostraca. These on egg characteristics relate to their drying resistance and their ability to colonize new temporary ponds. If we consider the egg-laying characteristic of each order, we find that Notostraca compensate for a restricted number of eggs per clutch by sticking them on gravel, sand or aquatic vegetation (Thiéry, 1985; Fryer, 1988) avoiding their scattering. Anostracan females expell their mature eggs out of the broodpouch into the water (Avery, 1940). Spinicaudata do not have an organised laying behaviour. They simply drop eggs with each moult (Thiéry & Mura, in preparation). These small-diameter eggs are very light and easily dispersed by wind. The potential loss is compensated by producing large clutches (>2500–3000 in *Cyzicus* for example).

These data are only three relevant aspects of the life history of phyllopods. They have to be completed by others, as total fertility, mortality rates, life span, and growth, to be interpreted in term of selection type in a r-K continuum, theory recently advanced by Thiéry (1987) and Hamer & Appleton (1991).

**Acknowledgements**

We thank Dr. J. Forest for loaning specimens kept in the collections of the National Museum of Natural History at Paris; C. Gril of the Languedoc University of Sciences and Techniques Electron Microscopy Center, Montpellier, for assistance with Scanning Electron Microscopy, M. Fargier for help with the translation, and L. Brendonck, D. Belk and H.J. Dumont for comments on the manuscript.

**References**

Alonso, M., 1985. A survey of the Spanish Euphyllopoda. Misc. Zool. 9: 179–208.

Alonso, M. & M. Alcaraz, 1984. Huevos resistentes de crustaceos eufilopodos no cladoceros de la peninsula Iberica: Observacion de la morfologia externa mediante tecnicas de microscopia electronica de barrido. Oecol. aquat. 7: 73–78.

Amat, F., 1982. Diferenciacion y distribucion de las poblaciones de *Artemia* de Espana. IV. Biometria de quistes y nauplius. Inv. Pesq. 46: 55–62.

Avery, J. L., 1940. Studies on the egg-laying habits of the fairy shrimp. J. Wash. Acad. Sci. 30: 31–33.

Belk, D., 1989. Identification of species in the Conchostracan genus *Eulimnadia* by egg shell morphology. J. Crust. Biol. 9: 115–125.

Botnariuc, N., 1947. Contribution à la connaissance des Phyllopodes Conchostracés de Roumanie. Not. Biol., Bucarest 5: 1–158.

Botnariuc, N. & T. Orghidan, 1953. Crustacea Phyllopoda. In Fauna Republicii Populare Romane. Acad. Rep. Pop. Romane 4: 99 pp.

Brendonck, L., A. Thiéry & A. Coomans, 1990. Taxonomy and biogeography of the Galapagos Branchiopod fauna (Anostraca, Notostraca, Spinicaudata). J. Crust. Biol. 10: 676–694.

Brtek, J., 1957. *Imnadia voitestii* Botn. & Org. 1941 (Imnadiidae Phyllopoda Conchostraca) na juhozapadnom Slovensku. Biologia 12: 63–67.

Castritsi Catharios, J., B. Konstantinidis & B. Kiortsis, 1984. Occurrence of *Artemia* in two solar saltworks near Missolonghi, Greece (Anostraca). Crustaceana 47: 321–323.

Cobolli Sbordoni, M., E. De Matthaeis, G. Mura & V. Sbordoni, 1985. Genetic variability and divergence among populations and species of *Chirocephalus* fairy shrimp from Italy. Third I.C.S.E.B. Brighton Sussex.

Cottarelli, V., 1968. Su *Branchipus alpinus* Colosi (Euphyllopoda, Anostraca). Arch. Zool. ital. 53: 353–365.

Cottarelli, V. & G. Mura, 1984. *Chirocephalus ruffoi* n. sp. (Crustacea Euphyllopoda Anostraca) di acque astatiche d'alta quota dell'Appenninbo Calabro-Lucano. Boll. Mus. civ. St. nat. Verona 11: 515–523.

Daday, E., 1914. Monographie systématique des Phyllopodes Conchostracés. Ann. Soc. Sci. Nat. Zool., Paris S 9, 20: 39–330.

Daday, E., 1925. Monographie systématique des Phyllopodes Conchostracés Troisième partie. Fam. Limnadiidae G.O. Sars. Ann. Soc. Sci. Nat. Zool. Paris, S 10, 8: 143–184 ( = 463–504).

De Walsche, C., N. Munuswamy & H. J. Dumont, 1991. Structural differences between the cyst walls of *Streptocephalus dichotomus* (Baird), *S. torvicornis* (Waga), and *Thamnocephalus platyurus* (Packard), and a comparison with other genera and species. Hydrobiologia, this volume.

Freiner, D. & O. Gruttner, 1984. Der Eichener Kiemen fubkrebs. Natur Museum 114: 274–286.

Fryer, G., 1987. A new classification of the branchiopod Crustacea. Zool. J. linn. soc. 91: 357–383.

Garreau de Loubresse, N., 1974. Etude chronologique de la mise en place des enveloppes de l'œuf d'un crustacé phyllopode: *Tanymastix lacunae*. J. Microsc. 20: 21–38.

Gasc, C., J. C. Vala & J. M. Reidenbach, 1984. Microstructure cuticulaires et récepteurs sensoriels des larves de *Sepedon sphegea* (F.) (Diptera, Sciomyzidae). Int. J. Insect. Morph. Embryol. 13: 275–281.

Gilchrist, B. M., 1978. Scanning electron microscope studies of the egg shell in some Anostraca (Crustacea: Branchiopoda). Cell Tiss. Res. 193: 337–351.

Hamer, M. L. & C. C. Appleton, 1991. Life history adaptations of phyllopods in response to predators, vegetation and habitat duration in north-eastern Natal. In: Belk, D., H. J. Dumont & N. Munuswamy (eds), Studies on Large Branchiopod Biology and Aquaculture. Developments in Hydrobiology 64. Kluwer Academic Publishers, Dordrecht: 105–116. Reprinted from Hydrobiologia 212.

Johnson, A. R., 1989. Movements of greater flamingos (*Phoenicopterus ruber roseus*) in the Western Palearctic. Rev. Ecol. (Terre Vie) 44: 75–94.

Kupka, E., 1940. Untersuchungen über die Schalenbildung und Schalenstruktur bei den Eiern von *Branchipus schäefferi* (Fischer). Zool. Anz. 132: 130–139.

Linder, H. J., 1960. Studies on the fresh water fairy shrimp *Chirocephalopsis bundyi* (Forbes). II. Histochemistry of egg-shell formation. J. Morphol. 107: 259–283.

Mac Donald, G., 1980. The use of *Artemia* cysts as food by the flamingo (*Phoenicopterus ruber roseus*) and the shelduck (*Tadorna tadorna*). In G. Persoone, P. Sorgeloos, O. Roels and E. Jaspers (eds), The Brine shrimp *Artemia*. Vol. 3: Ecology, Culturing, Use in aquaculture. Universa Press, Wetteren, Belgium pp. 97–104.

Martin, J. W., 1989. *Eulimnadia belki*, a new clam shrimp from Cozumel, Mexico (Conchostraca: Limnadiidae), with a review of Central and South American species of the genus *Eulimnadia*. J. Crust. Biol. 9: 104–114.

Mazzini, M., 1978. Scanning electron microscope morphology and Amino-acid analysis of the egg-shell of encysted brine shrimp, *Artemia salina* Leach (Crustacea Anostraca). Monit. zool. ital. (N.S.) 12: 243–252.

Munuswamy, N. & T. Subramoniam, 1983. Scanning electron microscope studies on the egg shell of two freshwater fairy shrimps, *Streptocephalus dichotomus* and *Branchinella kugenumaensis* (Branchiopoda: Anostraca). Cytobios 37: 181–186.

Munuswamy, N., T. Subramoniam & G. Mura, 1985. Ootaxonomic findings on Anostracan eggs: a scanning electron microscopic study. Cytobios 42: 93–97.

Mura, G., 1986. SEM morphological survey on the egg shell in the italian Anostracans (Crustacea, Branchiopoda). Hydrobiologia 134: 273–286.

Mura, G., 1988. Occurrence of *Artemia* in solar saltworks and coastal brine ponds in Sardinia, Italy. J. Crust. Biol. 7: 697–703.

Mura, G., F. Accordi & M. Rampini, 1978. Studies on the resting eggs of some fresh water fairy shrimps of the genus *Chirocephalus*: biometry and scanning electron microscopic morphology (Branchiopoda, Anostraca). Crustaceana 35: 190–194.

Mura, G. & A. Thiéry, 1986. Taxonomical significance of scanning electron microscopic morphology of the Euphyllopods' resting eggs from Morocco. Part I. Anostraca. Vie Milieu 36: 125–131.

Nourisson, M. & A. Thiéry, 1988. Crustacés Branchiopodes (Anostracés, Notostracés, Conchostracés). In Introduction pratique à la systématique des organismes des eaux continentales françaises. Bull. mens. Soc. linn. Lyon 57: 1–53 ( = 75–135).

Orghidan, T., 1945. Sur une nouvelle espèce du genre *Tanymastix* (Sim.) en Roumanie. Bull. Sec. Sci. Acad. Roumanie 27: 676–680.

Persoone, G. & P. Sorgeloos, 1980. General aspects of the ecology and biogeography of *Artemia*. In G. Persoone, P. Sorgeloos, O. Roels and E. Jaspers (eds), The Brine Shrimp *Artemia*. Vol. 3: Ecology, Culturing, Use in Aquaculture. Universa Press, Wetteren, Belgium, pp. 3–24.

Proctor, V. W. & C. R. Malone, 1965. Further evidence of the passive dispersal of small aquatic organisms via the intestinal tract of birds. Ecology 46: 728–729.

Saadi, A. & A. Champeau, 1987. Hydrobiologie de deux hydrosystèmes temporaires saumâtres: Zima et Sedd-El-Messjoun (Bassin de la Bahira, Maroc). Sci. Eau 6: 319–336.

Samyiah, N., K. Venkataraman & S. Krishnaswamy, 1985. Morphology of three species of Conchostraca using scanning electron microscope. Curr. Sci. 54: 869–871.

Sars, G. O., 1896. Phyllocarida and Phyllopoda. In Fauna Norvegiae Vol. I. Christiana.

Thiéry, A., 1985. Ponte et ultrastructure de l'œuf de *Triops granarius* Lucas (Crustacea, Notostraca): adaptations à l'assèchement de l'habitat. Verh. int. Ver. Limnol. 22: 3024–3028.

Thiéry, A., 1986. *Linderiella africana* n. sp., premier représentant des Linderiellidae (Branchiopoda Anostraca) en Afrique du Nord (Maroc). Crustaceana 51: 155–162.

Thiéry, A., 1987. Les Crustacés Branchiopodes Anostraca, Notostraca et Conchostraca des mares temporaires (Dayas) du Maroc occidental. Taxonomie, Biogéographie Ecologie. Thèse Doct. es Sciences, Université d'Aix-Marseille III, 405 pp.

Thiéry, A. & A. Champeau, 1988. *Linderiella massaliensis*, new species (Anostraca, Linderiellidae), a fairy shrimp from Southeastern France, its ecology and distribution. J. Crust. Biol. 8: 70–78.

Thiéry, A., F. Robert & C. Gabrion, 1991. Distribution des populations d'*Artemia* (Crustacea Anostraca) et de leur parasite *Flamingolepis liguloides* (Cestode Cyclophyllidae) dans les salins du littoral méditerranéen français. Can. J. Zool. 68: 2199–2204.

Tommasini, S. & F. Scanabissi Sabelli, 1989. Eggshell origin and structure in two species of Conchostraca (Crustacea, Phyllopoda). Zoomorphology 109: 33–37.

Tommasini, S., F. Scanabissi Sabelli & M. Trentini, 1989. Scanning electron microscope study of eggshell development in *Triops cancriformis* (Bosc) (Crustacea, Notostraca). Vie Milieu 39: 29–32.

Trentini, M. & F. Sabelli-Scanabissi, 1982. Follicle duct cell ultrastructure and egg-shell formation in *Triops cancriformis* (Crustacea, Notostraca). J. Morphol. 172: 113–121.

Van Ballaer, E., D. Versichele, P. Vanhaecke, P. Léger, N. Ben Abdelkader, S. Turki & P. Sorgeloos, 1987. Characterization of *Artemia* from different localities in Tunisia with regard to their use in local aquaculture. In P. Sorgeloos, D. A. Bengtson, W. Decleir & E. Jaspers (eds), *Artemia* research and its applications, Vol. I: Morphology, Genetics, Strain characterization, Toxicology. Universa Press, Wetteren, Belgium, pp. 199–209.

Vanhaecke, P. & P. Sorgeloos, 1980. International study on *Artemia*. IV. The biometrics of *Artemia* strains from different geographical origin. In G. Persoone, P. Sorgeloos, O. Roels and E. Jaspers (eds), The Brine shrimp *Artemia*, Vol. 3: Ecology, Culturing, Use in aquaculture. Universa Press, Wetteren, Belgium, pp. 393–405.

Vieira, M. N. & F. Amat, 1985. *Artemia* sp. from Aveiro: its characterization. Publ. Inst. Zool. Fac. Cien. Porto 191: 1–9.

Vilela, M. H. & M. A. Castelo Branco, 1987. Characterization of Portuguese *Artemia* strains. In P. Sorgeloos, D. A. Bengtson, W. Decleir and E. Jaspers (eds), *Artemia* Research and its Applications, Vol. I: Morphology, Genetics, Strain characterization, Toxicology. Universa Press, Wetteren, Belgium, pp. 211–217.

Webb, J. A. & G. D. Bell, 1977. A new species of *Limnadia* (Crustacea: Conchostraca) from the granite belt in southern Queensland and north New South Wales. Proc. linn. Soc. N.S.W. 103: 237–245.

Zaffagnini, F., 1967. Contributo alla conoscenza della biologia riproduttiva dei fillopodi *Concostraci* i osservazioni sull'accrescimento e sul ciclo riproduttivo di *Limnadia lenticularis* (L.). Mem. Ist. ital. Idrobiol. 21: 105–115.

*Hydrobiologia* **212**: 261–266, 1991.
*D. Belk, H. J. Dumont & N. Munuswamy (eds), Studies on Large Branchiopod Biology and Aquaculture.*



# The fairy shrimp *Chirocephalus brevipalpis* (Orghidan, 1953), (Crustacea: Anostraca) in Yugoslavia

Swetozar T. Petkovski
*Prirodonaučen muzej na Makedonija, Bulevar Ilinden 86, Yu-91000, Skopje 55, Yugoslavia*

*Key words:* Anostraca, taxonomy, zoogeography

## Abstract

The fairy shrimp *Chirocephalus brevipalpis* (Orghidan, 1953) has been found in the Srem-district, Northern Yugoslavia, at the southern margin of the Pannonian lowland. It is the second record of the species, which had been reported only from Banat province, Romania.

Since the specimens from Yugoslavia show some morphological differences with the typical form, a detailed description and illustration of essential taxonomic features in each sex are presented.

The specific characters of the population of Srem are not sufficient to establish a new species, and fall within the intraspecific variability of *Chirocephalus brevipalpis*.

## Introduction

During an investigation of the Anostraca from the collection of the Natural History Museum of Macedonia, Skopje, a species new to Yugoslavia was found. It corresponds with *Chirocephalus brevipalpis* (Orghidan, 1953), described from Western Romania near Timisoara. It was described as a new genus, *Palpicephalus* (Orghidan, 1952). To the same genus, Orghidan also assigned the earlier described species *Ceratocephalus recticornis* (Orghidan, 1948). In his taxonomical revision of the family Chirocephalidae, Brtek (1966) suppressed the genus *Palpicephalus*, and its species were transferred to *Chirocephalus* Prevost, 1803. At the same time, Brtek gave a short redescription of both species, based on new material from the type-locality in Romania.

Detailed analysis of our specimens of *C. brevipalpis*, collected from the Srem-district in the lower course of the Sava river near Kupinovo, revealed some morphological features not mentioned in previous papers. We here present a description of the material, consisting of 20 adult specimens, 10 of each sex.

## Distribution and habitat

Hitherto *Chirocephalus brevipalpis* had been recorded only from a restricted area in Western Romania. The Yugoslavian population was collected in a shallow, ephemeral pond on clayey soil in a flooded wasteland near the river Sava in the vicinity of the village Kupinovo, Sremska Mitrovica. It was associated with the notostracan *Triops cancriformis* Bosc 1801, (leg. Dr. T. Petkovski, 28.05.1975). In Limnofauna Europaea (Illies, 1978) both localities of *C. brevipalpis* are situated in area 11 (Hungarian lowland).

## Description

*Female*

Total length, from front to end of shaft of cercopods 14.0–16.4 mm. Average ratio of length of head and thorax to that of genital segments, abdomen and cercopods 1 : 1.14. Antennula 1.5–1.8 mm long, much longer than antenna, with 3 setae and 8–10 aesthetascs (Fig. 1. A). Antenna 1.1–1.4 mm long, length about 74% of antennula, uniarticulate, distal half narrowing to a tapering point (Fig. 1. A). Compound eye 0.65–0.75 mm (Fig. 1. A). Of the 11 thoracic segments, only the 7-th to 10-th bear a medio-dorsal thorn, while the 8-th and 11-th segments each have a dorsolateral thorn on each side. Thorns of 11-th segment much larger than those of 8-th segment and more laterally placed (Fig. 1. B). The first of the two genital segments with a pair of lateral outpocketings connected with the ovisac (Fig. 1. B,C,D), similar to *Chirocephalus diaphanus carinatus* (Daday, 1910). In adults, abdominal segments 5 to 7 each with dorsolateral thornlike outgrowths on their postero-lateral margins (Fig. 1. C,D). Ovisac extending backward to sixth abdominal segment (Fig. 1. D). Eggs of preserved specimens 0.22–0.24 mm in diameter, with surface structure as in Fig. 1. E. Cercopods 1.2–2.1 mm long, with long plumose setae on median and lateral borders.

*Male*

Total length, from front to end of shaft of cercopods 12.5–20.5 mm. Average ratio of length of head and thorax to that of genital segments, abdomen and cercopods 1 : 1.19.

Antennula 1.9–2.9 mm long, length about 64 percent of that of antenna, with 3 setae and 8–10 aesthetascs at the tip.

Antenna biarticulate, 3.0–4.9 mm long (Fig. 1. F; 2. A,C). Free part of proximal article of antenna 1.3–2.4 mm long, cylindrical, in the last third curving gently inward (Fig. 1. F). It bears an apophysis on its posterior surface, near its junction with the head. The apophysis is conical, strongly thickened at the basis, terminating in a bluntly rounded tip, covered with a pile of small, conical, sensory papillae (Fig. 1. F; 2. A,C). Proximal article of antenna with well-developed lamellar prolongation on its anterior surface. Antennal prolongation divided into a large upper lamella, 2.1–3.4 mm long, and small lower lamella 1.3–1.7 mm long (Fig. 2. A,B,C). Ratio of length of lower lamella to that of upper lamella 1 : 1.5–1 : 2.15. The long outgrowths on the lateral and median borders of the upper lamella become gradually shorter towards the tip. Relatively short fingerlike outgrowths are present at the basis of the lateral border of the upper lamella (Fig. 2. B). Lower lamella with weakly differentiated lateral and median lobes. Median lobe without carina (sac-like protuberance) (Fig. 2. A,B,C).

Distal article of antenna cylindrical, curving gently inward across its entire length. Length across the bend 1.6–2.5 mm. Terminal part with a hook-like expansion on its inner margin (Fig. 1. F,G,H,I,J; 2. A,C). Distal article with a thornlike outgrowth, 0.15–0.34 mm long, on inner surface of basal half. Length about 40% of that of apophysis of proximal article, medially directed, making an angle of 90° with the article. Thornlike outgrowth variable in shape: variations are shown on Fig. 1. G,H,I,J.

Thoracic appendages typical of the genus. Two praeepipodites with serrated margins, except those of the eleventh pair of legs, where praeepipodites are lacking. Epipodite tapering distally (Fig. 3. A,B,C).

All thoracic and abdominal segments without lateral or dorsolateral outgrowths or bosses.

Genital segments rather swollen. Construction of the rigid part of penes similar to the other species of the genus (Fig. 2. E,F). Retractile part of penis with a short, nearly cylindrical outgrowth on its lateral side near its proximal end (Fig. 2. E,G). Armature at the tip of the apical, retractile part of the penis a conical spur, with median surface chitinized and bearing small teeth (Fig. 2. G,H).

Cercopods 1.5–2.8 mm long, distinctly articulated with the telson, with long plumose setae on lateral and median borders (Fig. 3. D).

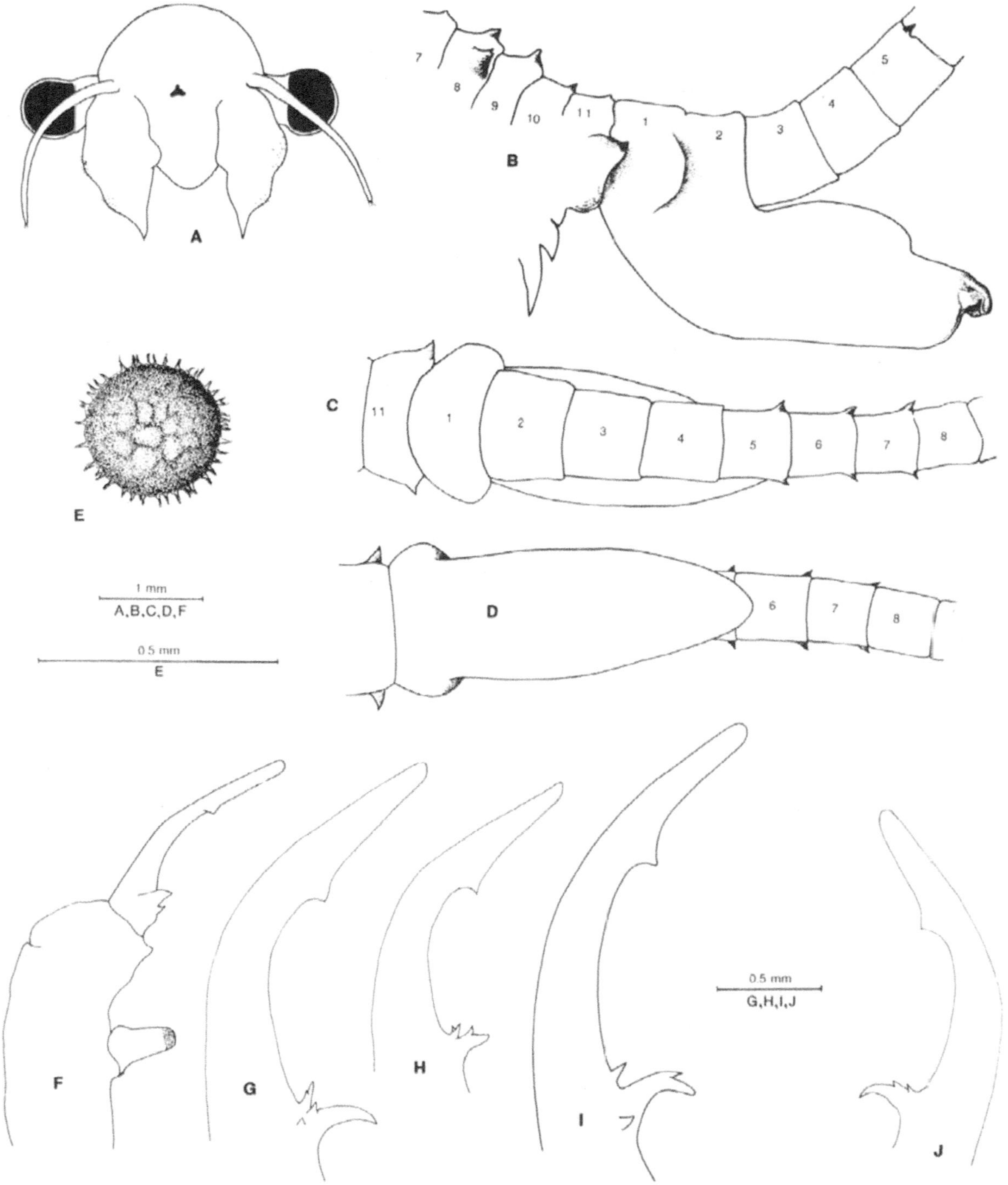


*Fig. 1.* *Chirocephalus brevipalpis* (Orghidan, 1953): A. Head of female, anterior view; **B**. Last thoracic segments, genital segments with ovisac and abdomen of female, lateral view; C. Last thoracic segment, genital segments and abdomen of female, dorsal view; D. Last thoracic segment, ovisac and abdomen of female, ventral view; E. Cyst of preserved specimen (external structure not clearly visible); F. Left antenna of male, median aspect; G,H,I,J. Distal article of antenna of male showing variation of the thornlike outgrowth at the base of the article.

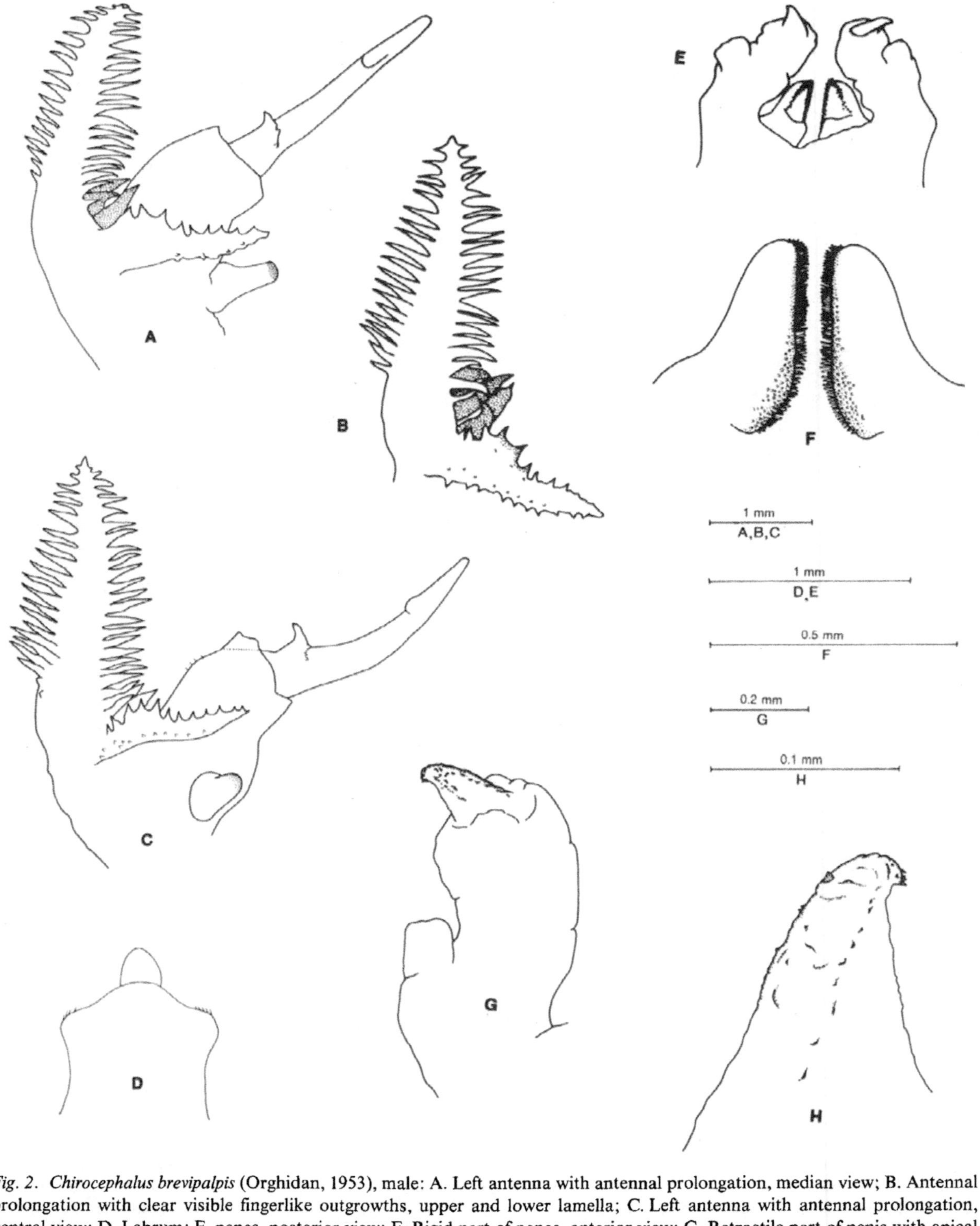


*Fig. 2. Chirocephalus brevipalpis* (Orghidan, 1953), male: A. Left antenna with antennal prolongation, median view; B. Antennal prolongation with clear visible fingerlike outgrowths, upper and lower lamella; C. Left antenna with antennal prolongation, ventral view; D. Labrum; E. penes, posterior view; F. Rigid part of penes, anterior view; G. Retractile part of penis with apical spur; H. Apical spur of the retractile part of penis.

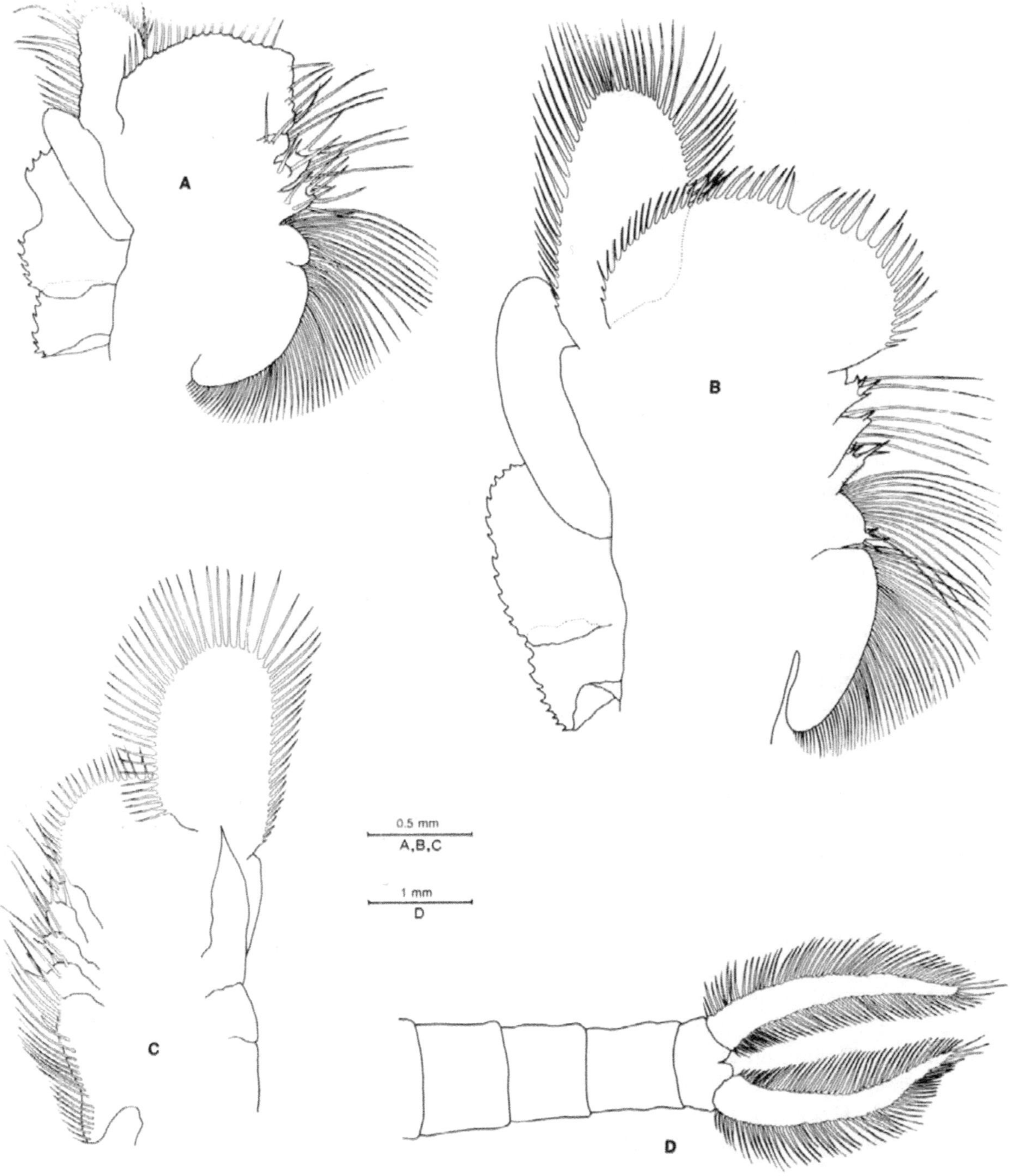


*Fig. 3. Chirocephalus brevipalpis* (Orghidan, 1953), male: A. Left first thoracic appendage; B. Left sixth thoracic appendage; C. Left eleventh thoracic appendage; D. Last three abdominal segments, telson and cercopods.

## Discussion

According to the above description, the Yugoslavian specimens differ in some features from those collected in Romania.

Male: (1) The apical part of the penis bears a short, nearly cylindrical outgrowth on its side, near its proximal end (Fig. 2. E,G), which is absent in specimens from Romania (see Brtek, 1966, Taf. I, Abb. 8,11). (2) The armature at the tip of the apical part of the penis has no knob-like protuberance oposite to the spur in the Yugoslavian specimens (Fig. 2. G). In the specimens from Romania this protuberance is conspicuous (see Brtek, 1966, Taf. I, Abb. 11). (3) The lateral lobe of lower lamella is considerably less developed than that in the specimens from Romania (see Orghidan, 1953, Fig. 30. a). (4) The apophysis of the proximal article of the antenna is conical and thickened at the basis. In specimens from Romania it is fingerlike (see Brtek, 1966, Taf. XI, Abb. 9).

Female: The females from Yugoslavia are morphologically similar to those from Romania. The only noticeable difference concerns the fifth, sixth and seventh abdominal segments, each of which bear a pair of dorsolateral thornlike outgrowths on their postero-lateral margin. In specimens from Romania, these outgrowths are lacking.

Since the described differences are obviously within the morphological variability of the species, the specimens from Yugoslavia have to be considered as representatives of *C. brevipalpis*.

According to present insights in the taxonomy of the genus *Chirocephalus* (see Brtek, 1968), *Chirocephalus brevipalpis* (Orghidan, 1953) belongs to the species group 'bairdi', which includes the following Ponto-Mediterranean species: *Chirocephalus bairdi* (Brauer, 1877), *C. appendicularis* Vavra, 1905, *C. kerkyrensis* Pesta, 1936, *C. kerkyrensis stellae* Brtek, 1966, *C. orghidani* Brtek, 1966 ( = *Palpicephalus recticornis* Orghidan, 1953), *C. vornatscheri* Brtek, 1968, *C. vornatscheri bulgaricus* Flössner, 1980 and *C. pelagonicus* Petkovski, 1986.

Of all these species and subspecies (Daday, 1910b; Brtek, 1966; Flössner, 1980; Petkovski, 1986), *C. orghidani* from South Romania is closest to *C. brevipalpis*.

## Acknowledgements

I thank Prof. Dr. Jan Brtek, Hornonitrianske Museum Prievidza, ČSSR, Prof. Dr. Heinz Löffler, Zoological Institute of the University of Vienna, Austria, and Prof. Dr. Geoffrey Fryer, Windermere, England, for helpful comments.

### References

Botnariuc, N. & T. Orghidan, 1953. Phyllopoda. In: Fauna Republici Populare Române, Vol. 4, Fasc. 2: 1–98.

Brtek, J., 1966. Einige Notizen zur Taxonomie der Familie Chirocephalidae Daday 1910. Annot. Zool. Bot., Bratislava 33: 1–65.

Brtek, J., 1968. *Chirocephalus vornatscheri* n. sp. aus der Türkei. Annot. Zool. Bot., Bratislava 51: 1–5.

Daday de Dées, E., 1910-b. Quelques Phyllopodes Anostracés nouveaux. Appendice a la Monographie Systematique des Phyllopodes Anostracés nouveaux. Ann. Sci. nat. Zool., Paris Ser. 9, 12: 241–264.

Flössner, D., 1980. Über zwei bemerkenswerte Branchiopoda aus Bulgarien – *Chirocephalus vornatscheri bulgaricus* subsp. n. (Crustacea, Anostraca) und *Daphnia chevreuxi* Richard (Crustacea, Cladocera). Acta Zool. Bulg. Acad. Sci. Bulg. Sofia 16: 63–68.

Illies, J., (Editor), 1978. Limnofauna Europaea, 2. Auflage. Stuttgart-New York: Gustav Fischer Verlag, 532 pp.

Orghidan, T., 1948. *Ceratocephalus recticornis* n.gen. n.sp. Ann. Sci. Univ. Jassy II-a, 31: 213–217.

Petkovski, S., 1986. The fairy shrimp *Chirocephalus pelagonicus* n. sp. (Crustacea: Branchiopoda) from Macedonia. Fragm. balc., Mus. maced. sci. nat., Skopje, 13, 3/281: 15–26.

*Hydrobiologia* **212**: 267–272, 1991.
*D. Belk, H. J. Dumont & N. Munuswamy (eds), Studies on Large Branchiopod Biology and Aquaculture.*



# On the Anostraca (Crustacea) of Yugoslavia

Brigita Petrov & Magdalena Marinček
*Institute of Zoology, Faculty of Biology, University of Belgrade, Studentski trg 16, 11000 Belgrade, Yugoslavia*

*Key words:* Anostraca, distribution, Yugoslavia, Branchipus serbicus n.sp.

## Abstract

During these investigations the following species of *Anostraca* have been found: *Chirocephalus diaphanus* Prevost 1803, *Chirocephalus brevipalpis* (Orghidan, 1953), *Streptocephalus torvicornis* Waga 1892, *Branchipus serbicus* Marinček & Petrov 1988, *Branchipus stagnalis* L. (1758) and *Branchipus sp.* Thus, the total number of species discovered in Yugoslavia is 13.

The shape of the frontal shield and the degree of flexure of the distal articles of antennae II in males are of no relevance to taxonomy in the genus *Branchipus* Schaeffer 1776. However, the apophyses on antennae II in males, shape of antennae II and of the ovisac in females, structure of eggs, and of appendages appear species-specific. We conclude that the differentiation of *B. stagnalis* into two forms (forma *typica* and forma *visnyai*) as well as the separation of *B. visnyai* Kertész 1956 as a distinct species are unfounded.

## Introduction

The *Anostraca* of Yugoslavia were studied by Daday (1910a, b), Marcus (1913), Marinček (1961, 1962), Brtek (1962, 1976), and S. Petkovski (1986, 1990), who described a new species *Chirocephalus pelagonicus*, from Macedonia. A list of 11 species known from Yugoslavia by the time our investigations started is presented in Table 1. The greatest number belongs to the *Chirocephalidae* Daday 1910. *Chirocephalus croaticus* (Steuer, 1899), *Chirocephalus reiseri* Marcus 1913 and *Chirocephalus pelagonicus* S. Petkovski 1986 are restricted to Yugoslavia. The majority of these species has been found in single localities only, because up to now no detailed studies on their distribution has been carried out. Still, the total number of species found in Yugoslavia is relatively great and similar to that of the species discovered in neighbouring countries Hungary (9) and Roumania (11) as well as in Czechoslovakia (9). Bearing in mind Yugoslavia's geographical situation, great variety of landscapes and ecological conditions, such a rich fauna was to be expected.

## Materials and methods

We initiated our studies in the *Anostraca* of Yugoslavia seven years ago. Our present investigations are mainly concerned with the north and east (i.e. areas of the southern border of the Pannonian Plain and with Serbia), and partly also with Montenegro.

Material was collected from ephemeral ponds and fixed in 70% ethanol, and is stored in the Institute of Zoology, Faculty of Biology, University of Belgrade.

*Table 1.* Species of Anostraca found in Yugoslavia before our investigations with first citation and collecting site.

| | | |
|---|---|---|
| Artemiidae | | |
| Artemia salina (Linnaeus, 1758) | Istra | Daday, 1910 |
| Branchinectidae | | |
| Branchinecta ferox Milne-Edwards, 1840 | Melenci | T. Petkovski (in Brtek, 1976) |
| Branchipodidae | | |
| Branchipus stagnalis (syn. schaefferi) Fischer, 1834 | Varaždin | Daday, 1910 |
| Tanyimastix stagnalis (Limnaeus, 1758) | Golemo Konjari | S. Petkovski, 1986 |
| Chirocephalidae | | |
| Chirocephalus carnuntanus (Brauer, 1877) | Melenci | T. Petkovski (in Brtek, 1976) |
| Chirocephalus croaticus (Steuer, 1899) | Blatačko Jezero | Daday, 1910 |
| Chirocephalus diaphanus Prevost, 1803 | Gospić, Široka Kula, Vrhovina, Crna Gora | Daday, 1910 |
| | Kočani, Sokolarci | T. Petkovski (in Brtek, 1976) |
| Chirocephalus reiseri Marcus, 1913 | Treskavica, Zelengora | Marcus, 1913 |
| Chirocephalus pelagonicus S. Petkovski, 1986 | Golemo Konjari | S. Petkovski, 1986 |
| Chirocephalus brevipalpis (Orghidan, 1953) | Kupinovo | S. Petkovski, 1990 |
| Streptocephalidae | Blatačko Jezero | Daday, 1910 |
| Streptocephalus torvicornis Waga, 1842 | Zrenjanin | Marinček, 1961 |

## Results and discussion

In the territory studies, three families occur: *Chirocephalidae* Daday 1910, *Streptocephalidae* Daday 1910 and *Branchipodidae* Daday 1910 (Fig. 1).

The family *Chirocephalidae* is represented by two species of *Chirocephalus* Prevost 1803. *Chirocephalus diaphanus* Prevost 1803 was found in Serbia in the vicinity of the village Ratari near Mladenovac, near Osipaonica as well as between Osipaonica and Požarevac. Besides, it was discovered in three localities in Montenegro (two localities in the vicinity of Vilusi and one on Durmitor Mt.), and in south Serbia near Preševo. All these localities are in the Dinaric West Balkan (according to Illies, 1978). In the Pannonian Plain we did not find this species. In addition to *Ch. diaphanus* we found one other species which is similar to *Chirocephalus brevipalpis* (Orghidan, 1953), but differs from the typical form in some morphological pecularities. It occurs in two localities west of Belgrade and in one locality somewhat more north, all in the south Pannonian Plain.

The family *Streptocephalidae* is represented by *Streptocephalus torvicornis* Waga 1892, a species most frequently found in plains north of the Danube and Sava near Čokadinci, Vrbanja, Jamena, Kuzmin and Jarak, as well as near Zrenjanin and Vojvoda Stepa in Banat. Besides, it was found south of the Danube and Sava in Serbia, in the vicinity of the Velika Morava near Osipaonica and Smederevska Palanka. This was the first report for the Dinaric West Balkan. *S. torvicornis* was also found near to the borders of Roumania and Bulgaria, on the Radujevac-Negotin route. This locality is in the East Balkan, and is the first citation of this species in this region.

The family *Branchipodidae* is most common in the territory studied.

Far in the northeast, along the road Vrbnica-Banatsko Arandjelovo, we recently discovered *Branchipus serbicus* Marinček & Petrov 1988. This has so far remained the only finding of this species. *B. serbicus* is characterized by the presence of posterolateral elongations on the abdominal segments of males and females (one pair of dorsal and one pair of ventral elongations), by the presence of a conical elevation on the dorsum of the head at the insertion of antennae I, by the apophyses on antennae II in males and by the terminal segment of apical article of antennae II in males, which is bent backwards (Fig. 2).

*Branchipus stagnalis* L. (1758) (syn. *schaefferi* Fischer, 1834) is common in flats north of the Danube and Sava (Fig. 1).

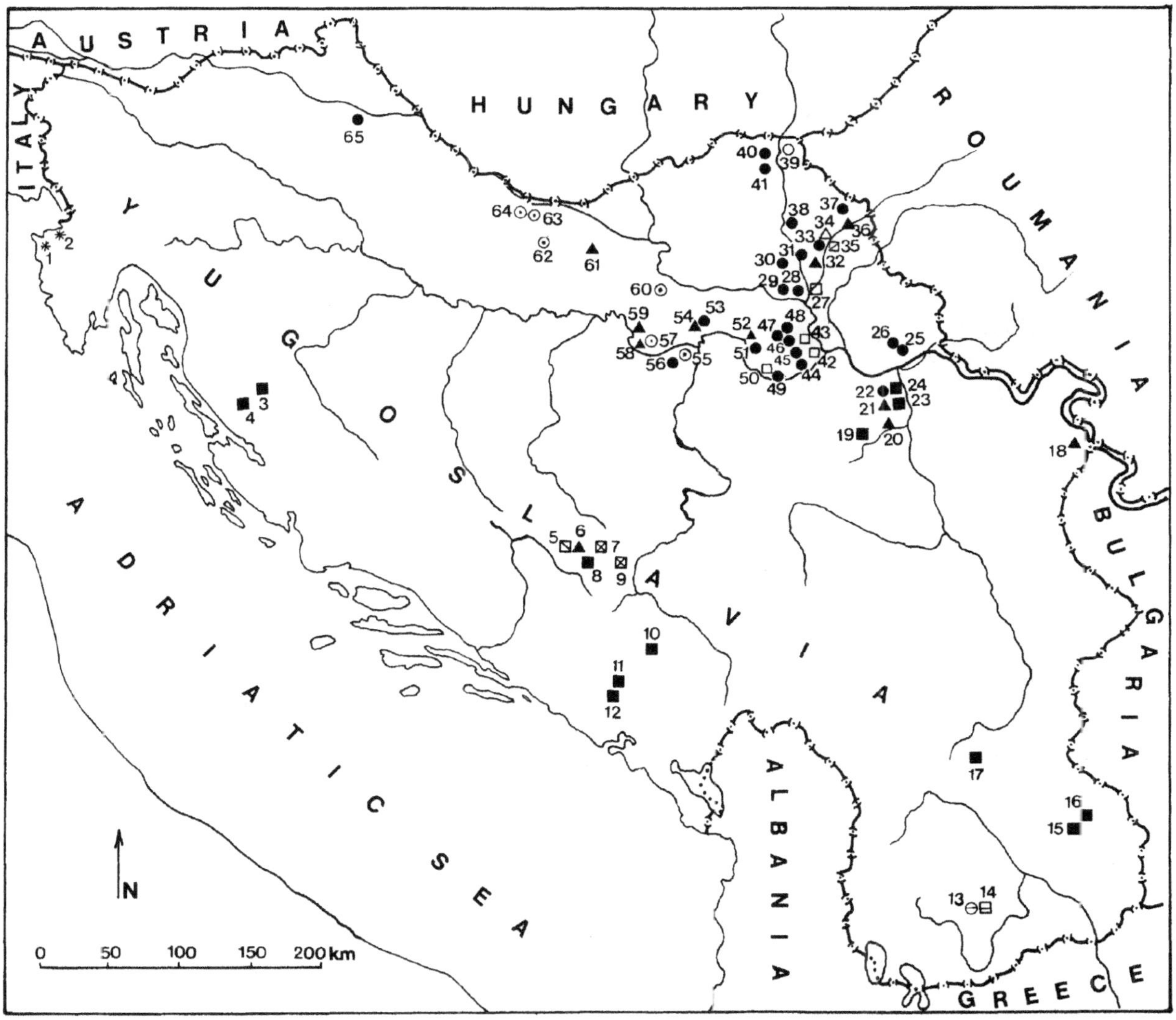


* Artemia salina
△ Branchinecta ferox
● Branchipus stagnalis
⊙ Branchipus sp.
○ Branchipus serbicus
⊖ Tanymastix stagnalis
⧄ Chirocephalus carnuntanus
⧅ Chirocephalus croaticus
⊠ Chirocephalus reiseri
■ Chirocephalus diaphanus
□ Chirocephalus brevipalpis
⊟ Chirocephalus pelagonicus
▲ Streptocephalus torvicornis

*Fig. 1.* Distribution map of Anostracan species in Yugoslavia (1 – Piran, 2 – Kopar, 3 – Široka Kula, 4 – Gospić, 5 – Blatačko Jezero, 6 – Blatačko Jezero, 7 – Treskavica, 8 – Vrhovina, 9 – Zelengora, 10 – Durmitor, 11 – Vilusi, 12 – Vilusi, 13 – Golemo Konjari, 14 – Golemo Konjari, 15 – Sokolarci, 16 – Kočani, 17 – Preševo, 18 – Radujevac-Negotin, 19 – Ratari, 20 – Smederevska Palanka, 21 – Osipaonica, 22 – Osipaonica, 23 – Osipaonica, 24 – Osipaonica-Požarevac, 25 – Deliblato, 26 – Volovska Paša, 27 – Perlez, 28 – Titel-Mošorin, 29 – Lok, 30 – Žabalj, 31 – Taraš, 32 – Zrenjanin, 33 – Melenci, 34 – Melenci, 35 – Melenci, 36 – Vojvoda Stepa, 37 – Srpska Crnja, 38 – Novi Bečej, 39 – Vrbnica-Banatsko Arandjelovo, 40 – Kanjiža, 41 – Nadrljan, 42 – Bečmen, 43 – Petrovčić, 44 – Progar, 45 – Bečmen, 46 – Deč, 47 – Prhovo, 48 – Šimanovci, 49 – Progar-Kupinovo, 50 – Kupinovo, 51 – Nikinci, 52 – Jarak, 53 – Kuzmin, 54 – Kuzmin, 55 – Bjeljina (Brodac), 56 – Bjeljina (Crnjeljevo Donje-Batković), 57 – Jamena, 58 – Jamena, 59 – Vrbanja, 60 – Vukovar-Tovarnik, 61 – Čokadinci, 62 – Podravska Slatina-Orahovica, 63 – Podravska Moslovina-Podravska Slatina, 64 – Kapinci, 65 – Varaždin)

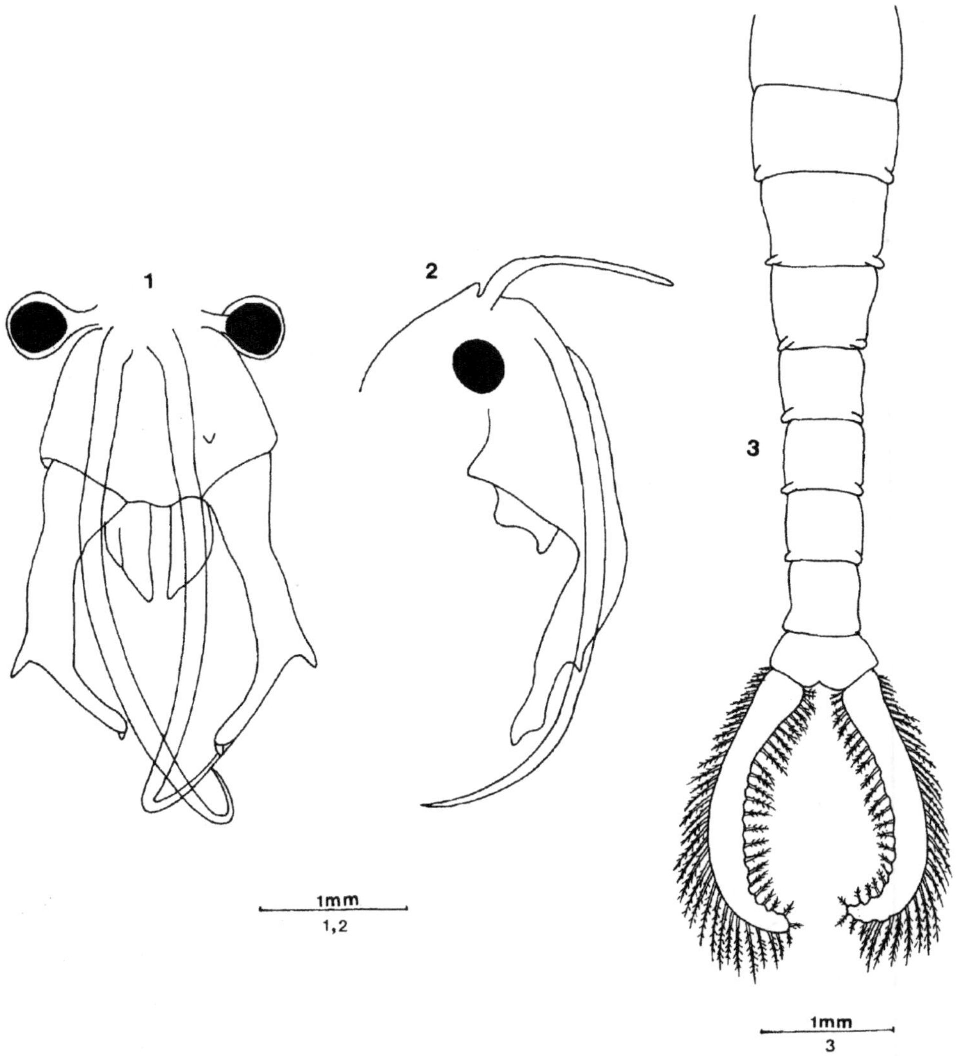


*Fig. 2.* Appearance of the head (1 – dorsal, 2 – lateral view) and the abdomen (3) of males in Branchipus serbicus Marinček et Petrov 1988 from the type locality Vrbnica-Banatsko Arandjelovo (24.6.1971.)

However, in several localities of Slavonija and Srem, between the Danube and Sava, near Jamena, in the vicinity of Bijeljina, between Vukovar and Tovarnik, between Podravska Moslovina and Podravska Slatina, at Kapinci near Podravska Moslovina, and between Podravska Slatina and Orahovica, populations of a *Branchipus* were found which differ from *B. stagnalis*. All these localities, except one (in the vicinity of Bijeljina), are in the Pannonian Plain, and they are to the west of those of *B. stagnalis*. Males from these localities differ from *B. stagnalis* by having more or less prominent posterolateral elongations on the abdominal segments. Besides, they are characterized by a little cone-like elevation on the dorsum of the head at the level of insertion of the antennae I. In females setae occur on the proximal part of the external edge of antennae II. By the presence of the posterolateral elongations, these populations are probably transitional between *B. stagnalis* without such elongations, and *Branchipus intermedius* Orghidan

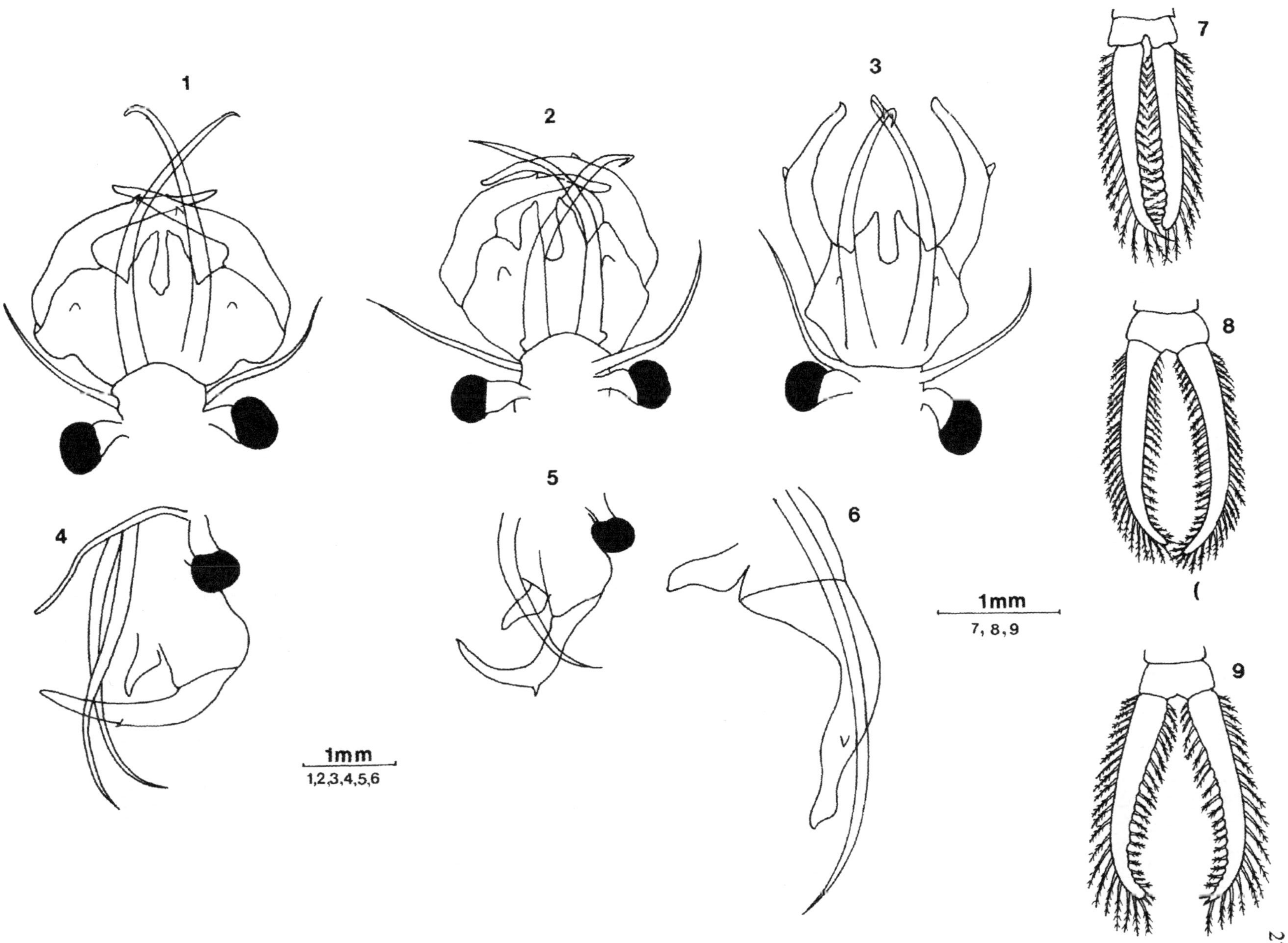


*Fig. 3.* Variability of head (1,2,3-dorsal, 4,5,6,-lateral view) and furca (7,8,9) in males of Branchipus stagnalis L. (1758). All specimens are from the same population collected near Kuzmin (1.7.1981.).

1953 and *Branchipus blanchardi* Daday 1910 which do have them. Our populations represent a distinct form, perhaps a new species.

A detailed study on the populations of the genus *Branchipus* from these localities revealed variability in some morphological characters (Marinček & Petrov, 1988b). This may contribute to the elucidation of taxonomic relationships within the genus *Branchipus*, and particularly to the *B. stagnalis* complex.

In most males, the frontal shield is hexagonal and less wide than the span between the eyes (Fig. 3). However, in several specimens it widens as to become rectangular, and considerably exceeds the distance between the eyes. Apical articles of antennae II in most males converge, but usually do not meet. However, in specimens with rectangle-like shield distal parts of antennae II are so strongly bent toward each other that they overcross. In side view, the majority of antennae II resemble the letter S, but in specimens with rectangular shield they are reminiscent of an L or even a V. The curvature of the male furca may also vary, but the furcal rami are turned inward only in their distal portions. The furcal setae and their distribution are stable.

Brtek (1962) reported similar variation of the frontal shield, the degree of bending of antennae II in males, and variations in the furca in *B. stagnalis* from Czechoslovakia.

Bearing in mind that all this variation occurs within populations, we can only conclude that this characters have no taxonomic value. In contrast, we have established that apophyses on antennae II in males, shape of antennae II and of the ovisac in females, structure of eggs and of appendages, are stable characters which may be of significance for the distinction of species within *Branchipus*.

Exactly on the basis of the characteristics showing greatest variability (shape of the frontal shield and the apical articles of antennae II, and partly the shape of the furca in males), Kertész (1956) differentiated two forms within the species *B. stagnalis*. Forma *typica* is characterized by a hexagonal shield, apical articles of antennae II in males slightly turned (antennae resemble the letter S), and furcal rami sicle-shaped and bent inwardly. Forma *visnyai* has a wide, rectangle-shaped frontal shield, sharply curved apical articles of male antennae II (antennae reminiscent of the letter L) and furcal rami which converge only in their terminal third.

Cottarelli (1969) maintains that full species rank should be given to forma *visnyai*: *Branchipus visnyai* Kertész 1956, a position which can no longer be maintained.

## References


Botnariuc, N. & T. Orghidan, 1953. Phyllopoda. Fauna Populare Romane 4, 2: 1–98.

Brtek, J., 1962. Anostraca. In Šramek-Hušek, R. *et al.*, Lupenonožci-Branchiopoda. Fauna ČSSR 16, 103–144.

Brtek, J., 1976. Anostraca, Notostraca, Conchostraca a Calanoida Slovenska (1. čast). Acta Rer. Natur. Mus. Nat. Slov., Bratislava 22, 19–80.

Cottarelli, V., 1969. Osservazioni tassonomiche su Branchipus visnyai Kertész e diagnosi preliminare di Branchipus passai n.sp. (Crustacea, Euphyllopoda, Anostraca). Fragm. ent. 6: 63–68.

Daday de Dées, E., 1910a. Monographie sistematique des Phyllopodes Anostracés. Ann. Sci. nat. Zool., Paris 9, 11: 91–489.

Daday de Dées, E., 1910b. Quelques Phyllopodes Anostracés nouveaux. Ann. Sci. nat. Zool., Paris 9, 12: 241–264.

Illies, J. (ed.), 1978. Limnofauna Europea, 2. Auflage. Gustav Fischer Verlag, Stuttgart-New York, 532 pp.

Kertész, G., 1956. Die Anostraca-Art Branchipus stagnalis (L.) und ihre Formen. Opusc. zool.: 33–40.

Marcus, K., 1913. Beiträge zur Kenntnis des Süsswasserfauna der nordwestlichen Balkanhalbinsel. Zool. Anz. 41: 409–412.

Marinček, M., 1961. Neke Eufilopode (Euphyllopoda) Panonske Nizije. Zbornik M.S. za prirodne nauke 20: 160–163.

Marinček, M., 1962. Prilog poznavanju faune Phyllopoda Panonske Nizije. Zbornik II kongresa biologa Jugoslavije, knjiga rezimea: 203–204.

Marinček, M. & B. Petrov, 1988a. On a new species of the genus Branchipus (Anostraca, Crustacea) discovered in Yugoslavia. Zbornik radova o fauni Srbije, SANU 5 (in press).

Marinček, M. & B. Petrov, 1988b. Taxonomical investigations within the genus Branchipus (Anostraca, Crustacea). Zbornik radova o fauni Srbije, SANU 5 (in press).

Petkovski, S., 1986. The fairy shrimp Chirocephalus pelagonicus n.sp. (Crustacea: Branchiopoda) from Macedonia. Mus. maced. sci. nat. 13, 3/281, 15–25.

Petkovski, S., 1991. The fairy shrimp *Chirocephalus brevipalpis* (Orghidan, 1953), (Crustacea: Anostraca) in Yugoslavia. In Belk, D., H. J. Dumont & N. Munuswany (eds), Studies on Large Branchiopod Biology and Aquaculture. Developments in Hydrobiology 64. Kluwer Academic Publishers, Dordrecht: 261–266. Reprinted from Hydrobiologia 212.

*Hydrobiologia* **212**: 273–282, 1991.
*D. Belk, H. J. Dumont & N. Munuswamy (eds), Studies on Large Branchiopod Biology and Aquaculture.*



# A review of the Conchostraca (Crustacea) of Yugoslavia

Madgalena Marinček & Brigita Petrov
*Institute of Zoology, Faculty of Biology, University of Belgrade, Studentski trg 16, 11000 Belgrade, Yugoslavia*

*Key words:* Conchostraca, Yugoslavia, taxonomical revision

## Abstract

We studied the taxonomy and distribution of the *Conchostraca* in Yugoslavia. A diverse fauna occurs in the country. However, the systematics of this group is still incomplete and has many problems. There is considerable variability of morphological characters. By a comparative-morphological analysis of the populations found, we re-evaluated the taxonomic significance of a series of structural characters, and offer comments on the results.

In Yugoslavia four families were found: *Cyzicidae* Stebbing, 1910; *Limnadiidae* Burmeister, 1843; *Imnadiidae* Botnariuc & Orghidan, 1941 and *Leptestheriidae* Daday, 1923, with a present total of 8 species and 4 subspecies. Within the genus *Imnadia* Hertzog, 1935, three new species were found: *Imnadia cristata* Marinček, 1972; *Imnadia banatica* Marinček & Valvajter, 1982 and *Imnadia panonica* Marinček & Petrov, 1984. Within the genus *Eoleptestheria* Daday, 1923, the species *Eoleptestheria spinosa* Marinček, 1978, with four distinct subspecies was discovered.

## Introduction

Around 1970, only Daday 1923 and Straškaba 1966 had recorded the presence of some *Conchostraca* (the genera *Eoleptestheria* Daday 1923 and *Leptestheria* Sars, 1896) from Yugoslavia, but all detail was lacking. We therefore started a program of extensive studies on the distribution and systematics of these crustaceans, which is still ongoing today.

## Materials and methods

Our investigations were performed mainly in the north of Yugoslavia, i.e. the Pannonian Plain, but also south of the Danube and Sava, concentrating on the fauna of ephemeral rain pools. During these investigations many ephemeral ponds (few hundreds) have been explored, and in 40 localities *Conchostraca* were found.

The material was collected with a hand net, with mesh size 1 mm or less, and fixed in 4% formalin or 70% ethyl-alcohol. For microscopic examination, appendages were dissected and mounted in glycerin.

## Results

We discovered representatives of 4 families with 5 genera, 8 species and some subspecies. Several were new to science (Marinček, 1972, 1978; Marinček & Petrov, 1983, 1984; Marinček & Valvajter, 1979, 1982a, b).

In the course of our investigations, particular emphasis was given to the comparative morphological analysis of populations, in order to

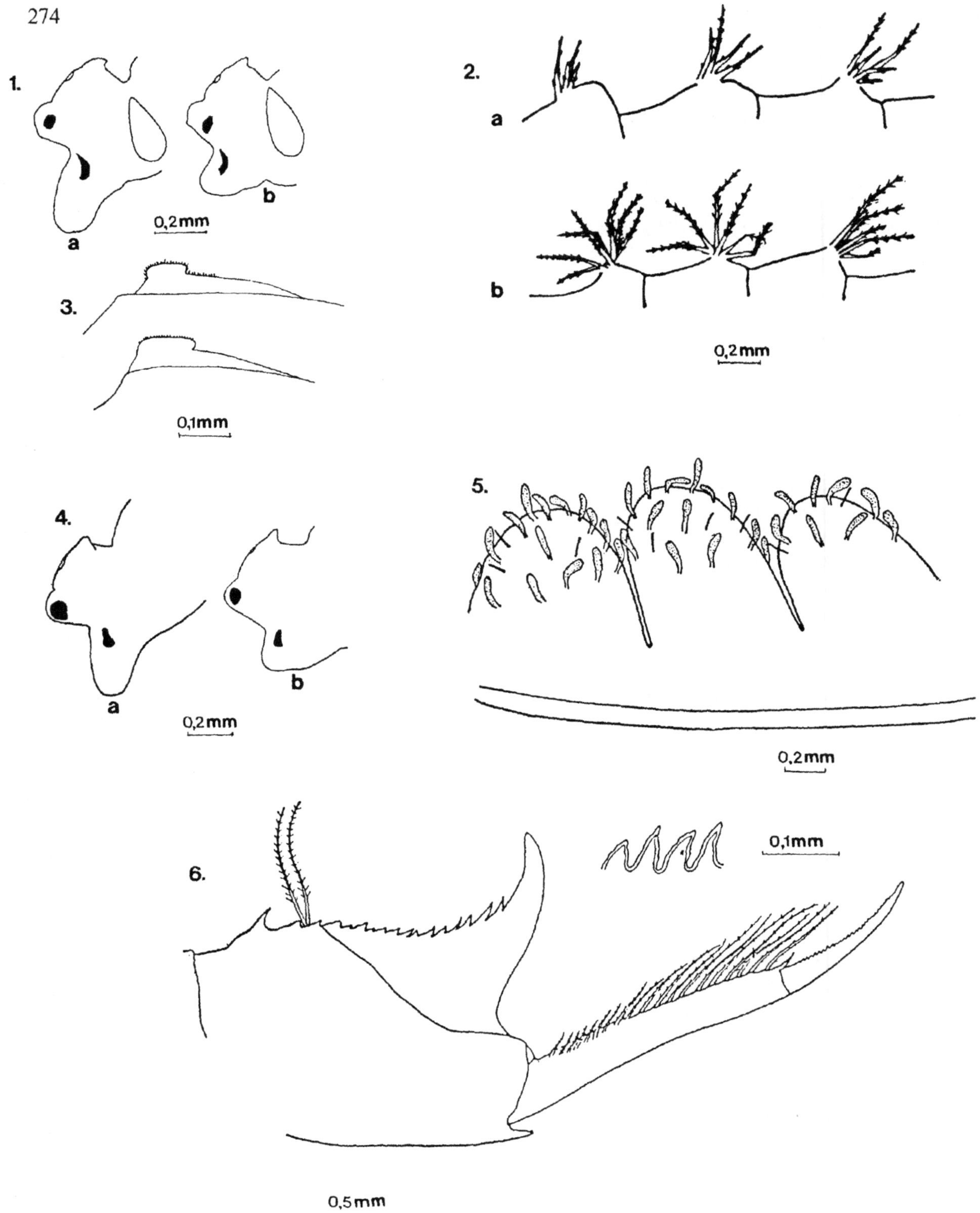


*Figs 1–6. Imnadia cristata*: Fig. 1. Head; a-male, b-female Fig. 2. Mediodorsal part of trunk segments; a-male, b-female Fig. 3. Ridge on the first pair of female appendages

*Imnadia banatica*: Fig. 4. Head; a-male, b-female Fig. 5. Part of the male antennule Fig. 6. Male telson

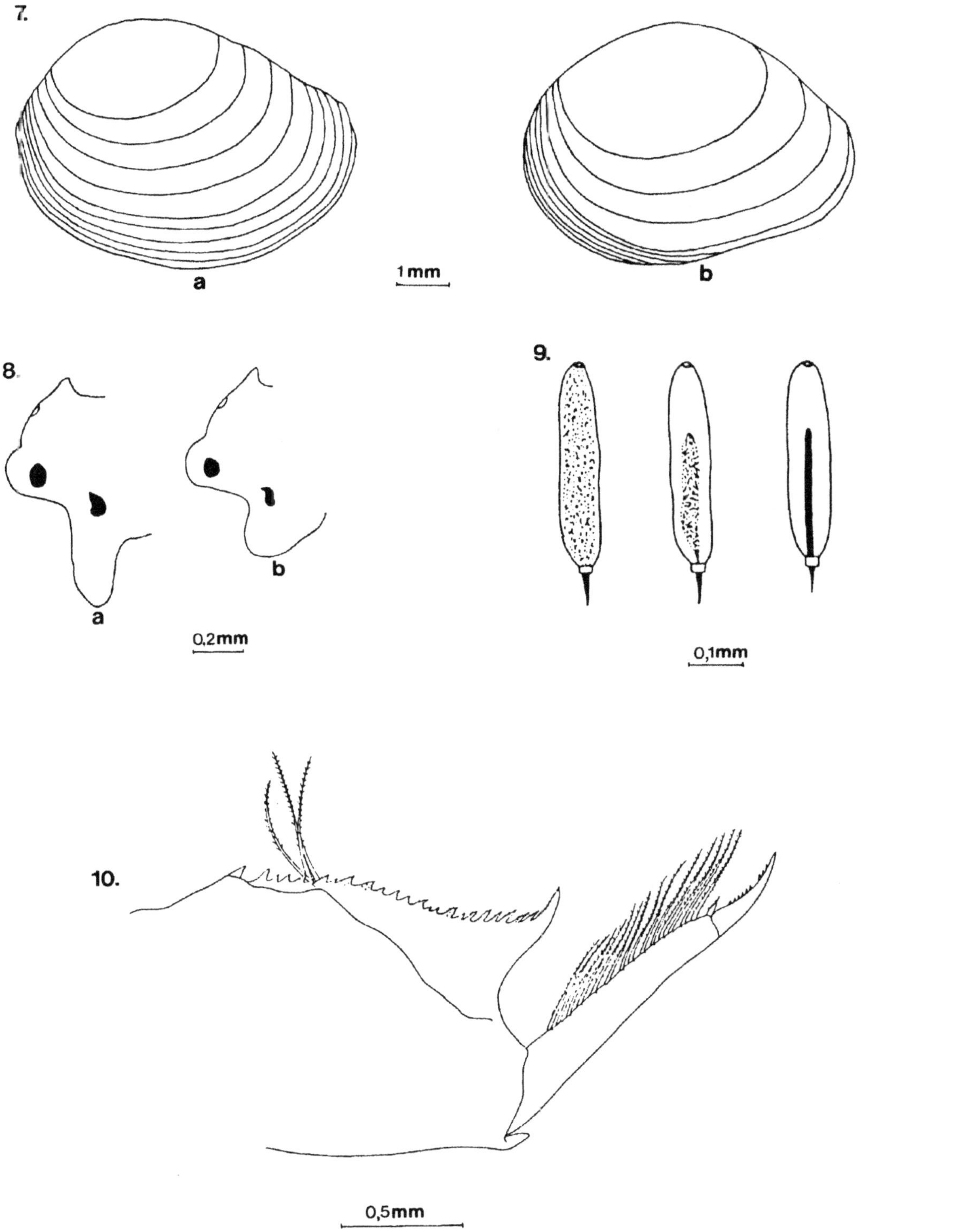


*Figs 7–10. Imnadia panonica*: Fig. 7. Shell; a-male, b-female Fig. 8. Head; a-male, b-female Fig. 9. Sensory organs on antennules Fig. 10. Female telson

11.

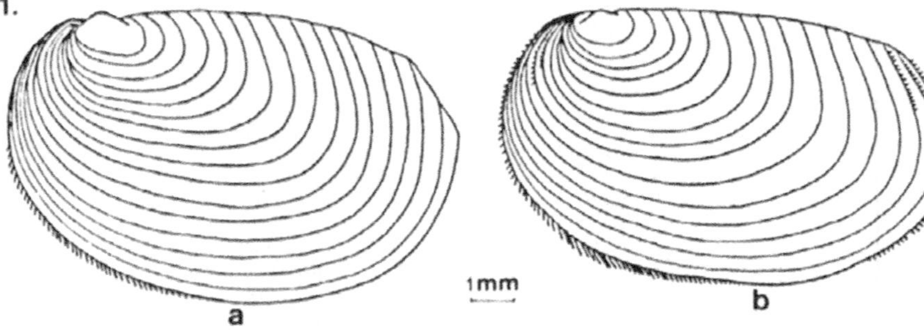


12.

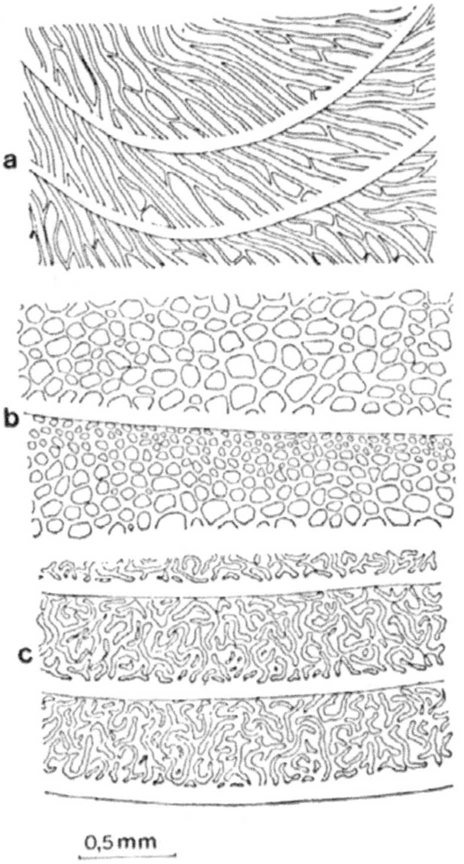


13.

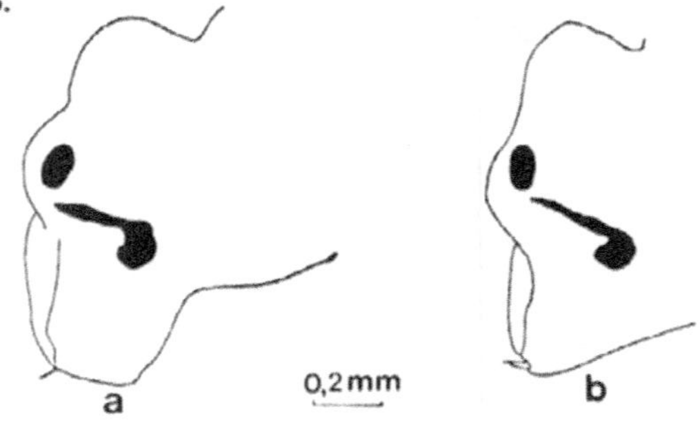


14.

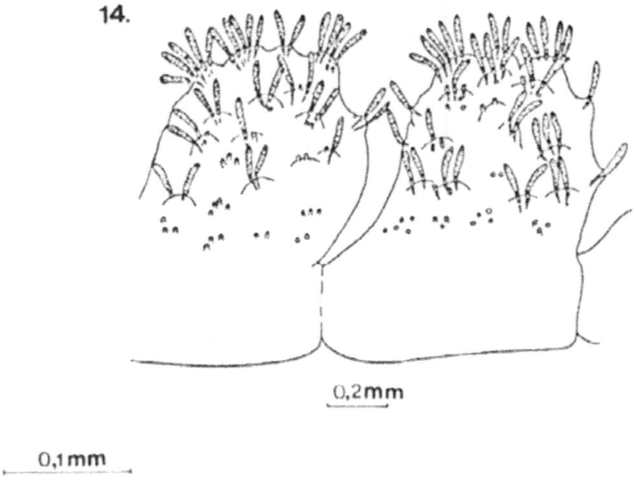


15.

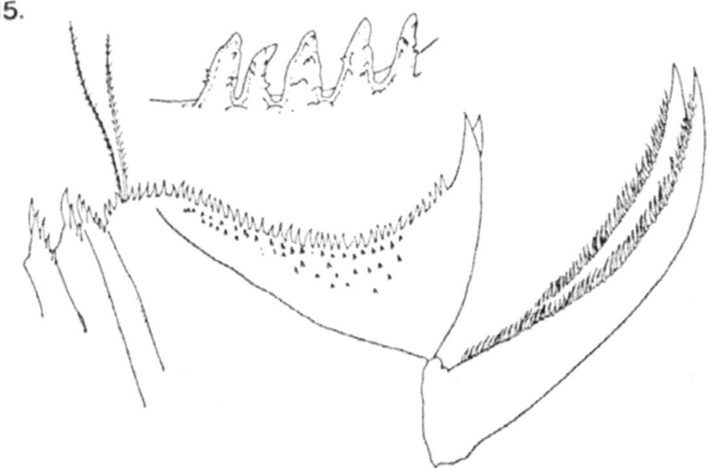

16.

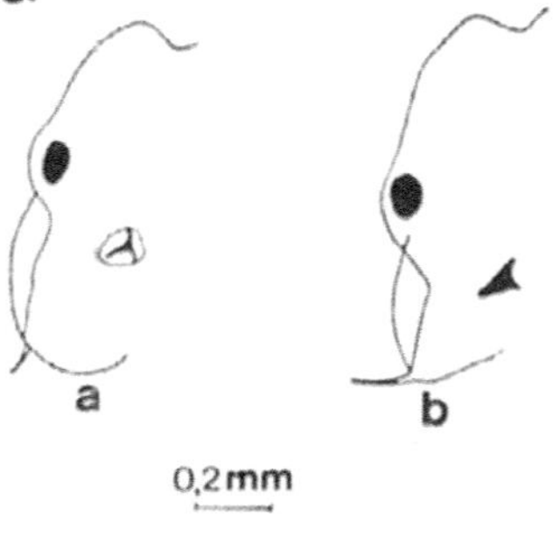


17.

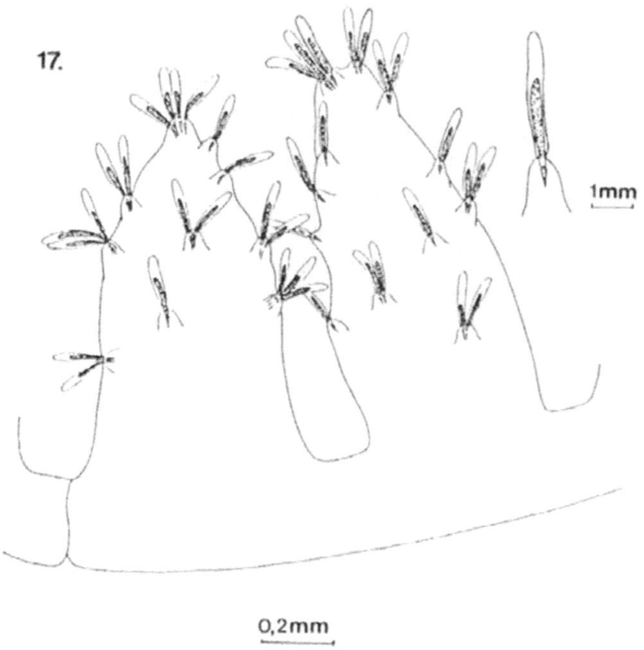


18.

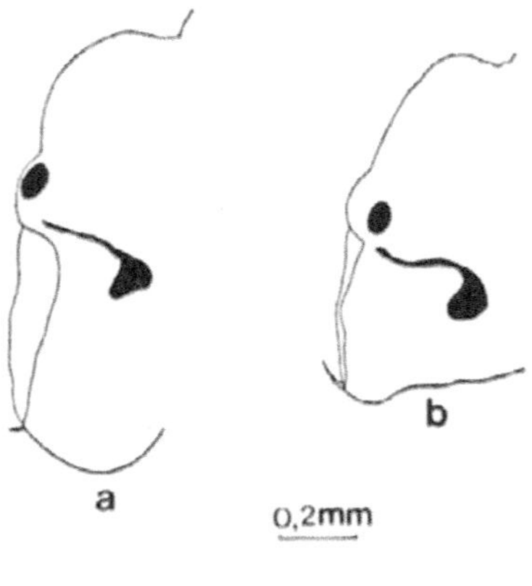


19.

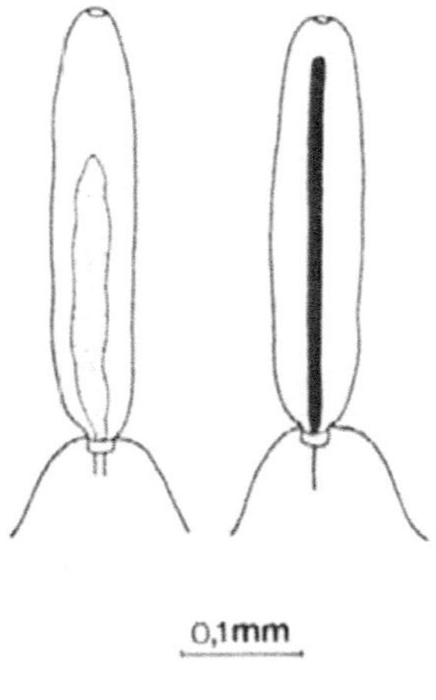


*Figs 16–19. Eoleptestheria spinosa tenuis*: Fig. 16. Head; a-male, b-female Fig. 17. Pappilae and sensory organs on male antennules

*Eoleptestheria spinosa magna*: Fig. 18. Head; a-male, b-female Fig. 19. Sensory organs on female antennules

---

*Figs 11–15. Eoleptestheria spinosa*: Fig. 11. Shell; a-male, b-female Fig. 12. Microscopic structure of shell; a-oldest zone, b-mid zone, c-peripheral zone Fig. 13. Head; a-male, b-female Fig. 14. Part of male antennule Fig. 15. Male telson check the taxonomic value of some much used structural characters. In addition to characters previously used, we also studied details of structure of the antennules and trunk limbs. Antennules were examined for presence or absence of segmentation, number of segments, number and shape of sensory papillae, and structure and distribution of sensory organs. As for appendages, we examined the number, distribution and shape of setae and spines on certain endites on the anterior and posterior side of the trunk appendages, and the presence and structure of a palpus endopoditalis and palpus enditalis. We also studied the shape of epipodites, the presence and shape of the cone at their base, of the suture at the foot of endite 5 or 6, and the presence and appearance of a longitudinal ridge.

We found that a broad variability occurs in many of the characters examined, while others are relatively stable, and of significance to the taxonomy of the group. As far as the valves are concerned, useful criteria are: general shape, distribution of growth lines, presence and distribution of setae along the edge and the growth lines, and microscopic structure of various regions of the shell. Significant characteristics of the head are: shape of the nauplius eye and the rostrum in the given age category. It should be noted that the shape of the occipital region is not a reliable character for the taxonomy of *Leptestheriidae* Daday 1923, nor is the presence of spines at the top of the rostrum. We also found that antennules of *Conchostraca* have a number of stable morphological characteristics, such as the shape of sensory papillae and the structure of sensory organs, which have not yet been sufficiently taken into account in taxonomic work. In contrast, the structure of the antennae is of no importance for the taxonomy of these forms. Among characteristics of the trunk, a particularly significant charac-

20.

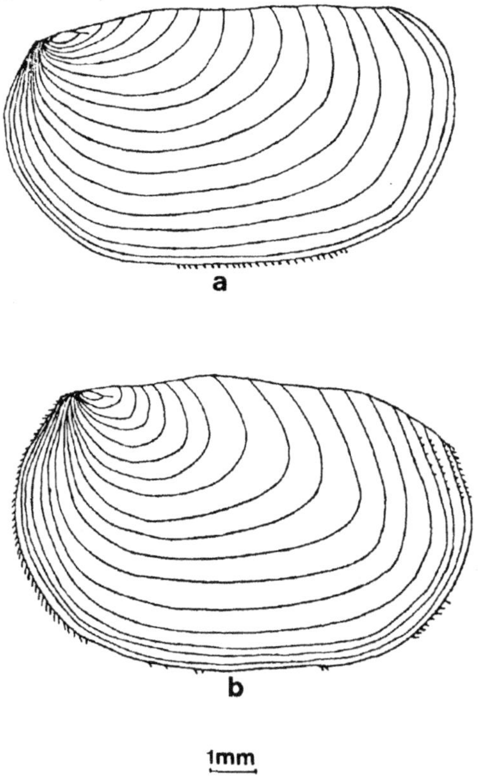


21.

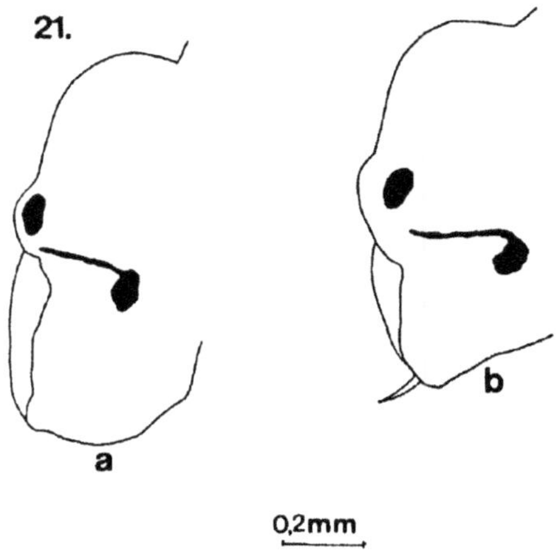


*Figs 20–21. Eoleptestheria spinosa mira*: Fig. 20. Shell; a-male, b-female Fig. 21. Head; a-male, b-female

ter is the appearance of the mediodorsal part of segments; this character too has not been given sufficient importance before. We concluded that the differentiation of genera within *Leptestheriidae* should be primarily based on the characteristics of the mediodorsal part of the trunk segments, and not on the appearance of the occipital region.

The detailed structure of the trunk limbs has not been sufficiently used for the taxonomy of these forms. Yet, almost all their structural features may be suitable criteria in this respect, except for the tuberculum on the first and second pairs of male appendages. The presence of a palpus endopoditalis and palpus enditalis has relative value. However, just these characters were previously given most taxonomic weight. All characteristics of the telson, especially by appearance of dorsal spines, may be useful for the evaluation of taxonomic relationships. The same holds true for the structure of eggs. However, within the family *Leptestheriidae* the number of appendages modified for egg bearing is not stable.

On the basis on these detailed comparative morphological analyses, we could establish the presence of representatives of families *Cyzicidae* Stebbing, 1910, *Limnadiidae* Burmeister 1843, *Imnadiidae* Botnariuc & Orghidan 1941 and *Leptestheriidae* Daday 1923 in Yugoslavia.

The *Cyzicidae* are represented by the genus *Cyzicus* Audoin, 1837. *Cyzicus tetracerus* Kryncki, 1830 was found in the surrounding of Zrenjanin, whereas *Cyzicus sp.* was discovered in several localities, both north and south of the Danube and Sava.

The *Limnadiidae* are represented by the species *Limnadia lenticularis* Linnaeus, 1761, which has been found in only two localities (near Zrenjanin and in the vicinity of Osijek), north of the Danube and Sava, and in a small number of specimens, all females.

The family *Imnadiidae* is of particular interest. By the time our investigations started, only two species were known one discovered in Roumania and other in France. We discovered three new species of this genus in Yugoslavia.

*Imnadia cristata* Marinček 1972 was found in central Banat near the village of Bočar. It has been given species rank on the grounds of the specific appearance, number and arrangement of spines and setae on the mediodorsal side of the trunk segments, characteristics of antennules and their segmentation, appearance of sensory organs and some structural specifities of the appendages (Figs 1–3). On its first pair of appendages, females have one elongated ridge, similar to that present in *Imnadia voitestii* Botnariuc & Orghidan 1941 (only their proximal apexes differ), plus one additional, prominent ridge formed by the extreme longitudinal bending of these appendages.

*Imnadia banatica* Marinček & Valvajter 1982 was discovered in the extreme northeast of Yugoslavia, near Novi Kneževac. It is characterized by the shape of the nauplius eye, and the antennules, and by having two types of sensory organs (Figs 4–6). In this respect it is intermediate between *I. voitestii* and *I. cristata.* However, *I. banatica* is also distinguished by specific spines on the telson and by structural specificities of the appendages.

*Imnadia panonica* Marinček & Petrov 1984 was discovered in northern Yugoslavia (surrounding of Kikinda). It is characterized by the appearance of the valves (Figs 7–10). Their dorsal part in both sexes is very prominent; maximum height in males is on the anterior half. Besides, it has a considerably greater number of growth lines (15) than observed in any other known species. However, this family was previously thought to be best characterized precisely by a small number of, marginnaly distributed, growth lines. This species is also distinguished by having three types of sensory organs on the antennules, by the shape of the female telson, and by structural characters on the appendages.

The *Leptestheriidae* are represented by two genera: *Eoleptestheria* Daday 1923, and *Leptestheria* Sars, 1896.

Within *Eoleptestheria*, the species *Eoleptestheria spinosa spinosa* Marinček 1978 was discovered between Kikinda and Senta. It has been diagnosed on evidence of shell characters, presence of setae on the shell edge, and growth lines, shape of the nauplius eye, some structural specificities of the appendages, and structure of

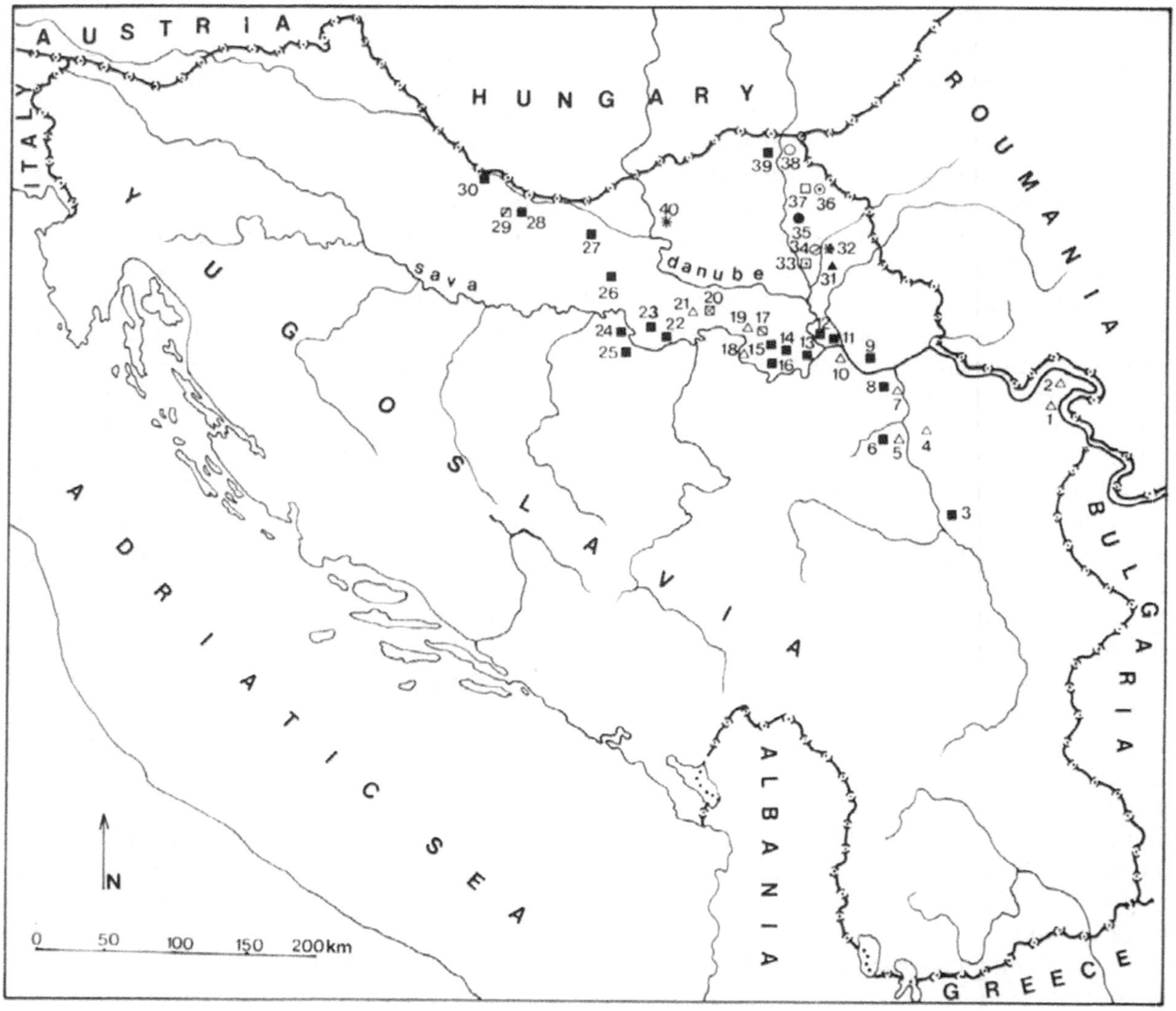


- ▲ Cyzicus tetracerus
- △ Cyzicus sp.
- ✳ Limnadia lenticularis
- ● Imnadia cristata
- ○ Imnadia banatica
- ⦶ Imnadia panonica
- ⊖ Imnadia sp.
- □ Eoleptestheria spinosa spinosa
- ▨ Eoleptestheria spinosa tenuis
- ⧅ Eoleptestheria spinosa magna
- ⊠ Eoleptestheria spinosa mira
- ⊡ Leptestheria dahalacensis
- ■ Leptestheria sp.

*Fig. 22.* Distribution of Conchostraca in Yugoslavia (1-Negotin, 2-Grabovica, 3-Donje Vidovo, 4-Žabari, 5-Velika Plana, 6-Velika Plana, 7-Osipaonica, 8-Osipaonica, 9-Omoljica, 10-Beograd (Ada Huja), 11-Krnjača, 12-Kovilovo, 13-Progar, 14-Petrovčić, 15-Donji Tovarnik, 16-Pećinci, 17-Jarak, 18-Klenak, 19-Jarak, 20-Kuzmin, 21-Kuzmin, 22-Jamena, 23-Vrbanja, 24-Krepšić, 25-Brčko, 26-Stari Mikanovci, 27-Čokadinci, 28-Podravska Slatina (Sopje), 29-Podravska Slatina (Sopje), 30-Terezino Polje, 31-Zrenjanin, 32-Zrenjanin, 33-Zrenjanin, 34-Zrenjanin, 35-Bočar, 36-Sajan, 37-Padej, 38-Novi Kneževac, 39-Kanjiža, 40-Osijek).

the telson, which has tiny teeth on the sides of its ridges (Figs 11–15). In addition to the nominal subspecies, three other subspecies were described.

*Eoleptestheria spinosa tenuis* Marinček & Valvajter 1979 was found considerably further to the west (near Podravska Slatina). It is characterized by a different distribution of setae on the shell, shape of the nauplius eye, appearance of sensory papillae and sensory organs, distribution of characteristic teeth on the telson, and by particular details of appendages (Figs 16, 17).

*Eoleptestheria spinosa magna* Marinček & Valvajter 1982 was also discovered north of the Sava, in the vicinity of Sremska Mitrovica (Jarak). It is characterized by having two types of sensory organs on the antennules, a smaller number of teeth on the female telson, which may be lacking, and by structural details of the appendages (Figs 18, 19).

*Eoleptestheria spinosa mira* Marinček & Petrov 1983 occurs near Kuzmin. Compared to the other subspecies, it is characterized by the particular arrangement of setae on the shell, the different arrangement of teeth on the male telson (they occur only on the internal side of ridges) and, again, by specific structural features of the appendages (Figs 20, 21).

The genus *Leptestheria*, the systematics of which certainly constitute the major problem of the *Conchostraca* in general, is most frequent in Yugoslavia. Thus, *Leptestheria dahalacensis* Rüppell, 1837 was found in the vicinity of Zrenjanin and *Leptestheria sp.* was discovered in many localities north and south of Danube and Sava. We compared the populations found in the northern regions (Terezino Polje) and a population from the central part of Yugoslavia, between Velika Plana and Žabari. We established that these populations, in terms of the presence of setae on the valve edge and growth lines, and to a certain extent, by the number of appendages modified for egg bearing, are similar to *Leptestheria intermedia* Botnariuc 1947, found in Roumania. Although both taxa differ by some details, we conclude that there is certainly a close phylogenetic relationship between them.

**Discussion**

Out of five families of *Conchostraca* occurring in Europe, four have been found in Yugoslavia. Only the family *Lynceidae* A.O. Sayce 1902 is lacking, although it occurs in neighbouring Roumania and Hungary. The family *Imnadiidae*, within which we discovered three new species, is certainly of special interest for our fauna. Before our investigations, only two species of this genus had been described: *Imnadia yeyeta* Hertzog 1935, found in France, and *Imnadia voitestii* Botnariuc & Orghidan 1941, which occurs in Roumania.

The family *Limnadiidae* is rare in Yugoslavia, and found only north of the Danube and Sava.

The *Leptestheriidae* is the most common family in Yugoslavia. Within the genus *Eoleptestheria*, three species used to be listed for Europe. Of these, *Eoleptestheria inopinata* Daday 1923 lives in Roumania. *Eoleptestheria ticinensis* Balsamo-Crivelli 1859 is distributed across Middle Europe. According to Daday (1923), it should occur in northwest Yugoslavia (Croatia). *Eoleptestheria variabilis* Botnariuc 1947 was discovered in Roumania. However, Straškraba (1966) is of the opinion that it is merely a synonym of *E. ticinensis*. *E. spinosa* discovered in Yugoslavia, is the fourth species found in Europe.

In Yugoslavia, *Leptestheria* is the most common Conchostracan genus. In Europe, five species and one subspecies of this genus have been described. *L. dahalacensis* is distributed in Europe, Asia Minor and in the southern parts of the Soviet Union. According to Daday (1923) it is present in northwest Yugoslavia (Croatia). The species *Leptestheria tenuis* Sars 1901 and *Leptestheria dives* Daday 1923, reported by Daday, occur in the Vardar Valley, but the specific sites mentioned are not situated in Yugoslavia. On the basis of specimens from Hungary and the Soviet Union, Daday (1923) differentiated the species *Leptestheria rotundirostris*. *L. intermedia*, finally, was discovered in Roumania. However, it should be mentioned that Straškraba (1966) believes that all species mentioned are identical to *L. dahalacensis*.

Our studies on populations of *Leptestheria*

showed that they are phyllogenetically closely related to the species *L. intermedia* from Roumania.

The systematics of the *Conchostraca* in general, and of the *Leptestheriidae* and the genus *Leptestheria* in particular, is incomplete. Thus, it should be mentioned that in *Leptestheria* the number of appendages modified for egg bearing is not yet stabilized, although it was used as criterion for taxonomy of this genus. This suggests that, within this genus, speciation is still in progress.

## References


Daday de Dées, E., 1923. Monographie systématique des Phyllopodes Conchostracés, II. Ann. Sci. nat. Zool., Paris 10: 255–310.

Marinček, M., 1972. Une nouvelle espèces du genre Imnadia, Hertzog (Phyllopoda, Conchostraca) découverte en Yougoslavie. Bull. Mus. Hist. nat. Belgrade B 27: 109–123.

Marinček, M., 1978. Eoleptestheria spinosa, a new species of Conchostraca (Phyllopoda) discovered in Yugoslavia. Bull. Mus. Hist. nat. Belgrade B 33: 103–118.

Marinček, M. & B. Petrov, 1983. On a new subspecies of Conchostraca (Phyllopoda). Bull. Must. Hist. nat. Belgrade B 38: 89–103.

Marinček, M. & B. Petrov, 1984. Taxonomic investigations within the genus Imnadia (Conchostraca, Crustacea). Bull. Mus. Hist. nat. Belgrade B 39: 105–122.

Marinček, M. & B. Valvajter, 1979. Eoleptestheria spinosa tenuis, a new subspecies of Conchostraca (Phyllopoda) found in Yugoslavia. Bull. Mus. Hist. nat. Belgrade B 34: 155–167.

Marinček, M. & B. Valvajter, 1982a. On a new species of the genus Imnadia (Conchostraca, Crustacea). Bull. Mus. Hist. nat. Belgrade B 37: 37–55.

Marinček, M. & B. Valvajter, 1982b. Eoleptestheria spinosa magna, a new subspecies of Conchostraca (Crustacea). Bull. Mus. Hist. nat. Belgrade B 37: 63–72.

Straškraba, M., 1966. Taxonomical studies on Czechoslovak Conchostraca. III, Family Leptestheriidae. Hydrobiologia 27: 571–589.

INDEX.

GPSR Compliance

*The European Union's (EU) General Product Safety Regulation (GPSR) is a set of rules that requires consumer products to be safe and our obligations to ensure this.*

*If you have any concerns about our products, you can contact us on ProductSafety@springernature.com*

In case Publisher is established outside the EU, the EU authorized representative is:

Springer Nature Customer Service Center GmbH
Europaplatz 3
69115 Heidelberg, Germany

**Batch number: 10809834**

Printed by Printforce, the Netherlands